Encyclopedia of Astrophysics

Encyclopedia of Astrophysics

Edited by **Jun Yu**

Callisto Reference

New York

Published by Callisto Reference,
106 Park Avenue, Suite 200,
New York, NY 10016, USA
www.callistoreference.com

Encyclopedia of Astrophysics
Edited by Jun Yu

International Standard Book Number: 978-1-63239-209-1 (Hardback)

Printed in China.

Contents

Preface

This book has been a concerted effort by a group of academicians, researchers and scientists, who have contributed their research works for the realization of the book. This book has materialized in the wake of emerging advancements and innovations in this field. Therefore, the need of the hour was to compile all the required researches and disseminate the knowledge to a broad spectrum of people comprising of students, researchers and specialists of the field.

The astronomical science dealing with the study of nature and physics of celestial bodies is termed as astrophysics. This book provides readers with a theoretical and observational approach to astrophysics. The invention of sophisticated telescopes and other equipment has brought the universe closer. This is the reason why the field of astrophysics is growing day by day. This book demonstrates the cosmic activities that occur in nature. This book covers different dimensions of astrophysics i.e. observational astrophysics, cosmology, the solar system, stars, planets, galaxies, spectroscopy, theoretical astrophysics, dark matter, neutron stars and high energy astrophysics.

At the end of the preface, I would like to thank the authors for their brilliant chapters and the publisher for guiding us all-through the making of the book till its final stage. Also, I would like to thank my family for providing the support and encouragement throughout my academic career and research projects.

Editor

Part 1

Nucleosynthesis

1

Diffuse Emission of ^{26}Al and ^{60}Fe in the Galaxy

Wei Wang
National Astronomical Observatories, Chinese Academy of Sciences, Beijing China

1. Introduction

The Universe is a huge factory of the elements. Just 3 minutes after the big bang, primordial nucleosynthesis occurred (original idea by Gamow 1946). The big bang nucleosynthesis (BBN) produces the present abundance of light elements: D (Deuterium), ^{3}He, ^{4}He, ^{7}Li. But the heavier elements (A>12) cannot be produced during BBN. Some light elements, e.g. B, Be and ^{6}Li, are produced by cosmic ray spallation processes. But the heavier elements are mainly produced by nucleosynthesis in stars (see details in a review by Burbidge et al. 1957; Wallerstein et al. 1997).

Cosmic element abundance could be measured precisely by astronomical observations over a wide range of the electromagnetic spectrum, specially in the optical and infrared bands. Anyway, optical, infrared, and ultraviolet observatories provide only a part of the information. Observations of gamma-ray lines from radioactive isotopes open a new window of observations and guide our understanding of nucleosynthesis in stars. Radioactive decays can been studied by measuring γ-ray line spectra of celestial sources. The γ-ray lines can identify the individual isotope, and the abundance of these isotopes can be quantified with the measurement of γ-ray line intensity of the sky. In particular, γ-ray line photons are nearly transparent for the universe, with no absorption by the very dense molecular clouds and interstellar medium. So detections of these γ-ray lines are a powerful tool to study the cosmic abundance of radioactive isotopes.

Two long-lived radioactive isotopes ^{26}Al and ^{60}Fe have the similar half-life (about million years) and astrophysical origins in the Galaxy. Their nucleosynthesis and ejection into the interstellar medium (ISM) are dominated by massive stars and the subsequent core-collapse supernova explosions (Diehl & Timmes, 1998; Prantzos & Diehl, 1996). Detections of these isotopes provide direct evidence that nucleosynthesis is ongoing in the Galaxy. Their line shapes reflect the dynamics of the ejected isotopes in the interstellar medium and then probe properties of ISM and Galactic rotation effect. Measurements of gamma-ray fluxes of ^{26}Al and ^{60}Fe in the Galaxy provide the unique way to constrain the nuclear reaction rates and stellar evolution models.

Cosmic gamma-ray line signals are very weak, and the gamma-ray photons are absorbed by the atmosphere. The detection of Galactic radioactivity must be based on the gamma-ray detectors aboard balloons and spacecrafts. The first gamma-ray line observations were from OSO-3, OSO-7 in 1960's, which revealed the strong 2.223 MeV line from solar flares (Brandt, 1969). This line results from the formation of deuterium via the union of a neutron and proton; in a solar flare the neutrons appear as secondaries from interactions of high-energy ions accelerated in the flare process. And the field of gamma-ray line astronomy took great leaps forward with the HEAO C, the COS-B (1975-1982) and the Solar Maximum Mission (launched in 1980) satellites. The HEAO C first detected the 1809 keV line from radioactive ^{26}Al in the inner Galaxy (Mahoney et al., 1982). In 1977, NASA announced plans to build a "great observatory" for gamma-ray astronomy. The Compton Gamma-Ray Observatory (CGRO) was designed to take advantage of the major advances in detector technology during the 1980s, and was launched in 1991. The COMPTEL telescope aboard CGRO first mapped the 1809 keV γ-ray line emission of the whole sky (Diehl et al., 1995a; Plüschke et al., 2001), and first detected the strong and broad ^{44}Ti line around 1157 keV from a young supernova remnant (SNR) Cas A (Iyudin et al., 1994). The OSSE telescope aboard CGRO also firstly mapped the 511 keV annihilation line emission in the inner Galaxy (Purcell et al., 1997). Currently, the main space-based gamma ray observatory is the *INTE*rnational *G*amma-*R*ay *A*strophysics *L*aboratory (INTEGRAL). The spectrometer aboard INTEGRAL (SPI) provides the high spectral resolution enough to resolve astrophysical lines and allow spectroscopy in the regime of gamma-rays.

Gamma-ray emissions of ^{26}Al and ^{60}Fe in the Galaxy are studied with the high spectral resolution INTEGRAL spectrometer (SPI). Observations of diffuse ^{26}Al and ^{60}Fe emission and their line shapes in the Galaxy and active nearby star-formation regions with SPI are main scientific objectives in this chapter.

2. Origin of Galactic ^{26}Al and ^{60}Fe

2.1 ^{26}Al

^{26}Al is an unstable nucleus, produced almost exclusively by proton capture on ^{25}Mg in a sufficiently hot environment (Woosley, 1986), mainly destroyed by the β^+ decay into ^{26}Mg since the competing destruction process, i.e., the ^{26}Al $(p, \gamma)^{27}$Si reaction, becomes efficient for $T > 5 \times 10^7$ K, and the end of central the H burning barely reaches such a temperature (see the reaction chains shown in Fig. 1). In addition, the ^{26}Al freshly synthesized must be ejected into the interstellar medium before it is destroyed *in situ*. So its synthesis occurs, essentially in three different specific environments: i.e., the core H burning, the C and Ne convective shells, and the explosive Ne burning. For the hydrostatic nucleosynthesis in the core of stars with convective envelopes, the fresh ^{26}Al requires to be convected away from the hot inner burning region sufficiently fast to prevent destruction, and ejected by strong stellar winds. The present theoretical knowledge of ^{26}Al origin in the Galaxy will be presented. Five possible origins of ^{26}Al are discussed separately: core-collapse supernovae; Wolf-Rayet stars; novae; asymptotic giant branch (AGB) stars; and cosmic-ray nuclear reactions in the interstellar medium.

2.1.1 Core-collapse supernovae

Massive stars (e.g. $M > 8M_{\odot}$) end up as core-collapse supernovae (Type II and Type Ib/c events). Thirty years ago, it has been suggested that ^{26}Al is created in core-collapse supernovae (Arnett, 1977; Ramaty & Lingenfelter, 1977). Explosive ^{26}Al nucleosynthesis is triggered by the shock wave in the Ne burning shell (Woosley & Weaver 1980) and ^{26}Al production can be enhanced by neutrino-induced nuclear reactions (Woosley et al., 1990).

Significant ^{26}Al production occurs in both the late pre-supernova phases and explosion processes. In general, the larger the mass, the lager amount of ^{26}Al survives the explosion

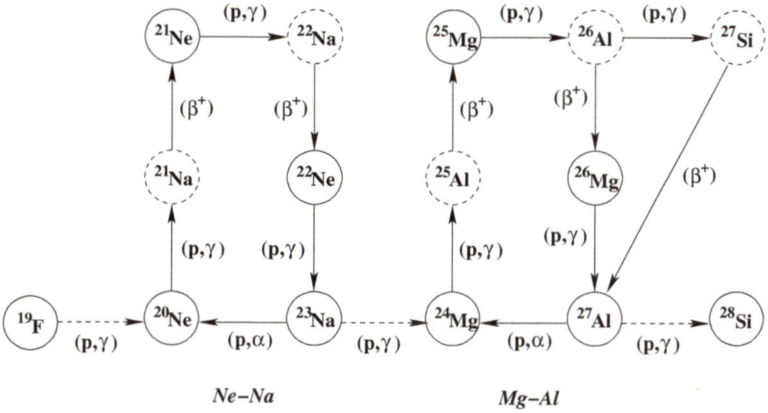

Fig. 1. Reactions of the Ne-Na and Mg-Al chains. Unstable isotopes are denoted by dashed circles (from Rolfs & Rodney 1994).

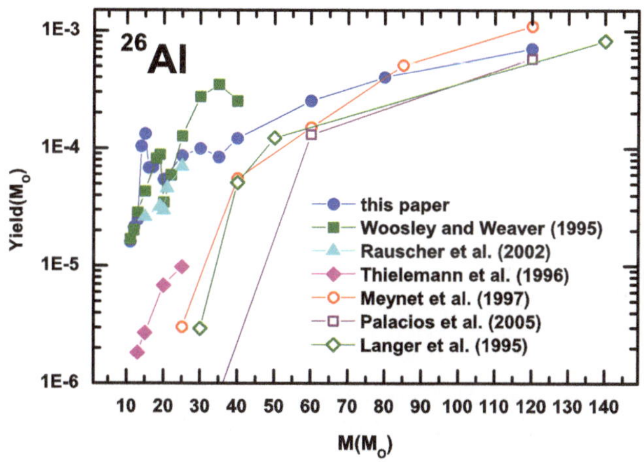

Fig. 2. ^{26}Al yields in different processes (C/Ne convective shells, explosion, winds in the WR star phase) as a function of the initial stellar mass (from recent calculations by Limongi & Chieffi 2006).

(Woosley & Weaver 1995; Rauscher et al. 2002; Limongi & Chieffi 2006). The triangles in Fig. 2 show the ^{26}Al yield produced in C/Ne convective shells that survived to the explosion as a function of the initial mass. The explosive ^{26}Al yield is generally a little higher than pre-supernova production with the different initial masses (Limongi & Chieffi 2006). But different models may have different predictions, e.g., the explosive ^{26}Al is low compared to pre-supernova production for stars with masses higher than $35M_\odot$ (Weaver & Woosley, 1993). Typically, the ^{26}Al yields from core-collapse supernovae range from $2\times10^{-5} - 5\times10^{-4}M_\odot$ with initial star masses of $12 - 120M_\odot$ (see Fig. 2).

2.1.2 Wolf-Rayet stars

Hydrostatic core H burning in the main sequence stars can produce large amounts of ^{26}Al . At central H exhaustion, the ^{26}Al is located in the He core and in the region of variable H left behind by the receding convective core. Since the He burning easily and quickly destroys the ^{26}Al (via the (n, α) and (n, p) reactions), the amount of ^{26}Al synthesized by central H burning and possible preserved up to the explosion is just the one located in the H-rich layers plus the one locked in the fraction of the He core that would not be affected by the He burning. If the dredge-up and the mass-loss were not effective, ^{26}Al would mostly decay before it could be ejected by the explosion.

Stars with masses between $10 - 35M_{\odot}$ will undergo a dredge-up episode that does not enter into the He core, and the mass-loss is weak. Since the He convective shell extends almost up to the base of the H burning shell, only a tiny amount of ^{26}Al which is present in the region of variable H left by the receding H convective core and engulfed in the convective envelope, would be ejected into the interstellar medium. Stars more massive than 35 $M_{\odot}$ do not show dredge-up episodes, but the mass-loss is so strong that a substantial fraction of the He core is ejected through stellar winds, i.e., the Wolf-Rayet phase (van der Hucht et al., 1988). For the Wolf-Rayet stars, a large amount of ^{26}Al present in the He core is thus preserved from the destruction and ejected into the interstellar medium. So the main ^{26}Al production in the ISM before supernova explosions comes from the stellar winds of Wolf-Rayet stars. The average yield of ^{26}Al production during the Wolf-Rayet phase ranges from $1 \times 10^{-5} - 3 \times 10^{-4} M_{\odot}$ for with initial star masses of $35 - 120M_{\odot}$ (Limongi & Chieffi, 2006; Palacios et al., 2005).

From the all-sky survey of 1809 keV emission by COMPTEL (Plüschke et al. 2001), the ^{26}Al in the interstellar medium is predominantly synthesized in massive stars, through the strong stellar winds of Wolf-Rayet stars and core-collapse supernova explosions. The total ^{26}Al yields due to two processes have been calculated by different work (e.g., Limongi & Chieffi 2006; Woosley & Weaver 1995; Langer et al. 1995; Meynet et al. 1997; Thielemann et al. 1996; Rauscher et al. 2002; Palacios et al. 2005). The average ^{26}Al yields provided by various work range from $(0.3 - 30) \times 10^{-5} M_{\odot}$ for $10 - 35$ $M_{\odot}$ stars, and from $(1 - 10) \times 10^{-4}$ $M_{\odot}$ for stars above 35 $M_{\odot}$ (also see Fig. 2). The calculations of the total ^{26}Al yield would be directly compared with the observational limits. Knödlseder (1999) argued that most of ^{26}Al in the Galaxy comes from WR stars, and the study of ^{26}Al in the Cygnus region is a way to resolve this argument.

2.1.3 Novae

^{26}Al production requires moderate peak temperatures, e.g., $T_{\rm peak} < 2 \times 10^8$ K, and a fast decline from maximum temperature (Ward & Fowler, 1980). These conditions are commonly achieved in nova outbursts. In 1980s, one-zone model calculations of explosive H-burning nucleosynthesis with solar or CNO-enhanced envelopes (Hillebrandt & Thielemann, 1982; Wiescher et al., 1986) suggested that classical novae might produce sufficient amounts of ^{26}Al to account for some of the observed meteoritic anomalies but would not represent major Galactic sources. New calculations on the basis of ONeMg white dwarf stars (Nofar et al., 1991; Weiss & Truran, 1990) produced large amounts of long-lived radioactive nuclei, such as ^{22}Na and ^{26}Al , concluding that the ONe novae might be important sources of the Galactic ^{26}Al . Furthermore, the production of ^{26}Al by novae is very sensitive to the initial composition of the envelope and to the nuclear reaction rates adopted. ONe novae should be more important ^{26}Al sources than CO novae, because seed nuclei for the Ne-Na and Mg-Al cycles are almost absent in CO novae. For the same reason, the amount of ^{26}Al synthesized in ONe novae depends on the initial composition of the white dwarf core. Some improvements in the nuclear reaction rates since Caughlan & Fowler (1988) would lead to a lower ^{26}Al production. Recent hydrodynamic calculations of nova outbursts (Jose et al., 1997) predict that the ejected ^{26}Al mass by ONe novae ranges from $(0.3 - 1.7) \times 10^{-8}\ M_\odot$ considering various white dwarf masses and accretion rates.

The amount of ^{26}Al ejected into the interstellar medium by ONe novae decreases as the mass of the underlying white dwarf increases (Jose et al., 1997). So the low-mass white dwarfs are most likely candidates for ^{26}Al production, with the higher ^{26}Al production and higher ejected mass. But white dwarfs lower than $\sim 1.1 M_\odot$ are expected to be CO white dwarf, which are unable to produce important quantities of ^{26}Al . Then the maximum ejection mass of ^{26}Al by one ONe nova event would be $\sim 2 \times 10^{-8} M_\odot$. The predicted contribution of nova outbursts to the Galactic ^{26}Al ranges $(0.1 - 0.4)\ M_\odot$, which is small compared with the present observational limits ($M_{gal}(^{26}\mathrm{Al}) \sim 2\ M_\odot$) derived by COMPTEL measurements (Diehl et al., 1995a) and INTEGRAL/SPI (Diehl et al., 2006a; Wang et al., 2009). Hence, novae represent important ^{26}Al sources in the Galaxy, but cannot be the dominant ones, which is consistent with the accepted hypothesis of young populations as major sources of the Galactic ^{26}Al .

2.1.4 Asymptotic giant branch (AGB) stars

The intermediate-mass stars ($1.5 \leq M/M_\odot \leq 6$) can evolve through a thermally-pulsing phase with the strong stellar wind, so-called asymptotic giant branch (AGB) stars. AGB stars are characterized by two burning shells, one of helium, and one of hydrogen, and by a deep convective envelope extending from above the H-burning shell up to the surface. ^{26}Al could be produced in three sites of AGB stars: the H-burning shell via Mg-Al chain; the He-burning shell via α-capture on ^{22}Ne, and at the base of the convective envelope in the most massive AGB stars that experience H-burning. In AGB stars, ^{26}Al is efficiently produced by H-burning, but destruction by n-capture reaction during the interpulse and pulse phases becomes increasingly more efficient (Mowlavi & Meynet, 2000).

The ejected ^{26}Al masses by AGB stars depend on the temperature which directly relates to the initial mass, and initial composition. Recent calculations show that the low mass AGB stars ($< 4M_\odot$) cannot significantly contribute to ^{26}Al production (Karakas & Lattanzio, 2003; Mowlavi & Meynet, 2000). The AGB star with masses of $4 - 6M_\odot$ can yield ^{26}Al in the regime of $(0.2 - 8) \times 10^{-8}\ M_\odot$ (Karakas & Lattanzio 2003). The AGB stars with lower metallicity produce a higher amount of ^{26}Al . The rough estimation of the contribution by AGB stars to the Galactic ^{26}Al varies from $0.01 - 0.4\ M_\odot$ (Mowlavi & Meynet, 2000). Though the large uncertainties exist for the prediction of ^{26}Al by AGB stars, AGB stars cannot be the main ^{26}Al sources in the Galaxy. But the AGB stars could be best candidates to explain the inferred ^{26}Al /^{27}Al ratios ranging from $\sim 10^{-4} - 10^{-2}$ observed in meteoritic grains (Clayton & Leising, 1987). As an interesting science, models predict a higher amount of ^{26}Al ($\sim (1-2) \times 10^{-7}\ M_\odot$, Mowlavi & Meynet, 2000) around planetary nebulae, which could be possible candidates for direct ^{26}Al detection in future.

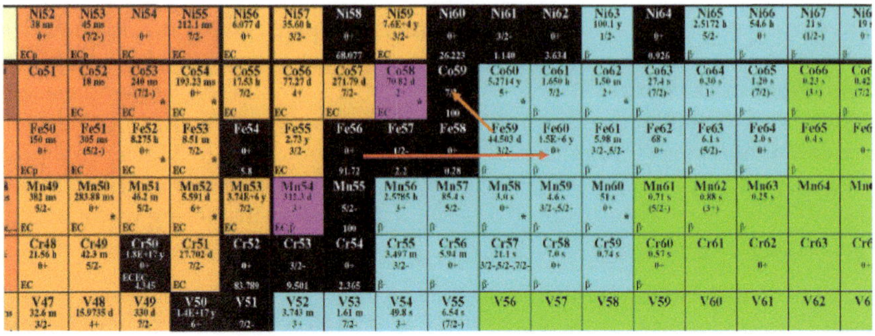

Fig. 3. The caption of ^{60}Fe produced by neutron capture processes of iron isotopes e.g., ^{56}Fe, and ^{59}Fe β-decay will generate a leak in the ^{60}Fe production.

2.1.5 Cosmic-ray nuclear reactions in the interstellar medium

The interactions of accelerated particles with ambient matter can produce a variety of gamma-ray lines following the de-excitation of excited nuclei in both the ambient matter and the accelerated particles. Nuclear reaction of low energy heavy cosmic-ray particles have been proposed as another ^{26}Al source process (Clayton, 1994). Based on the COMPTEL measurement of excited ^{12}C in the Orion molecular cloud complex (Bloemen et al., 1994), Clayton (1994) suggested that this could be an efficient ^{26}Al source process. The cross sections for the ^{26}Mg(H,n)^{26}Al and ^{28}Si(H,ppn)^{26}Al reactions are of the similar magnitude as the one for the ^{12}C(H,p)^{*12}C reaction. Therefore, the estimation for the Orion region of active star formation corresponds to an ^{26}Al yield of $\sim 10^{-4}\ M_{\odot}$.

The Galactic ^{26}Al yield is quite uncertain, depending on the fractions of molecular clouds irradiated by low-energy cosmic rays. And the absence of substantial Galactic plane 4.4 MeV emission due to ^{12}C de-excitation suggests that this process is probably negligible as a Galactic ^{26}Al source (Ramaty, 1996).

2.2 ^{60}Fe

The radioactive isotope ^{60}Fe is believed to be synthesized through successive neutron captures on Fe isotopes (e.g., ^{56}Fe, see Fig. 3) in a neutron-rich environment inside He shells in AGB stars (^{60}Fe is stored in white dwarfs and cannot be ejected), and massive stars, before or during their final evolution to core collapse supernovae. ^{60}Fe can be also synthesized in Type Ia SNe (Woosley, 1997). It is destroyed by the ^{60}Fe (n, γ) process. Since its closest parent, ^{59}Fe is unstable, the ^{59}Fe(n, γ) process must compete with the ^{59}Fe(γ^-) decay to produce an appreciate amount of ^{60}Fe .

2.2.1 Massive stars

A neutron-rich environment in massive stars is required to produce ^{60}Fe . And a temperature of the order of 2 billion degrees represents an upper limit for the synthesis of ^{60}Fe because above this temperature the (γ, n) and the (γ, p) photon disintegrations of both ^{59}Fe and ^{60}Fe become tremendously efficient. Such an occurrence limits a possible ^{60}Fe production to the He, C, Ne shell burning phases.

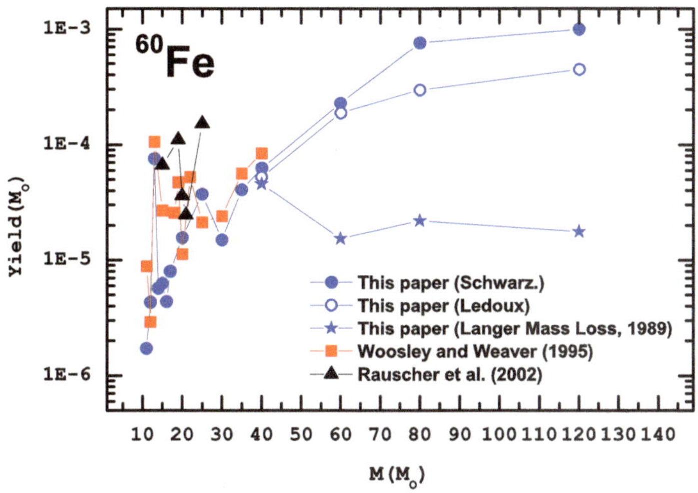

Fig. 4. Comparison among the ^{60}Fe yields as a function of the initial stellar mass provided by various work (from Limongi & Chieffi 2006). The solid and open circles refer to the amount of ejected ^{60}Fe for two different choices of the stability criterion in the He convective shell, i.e., the Schwarzschild and the Ledoux criteria, respectively. The stars represent the amount of ejected ^{60}Fe obtained by adopting the Langer (1989) mass-loss rate for massive stars (WR stars of masses from 40 - 120 $M_{\odot}$). Solid squares and triangles are the results from Woosley & Weaver (1995) and Rauscher et al. (2002) in the initial mass range of 12 - 40 $M_{\odot}$.

In He shell burning phases, significant ^{60}Fe production occurs when the star becomes a WR star which experiences such strong mass loss that it first loses all H-rich envelope and then continues eroding the He core up to the moment of the core collapse. In these stars, the He convective shell forms within the region of variable He abundance, then a problem arises of whether the Schwarzschild criterion or the Ledoux one is used to determine if a convective region forms. In Fig. 4, The solid and open circles refer to the amount of ejected ^{60}Fe for two different choices of the stability criterion in the He convective shell. In C shell burning, the high temperature ($> 10^9$ K) allows a large production of α-particles which translates into a high neutron density and hence a large yield of ^{60}Fe . C convective shell could produce a conspicuous amount of ^{60}Fe in stars of the initial mass below 40 $M_{\odot}$. Ne burning may produce ^{60}Fe , but the lack of an extended and stable convective shell lasting up to the explosion prevents the build up of a significant amount of ^{60}Fe . The average yield of ejected ^{60}Fe in convective shells varies from $10^{-6} - 7 \times 10^{-4}$ $M_{\odot}$, increasing with the initial mass from 11 -120 $M_{\odot}$ (Limongi & Chieffi, 2006). Calculations by Limongi & Chieffi (2006) suggested the dominant yield of ^{60}Fe in massive stars above 40 $M_{\odot}$ comes from the He convective shell burning, which sensitively depends on the mass-loss rate. Models with a strong mass-loss rate (Langer, 1989) would reduce the ^{60}Fe production during the WR phases into a level of $\sim 10^{-5}$ $M_{\odot}$ (see Fig. 4).

2.2.2 Supernovae

The last episode of synthesis of ^{60}Fe occurs when the blast wave crosses the mantle of the star on its way out during the core-collapse supernova explosions. The peak temperature is of the order of 2.2×10^9 K, and hence roughly in the same region where the explosive synthesis of ^{26}Al occurs. The average yield of ^{60}Fe in explosion varies from 10^{-5} $M_\odot$ (initial mass region of $11-50$ $M_\odot$) to 4×10^{-5} $M_\odot$ ($50-120$ $M_\odot$).

Fig. 4 shows the total amount of ejected ^{60}Fe as a function of the initial stellar masses (from 11 -120 $M_\odot$) calculated by Limongi & Chieffi (2006). The results by the previous work (initial masses from 12 -40 $M_\odot$, Woosley & Weaver 1995; Rauscher et al. 2002) are also presented for a comparison. In the initial mass range < 40 $M_\odot$, the ^{60}Fe yield by Rauscher et al. (2002) are significantly larger than the yields by both Woosley & Weaver (1995) and Limongi & Chieffi (2006). In Fig. 4, the solid and open circles are obtained by adopting the mass-loss rate by Nugis & Lamers (2000); the stars are obtained by taking the rate by Langer (1989). The mass-loss rate proposed by Langer (1989) is much stronger than the Nugis & Lamers (2000) one. With adopting a stronger mass-loss rate, they have significantly reduce the ^{60}Fe production during the WR phases with the initial masses above 40 $M_\odot$.

In addition, ^{60}Fe could also be produced in substantial amounts by rare subtypes of SN Ia (Woosley, 1997), which would then be point sources of ^{60}Fe gamma-rays. The average yield of ^{60}Fe in SN Ia is $(1-5) \times 10^{-3}$ $M_\odot$ depending on models. The ^{60}Fe per event is typically 100 times greater in a high-density white dwarf explosion than in a Type II supernova. With an event rate 10^{-4} yr^{-1}, the composite signal here comes from ~ 100 point sources. A single source at 10 kpc that made 0.005 $M_\odot$ of ^{60}Fe would be visible for several million years at a flux level of 10^{-7} ph cm^{-2} s^{-1}. However, this is well beyond the capability of present gamma-ray telescopes.

3. All-sky observations of ^{26}Al by CGRO/COMPTEL

The 1809 keV gamma-ray line from radioactive ^{26}Al with its decay time of 1 million years can be used as the tracer of the recent nucleosynthesis activity in the Galaxy. Ramaty & Lingenfelter (1979) first predicted the 1809 keV γ-ray line flux of $\sim 10^{-4}$ph cm^{-2} s^{-1} rad^{-1} from the inner Galaxy, combining the solar ^{27}Al abundance with an estimate of the isotope ratio ^{26}Al /^{27}Al ($\sim 10^{-5}$) from supernovae.

The 1809 keV γ-ray line emission was first detected with the Ge spectrometer on the HEAO-C spacecraft (Mahoney et al. 1982). This detection was confirmed by the measurement of Galactic transits through the field of view by the NaI spectrometer on the SMM spacecraft (Share et al., 1985). The following balloon-borne experiments measured the intensity of the ^{26}Al emission and derived some information on its angular distribution (see Durouchoux et al. 1993; Prantzos & Diehl 1996; Schoenfelder et al. 1991; von Ballmoos et al. 1987). And the ^{26}Al fluxes derived from these different measurements depend on the assumed Galactic distribution of ^{26}Al line emission. For example, they found $(1.1-4.6) \times 10^{-4}$ph cm^{-2} s^{-1} for a point source at the Galactic center, and $(3.9-5.4) \times 10^{-4}$ph cm^{-2} s^{-1} from the inner Galaxy for a flat supernova model. So mapping the 1809 keV γ-ray line emission of the Galaxy can provide insight into the nature of ^{26}Al sources, and precise determination of the ^{26}Al flux.

The COMPTEL imaging telescope aboard the Compton Observatory performed the first survey of ^{26}Al γ-ray line emission in the whole Galaxy. COMPTEL covered the energy range of 1 – 30 MeV, with an energy resolution of 140 keV (FWHM) around 1809 keV and an angular resolution of 3.8° (see Schoenfelder et al. 1993). COMPTEL has the enough sensitivity to study the origin of ^{26}Al .

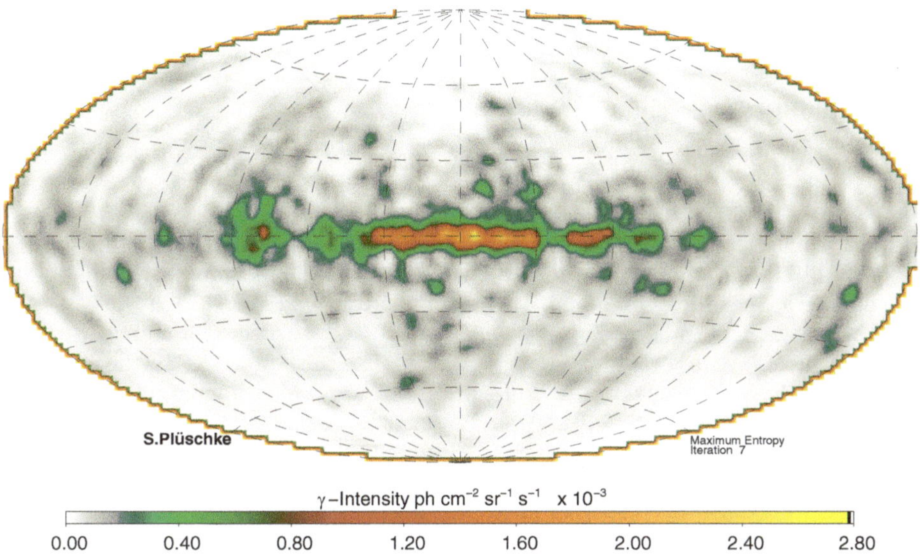

Fig. 5. The maximum-entropy all-sky images of the Galactic 1809 keV line emission observed with COMPTEL over 9 years (from Plüschke et al., 2001).

Diehl et al. (1995a) analyzed the first-year COMPTEL survey data, which has covered the whole Galactic plane. Overlays of the ^{26}Al emission map with the positions of supernova remnants and Wolf-Rayet (WR) stars in the Galaxy suggest that the large-scale structure of ^{26}Al emission in the Galaxy may be contributed to massive stars. And the Vela region (the Vela SNR) shows evidence for a single identified close-by ^{26}Al source (Diehl et al., 1995b). So the first ^{26}Al emission map of the Galactic Plane favors the dominant ^{26}Al origin from massive stars, presumedly core-collapse supernovae or Wolf-Rayet stars.

The first 1809 keV all-sky map was presented based on the first three years of COMPTEL observations (Oberlack et al., 1996). The all-sky map confirmed the non-local character of the detected ^{26}Al emission in the first Galactic plane survey (Diehl et al. 1995a). Most of the ^{26}Al emission is attributed to young, massive stars and star-formation regions.

Based on the 9-year COMPTEL observations (1991 – 2000), Plüschke et al. (2001) obtained 1809 keV all-sky map with best significance level. The image (Fig. 5) reconstructions show an extended Galactic ridge emission mostly concentrated towards the Galactic center region ($-30^\circ < l < 30^\circ$), plus an emission feature in the Cygnus region, and a low-intensity ridge along the Carina and Vela regions. These features confirm the previously reported emission structures. In addition, the image shows some low-intensity features in the longitude range between 110° and 270°, e.g. the Orion region (Fig. 5). Near the Galactic center region, the image shows a possible emission from the nearby Sco-Cen region. Also at latitudes beyond $\pm 30^\circ$, some of these low-intensity structures are visible, which may be artifacts, subject to further studies.

In summary, from the all-sky ^{26}Al emission image by COMPTEL, the observed 1809 keV γ-ray line is ascribed to the radioactive decay of ^{26}Al in the interstellar medium. ^{26}Al has been found to be predominantly synthesized in massive stars and their subsequent core-collapse supernovae. Furthermore, ^{26}Al flux enhancements are detected aligned with regions of recent star formation, such as apparently observed in the Cygnus and Vela regions.

4. INTEGRAL/SPI studies on Galactic ^{26}Al

The INTEGRAL Observatory is an European (ESA) Gamma-Ray Observatory Satellite Mission for the study of cosmic gamma-ray sources in the keV to MeV energy range (Winkler et al., 2003). INTEGRAL was successfully launched from Baikonur Cosmodrome (Kazakhstan) on October 17, 2002 using a Proton rocket provided by the Russian Space Agency. The INTEGRAL orbit is eccentric, with an apogee of 153 000 km, a perigee of 9000 km, and a 3 day period (Jensen et al., 2003). Two main instruments are on board INTEGRAL: the INTEGRAL imager (IBIS) with an angular resolution of 12′, allowing for source localization with arcmin precise with a field of view of $9° \times 9°$ (Ubertini et al., 2003); the INTEGRAL spectrometer (SPI) with an energy resolution of 2.5 keV at 1.3 MeV and angular resolution of 2.5° within a field of view of $16° \times 16°$ (Vedrenne et al. 2003).

The SPI composed of composed of 19 high purity germanium detectors (GeD) and their associated electronics can allow for high spectral resolution of $\sim$ 2.5 keV at 1 MeV, suitable for astrophysical studies of individual gamma-ray lines and their shapes, e.g. 511 keV line emission, γ-ray lines from radioactivities of ^{26}Al and ^{60}Fe .

^{26}Al is an unstable isotope with a mean lifetime of 1.04 Myr. ^{26}Al first decays into an excited state of ^{26}Mg, which de-excites into the ^{26}Mg ground state by emitting gamma-ray photons with the characteristic energy of 1809 keV (Fig. 6). The study of ^{26}Al line emission from the Galaxy is one of the main science goals of the INTEGRAL mission. SPI aboard INTEGRAL is a high resolution spectrometer with energy resolution of 3 keV (FWHM) at 1809 keV, which therefore adds high-resolution spectroscopic information to ^{26}Al astronomy. The detailed measurement of ^{26}Al line position and shape is expected to reveal more information beyond the COMPTEL imaging survey about the ^{26}Al sources and their location through the Doppler effect, induced from Galactic rotation and dynamics of the ejected ^{26}Al as it propagates in the interstellar medium around its stellar sources.

4.1 Diffuse ^{26}Al emission of the inner Galaxy

The spectral characteristics of ^{26}Al emission in the inner Galaxy serve to study the current nucleosynthesis activity and the properties of the interstellar medium near the ^{26}Al sources on a large-scale averaged scale. We define the "inner Galaxy" as the region $-30° < l < 30°$, $-10° < b < 10°$), and may use this as a representative region for this purpose, since it coincides with the bright ridge of observed 1809 keV emission as observed along the plane of the Galaxy.

Earlier INTEGRAL analysis had used 1.5 years of SPI data to first explore the large-scale spectral characteristics of ^{26}Al emission in the inner Galaxy (Diehl et al. 2006a, 2006b). A detection of the ^{26}Al line from the inner Galaxy with a significance of $\sim 16\sigma$ had confirmed the narrowness of the ^{26}Al line (FWHM < 2.8 keV, 2σ), which had already been seen by RHESSI (Smith et al. 2003) and HEAO-C (Mahoney et al. 1984) earlier.

At present, INTEGRAL/SPI data are accumulated from more than five years to extend this study towards spatially-resolved details of ^{26}Al line spectroscopy across the inner regions of

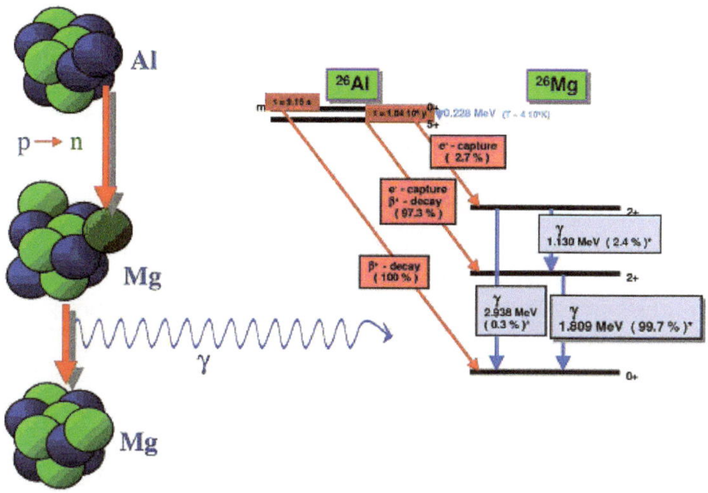

Fig. 6. The decay chain of the radioactive isotope ^{26}Al . ^{26}Al decays with a characteristic lifetime of 1.04 Myr into an excited state of ^{26}Mg, then de-excites into the Mg ground state by emitting γ-ray lines with the energies of 1809 keV (99.7%) and 2.936 keV (0.3%).

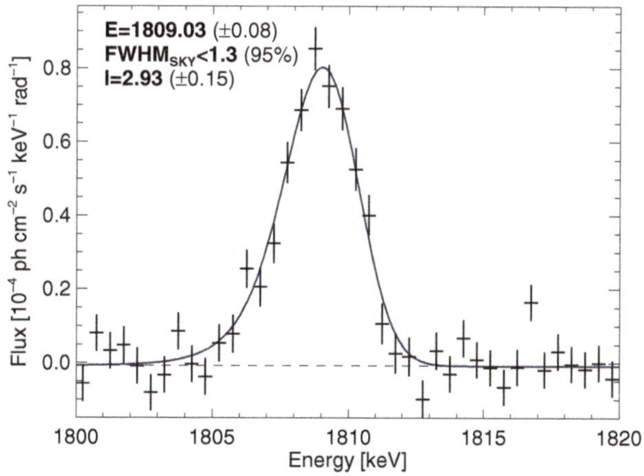

Fig. 7. ^{26}Al spectrum of the inner Galaxy from INTEGRAL/SPI observations (from Wang et al., 2009). The ^{26}Al line width is found to be (< 1.3 keV, 2σ). Fluxes are quoted in units of 10^{-4} ph cm^{-2} s^{-1} rad^{-1}.

the Galaxy. We characterize the ^{26}Al line details through two different approaches. When the sky signal is weak, we fit the spectra with Gaussians plus a linear residual background, and use the Gaussian intensities, centroid energies, and FWHM widths for relative comparisons, such as trends along the plane of the Galaxy. When the signal is sufficiently strong so that we are sensitive to line width details, we describe the line component not by a single Gaussian any more, but rather by the convolution of a Gaussian with the asymmetric instrumental line response. This allows us to infer immediate information about the celestial ^{26}Al dynamics, which we identify with this Gaussian, and in particular its width, which arises from Doppler shifting of the line energies with motion of decaying ^{26}Al nuclei relative to the observer.

Using more than five years of SPI data, the inner-Galaxy ^{26}Al emission spectrum is shown in Fig. 7. The ^{26}Al line is detected at $\sim 28\sigma$ significance. The ^{26}Al gamma-ray flux from the inner Galaxy turns out as $(2.93 \pm 0.15) \times 10^{-4}$ ph cm^{-2} s^{-1} rad^{-1}. This is consistent with our earlier values $(3.3 \pm 0.4) \times 10^{-4}$ ph cm^{-2} s^{-1} rad^{-1} (Diehl et al. 2006a, 2006b), and also with the COMPTEL imaging-analysis value of $(2.8 \pm 0.4) \times 10^{-4}$ ph cm^{-2} s^{-1} rad^{-1} (Plüschke et al. 2001). The above value is derived using an asymmetric line shape as best matching our expectations from SPI's spectral response and eventual additional celestial line broadening. The total ^{26}Al gamma-ray flux of the Gaussian fit as determined for the inner Galaxy region is $(2.73 \pm 0.17) \times 10^{-4}$ ph cm^{-2} s^{-1} rad^{-1}.

The measured ^{26}Al flux of $(2.93 \pm 0.15) \times 10^{-4}$ ph cm^{-2} s^{-1} rad^{-1} for the inner Galaxy thus translates into a Galactic ^{26}Al mass of (2.6 ± 0.6) $M_\odot$ using a plausible scale height of 180 pc. If we ignored the ejection of ^{26}Al into surrounding cavities and corresponding champagne flows, and used the lower scale heights of O stars or of the molecular disk, we would obtain lower total amounts around or even below 2 $M_\odot$.

The amount of ^{26}Al is maintained in steady state by a core-collapse supernova rate via $M_{eq} = SNRate \cdot \tau \cdot Y$, where the rate is measured in events per year, τ is the mean life of ^{26}Al , and Y is the IMF-averaged ^{26}Al yield in units of $M_\odot$ per supernova. The yield in this context must include the explosive yields from the supernova model as well as any ^{26}Al ejected in the Wolf-Rayet wind phase. Yields are moderated by the steep initial mass function (IMF), $\xi \propto m^{-\alpha}$, in our relevant mass range 10–120 $M_\odot$. We use the Miller-Scalo IMF ($\xi \propto m^{-2.7}$) for this higher-mass range (Miller & Scalo, 1979), supported by a wide range of astronomical constraints. Y is obtained from the high-mass initial mass function (IMF) and the nucleosynthesis yields of models. The resulting ^{26}Al yield per massive star is $(1.4 \pm 0.7) \times 10^{-4}$ $M_\odot$ based on various published yields as a function of progenitor mass (see Fig. 2). The corresponding supernova rate is $SNRate = 1.90 \pm 0.95$ events per century (this does not include type Ia supernovae, which have been found to be negligible sources of ^{26}Al). The resulting range of one to three core collapses per century coincides with the recent values obtained from on a survey of local O3-B2 dwarfs (Reed, 2005), extrapolated to the Galaxy as whole with spatial distribution models, and the study of the luminosity function of OB associations (McKee & Williams, 1997).

The intrinsic line width is now constrained to < 1.3 keV ($2\,\sigma$). In the inner Galaxy, Galactic differential rotation alone can lead to significant Doppler shifts towards specific longitudes where the projected-velocity differences with respect to the solar orbit reach maxima; line broadening results if we integrate over a larger longitude range with different bulk velocity differences. Kretschmer et al. (2003) have simulated the ^{26}Al line shape diagnostics in the inner Galaxy due to Galactic rotation and ^{26}Al ejection from sources, and find that line broadening

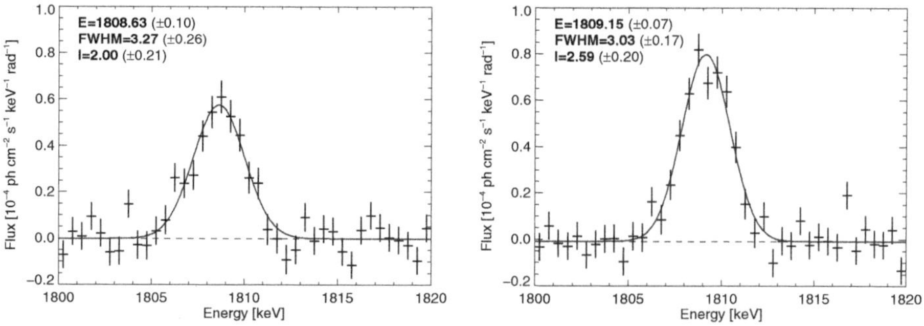

Fig. 8. ^{26}Al spectra for two Galactic quadrants (**left** $0° < l < 60°$, and **right** $-60° < l < 0°$, from Wang et al., 2009). Line centroids relative to the centroid energy of ^{26}Al line in the laboratory (1808.65 keV) show a significant blueshift in the 4th quadrant. Both the spectra have width values near the value of the instrumental line width, implying that ^{26}Al emissions from two quadrants are the narrow lines. In addition, the ^{26}Al flux of the 4th quadrant is higher than that of the 1st quadrant, and the flux ratio is ~ 1.3.

of up to 1 keV is expected if the signal is integrated over the inner region of the Galaxy. Our present large-scale line-shape constraints are consistent with these expectations. If we interpret line broadening of the ^{26}Al line from the inner Galaxy in terms of interstellar-medium characteristics, the intrinsic-width constraint of < 1.3 keV corresponds to 160 km s^{-1} as a corresponding 2σ limit on ISM velocities. This is well within the plausible and acceptable range for the environment of normal interstellar-medium turbulence (Chen et al. 1997). As a summary, the measured line width of ^{26}Al from the inner Galaxy is consistent with Galactic rotation and modest interstellar-medium turbulence around the sources of ^{26}Al .

4.2 Spectral variations of ^{26}Al gamma-ray line along the Galactic plane

The 9-year COMPTEL imaging of ^{26}Al line emission has already suggested some asymmetry in the inner Galaxy: the fourth Galactic quadrant appears somewhat brighter than the first quadrant (Plüschke (2001) found a significance of 2.5σ for a brightness difference). COMPTEL could provide the image details of ^{26}Al in the Galaxy, but no significant spectral information due to its spectral resolution of about 150 keV near the ^{26}Al line. SPI with its Ge detectors features sufficiently-high spectral resolution to allow astrophysical constraints from ^{26}Al line shapes, averaged over the Galaxy as discussed above, but also for different regions along the Galactic plane due to its imaging properties as a coded-mask telescope.

In this part, we will proceed towards increasing spatial resolution along the Galactic plane, starting out from testing Galactic asymmetries between the first and fourth quadrant. ^{26}Al line parameters toward the different directions of the Galactic plane are determined using separate sky maps covering each sky region, simultaneously fitting these together with our background model to the entire sky survey database. ^{26}Al line fluxes, centroid energies, and line widths then are derived by a simple Gaussian fit to the ^{26}Al line in the resulting spectra, as we are interested in relative changes between different portions of the sky. This will allow us to identify line shifts from bulk motion such as expected from large-scale Galactic rotation,

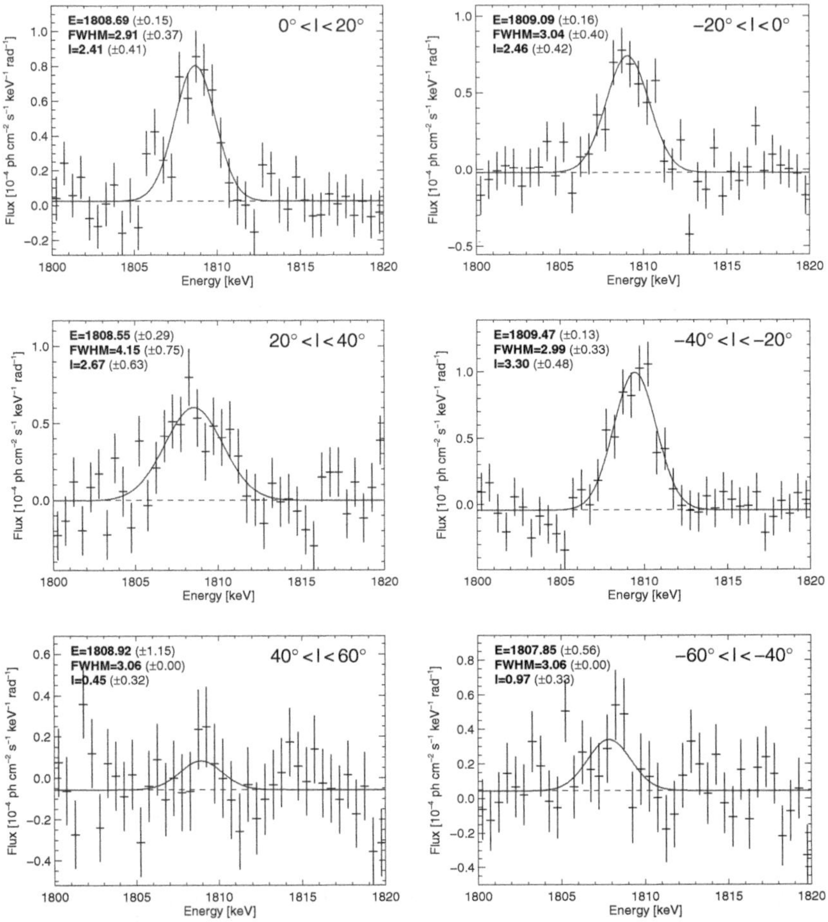

Fig. 9. ^{26}Al spectra of six segments along the Galactic plane ($-60^\circ < l < 60^\circ$, from Wang et al., 2009). Small longitude degree bin (20°) makes the detections of ^{26}Al not significant in the regions of $40^\circ < |l| < 60^\circ$.

and hints for additional line broadenings in particular regions, which would reflect increased ^{26}Al velocities in such regions.

^{26}Al spectra for the 1st ($0^\circ < l < 60^\circ$) and 4th quadrant ($-60^\circ < l < 0^\circ$) are presented in Fig. 8. In the 4th quadrant we note a blueshift of 0.49 ± 0.07 keV relative to the centroid energy of ^{26}Al line in the laboratory, but no significant redshift in the 1st quadrant is apparent ($\sim 0.04 \pm 0.10$ keV). Both spectra have width values compatible with no significant ^{26}Al line broadenings. The indicated ^{26}Al asymmetry between the two inner Galactic quadrants appears again, with a flux ratio of $\sim 1.3 \pm 0.2$.

Challenging the imaging capability of SPI for diffuse and extended emission, we refine spatial structure even more towards smaller longitude intervals. In Fig. 9, we present the spectra

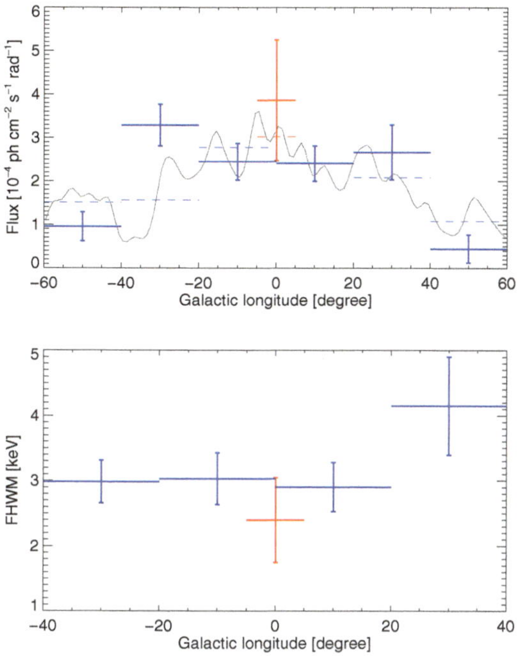

Fig. 10. **Top** ^{26}Al intensity distribution along the Galactic plane (from Wang et al., 2009). For comparison, the COMPTEL-derived ^{26}Al intensity profile is shown (solid line, and dashed lines when integrated over the same longitude bins). **Bottom** ^{26}Al FWHM (Gaussian fitting) variation along the Galactic longitudes, a broad ^{26}Al line feature is detected toward the longitudes $20° < l < 40°$ (from Wang et al., 2009).

of ^{26}Al emission line for six smaller longitude intervals of 20 degree width along the Galactic plane ($-60° < l < 60°$). The ^{26}Al line is still detected for the inner Galaxy ($-40° < l < 40°$, $> 6\sigma$ for each 20° bin region), but only marginal for the two outer regions ($40° < l < 60°$ and $-60° < l < -40°$, $< 4\sigma$). This may be attributed to both less ^{26}Al brightness and to less exposure in these regions, compared to the inner Galaxy.

Fig. 10 shows the intensity distribution of ^{26}Al emission (from the same 20° regions) along the Galactic plane, adding the (longitude-range normalized) Galactic-Center region ^{26}Al intensity ($|l| < 5°$) for comparison. The variability of ^{26}Al intensity along the Galactic plane again is evident.

We note that the ^{26}Al line in the region of $20° < l < 40°$ appears somewhat broadened, with a Gaussian width of FWHM$\sim 4.15 \pm 0.75$ keV. This may hint towards a peculiar ^{26}Al source region towards this direction, which could be associated with the Aquila region (Rice et al. 2006). Broadening could result from higher turbulence if the ^{26}Al source region is younger than average and dominated by the ^{26}Al ejection from more massive stars (see Knödlseder et al. 2004). Further studies would be interesting, and have the potential to identify star formation otherwise occulted by foreground molecular clouds.

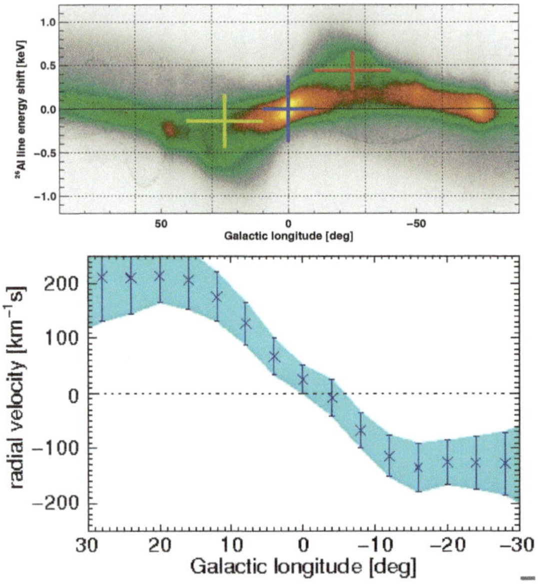

Fig. 11. ^{26}Al line energy shifts along the Galactic plane. Galactic rotation will shift the observed ^{26}Al line energy due to the Doppler effect, to appear blueshifted at negative longitudes and redshifted at positive longitudes. (**Top**) The ^{26}Al line shifts according to the 1.5-year SPI data ($-40° < l < 40°$, $-10° < b < 10°$, Diehl et al. 2006a). Colour scales are expectations of ^{26}Al line positions and intensity modelled from the Galactic rotation curve and a three-dimensional distribution of ^{26}Al sources (Kretschmer et al., 2003). Using the central Galactic region ($-10° < l < 10°$) as a reference (fitted line energy 1808.72 keV), we found centroid energy shifts of -0.14 keV ($10° < l < 40°$) and +0.44 keV ($-40° < l < -10°$), respectively. (**Bottom**) Radial velocity curve of ^{26}Al ejecta in the stellar medium along Galactic longitudes which shows Doppler shifts of the 1809 keV gamma-ray line from ^{26}Al due to Galactic rotation according to more than 6 years of SPI data on the Galactic plane ($-40° < l < 40°$, $-5° < b < 5°$, Kretschmer 2011).

4.3 ^{26}Al line centroid energy as probe of Galactic rotation

Galactic differential rotation can result in not only the broadening of ^{26}Al line but also the shifts of the line centroid energy due to the Doppler effect (Gehrels & Chen, 1996; Kretschmer et al., 2003).Galactic rotation would induce the ^{26}Al line centroid energy redshifts in the 1st quadrant and blueshifts in the 4th quadrant. The Galactic rotation curve can be also determined from observations of various objects (e.g., HI, CO, HII, Brand & Blitz 1993).

The first study on the Doppler shifts of ^{26}Al line due to Galactic rotation using the 1.5-year SPI data was present in Fig. 11 top panel (Diehl et al. 2006a). Using the central Galactic region ($-10° < l < 10°$) as a reference (fitted line energy 1808.72 keV), we found centroid energy shifts of -0.14 keV ($10° < l < 40°$) and +0.44 keV ($-40° < l < -10°$), respectively. Though the error bars for the line positions are very large, we concluded that the results of the line energy shifts are consistent with the Galactic rotation effect (Diehl et al. 2006a).

Recently, using more than 6 years of INTEGRAL/SPI data, Kretschmer (2011) obtained the radial velocity curve of ^{26}Al ejecta along Galactic longitudes with much better resolution (see Fig. 11 bottom panel). The measurements show the characteristic wave-like shape of velocity offset versus Galactic longitude due to the large-scale Galactic rotation. This has been observed for molecular and atomic gas in our and other galaxies. But with ^{26}Al gamma-rays, we get unique access to kinematics of the tenuous, diluted, and presumably hot phase of the interstellar medium, as it is expected around the massive stars which eject radioactive ^{26}Al in the Galaxy. We should note that comparing the velocity curves from ^{26}Al line shape measurements along the Galactic plane and other observations and theory found that the hot gas of ^{26}Al ejecta undergoes additional acceleration in the direction of Galactic rotation.

4.4 Latitudinal variations of ^{26}Al emission

The interpretation of ^{26}Al imaging and spectral results relies on (uncertain) distances of ^{26}Al sources. Along the line-of-sight, the detected ^{26}Al signal could originate from local star-formation complexes ($\sim$ 100 pc), or from the nearest part of the Sagittarius-Carina arm ($1-2$ kpc), or from the Galactic center region ($\sim$ 8 kpc), or even from the distant side of the Galaxy ($>$ 10 kpc). In this section, we try a possible way to resolve the ^{26}Al signals for the local complexes from the large scales of the Galactic plane by probing the latitudinal variation of ^{26}Al emission.

^{26}Al emission for low latitudes ($|b| < 5°$) would be dominated by the large-scale origin in the Galactic disk. While ^{26}Al sources for high latitudes ($|b| > 5°$) should originate from local star-formation systems in the Gould Belt. The Gould Belt appears as an ellipsoidal shaped ring with semi-major and minor axes equal to $\sim$ 500 pc and 340 pc, respectively (Perrot & Grenier, 2003). The Sun is displaced from the center of the Gould Belt about 200 pc towards $l = 130°$ (Guillout et al., 1998). The Vela region is located near the boundary of the Gould Belt towards $l \sim -90°$. The nearby Sco-Cen region also belongs to the Gould Belt, extending from $(l, b) = (0°, 20°)$ towards $(l, b) = (-30°, 0°)$ (Sartori et al., 2003).

In Fig. 12, we present ^{26}Al spectra along Galactic latitudes for the 1st and 4th quadrants, respectively. No ^{26}Al signals are detected for two intermediate latitude regions of the 1st quadrant. While in the 4th quadrant, ^{26}Al emission is clearly detected in the latitude region of $5° < b < 20°$, which may originate toward the nearby Sco-Cen OB associations at a distance of $\sim$ 140 pc. Additionally, a very weak ^{26}Al signal may be hinted toward the latitude region of $-60° < l < 0°$, $-20° < b < -5°$, which need a further check in future.

Finally, we determine the scale height of the Galactic plane in ^{26}Al emission by comparing the fit quality for sets of two different plausible geometrical models for the ^{26}Al source density distribution in the Galaxy (a Glactocentric double-exponential disk model (Wang et al., 2009) and a spiral-arm structure model based on Taylor & Cordes (1993), varying their scale height parameters). In order to avoid a bias from bright special regions such as Cygnus/Vela/Carina, we restrict this analysis to data within $|l| < 60°$ and $|b| < 30°$. Fig. 13 shows the variation of log-likelihood values with different scale heights for both model types. Here, the values for the exponential-disk models have been shifted by $+16$, as the spiral-arm model systematically provides a better description of our data. With $-2\,\log\mathrm{L}$ being asymptotically χ^2 distributed, we derive a scale height of 130^{+120}_{-70} pc (1σ) for the ^{26}Al emission in the inner Galactic disk, from the spiral arm model constraints. This confirms previous such studies based on COMPTEL data (Diehl et al., 1998).

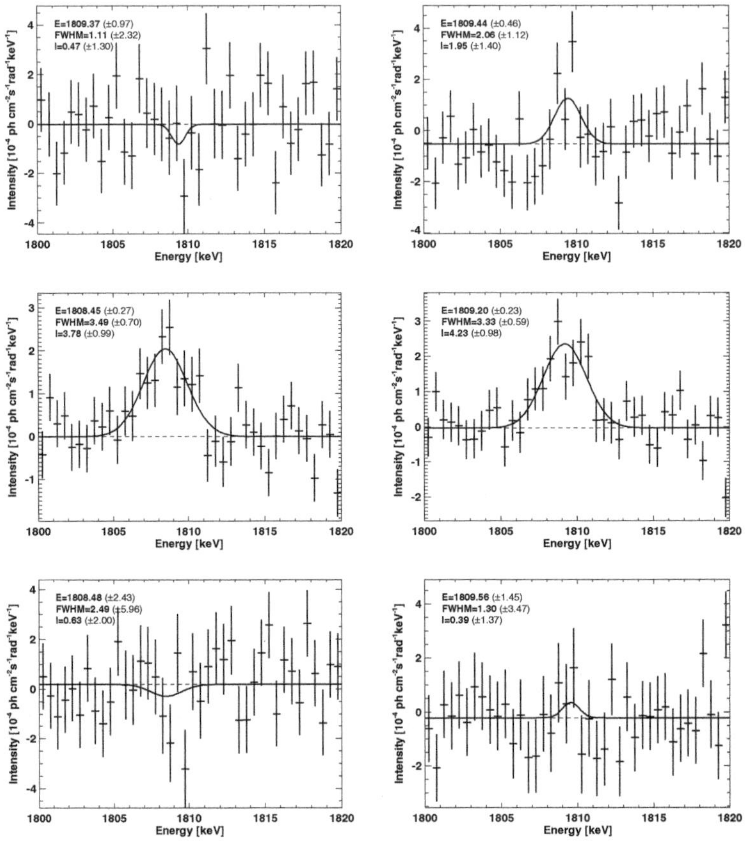

Fig. 12. ^{26}Al spectra for different latitude intervals of the 1st (left) and 4th (right) quadrants (taken from Wang, 2007): $5° < b < 20°$ (Top), $-5° < b < 5°$ (Middle), $-20° < b < -5°$ (Bottom).

4.5 ^{26}Al line shapes of the Cygnus and Sco-Cen regions

The COMPTEL 1809 keV all-sky map attributed to radioactive decay of Galactic ^{26}Al confirmed the diffuse emission along the inner Galaxy. Several significant features of the reconstructed intensity pattern are flux enhancements in the directions of nearby star-formation regions, e.g., the Cygnus region (del Rio et al., 1996), the Vela region (Diehl et al., 1995b), and a possible feature above the Galactic center which is attributed to the nearby Sco-Cen region. Detections of ^{26}Al in these regions strongly support the hypothesis of massive stars and their descendent supernovae being the dominant sources of interstellar ^{26}Al . ^{26}Al near these young populations may be located in the different medium environment from the large scales in inner Galaxy, e.g. high turbulent velocities of interstellar medium in these star-formation regions due to stellar winds and supernova explosions.

The INTEGRAL/SPI is a powerful spectrometer to probe the ^{26}Al line shapes. So spectral studies of ^{26}Al line in these star-formation regions can provide new information of ^{26}Al sources

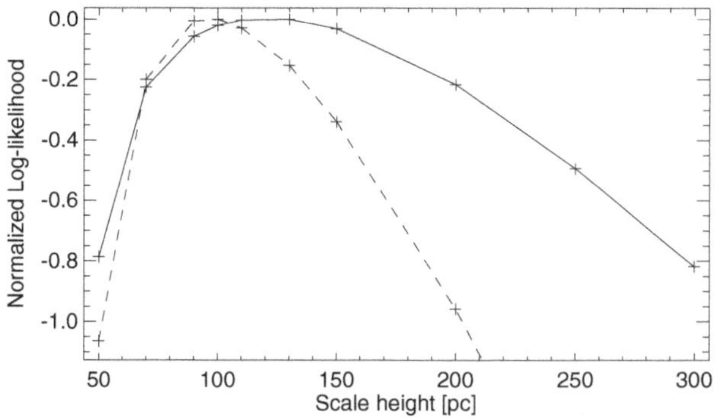

Fig. 13. Determination of the Galactic-disk scale height (from Wang et al., 2009). Shown is the logarithmic likelihood-ratio for the exponential-disk (dashed line) and spiral-arm structure models (solid line) for different scale height values. Log-likelihood values for the exponential-disk model have been offset by +16.

and their environments. In this section, we will show recent SPI results on ^{26}Al line shapes of two nearby star-formation regions: Cygnus and Sco-Cen.

4.5.1 Cygnus

The Cygnus region is one of the most active nearby star forming regions in the Galaxy with a mean age of $\sim$ 3 Myr (Plüschke, 2001). The region as defined by the Cygnus 1.8 MeV emission feature contains numerous massive stars. The galactic O star catalogue lists 96 O stars in this field (Garmany et al., 1982). In addition, one finds 23 Wolf-Rayet stars in this region of which 14 are of WN-type, 8 of WC-type and one is classified as WO-star (van der Hucht, 2001). The Galactic SNR Catalogue lists 19 remnants in this region, for 9 of those age and distance have been estimated with sufficient accuracy (Green, 2011). Beside numerous open clusters, the region contains nine OB associations (Alter et al., 1970; Plüschke, 2001). Fig. 14 shows the distribution of O and WR stars, SNRs and OB associations in the direction of Cygnus as viewed from above the galactic plane. Therefore, the Cygnus region provides a good opportunity to study the physics of massive stars, specially on the nucleosynthesis processes which will enrich the interstellar medium with new elements.

The ^{26}Al emission from the Cygnus region was first reported by the COMPTEL imaging observations (see Fig. 5). With two years of COMPTEL data, del Rio et al. (1996) derived a flux of $(7.0 \pm 1.4) \times 10^{-5}$ph cm^{-2} s^{-1} from a sky region defined as $73^\circ < l < 93^\circ$, $-9^\circ < b < 9^\circ$. Using 9 years of COMPTEL data, Plüschke (2001) found a flux of $(10.3 \pm 2.0) \times 10^{-5}$ph cm^{-2} s^{-1} from a slightly larger region covering $70^\circ < l < 96^\circ$, $-9^\circ < b < 25^\circ$.

INTEGRAL/SPI has done deep surveys around the Cygnus region. ^{26}Al line shape of Cygnus has been detailedly studied by SPI. With early SPI data, the first result of ^{26}Al line emission from the Cygnus region showed a 1809 keV line flux of $(7.3 \pm 1.8) \times 10^{-5}$ph cm^{-2} s^{-1} from a sky region defined as $73^\circ < l < 93^\circ$, $|b| < 9^\circ$ (Knödlseder et al., 2004). The ^{26}Al line from Cygnus appears moderately broadening, with an intrinsic FWHM of (3.3 ± 1.3) keV. Recently, new SPI results show the ^{26}Al flux of $\sim (7.2 \pm 1.2) \times 10^{-5}$ph cm^{-2} s^{-1} for the region

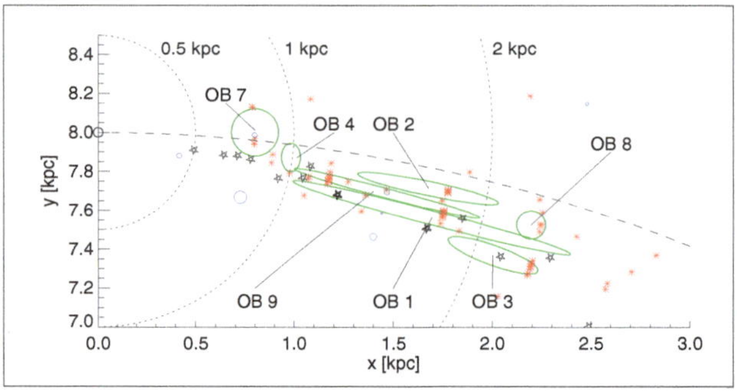

Fig. 14. Distribution of O and WR stars (red and black stars) as well as SNR (blue circles) and OB associations (green ellipses) in the direction of Cygnus as viewed from above the galactic plane (from Plüschke 2001). The sun is allocated along the y-axis at 8.0 kpc.

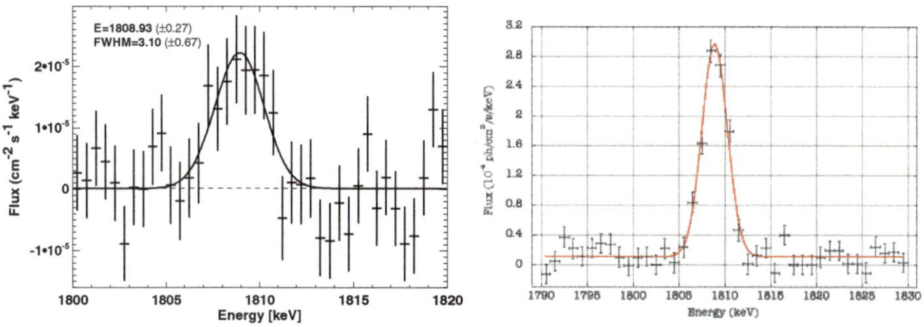

Fig. 15. Spectra of ^{26}Al line emission from the Cygnus region: left panel from Wang (2007); right one from Martin et al. (2009a). Both analyses show the ^{26}Al line flux around $(6-7)\times10^{-5}$ph cm^{-2} s^{-1} with a narrow line feature.

of $65^\circ < l < 95^\circ$, $-13^\circ < b < 17^\circ$ (Fig. 15 left panel), but found a narrow line feature with an intrinsic FWHM of 0.9 ± 0.8 keV (Wang 2007; 2008). The narrow line feature was confirmed by re-analysis SPI data by Martin et al. (2009), in which the ^{26}Al decay emission from Cygnus was represented as a $3^\circ\times3^\circ$ Gaussian centred on $(l,b)=(81^\circ,0.1^\circ)$, a position consistent with that of the massive Cyg OB2 cluster thought to dominate the energetics and nucleosynthesis of the Cygnus complex (Fig. 15 right panel). They obtained an ^{26}Al flux of $(6.0\pm1.0)\times10^{-5}$ph cm^{-2} s^{-1}. In addition, they considered the possible Galactic background and foreground contributions, reducing the ^{26}Al flux of the Cygnus region to $\sim(3.9\pm1.1)\times10^{-5}$ph cm^{-2} s^{-1}. This reduced ^{26}Al flux is comparable with the predicted one based on the latest models of stellar nucleosysnthesis (Martin et al. 2010).

4.5.2 Sco-Cen

The nearest site of massive star formation, the Scorpius-Centaurus-Lupus region (hereafter the Sco-Cen region), is about 100 – 150 pc from the Sun. Nearby young stars are seen mostly in the Southern Hemisphere, which is related to recent massive star formation in Sco-Cen OB association that consists of three sub-regions: Upper Scorpius (US), Upper Centaurus Lupus

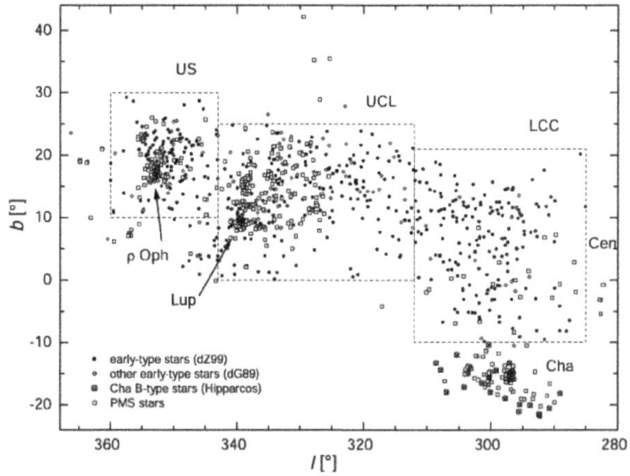

Fig. 16. Distribution of young stars including pre-main sequence stars (PMS) and young early-type stars in Sco-Cen region (from Sartori et al. 2003). Three sub-groups, UC, UCL and LCC are notified with the dashed-line boxes. The Chamaeleon OB association could be an extension of the Sco-Cen one. The positions of the rho Ophiuchus active star formation region and the Lupus cloud complex are also marked.

(UCL), and Lower Centaurus Crux (LCC), each distinguishable by different sky positions, age (de Geus et al., 1989), and kinematics (de Zeeuw et al. 1999, see Fig. 16).

The closest of the three sub-regions is Lower Centaurus-Crux (LCC) with an estimated distance of $\sim$ 120 pc (de Zeeuw et al., 1999). The age of LCC is thought to be 10– 20 Myr (de Geus et al., 1989). LCC is located toward the direction of the Galactic plane ($-70^\circ < l < -40^\circ$), and may contribute to ^{26}Al line emission observed by COMPTEL and SPI in the plane.

US and UCL are located at latitudes above the Galactic plane. The mean distance of UCL is $\sim$ 140 pc, with an age of $\sim$ 10 Myr (de Geus et al., 1989; de Zeeuw et al., 1999). UC is the youngest one with an age of $\sim 5-10$ Myr. Its mean distance is $\sim$ 145 pc (de Geus et al., 1989; de Zeeuw et al., 1999). The rho Ophiuchus star forming region is located near the center of this group (Fig. 16). UC is just located in the direction above the Galactic centre ($l \sim -5^\circ, b \sim 18^\circ$), so the ^{26}Al emission structure in this direction observed by COMPTEL (Fig. 5) is most probably attributed to the UC OB association.

In Fig. 12, the significant excess of ^{26}Al emission in the intermediate latitudes $5^\circ < b < 20^\circ$, $l < 0^\circ$ suggests that this extra ^{26}Al emission should be attributed to nearby Sco-Cen OB associations. With more than 5 years of SPI observations, ^{26}Al gamma-ray signal toward Sco-Cen is detected with a significance level of $\sim 5\sigma$ (Diehl et al., 2010). The ^{26}Al line spectrum toward the Sco-Cen region is presented in Fig. 17. The observed flux of Sco-Cen is about 6×10^{-5}ph cm^{-2} s^{-1} for the region of $12^\circ < b < 32^\circ$, $-16^\circ < l < -2^\circ$. This ^{26}Al flux corresponds to 1.1×10^{-4} $M_\odot$ at a distance of 150 pc. A typical ^{26}Al yield from a massive star in mass range 8–40 $M_\odot$ is about 10^{-4} $M_\odot$ (Limongi & Chieffi, 2006; Woosley & Heger, 2007), which is consistent with the derived ^{26}Al mass from observations. In Upper Sco, the most massive star is presumably a 50 $M_\odot$ O5–O6 star, which may have exploded as a supernova about 1.5 Myr ago; the pulsar PSR J1932+1059 may be its compact remnant (Hoogerwerf et

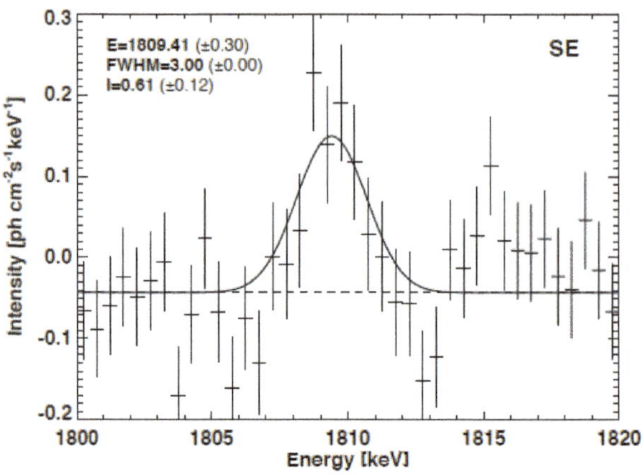

Fig. 17. ^{26}Al spectra for Sco-Cen by the 5-year SPI data. The ^{26}Al flux is derived as $F \sim (6.0 \pm 1.1) \times 10^{-5}$ph cm^{-2} s^{-1} (from Diehl et al., 2010). A significant blueshift of line centroid energy is found.

al., 2000; Chatterjee et al., 2004). But because of uncertainty of this event and no obvious supernova activity being evident in Sco-Cen less than 1 My ago, the observed ^{26}Al could be a product of massive-star nucleosynthesis from a number of sources during an earlier epoch, rather than from a single event, like a supernova explosion.

The line width of $\sim$ 3 keV is consistent with the instrumental line width of SPI at 1809 keV, implying no additional astrophysical line broadening in the Sco-Cen region. If we compare the centroid line energy with the ^{26}Al -decay laboratory value for the γ-ray energy of 1808.63 keV, a blueshift of 0.8 keV is evident. This blueshift suggests a kinematic Doppler effect from bulk motion of ^{26}Al ejecta corresponding to about 140 km s^{-1} toward us. The bulk motions of ^{26}Al ejecta in the Sco-Cen region suggest that the very nearby ^{26}Al sources have a quite different behavior in dynamics from the large scale feature in the inner Galaxy.

The origin of this bulk motions is still uncertain. The cavity surrounding Upper Sco appears to expand with a velocity of about 10±2 km s^{-1}, together with the relative motion of Sco-Cen stars with respect to the Sun of about 5 km s^{-1}, an approaching relative velocity of the entire complex is only few tens of km s^{-1} at most. But decaying ^{26}Al moves at higher velocities. So we exclude the association between the ^{26}Al decays and the cavity walls. The initial velocity of ^{26}Al ejecta from massive star winds and supernova explosion is about 1000 –1500 km s^{-1}, which will slow down due to the interactions with the interstellar medium. Generally, if the medium is homogenous, the expanding ^{26}Al ejecta would be isotropic. In this case, the isotropic motions of ^{26}Al ejecta ($\sim$ 140 km s^{-1}) would induce a significant line broadening, but this Doppler broadening is not detected in SPI observations (Fig. 17). The possibility of inhomogenous ISM around the Sco-Cen region leads to the fast flowing ^{26}Al materials toward us. The other possibility is related to the assumed scenario that the ^{26}Al ejecta originated in Sco-Cen stream into the pre-blown cavity around it, decelerating through turbulence and interactions with the cavity walls (Diehl et al. 2010). The size of the cavity walls will be very large. Thus the whole ^{26}Al emission from the cavity walls is more extended on the sky. The detected region ($12° < b < 32°$, $-16° < l < -2°$) is small but bright compared with the total

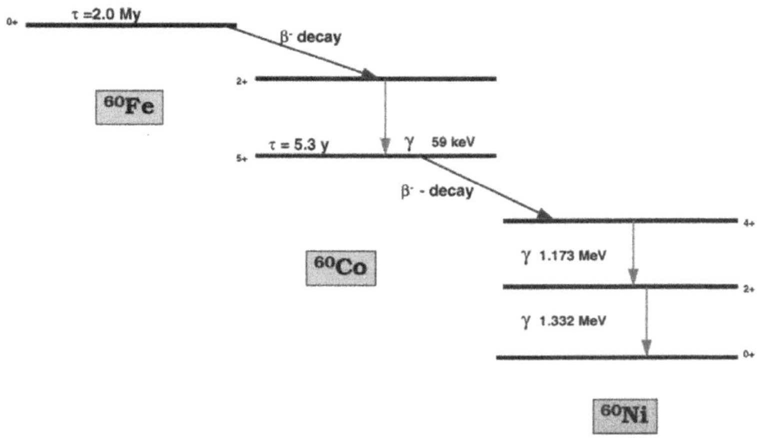

Fig. 18. The decay scheme of ^{60}Fe . The mean lifetime is about $(2-3) \times 10^6$ years. The gamma-ray flux at 59 kev line is $\sim 2\%$ of those at 1173 and 1332 keV.

extended ^{26}Al emission region. However, we cannot detect such more extended emission yet, both because it may be weaker than the main feature we show here, and because other locations of Sco-Cen stars overlap with Galactic-disk viewing directions.

5. Diffuse ^{60}Fe emission of the Galaxy

^{60}Fe is an unstable nucleus whose terrestrial half-life is $\simeq 1.5 \times 10^6$ years (recent measurements suggest a longer half-life time of $\sim 2.6 \times 10^6$ yr, Rugel et al. 2009), and it is located at the neutron rich side of the "valley of stable isotopes". The ^{60}Fe isotope is synthesized in neutron capture reactions from the ^{56}Fe isotope which is abundant from former equilibrium nucleosynthesis of ^{56}Ni and its decay, Such *s*-process is expected to occur in stellar regions with efficient neutron-liberating reactions, e.g., ^{13}C and ^{20}Ne α captures, hence in the O/Ne burning shell and bottom of He burning shell of core-collapse supernovae, and the He burning shell inside massive stars.

The decay chains of ^{60}Fe are shown in Fig. 18. ^{60}Fe firstly decays to ^{60}Co, with emitting γ-ray photons at 59 keV, and then decays to ^{60}Ni, with emitting γ-ray photons at 1173 and 1332 keV. The gamma-ray efficiency of the 59 keV transition is only $\sim 2\%$ of those at 1173 and 1332 keV, so the gamma-ray flux at 59 keV is much lower than the fluxes of the high energy lines. The 59 keV gamma-ray line is very difficult to be detected with present missions. Measurements of the two high energy lines have been the main scientific target to study the radioactive ^{60}Fe isotope in the Galaxy.

^{60}Fe has been found to be part of meteorites formed in the early solar system (Shukolyukov & Lugmair, 1993). The inferred ^{60}Fe /^{56}Fe ratio for these meteorites exceeded the interstellar-medium estimates from nucleosynthesis models, which led to suggestion that the late supernova ejection of ^{60}Fe occurred before formation of the solar system (Tachibana & Huss, 2003; Tachibana et al., 2006). Yet, this is a proof for cosmic ^{60}Fe production, accelerator-mass spectroscopy of seafloor crust material from the southern Pacific ocean has revealed an ^{60}Fe excess in a crust depth corresponding to an age of 2.8 Myr (Knie et al., 2004). From this interesting measurement, it is concluded that a supernova explosion event near the solar system occurred about 3 Myr ago, depositing some of its debris directly in the earth's atmosphere. All these measurements based on material samples demonstrate that ^{60}Fe necleosynthesis does occur in nature. It is now interesting to search for current ^{60}Fe production in the Galaxy through detecting radioactive-decay γ-ray lines.

Many experiments and efforts were made to measure the ^{60}Fe gamma-ray emission. The first detection of ^{60}Fe lines was provided by HEAO-3, the ^{60}Fe flux from the inner Galaxy region ($-30^\circ < l < 30^\circ$) is $\sim (5.3 \pm 4.3) \times 10^{-5}$ ph cm^{-2} s^{-1} rad^{-1} (Mahoney et al., 1982). The SMM Gamma-Ray Spectrometer reported a ^{60}Fe flux of $(2.9 \pm 2.5) \times 10^{-5}$ ph cm^{-2} s^{-1} rad^{-1} (Leising & Share, 1994). OSSE aboard the COMPTON Observatory gave a ^{60}Fe flux of $(6.3 \pm 4.5) \times 10^{-5}$ ph cm^{-2} s^{-1} rad^{-1} (Harris et al., 1994; 1997). And COMPTEL aboard the COMPTON Observatory also reported a ^{60}Fe flux limit of 1.2$\times$ 10^{-4} ph cm^{-2} s^{-1} rad^{-1} (2σ, Diehl et al. 1997). The Gamma-Ray Imaging Spectrometer (GRIS) reported an upper limit for the ^{60}Fe flux of 6.8$\times$ 10^{-5} ph cm^{-2} s^{-1} rad^{-1} (2σ, Naya et al. 1998). Recently, RHESSI reported observations of the gamma-ray lines from ^{60}Fe with a signal of 2.6 σ significance, and an average flux of $(3.6 \pm 1.4) \times 10^{-5}$ph cm^{-2} s^{-1} (Smith, 2004b) from the inner Galaxy.

The analysis of the first year of data from the SPI data resulted in a similarly marginally-significant detection of these γ-ray lines from ^{60}Fe ($\sim 3\sigma$, Harris et al. 2005), with an average line flux of $(3.7 \pm 1.1) \times 10^{-5}$ph cm^{-2} s^{-1} from the inner Galaxy. With more than 3-year SPI data, we detected the diffuse ^{60}Fe emission in the Galaxy with a significance level of $\sim 5\sigma$ by superposing two gamma-ray lines at 1173 keV and 1332 keV (Fig. 19, Wang et al., 2007). Line energies of the ^{60}Fe lines in the laboratory are 1173.23 and 1332.49 keV. For this superposition, we therefore define the zero of the relative energy axis at 1173 and 1333 keV, to derive the summed spectrum of all ^{60}Fe signals. The line flux estimated from the combined spectrum is $(4.4 \pm 0.9) \times 10^{-5}$ph cm^{-2} s^{-1} rad^{-1}.

We also done searching for the possible ^{60}Fe signal in nearby star-formation regions, like the Cygnus and Vela regions. Since the majority of Cygnus region star-clusters are young ($\sim$ 3 Myr), from population synthesis studies of the massive stars in the Cygnus region it has been suggested that the ^{60}Fe production is low, consistent with the small number of recent supernova events inferred for this region ($F(1173$ keV$) \sim 2 \times 10^{-6}$ph cm^{-2} s^{-1}, see Knödlseder et al., 2002). The Vela region including a core-collapse supernova remnant Vela SNR at $d \sim 250$ pc (Cha et al., 1999) should be a good candidate for gamma-ray line emission from ^{60}Fe. With the SPI data, no significant ^{60}Fe signal was detected in theses regions with an upper limit of $\sim 1.1 \times 10^{-5}$ph cm^{-2} s^{-1} (2σ) (Wang et al. 2007; Martin et al., 2009a).

6. The ratio of ^{60}Fe/^{26}Al

^{26}Al and ^{60}Fe would share at least some of the same production sites, i.e. massive stars and supernovae (Timmes et al. 1995; Limongi & Chieffi 2006). In addition both are long-lived radioactive isotopes, so we have good reasons to believe their gamma-ray distributions are similar as well. Therefore we adopt the sky distribution of ^{26}Al gamma-rays as our best model for celestial ^{60}Fe gamma-ray distribution. And we use an ^{26}Al distribution obtained in direct observations, from the 9-year COMPTEL data (Plüschke et al., 2001).

Different theoretical models have predicted the ratio of ^{60}Fe/^{26}Al (Limongi & Chieffi, 2006; Prantzos, 2004; Timmes et al., 1995). Gamma-ray observations could detect these two isotopes and report the flux ratio of ^{60}Fe/^{26}Al which can be directly compared with theories. Therefore,

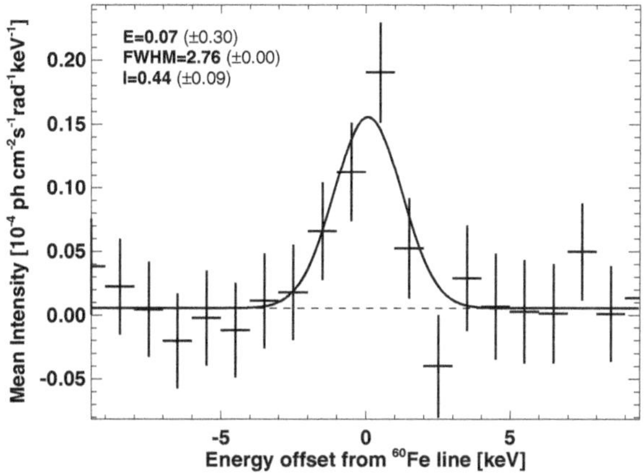

Fig. 19. The combined spectrum of the ^{60}Fe signal in the inner Galaxy by SPI, superimposing two gamma-ray lines (from Wang et al., 2007). In the laboratory, the line energies are 1173.23 and 1332.49 keV; here superimposed bins are zero at 1173 and 1332 keV. We find a detection significance of 4.9σ. The solid line represents a fitted Gaussian profile of fixed instrumental width (2.76 keV), and a flat continuum. The average line flux is estimated as $(4.4 \pm 0.9) \times 10^{-5}$ph cm^{-2} s^{-1} rad^{-1}.

Table 1. Different measurements of ^{60}Fe flux from the inner Galaxy and ^{60}Fe/^{26}Al flux ratio

Experiments	^{60}Fe flux (10^{-5} ph cm^{-2} s^{-1} rad^{-1})	$F(^{60}\mathrm{Fe})/F(^{26}\mathrm{Al})$	references
HEAO-3	5.3 ± 4.3	0.09 ± 0.08	Mahoney et al. 1982
SMM	2.9 ± 2.5	0.1 ± 0.08	Leising & Share 1994
OSSE	6.3 ± 4.5	0.21 ± 0.15	Harris et al. 1997
COMPTEL	$< 12(2\sigma)$	0.17 ± 0.135	Diehl et al. 1997
GRIS	$< 6.8(2\sigma)$	$< 0.14(2\sigma)$	Naya et al. 1998
RHESSI	6.3 ± 5.0	0.16 ± 0.13	Smith 2004a
RHESSI	3.6 ± 1.4	0.10 ± 0.04	Smith 2004b
SPI	3.7 ± 1.1	0.11 ± 0.07	Harris et al. 2005
SPI	4.4 ± 0.9	0.148 ± 0.06	Wang et al. 2007

the measurement of the gamma-ray flux ratio ^{60}Fe/^{26}Al is important for discussions of the astrophysical origins of the two radioactive isotopes, and the nuclear physics involved in models for their production (addressing the uncertain nuclear reaction cross sections and half-lives).

Many experiments and efforts were made (see Table 1) to measure the ^{60}Fe/^{26}Al flux ratio that was predicted by theory – we now provide the most significant detection to date. In Figure 20, we show the previous constraints on the flux ratio of ^{60}Fe/ ^{26}Al together with the present SPI result, and compare the observational results with different theoretical predictions. The earliest observational limit was given from HEAO-3, $F(^{60}\mathrm{Fe})/F(^{26}\mathrm{Al}) = 0.09 \pm 0.08$, an upper limit being 0.27 (Mahoney et al., 1982) (in Fig. 20, we chose to give limits at 2σ for all reported values below a significance of 3σ). Another limit was obtained with the SMM Gamma-Ray Spectrometer, a flux ratio of 0.1 ± 0.08, the upper limit being ~ 0.27 (Leising & Share, 1994). OSSE aboard the COMPTON Observatory gave a flux ratio of 0.21 ± 0.15, and the upper limit is ~ 0.51 (Harris et al., 1994; 1997). COMPTEL aboard the COMPTON Observatory also found ^{60}Fe gamma-rays, and reported a flux ratio value of ^{60}Fe/ ^{26}Al of 0.17 ± 0.135, which translates into an upper limit ~ 0.44 (Diehl et al., 1997). The Gamma-Ray Imaging Spectrometer (GRIS) reported an upper limit for the ratio of < 0.14 (2σ, Naya et al. 1998). RHESSI has reported the first detection of ^{60}Fe gamma-ray lines, and gave a flux ratio 0.16 ± 0.13 for two-year data (Smith, 2004a). The first year data of SPI gave a flux ratio 0.11 ± 0.07 (Harris et al., 2005), and the present analysis of the 3-year SPI data finds a flux ratio 0.148 ± 0.06 (Wang et al. 2007).

Theoretical predictions of the ratio of ^{60}Fe/ ^{26}Al have undergone some changes since Timmes et al. (1995) published the first detailed theoretical prediction. In their paper, they combine a model for ^{26}Al and ^{60}Fe nucleosynthesis in supernova explosions with a model of chemical evolution, to predict that the steady production rates are $(2.0 \pm 1.0) M_{\odot}$ Myr^{-1} for ^{26}Al, and $(0.75 \pm 0.4) M_{\odot}$ Myr^{-1} for ^{60}Fe, which corresponds to a gamma-ray flux ratio $F(^{60}\text{Fe})/F(^{26}\text{Al}) = 0.16 \pm 0.12$. This prediction would be consistent with our present measurements. Since 2002, theoreticians have improved various aspects of the stellar-evolution models, including improved stellar wind models and the corresponding mass loss effects on stellar structure and evolution, of mixing effects from rotation, and also updated nuclear cross sections in the nucleosynthesis parts of the models. As a result, predicted flux ratios ^{60}Fe/ ^{26}Al rather fell into the range 0.8 ± 0.4 (see Prantzos 2004, based on, e.g. Rauscher et al. 2002, Limongi & Chieffi 2003) – such high values would be inconsistent with several observational limits and our SPI result. Recently, new ^{26}Al and ^{60}Fe yield models are presented (Limongi & Chieffi, 2006; Woosley & Heger, 2007), for stars of solar metallicity ranging in mass from $(11 - 120) M_{\odot}$. Limongi & Chieffi (2006) then combined their models of full stellar evolution up to and including the supernova with an adopted mass function to obtain a new prediction with latest nuclear reaction rate inputs. Their calculations yield a lower prediction for the ^{60}Fe/ ^{26}Al flux ratio of 0.185 ± 0.0625, which is again consistent with the observational constraints (see Fig. 20).

In summary, uncertainties still exist, both in models and measurements of ^{60}Fe. On the theory side, stellar evolution in late stages is complex, nuclear reactions include neutron capture on unstable Fe isotopes, and explosive nucleosynthesis adds yet another complex ingredient. On the experimental side, the half-lives of ^{26}Al and ^{60}Fe , cosmic ray induced ^{60}Co radioactivity in the instrument and spacecraft and the limitations of spatial resolutions and sensitivity are issues reflected in the substantial uncertainties in experimental values. With more INTEGRAL/SPI data to come, and also with the development of next-generation gamma-ray spectrometers/telescopes, gamma-ray observations hopefully can help with an independent view on the astrophysical model components.

7. Summary and prospective

^{26}Al and ^{60}Fe have similar astrophysical origins in the Galaxy. Their nucleosynthesis and ejection into the interstellar medium are dominated by massive stars, their evolution, winds, and their subsequent core-collapse supernova explosions. With their much longer half-life (more than 1 million years), ^{26}Al and ^{60}Fe may propagate over significant distances of $\simeq$ few hundred pc, and accumulates in the interstellar medium from many stars and supernovae until injection and β-decay are in balance in the ISM. ^{26}Al emission shows the diffuse

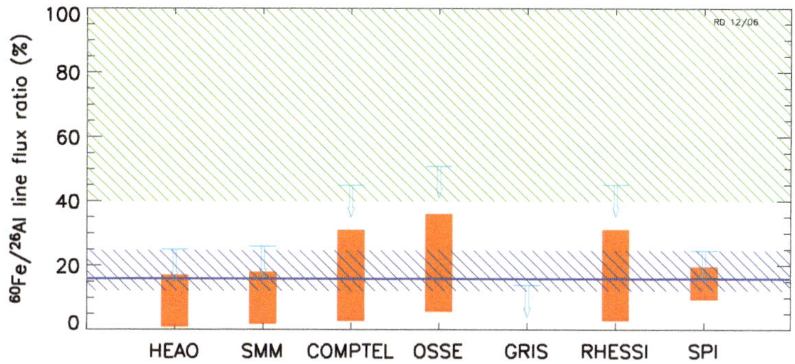

Fig. 20. Flux ratio of the gamma-ray lines from the two long-lived radioactive isotopes ^{60}Fe/^{26}Al from several observations, including our SPI result (also see Table 1; from Wang et al., 2007), with upper limits shown at 2σ for all reported values, and comparison with the recent theoretical estimates (the upper hatched region from Prantzos 2004; the horizontal line taken from Timmes et al. 1995; the lower hatched region, see Limongi & Chieffi 2006). The present SPI results find the line flux ratio to be $(14.8 \pm 6.0)\%$. Note that the primary instrument results on ^{60}Fe is compromised in the figure by different and uncertain ^{26}Al fluxes used for the ^{60}Fe/^{26}Al ratio.

Galaxy-wide feature which has been confirmed by the COMPTEL imaging observations, and Galactic ^{60}Fe emission may follow the similar diffuse distribution.

Diffuse γ-ray emissions from two long-lived radioactive isotopes ^{26}Al and ^{60}Fe in the Galaxy are significantly detected by the INTEGRAL/SPI. With high spectral resolution in MeV energy band, SPI has made the great progress on detecting diffuse ^{26}Al and ^{60}Fe emission and studying their line shapes. Details of line shapes help us to understand the dynamics of the ejected isotopes in the interstellar medium (e.g., nearby star formation regions) and in large scales of the Galaxy.

The new results on diffuse emissions of Galactic ^{26}Al and ^{60}Fe from more than five years of INTEGRAL/SPI observations and their astrophysical implications are briefly summarized here as follows:

(1) The study of the ^{26}Al line and its details in the large-scale of the Galaxy is one of the main goals of the SPI spectrometer on INTEGRAL. Using SPI data, we detect ^{26}Al γ-ray emission signal in the Galaxy with a significance level of $\sim 30\sigma$. The ^{26}Al flux integrated over the inner Galaxy of $(2.9 \pm 0.2) \times 10^{-4}$ ph cm^{-2} s^{-1} rad^{-1}. Taking the distance of the Sun to the Galactic center as $R_0 = 8.5$ kpc, we convert the measured ^{26}Al flux to a Galactic ^{26}Al mass of (2.6 ± 0.6) $M_\odot$. With detecting diffuse ^{26}Al emission in the Galaxy, we can also independently derive the core-collapse supernova rate of 1.9 ± 0.9 per century.

(2) The ^{26}Al line centroid energy appears blue-shifted relative to the laboratory value of 1808.65 ± 0.07 keV if integrated over the inner Galaxy ($\sim 1809.0 \pm 0.1$ keV). With refined spatial resolution this turns out to be mostly due to asymmetric bulk motion which we find along the plane of the Galaxy, and attribute to asymmetries in inner spiral arms and to the Galaxy's bar.

(3) The measured line width of ^{26}Al from the large-scale integrated inner Galaxy with an upper limit of 1.3 keV (2σ) is consistent with expectations from both Galactic rotation and modest interstellar-medium turbulence around the sources of ^{26}Al (turbulent velocities constrained below 160 km s^{-1} even when disregarding the effects of galactic rotation).

(4) The brightness asymmetry of the ^{26}Al line emission for the 1st and 4th quadrants is discovered with a flux ratio of $F^{4th}/F^{1st} \sim 1.3$. The ^{26}Al emission asymmetry between fourth and first quadrant of the Galaxy seems to be lower than the intensity contrast of $1.8^{+0.5}_{-0.3}$ reported for positron annihilation emission from the disk of the Galaxy (Weidenspointner et al. 2008). ^{26}Al by itself releases a positron in 82% of its decays; it is uncertain, however, how this translates into annihilation photons, from the variety of slowing down and annihilation processes which determine the fate of positrons in interstellar space. Both spatial and temporal variations and non-linearities scaling with gas density may occur.

(5) The study of ^{26}Al emission in spatially-restricted regions along the plane of the Galaxy found the ^{26}Al line centroid energy shifts along the Galactic plane: redshifts in the position longitudes and blueshifts in the negative ones. The centroid line shift curve is consistent with the Galactic rotation curve. What's more, comparing the results for ^{26}Al ejecta velocity along the Galactic longitudes with other observations like HI, CO detections and theory still suggests that the hot gas (^{26}Al) undergoes additional acceleration in the direction of Galactic rotation (Fig. 11 bottom panel). Further analysis suggested that ^{26}Al velocity curves are incompatible with the circular orbits (Kretschmer, 2011). This implies that the Galactic bar would have a major influence on the kinematics of ^{26}Al ejecta in the inner Galaxy. Anther possibility would be that ^{26}Al ejecta from the young star populations in molecular clouds would be emitted in the direction of Galactic rotation, possibly as a result of interaction with nearby clouds, thereby reaching the high radial velocities observed by SPI (Kretschmer, 2011).

(6) The ^{26}Al line towards the direction of $20^\circ < l < 40^\circ$ shows a hint for additional line broadening. This line broadening may be attributed to the Aquila region which shows increased interstellar turbulence from stellar-wind and supernova activity at the characteristic age of stellar groups of that region, which may have created a supershell of substantial size (320×550 pc, see Maciejewski et al. 1996).

(7) The scale height of Galactic-plane ^{26}Al emission is 130^{+120}_{-70} pc (1σ), determined towards the inner Galaxy ($|l| < 60^\circ$). The scale height of Galactic ^{26}Al distributions is larger than the distributions of molecular clouds which is generally around 50 pc. However, this scale height is still significantly smaller than the "thick disk" part of the Galaxy. It is consistent with ^{26}Al being ejected from star-forming regions, and partly extending more towards the Galactic halo where gas pressure is lower than within the plane of the Galaxy ("champagne flows").

(8) ^{26}Al emission brightness decreases significantly along the Galactic latitudes. But an additional ^{26}Al emission is discovered toward the intermediate latitude region of $5^\circ < b < 30^\circ$, $l < 0^\circ$. This component shows a hint for the ^{26}Al emission originated in nearby star-formation region Sco-Cen.

(9) ^{26}Al line shapes from nearby star-formation regions (e.g., Cygnus and Sco-Cen) are obtained. The ^{26}Al line flux from the Cygnus region is $\sim 6 \times 10^{-5}$ ph cm^{-2} s^{-1}, consistent with the previous COMPTEL result. No significant line broadening and shifts are detected. The Sco-Cen region is the nearest star-formation region. INTEGRAL/SPI first discovered the significant ^{26}Al line emission from this region and derived the ^{26}Al line shape. The ^{26}Al flux

from the Sco-Cen region is found to be 6×10^{-5} ph cm^{-2} s^{-1}. In addition a blueshift of ~ 0.8 keV for ^{26}Al line centroid energy was discovered. The blueshift suggested that the observed ^{26}Al ejecta in Sco-Cen have a bulk velocity of ~ 140 km s^{-1} toward us. The origin of the ^{26}Al ejecta bulk motions is still unknown, which would require more observational constraint.

(10) INTEGRAL/SPI first confirmed the existence of Galactic ^{60}Fe by detecting the γ-ray emission lines at 1173 and 1332 keV from the sky. ^{60}Fe signal was detected with a significance level of $\sim 5\sigma$ by combining the two γ-ray emission lines. The average line flux is derived to be 4.4×10^{-5} ph cm^{-2} s^{-1}.

(11) The flux ratio of ^{60}Fe /^{26}Al is derived to be $\sim 15\%$. This ratio provides a good constraint on the present nuclear reaction rate and massive star evolution models. There exist uncertainties in both observations and theories. On the theory side, stellar evolution in late stages is complex, nuclear reactions include neutron capture on unstable Fe isotopes, and explosive nucleosynthesis adds yet another complex ingredient. On the experimental side, half-lives of isotopes, cosmic ray induced ^{60}Co radioactivity in the instrument and spacecraft and the limitations of spatial resolutions and sensitivity are issues reflected in the substantial uncertainties in experimental values.

INTEGRAL still continues to accumulate the observational data for several years, with more exposure on the Galactic plane, covering more sky regions, specially in the intermediate latitudes. The deeper observations along the Galactic plane will increase the significance of the present results on ^{26}Al and ^{60}Fe . The further studies with high-resolution ^{26}Al line shapes will help to better understand the properties of ISM near groups of massive stars, and the bulk motion of gas in the inner Galaxy from Galactic rotation and other peculiar kinematics. The possible detection of ^{60}Fe signal variations along the Galactic plane can be carried out with nearly ten years of INTEGRAL/SPI data. Nearby ^{26}Al sources may be also discriminated with better exposure towards candidate sky regions, improving our determination of the ^{26}Al mass in the Galaxy and towards regions where the stellar census is known to a better degree. Studies of ^{26}Al from nearby star-formation regions (e.g., Cygnus, Vela, Sco-Cen, Orion) are a promising diagnostic for the massive star origin of Galactic ^{26}Al and kinematics of ^{26}Al ejecta in ISM. Presently ^{26}Al line shapes are derived only in two nearby regions Cygnus and Sco-Cen. The ^{26}Al emission signal from Vela and Orion regions and their line shapes would be studied with INTEGRAL/SPI. These studies will provide the important constraints on the nucleosynthesis of supernovae and massive stars.

SPI aboard INTEGRAL has a high spectral resolution but still a very limited sensitivity over 0.1 – 8 MeV, e.g., ^{26}Al emission can be only detected in the inner Galaxy and nearby star-formation regions, but no signal is detected from individual sources, like supernova remnants. The MeV gamma-ray astronomy retards for a long time for the lack of a sensitive instrument from 0.1 – 10 MeV compared to the present keV and GeV band instruments (Chandra, XMM-Newton, Fermi/LAT). Therefore, the development of the next-generation MeV gamma-ray telescope is very necessary and urgent in multi-wavelength instrument sensitivity for full understanding of the high energy astrophysics, specially in the new observational window for the nuclear radioactivity gamma-ray line emissions, e.g., ^{26}Al , ^{60}Fe , ^{44}Ti, ^{56}Co and 511 keV annihilation line (also see Fig. 21). The potential γ-ray line sources: the 511 keV line from Galactic novae and nearby supernova remnants; ^{60}Fe from the Vela SNR; ^{56}Co from SN Ia at a distance up to 80 Mpc; ^{44}Ti from more young SNRs, e.g. Tycho, Kepler, SN 1987A; ^{26}Al from individual sources, like the Vela SNR, the closest Wolf-Rayet star WR

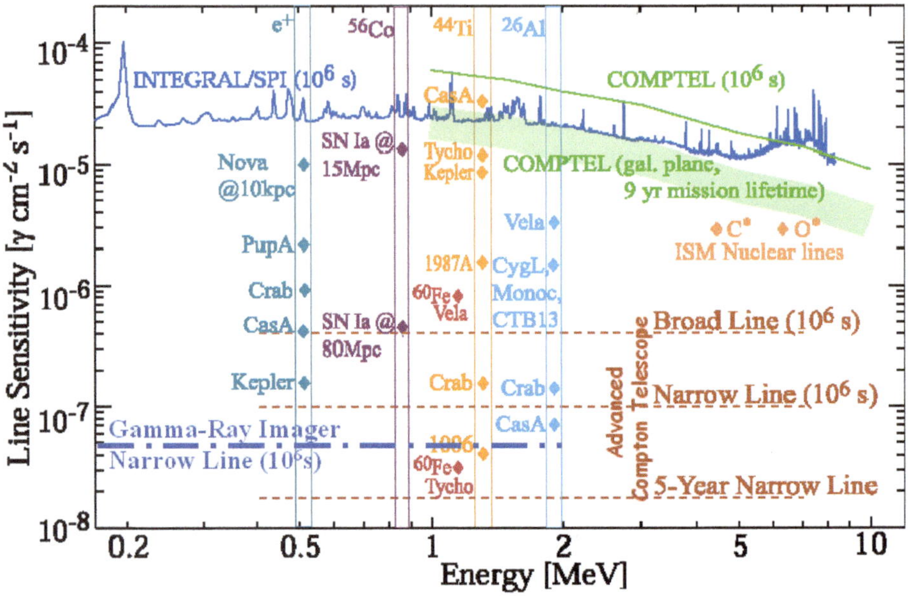

Fig. 21. Future goals for γ-ray line astronomy (0.2 – 10 MeV): possible next-generation instruments (e.g., Advanced-Compton Telescope, Gamma-Ray Imager) and potential candidate γ-ray line sources for e^-e^+ annihilation line, ^{56}Co, ^{44}Ti, ^{60}Fe and ^{26}Al (from Boggs et al. 2003). Sensitivities of SPI and COMPTEL are also shown here for comparison.

11 in the Vela star-formation region, even from Crab and Cas A; and the nuclear exciting lines in ISM from C and O, are well below the sensitivity limits of present missions, but could be detected by the future advanced γ-ray telescopes, i.e., visionary Advanced Compton Telescope (ACT) and European Gamma-Ray Imager (GRI) mission. Detections of the 511 keV line, ^{44}Ti, ^{26}Al and ^{60}Fe from some individual SNRs will put the connection with yields of these radioactivity isotopes from stellar nucleosynthesis on solid ground. Era for accurate γ-ray line astronomy and nuclear astrophysics may come near with the launch of next-generation MeV γ-ray missions.

8. Acknowledgments

The author thanks the book editor for inviting me to write a book chapter in addressing the issues of long-lived radioactivity isotopes in the Galaxy. This work is based on observations of INTEGRAL, an ESA project with instrument and science data center funded by ESA member states (principle investigator countries: Demark, France, Germany, Italy, Switzerland and Spain), the Czech Republic and Poland, and with participation of Russia and US. We are also grateful to the support by the National Natural Science Foundation of China under grants 10803009, 10833003, 11073030.

9. References

Alter, G. and Balazs, B. and Ruprecht, J. and Vanysek, J. (1970). Catalogue of star clusters and associations, Budapest: Akademiai Kiado, 2nd ed., edited by G. Alter, B. Balazs,

J. Ruprecht
Arnett, W. D. (1977). Eight Texas Symposium on Relativistic Astrophysics, Papagiannis, M. D., ed., 90
Bloemen, H. et al. (1994). Astrophysical Journal Supplements, Vol. 92, 419-423
Boggs, S. E. et al. (2003). X-Ray and Gamma-Ray Telescopes and Instruments for Astronomy, Proceedings of the Society of Photo-Optical Instrumentation Engineers (SPIE) Conference, Truemper, J. E. and Tananbaum, H. D. ed., Vol. 4851, 1221-1227
Brandt, J. C. (1969). Solar Physics, Vol. 6, 171
Brand, J. and Blitz, L. (1993). Astronomy & Astrophysics, Vol. 275, 67
Burbidge, E. M., Burbidge, G. R., Fowler, W. A. and Hoyle, F. (1957). Synthesis of the Elements in Stars, Reviews of Modern Physics, Vol. 29, Issue 4, 547-650
Caughlan, G. R. and Fowler, W. A. (1988). Atomic Data and Nuclear Data Tables, Vol. 40, 283
Cha, A. N. and Sembach, K. R. and Danks, A. C. (1999). Astrophysical Journal Letters, Vol. 515, L25-L28
Chatterjee, S. et al. (2004). Pulsar Parallaxes at 5 GHz with the Very Long Baseline Array, Astrophysical Journal, Vol. 604, 339-345
Chen, W. and Diehl, R. and Gehrels, N. and Hartmann, D. and Leising, M. and Naya, J. E. and Prantzos, N. and Tueller, J. and von Ballmoos, P. (1997). ESA SP-382: The Transparent Universe, Winkler, C. and Courvoisier, T. J.-L. and Durouchoux, P., ed., 105
Clayton, D. D. & Leising, M. D. (1987). 26Al in the interstellar medium, Physics Report, Vol. 144, 1-50
Clayton, D. D. (1994). Nature, Vol. 368, 222
de Geus, E. J. and de Zeeuw, P. T. and Lub, J. (1989). Astronomy & Astrophysics, Vol. 216, 44-61
del Rio, E. et al. (1996). Astronomy & Astrophysics, Vol. 315, 237-242
de Zeeuw, P. T. and Hoogerwerf, R. and de Bruijne, J. H. J. and Brown, A. G. A. and Blaauw, A. (1999). Astronomical Journal, Vol. 117, 354-399
Diehl, R. et al. (1995a). Astronomy & Astrophysics, Vol. 298, 445
Diehl, R. et al. (1995b). Astronomy & Astrophysics, Vol. 298, L25-L28
Diehl, R. et al. (1997). AIP Conf. Proc. 410: Proceedings of the Fourth Compton Symposium, Dermer, C. D. and Strickman, M. S. and Kurfess, J. D. ed., 1109
Diehl, R. et al. (1998). IAU Colloq. 166, D. Breitschwerdt, M. J. Freyberg and J. Truemper, ed., Lecture Notes in Physics, Berlin Springer Verlag, 506, 393
Diehl, R. and Timmes, F. X. (1998). Gamma-Ray Line Emission from Radioactive Isotopes in Stars and Galaxies The Publications of the Astronomical Society of the Pacific, Vol. 110, 637-659
Diehl, R. and Halloin, H. and Kretschmer, K. and Lichti, G. G. and Schönfelder, V. and Strong, A. W. and von Kienlin, A. and Wang, W. and Jean, P. and Knödlseder, J. and Roques, J.-P. and Weidenspointner, G. and Schanne, S. and Hartmann, D. H. and Winkler, C. and Wunderer, C. (2006a). Nature, Vol. 439, 45-47
Diehl, R. and Halloin, H. and Kretschmer, K. and Strong, A. W. and Wang, W. and Jean, P. and Lichti, G. G. and Knödlseder, J. and Roques, J.-P. and Schanne, S. and Schönfelder, V. and von Kienlin, A. and Weidenspointner, G. and Winkler, C. and Wunderer, C. (2006b). Astronomy & Astrophysics, Vol. 449, 1025-1031
Diehl, R.; Lang, M. G.; Martin, P.; Ohlendorf, H.; Preibisch, Th.; Voss, R.; Jean, P.; Roques, J.-P.; von Ballmoos, P.; Wang, W. (2010). Radioactive 26Al from the Scorpius-Centaurus association, Astronomy & Astrophysics, Vol. 522, 51

Durouchoux, P. et al. (1993). Astrophysics & Space Science, Vol. 97, 185-187
Gamow, G. (1946). Expanding Universe and the Origin of Elements, Physical Review, Vol. 70, 572-573
Garmany, C. D. and Conti, P. S. and Chiosi, C. (1982). Astrophysical Journal, Vol. 263, 777-790
Gehrels, N. and Chen, W. (1996). Astronomy & Astrophysics Supplements, Vol. 120, 331
Green, D. A. (2011). Catalogue of Galactic Supernova Remnants, URL: "http://www.mrao.cam.ac.uk/surveys/snrs/"
Guillout, P. and Sterzik, M. F. and Schmitt, J. H. M. M. and Motch, C. and Neuhaeuser, R. (1998). Astronomy & Astrophysics, Vol. 337, 113-124
Harris, M. J. et al. (1994). Bulletin of the American Astronomical Society, 968
Harris, M. J. et al. (1997). AIP Conf. Proc. 410: Proceedings of the Fourth Compton Symposium, Dermer, C. D. and Strickman, M. S. and Kurfess, J. D. ed., 1079
Harris, M. J. et al. (2005). Astronomy & Astrophysics, Vol. 433, L49-L52
Hillebrandt, W. and Thielemann, F.-K. (1982). Astrophysical Journal, Vol. 255, 617-623
Hoogerwerf, R.; de Bruijne, J. H. J.; de Zeeuw, P. T. (2000). The Origin of Runaway Stars, Astrophysical Journal Letters, Vol. 544, L133-L136
Iyudin, A. F. and Diehl, R. and Bloemen, H. and Hermsen, W. and Lichti, G. G. and Morris, D. and Ryan, J. and Schoenfelder, V. and Steinle, H. and Varendorff, M. and de Vries, C. and Winkler, C. (1994). Astronomy & Astrophysics, Vol. 284, L1-L4
Jensen, P. L. and Clausen, K. and Cassi, C. and Ravera, F. and Janin, G. and Winkler, C. and Much, R. (2003). Astronomy & Astrophysics, Vol. 411, L7-L17
Jose, J. and Hernanz, M. and Coc, A. (1997). Astrophysical Journal Letters, Vol. 479, L55-L58
Karakas, A. I. and Lattanzio, J. C. (2003). Publications of the Astronomical Society of Australia, Vol. 20, 279-293
Knie, K. and Korschinek, G. and Faestermann, T. and Dorfi, E. A. and Rugel, G. and Wallner, A. (2004). Physical Review Letters, Vol. 93, 171103
Knödlseder, J. (1999). Astrophysical Journal, Vol. 510, 915-929
Knödlseder, J. and Cerviño, M. and Le Duigou, J.-M. and Meynet, G. and Schaerer, D. and von Ballmoos, P. (2002). Astronomy & Astrophysics, Vol. 390, 945-960
Knödlseder, J. et al. (2004). ESA SP-552: 5th INTEGRAL Workshop on the INTEGRAL Universe, Schoenfelder, V. and Lichti, G. and Winkler, C. ed., 33
Kretschmer, K. and Diehl, R. and Hartmann, D. H. (2003). Astronomy & Astrophysics, Vol. 412, L47-L51
Kretschmer, K. (2011). PhD Thesis, Technische Universität München
Langer, N. (1989). Astronomy & Astrophysics, Vol. 220, 135-143
Langer, N. and Braun, H. and Fliegner, J. (1995). Astrophysics & Space Science, Vol. 224, 275-278
Leising, M. D. and Share, G. H. (1994). Astrophysical Journal, Vol. 424, 200-207
Limongi, M. and Chieffi, A. (2006). Astrophysical Journal, Vol. 647, 483-500
Maciejewski, W. et al. (1996). Astrophysical Journal, Vol. 469, 238
Mahoney, W. A. and Ling, J. C. and Jacobson, A. S. and Lingenfelter, R. E. (1982). Astrophysical Journal, Vol. 262, 742
Mahoney, W. A. and Ling, J. C. and Wheaton, W. A. and Jacobson, A. S. (1984). Astrophysical Journal, Vol. 286, 578-585
Martin, P.; Knödlseder, J.; Diehl, R.; Meynet, G. (2009). New estimates of the gamma-ray line emission of the Cygnus region from INTEGRAL/SPI observations, Astronomy & Astrophysics, Vol. 506, 703-710

Martin, P.; Knödlseder, J.; Meynet, G.; Diehl, R. (2010). Predicted gamma-ray line emission from the Cygnus complex, Astronomy & Astrophysics, Vol. 511, 86
McKee, C. F. and Williams, J. P. (1997). Astrophysical Journal, Vol. 476, 144
Meynet, G. and Arnould, M. and Prantzos, N. and Paulus, G. (1997). Astronomy & Astrophysics, Vol. 320, 460-468
Miller, G. E. and Scalo, J. M. (1979). Astrophysical Journal Supplements, Vol. 41, 513-547
Mowlavi, N. and Meynet, G. (2000). Astronomy & Astrophysics, Vol. 361, 959-976
Naya, J. E. et al. (1998). Gamma-Ray Limits on Galactic 60Fe Nucleosynthesis and Implications on the Origin of the 26Al Emission, Astrophysical Journal Letters, Vol. 499, L169-L172
Nofar, I. and Shaviv, G. and Starrfield, S. (1991). Astrophysical Journal, Vol. 369, 440-450
Nugis, T. and Lamers, H. J. G. L. M. (2000). Astronomy & Astrophysics, Vol. 360, 227-244
Oberlack, U. et al. (1996). Astronomy & Astrophysics Supplements, Vol. 120, 311
Palacios, A. and Meynet, G. and Vuissoz, C. and Knödlseder, J. and Schaerer, D. and Cerviño, M. and Mowlavi, N. (2005). Astronomy & Astrophysics, Vol. 429, 613-624
Perrot, C. A. and Grenier, I. A. (2003). Astronomy & Astrophysics, Vol. 404, 519-531
Plüschke, S. (2001). PhD Thesis, Technische Universität München
Plüschke, S. and Diehl, R. and Schönfelder, V. and Bloemen, H. and Hermsen, W. and Bennett, K. and Winkler, C. and McConnell, M. and Ryan, J. and Oberlack, U. and Knödlseder, J. (2001). ESA SP-459: Exploring the Gamma-Ray Universe, Gimenez, A. and Reglero, V. and Winkler, C. ed., 55-58
Prantzos, N. and Diehl, R. (1996). Radioactive 26Al in the galaxy: observations versus theory, Physics Reports, Vol. 267, 1-69
Prantzos, N. (2004). Astronomy & Astrophysics, Vol. 420, 1033-1037
Purcell, W. R. and Cheng, L.-X. and Dixon, D. D. and Kinzer, R. L. and Kurfess, J. D. and Leventhal, M. and Saunders, M. A. and Skibo, J. G. and Smith, D. M. and Tueller, J. (1997). Astrophysical Journal, Vol. 491, 725
Ramaty, R. and Lingenfelter, R. E. (1977), Astrophysical Journal Letters, Vol. 213, L5-L7
Ramaty, R. (1996). Astrophysics & Space Science, Vol. 120, 373
Ramaty, R. and Lingenfelter, R. E. (1979). Nature, 278, 127-132
Rauscher, T. and Heger, A. and Hoffman, R. D. and Woosley, S. E. (2002). Astrophysical Journal, Vol. 576, 323-348
Reed, B. C. (2005). Astronomical Journal, Vol. 130, 1652-1657
Rice, E. L., Prato, L. & McLean, I. S. 2006, Astrophysical Journal, Vol. 647, 432
Rolfs, C. E. and Rodney, W. S. (1994). Cauldrons in the Cosmos, University of Chicago Press
Rugel, G.; Faestermann, T.; Knie, K.; Korschinek, G.; Poutivtsev, M.; Schumann, D.; Kivel, N.; Günther-Leopold, I.; Weinreich, R.; Wohlmuther, M. (2009). Physical Review Letters, Vol. 103, Issue 7, id. 072502
Sartori, M. J. and Lépine, J. R. D. and Dias, W. S. (2003). Astronomy & Astrophysics, Vol. 404, 913-926
Schoenfelder, V. and Varendorff, M. (1991). AIP Conf. Proc. 232: Gamma-Ray Line Astrophysics, Durouchoux, P. and Prantzos, N. ed., 101-115
Schoenfelder, V. et al. (1993). Astrophysical Journal Supplements, Vol. 86, 657-692
Share, G. H. and Kinzer, R. L. and Kurfess, J. D. and Forrest, D. J. and Chupp, E. L. and Rieger, E.. (1985). Astrophysical Journal Letters, Vol. 292, L61-L65
Shukolyukov, A. and Lugmair, G. W. (1993). Science, Vol. 259, 1138-1142
Smith, D. M. (2003). Astrophysical Journal Letters, Vol. 589, L55-L58
Smith, D. M. (2004a). New Astronomy Review, Vol. 48, 87-91

Smith, D. M. (2004b). ESA SP-552: 5th INTEGRAL Workshop on the INTEGRAL Universe, Schoenfelder, V. and Lichti, G. and Winkler, C. ed., 45
Tachibana, S. and Huss, G. R. (2003). Astrophysical Journal Letters, Vol. 588, L41-L44
Tachibana, S. and Huss, G. R. and Kita, N. T. and Shimoda, G. and Morishita, Y. (2006). Astrophysical Journal Letters, Vol. 639, L87-L90
Taylor, J. H. and Cordes, J. M. (1993). Astrophysical Journal, Vol. 411, 674-684
Thielemann, F.-K. and Nomoto, K. and Hashimoto, M.-A. (1996). Astrophysical Journal, Vol. 460, 408
Timmes, F. X. and Woosley, S. E. and Hartmann, D. H. and Hoffman, R. D. and Weaver, T. A. and Matteucci, F. (1995). Astrophysical Journal, Vol. 449, 204
Ubertini, P. et al. (2003). Astronomy & Astrophysics, Vol. 411, L131-L139
van der Hucht, K. A. and Hidayat, B. and Admiranto, A. G. and Supelli, K. R. and Doom, C. (1988). Astronomy & Astrophysics, Vol. 199, 217-234
van der Hucht, K. A. (2001). The VIIth catalogue of galactic Wolf-Rayet stars, New Astronomy Review, Vol. 45, 135-232
Vedrenne, G. et al. (2003). Astronomy & Astrophysics, Vol. 411, L63-L70
von Ballmoos, P. and Diehl, R. and Schoenfelder, V. (1987). Astrophysical Journal, Vol. 318, 654-663
Wallerstein, G. et al. (1997). Synthesis of the elements in stars: forty years of progress, Reviews of Modern Physics, Vol. 69, Issue 4, 995-1084
Wang, W. (2007). PhD Thesis, Technische Universität München
Wang, W. et al. (2007). SPI observations of the diffuse 60Fe emission in the Galaxy, Astronomy & Astrophysics, Vol. 469, 1005-1012
Wang, W. (2008). Study of Long-lived Radioactive Sources in the Galaxy with INTEGRAL/SPI, Publications of the Astronomical Society of the Pacific, Vol. 120, 118-119
Wang, W. et al. (2009). Spectral and intensity variations of Galactic 26Al emission, Astronomy & Astrophysics, Vol. 496, 713-724
Ward, R. A. and Fowler, W. A. (1980). Astrophysical Journal, Vol. 238, 266-286
Weaver, T. A. and Woosley, S. E. (1993). Physics Reports, Vol. 227, 65-96
Weidenspointner, G. et al. (2008). Nature, Vol. 451, 159
Weiss, A. and Truran, J. W. (1990). Astronomy & Astrophysics, Vol. 238, 178-186
Wiescher, M. and Gorres, J. and Thielemann, F.-K. and Ritter, H. (1986). Astronomy & Astrophysics, Vol. 160, 56-72
Winkler, C. et al. (2003). Astronomy & Astrophysics, Vol. 411, L1-L6
Woosley, S. E. and Weaver, T. A. (1980). Astrophysical Journal, Vol. 238, 1017-1025
Woosley, S. E. (1986). Saas-Fee Advanced Course 16: Nucleosynthesis and Chemical Evolution, Audouze, J. and Chiosi, C. and Woosley, S. E., ed., 1
Woosley, S. E., Hartmann, D. H., Hoffman, R. D. and Haxton, W. C. (1990). Astrophysical Journal, Vol. 356, 272-301
Woosley, S. E. and Weaver, T. A. (1995). Astrophysical Journal Supplements, Vol. 101, 181
Woosley, S. E. (1997). Astrophysical Journal, Vol. 476, 801
Woosley, S. E. & Heger, A. (2007). Nucleosynthesis and remnants in massive stars of solar metallicity, Physics Reports, Volume 442, Issue 1-6, p. 269-283

2

Nuclear Excitation Processes in Astrophysical Plasmas

G. Gosselin[1], P. Mohr[2,3], V. Méot[1] and P. Morel[1]
[1]CEA,DAM,DIF, Arpajon
[2]Diakonie-Klinikum, Schwäbisch Hall
[3]ATOMKI, Debrecen
[1]France
[2]Germany
[3]Hungary

1. Introduction

In general, nuclear transitions are almost independent of the atomic environment of the nucleus. This feature is a basic prerequisite for the widely used nuclear chronometers (with the most famous example of ^{14}C, the so-called radiocarbon dating). However, a closer look at the details of nuclear transitions shows that under special circumstances the atomic environment may affect nuclear transitions. This is most obvious for electron capture decays where the nucleus captures an electron (typically from the lowest K-shell). A nice example for the experimental verification of this effect is the dependence of the electron capture half-life of ^{7}Be on the chemical form of the beryllium sample (Ohtsuki et al., 2004). Also the half-lives of β^--decays may be affected by the environment: for fully ionized nuclei the emitted electron may remain in the (otherwise completely occupied) K-shell, thus enhancing the decay Q-value and decay rate. An experimental verification was found for ^{187}Re (Bosch et al., 1996). As electron densities in solids may also vary with temperature (e.g. in the Debye-Hückel model), β-decay half-lives may also depend on temperature. However, the latest study of the decay branching between β^--decay and β^+-decay/electron capture in ^{74}As could not confirm earlier claims in this direction (Farkas et al., 2009). The relevance of temperature and density dependence of β^--decay has been studied in detail in the review (Takahashi and Yokoi, 1987).

Contrary to the above mentioned β-decays where the role of electrons in the environment of the nucleus is obvious, the present study investigates electromagnetic transitions in nuclei. We also do not analyze electron screening where stellar reaction rates between charged particles at extremely low energies are enhanced because the repulsive Coulomb force between the positively charged nuclei is screened by the electrons in the stellar plasma. Details on electron screening can also be found in this book (Kücük, 2012) and in the latest review of solar fusion reactions (Adelberger et al., 2011).

The electromagnetic transitions under study in this chapter are extremely important in almost any astrophysical scenario. Capture reactions like (p,γ), (n,γ), and (α,γ) play key roles in hydrostatic and explosive burning of stars, in the neutrino production of our sun, and in the synthesis of heavy elements in the so-called s-process and r-process. Photodisintegration reactions like (γ,p), (γ,n) and (γ,α) define the reaction path in the so-called γ-process which produces a significant amount of the rare p-nuclei. In addition, half-lives of isomeric states may be affected under stellar conditions via photon-induced excitation of so-called intermediate states.

As we will show in this chapter, γ-transitions which are most affected by the electronic environment are found in heavy nuclei and are characterized by relatively low γ-transition energies below approximately 100 keV. First of all, astrophysical processes have to be identified where such γ-transitions play an important role.

In early burning stages of stars from hydrogen burning up to silicon burning heavier nuclei are synthesized mainly by capture reactions along the valley of stability. Typical Q-values of these capture reactions between light nuclei are of the order of several MeV. In these scenarios γ-transitions are practically not affected by the surrounding plasma. A possible exception in the ^{7}Be(p,γ)^{8}B reaction will be discussed separately as a special example later in this chapter.

The synthesis of heavy nuclei proceeds mainly via neutron capture reactions in the slow and rapid neutron capture processes (s-process, r-process). The s-process path is located close to stability, and typical Q-values for neutron capture reactions are again of the order of several MeV. The corresponding capture γ-rays are also not significantly affected by the environment. As the r-process operates close to the neutron dripline, typical Q-values decrease down to about 2-3 MeV or even below. However, under typical r-process conditions an equilibrium between the (n,γ) capture and (γ,n) photodisintegration reaction is found, and the r-process path becomes mainly sensitive to the neutron separation energies, but almost independent of the corresponding (n,γ) and (γ,n) cross sections. Although there may be some influence of the plasma environment on the low-energy γ-transitions in the r-process, there is no significant influence on the outcome of the r-process.

Contrary to the s-process and the r-process, the so-called rp-process proceeds via proton captures on the neutron-deficient side of the chart of nuclides close to the proton dripline. The Q-values of these (p,γ) reactions may become small. However, in general not much is known on nuclei on the path of the rp-process, and thus any discussion of the influence of the surrounding plasma on low-energy γ-transitions in the rp-process must remain quite speculative and is omitted in this chapter.

In the so-called p-process or γ-process existing heavy seed nuclei are destroyed in the thermal photon bath of a hot environment by (γ,p), (γ,n) and (γ,α) reactions leading to the production of the low-abundance p-nuclei. Again, the required γ-ray energy is of the order of several MeV, by far too high for a significant influence of the plasma environment.

Further details on the various nucleosynthesis processes can be found in the latest textbooks (Iliadis, 2007; Rolfs and Rodney, 1988) and in several contributions to this book (Matteucci, 2012; Pumo, 2012; Arnould and Goriely, 2012).

At first view, it seems that the plasma environment is not able to play a significant role in any of the above processes. However, a closer look at the s-process nucleosynthesis shows that there are a number of cases where low-energy γ-transitions turn out to be extremely important because such transitions may be able to produce and/or destroy isomers and thus affect s-process nucleosynthesis. The s-process path is located along the valley of stability. The level schemes of the nuclei under study are well-known; this holds in particular for the excitation energies and spins. This information allows for a careful study of the influence of the plasma.

Details of the s-process are given in this book (Pumo, 2012) and in a recent review (Käppeler et al., 2011). Here we repeat very briefly the most important properties of s-process nucleosynthesis. The main component of the s-process operates in low-mass AGB stars. Two alternating neutron sources are active. The $^{13}C(\alpha,n)^{16}O$ reaction operates for about 10^4 to 10^5 years at low temperatures below 10 keV; in most cases this temperature is too low to affect isomer production or destruction. The $^{22}Ne(\alpha,n)^{25}Mg$ reaction is activated for a few years during so-called helium shell flashes at temperatures around 25 keV and densities of about 10^3 g/cm^3. (Gallino et al., 1998). Under these conditions dramatic variations of isomer production and destruction rates can be expected (Ward & Fowler, 1980; Mohr et al. 2007; Gintautas et al., 2009; Mohr et al., 2009; Hayakawa et al., 2010). It has to be noted that the stellar transition rates may exceed the experimentally accessible ground state contribution (Belic et al., 2002; Mohr et al., 2007b; Rauscher et al., 2011) by orders of magnitude.

Two different temperature dependencies can be found for such low-energy γ-transitions which should not be mixed up. First, the total transition rate between states is given by the sum over all contributing branchings; all these different contributions vary strongly with temperature because of the exponential temperature dependence of the surrounding blackbody radiation. Second, each individual transition may additionally be modified by the plasma environment; this may lead to an additional temperature and density dependence of individual transitions with low γ-ray energy. The latter effect is the main subject of this chapter.

As example we have chosen the nucleus ^{171}Tm. It has a ground state with $J^\pi = 1/2^+$ and a low-lying first excited state with $J^\pi = 3/2^+$ at the excitation energy of 5.04 keV. This state decays by a M1 transition with small E2 admixture (mixing $\delta = 0.021\pm0.001$) with a half-life of $T_{1/2} = 4.77\pm0.08$ ns. Because of its low energy, this transition is highly converted (internal conversion coefficient $\alpha = 1408\pm55$). All data have been taken from the latest data evaluation (Baglin, 2002). ^{171}Tm is located on a neutron-rich branch of the s-process and may be reached either via the branching at ^{169}Er in the $^{168}Er(n,\gamma)^{169}Er(n,\gamma)$ $^{170}Er(n,\gamma)$ $^{171}Er(\beta^-)^{171}Tm$ reaction chain or via the branching at ^{170}Tm in the $^{169}Tm(n,\gamma)^{170}Tm(n,\gamma)^{171}Tm$ chain. It is interesting to note that the destruction of ^{171}Tm in the $^{171}Tm(n,\gamma)^{172}Tm$ capture reaction proceeds mainly via neutron capture in the thermally excited $3/2^+$ state (Rauscher et al., 2011).

2. Modification of a particular transition

As mentioned above, nuclear excitation in astrophysical plasmas may be significantly modified by the electronic environment γ-transitions of relatively low energy. Particles from the plasma other than photons interact with the nucleus and may excite it to an upper level. Thermodynamic conditions in the plasma may substantially alter the electronic environment of the nucleus and also perturb the de-excitation process of internal conversion. As a rule of thumb, every transition with an energy lower than 100 keV must be looked at, as internal conversion can significantly contribute to the transition rate. Some examples for several γ-transition energies have already been shown earlier (Gosselin et al., 2010). The two levels involved need not include the ground state, but can also be built on an isomeric state.

At least four different electromagnetic excitation processes may be able to excite a nucleus under typical astrophysical plasma conditions (Gosselin et al., 2010):

- Radiative excitation. A photon from the blackbody spectrum in the plasma is absorbed by the nucleus (Ward & Fowler, 1980).
- Nuclear Excitation by Electron Capture (NEEC). A free electron from the plasma is captured onto an empty atomic shell, giving its energy to the nucleus. This is also known as Inverse Internal Conversion (Gosselin & Morel, 2004).
- Nuclear Excitation by Electron Transition (NEET). A loosely bound electron makes a transition to a deeper atomic shell and gives its energy to the nucleus (Morel et al., 2004).
- Inelastic scattering of electrons (Gosselin et al., 2009).

Another non-electromagnetic excitation process is inelastic neutron scattering. It will not be dealt with in here, as it is strongly dependent on the specific astrophysical plasma in which it occurs. As the neutron spectrum is not directly related to the thermodynamic conditions of the plasma, but rather to the astrophysical site under study, it is impossible to plot an excitation rate as a function of the temperature as for the other processes.

All these electromagnetic processes must be dealt with along with their inverse processes, respectively photon emission, internal conversion, bound internal conversion (BIC), super-elastic scattering of electrons (a scattering process where the scattered electron gains some energy from the nucleus) and neutrons.

Describing each process is a two-step undertaking. The first step is a microscopic description of the excitation process which uses quantum mechanics formalism and aims at calculating a cross section (when there is an incident particle) or a transition rate (when there is none, such as with NEET). The electronic environment of the nucleus is described with a relativistic average atom model (RAAM) (Rozsnyai, 1972) from which an atomic potential can be extracted which depends on the density and the temperature of the plasma.

The second step is to derive a macroscopic plasma transition rate for all processes. Thermodynamics and plasma physics in the RAAM model are used to get distribution functions of photons or electrons and build the corresponding transition rate. Except for NEET, such a model is able to provide reliable values of electronic shells binding energies and occupancies, as well as a distribution function of free electrons.

If the plasma can be considered to be at local thermodynamic equilibrium (LTE), the excitation and de-excitation rate of each process are related to each other by:

$$\frac{\lambda_e}{\lambda_d} = \frac{2J_f + 1}{2J_i + 1} e^{-\frac{\Delta E}{kT}} \tag{1}$$

In this expression, λ_e is the excitation rate, λ_d the de-excitation rate, J_i and J_f the spins of the initial and final states in the nucleus and ΔE the nuclear transition energy. This relation is known as the principle of detailed balance and expresses the micro-reversibility of excitation processes. Here, we consider that electrons and photons are at LTE, but obviously nuclei are not.

2.1 Radiative excitation

Radiative excitation occurs through the resonant absorption of a photon. Most astrophysical plasmas have a large blackbody component in their photon spectrum. A blackbody photon can easily be absorbed by a nucleus if its energy is very close to a nuclear transition energy. However, the huge number of blackbody photons ensures that the resulting rate is significant. Two de-excitation processes compete with photon absorption: spontaneous and induced emission.

The microscopic cross section is a Breit and Wigner resonant capture cross section (Hamilton, 1975):

$$\sigma(h\nu) = \frac{2J_f+1}{2(2J_i+1)}\pi\hbar^2c^2\frac{\Gamma_\gamma\Gamma}{(h\nu)^2\left[(h\nu-\Delta E)^2+\frac{\Gamma^2}{4}\right]} \quad (2)$$

Here ν is the photon frequency, Γ_γ the gamma width of the excited nuclear level and Γ its total width. By folding this cross section with the blackbody distribution, we deduce an excitation rate:

$$\lambda_e = \frac{2J_f+1}{2J_i+1}\frac{\ln 2}{T^\gamma_{J_f\to J_i}}\frac{1}{e^{\frac{\Delta E}{kT_r}}-1} \quad (3)$$

where T_r is the radiative temperature and $T^\gamma_{J_f\to J_i}$ the transition radiative half-life.

The induced de-excitation rate can be deduced from the spontaneous rate by multiplying by the number of modes:

$$\lambda_d^{ind} = n_\nu\lambda_d^{sp} = \frac{c^2 I_\nu}{2h\nu^3}\frac{\ln 2}{T^\gamma_{J_f\to J_i}} \quad (4)$$

where I_ν is the blackbody radiative intensity. This gives a total de-excitation rate:

$$\lambda_d = \frac{\ln 2}{T^\gamma_{J_f\to J_i}}\frac{e^{\frac{\Delta E}{kT_r}}}{e^{\frac{\Delta E}{kT_r}}-1} \quad (5)$$

This total de-excitation rate and the excitation rate satisfy the principle of detailed balance.

As mentioned above, all excitation processes were calculated for ^{171}Tm. Fig. 1 shows excitation and induced emission rates becoming significant when the plasma temperature reaches a value in the same order of magnitude as the nuclear transition energy.

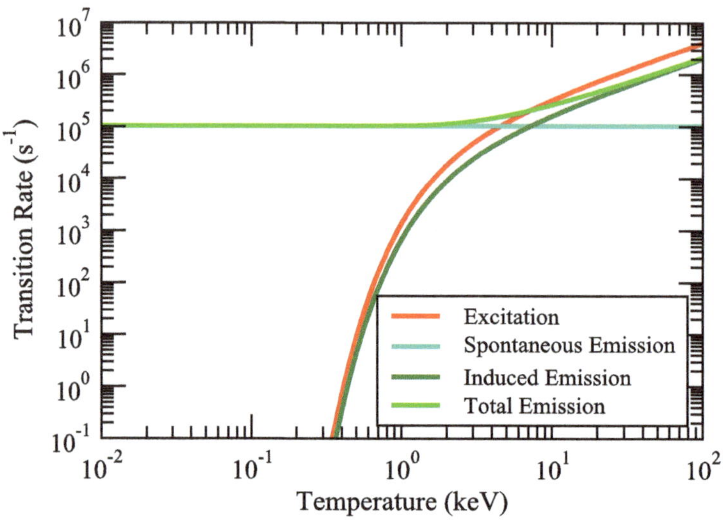

Fig. 1. Radiative excitation and de-excitation of ^{171}Tm.

2.2 Nuclear excitation by electron capture

NEEC is the inverse process of internal conversion. A free electron from the plasma is captured onto an atomic shell and the excess energy is used to excite the nucleus. It still has not been observed in the laboratory despite some attempts in channeling experiments (Kimball et al., 1991) and some projects with EBIT or EBIS (Marss, 2010). Considering NEEC in plasmas has first been proposed by Doolen (Doolen, 1978) in plasma at LTE.

If an electron has a kinetic energy lower than the nuclear excitation energy, the NEEC cross section can be expressed by the Fermi golden rule (Messiah, 1961):

$$\sigma_{NEEC}(E) = \frac{2\pi}{\hbar v_e} \left| \langle \psi_f \varphi_b | H | \psi_i \varphi_r \rangle \right|^2 \rho_b(E) \qquad (6)$$

where v_e is the incident electron speed, Ψ_i and Ψ_f the nuclear initial and final states wave functions, φ_b and φ_r the bound and free electron wave functions and $\rho_b(E)$ the total final state density. The matrix element is directly related to the internal conversion coefficient (Hamilton, 1975), which gives a resonant electron capture cross section:

$$\sigma_{NEEC}(E) = \frac{\pi\hbar^2}{2m_e E} \frac{2J_f+1}{2J_i+1} \frac{\ln 2}{T^{\gamma}_{J_f \to J_i}} \alpha \frac{\hbar\Gamma}{(E-E_r)^2 + \left(\frac{\Gamma}{2}\right)^2} \qquad (7)$$

where $T^{\gamma}_{J_f \to J_i}$ is the radiative half-life of the transition, α the internal conversion coefficient, E_r the resonance energy and Γ the nuclear level width.

A NEEC rate in plasma can then be derived by folding this cross section with the free electron distribution:

$$\lambda_e = \frac{2J_f+1}{2J_i+1}\frac{\ln 2}{T^{\gamma}_{J_f \to J_i}}\alpha(T_e) f_{FD}(E_r)\left[1 - f_{FD}(E_b)\right] \tag{8}$$

with E_b the binding energy of the bound electron and f_{FD} is the Fermi-Dirac distribution.

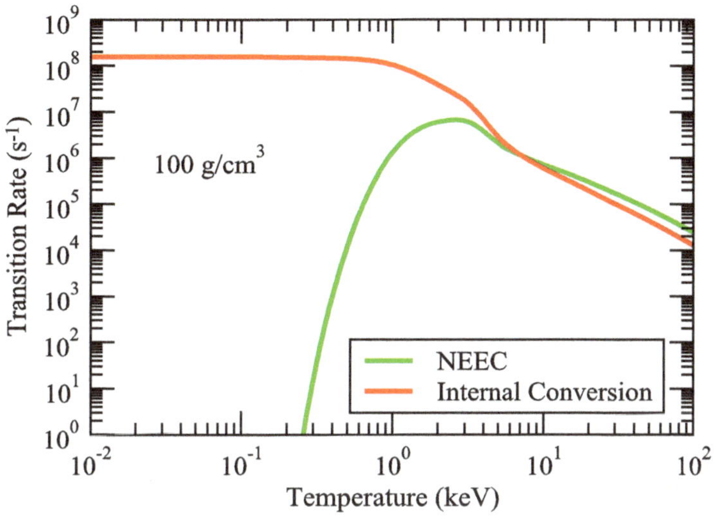

Fig. 2. NEEC excitation and Internal Conversion de-excitation of ^{171}Tm.

At low temperatures, Fig. 2 shows a near constant internal conversion rate which is very close to the laboratory value. For higher temperatures, the number of bound electrons decreases as the atom is ionized, and the number of allowed conversions must decrease. At low temperatures, the NEEC rate is very small as there are few free electrons to be captured and few vacant atomic states on which they could be captured. The NEEC rate then rises as the temperature increases to reach a maximum. The decrease at the higher temperatures can be attributed to the rising kinetic energy of the free electrons, the fraction of which below the nuclear transition threshold becoming less and less important.

2.3 Nuclear excitation by electron transition

Nuclear Excitation by Electron Transition (NEET) occurs when a loosely bound electron makes a transition to a deeper atomic shell and gives its energy to the nucleus. This may happen when the electronic and nuclear transition energies are very close to each other (separated by less than the atomic widths). NEET requires at least one electron on the outer atomic shell and at least a vacancy on the inner atomic shell. The energy difference between the atomic and the nuclear transition energy is called the mismatch and is denoted by δ. In the laboratory, such restrictive conditions can only be achieved for a very small number of nuclei. In astrophysical plasma, various conditions of density and temperature can be encountered, with a huge number of different electronic configurations at various charge states. The electronic shell binding energies are modified and the energy resonance condition ($\delta \approx 0$) can more often be achieved.

The probability of NEET occurring on an isolated and excited atom is given by:

$$P_{NEET}(\delta) = \frac{|R_{1,2}|^2}{\delta^2 + \left(\frac{\Gamma_1 + \Gamma_2}{2}\right)^2}\left(1 + \frac{\Gamma_2}{\Gamma_1}\right) \tag{9}$$

where Γ_1 and Γ_2 represent the total width (atomic and nuclear) of the initial and final configurations, respectively, and $|R_{1,2}|$ is the atom-nucleus matrix coupling element (Morel et al., 2004). The NEET probability reaches a maximum when the mismatch is zero. Mismatch variations for a 3s-6s atomic transition on Fig. 3 exhibit matching conditions around a temperature of 4 keV for a plasma density of 100 g/cm^3.

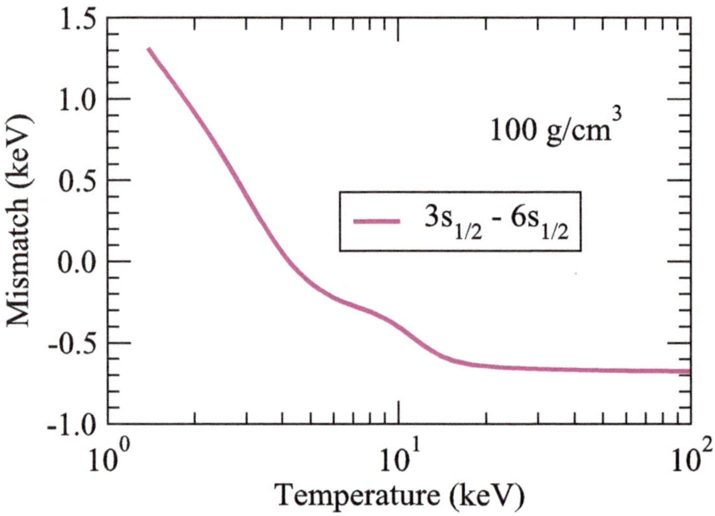

Fig. 3. Mismatch of ($3s_{1/}$ $-6s_{1/2}$) atomic transition for ^{171}Tm.

The atom-nucleus coupling matrix element is little sensitive to temperature as illustrated on Fig. 4. As the temperature increases, there are less bound electrons. This reduces screening of the nucleus by the remaining bound electrons, whose orbitals are closer to the nucleus. The overlap between the electron and the nucleus wave functions is larger, which increases the coupling.

In plasma, the NEET rate can be estimated as a summation over all initial configurations of the rate of creation of such a configuration multiplied by the NEET probability:

$$\lambda^{NEET}(\rho, T_e) = \sum_{\alpha} P_\alpha(\rho, T_e)\frac{\Gamma_\alpha}{\hbar} N_1\left(1 - \frac{N_2}{\Omega_2}\right)\frac{|R_{\alpha,\beta}|^2}{\delta_{\alpha,\beta}^2 + \left(\frac{\Gamma_\alpha + \Gamma_\beta}{2}\right)^2} \tag{10}$$

Here, N_1 and N_2 are the initial electronic occupations of the two atomic shells involved in the transition, of degeneracy Ω_1 and Ω_2, corresponding to outer and inner shell, respectively.

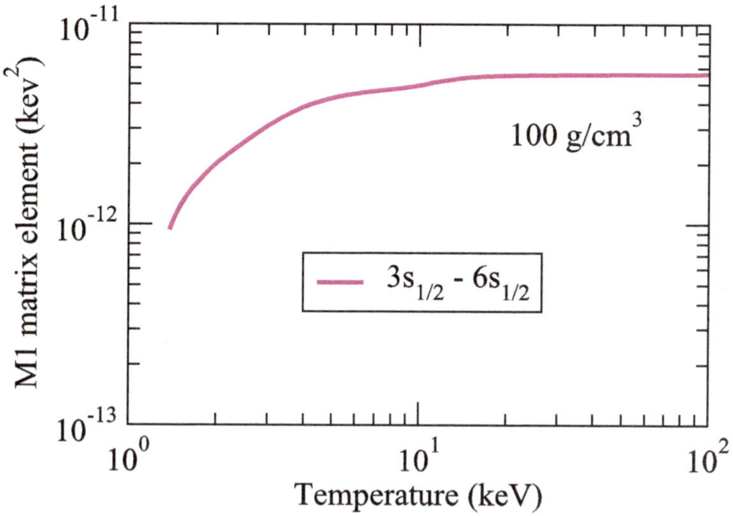

Fig. 4. Atom-Nucleus coupling matrix element of ($3s_{1/2}$–$6s_{1/2}$) atomic transition for ^{171}Tm.

An accurate calculation requires a good knowledge of nuclear and atomic wave functions for every configuration. However, the huge number of electronic configurations in this approach, which mirrors the DCA (Detailed Configuration Accounting) approach used to determine atomic spectra (Abdallah et al., 2008), makes such a calculation a prohibitive task. Therefore, we replace the detailed spectrum of atomic transitions by a Gaussian envelope, whose mean energy and statistical standard deviation σ are extracted from the RAAM (Faussurier et al., 1997). This quicker approach works the best when only outer atomic shells are involved. The NEET rate becomes:

$$\lambda^{NEET}(\rho,T_e)=\frac{2\pi}{\hbar}N_1\left(1-\frac{N_2}{\Omega_2}\right)\left|R_{1,2}\right|^2\frac{1}{\sqrt{2\pi\sigma^2}}e^{-\frac{\delta^2}{2\sigma^2}} \tag{11}$$

The NEET rate as a function of electronic temperature is presented in Fig. 5. Many atomic transitions contribute to the total NEET rate, the $3s_{1/2}$-$6s_{1/2}$ and $3p_{1/2}$-$6p_{3/2}$ transitions being the two dominant ones around 4 keV (only the M1 and E2 transitions with a maximum NEET rate higher than 10^3 s^{-1} are shown).

The average atom model cannot be steadily applied for atomic transitions involving deep shells because their average occupation numbers differ highly from the NEET requirements of at least a vacancy in the inner shell and an electron in the outer shell. In such cases, the mismatch of the RAAM mean configuration can be very different from the mismatches of the real configurations on which NEET is possible. When this discrepancy gets higher than the statistical standard deviation, a detailed configuration approach such as DCA can then not be avoided.

Fully comprehensive DCA calculations are still out of reach of the fastest available supercomputers. However, a careful selection of the atomic transitions may significantly reduce the number of electronic configurations and may soon be an accessible goal.

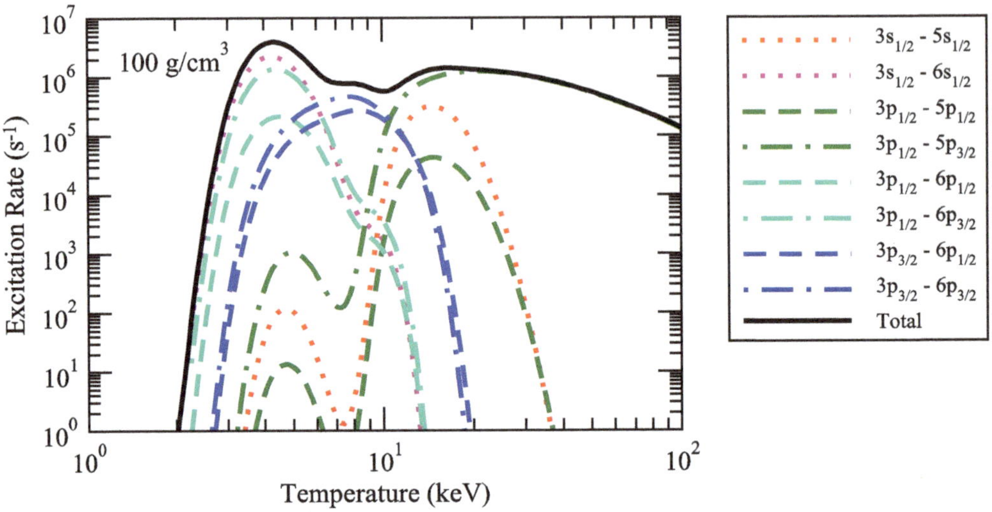

Fig. 5. NEET excitation of ^{171}Tm.

2.4 Inelastic electron scattering

In astrophysical plasmas, there usually exists a huge number of free electrons. If the nuclear transition energy is low enough, a significant part of these electrons might be able to excite the nucleus through Coulomb excitation. A semi-classical theory works well for high -above 1 MeV- energy ions (Alder et al., 1956), whereas a more sophisticated quantum model is required to deal with electrons in the keV range. With an unscreened atomic potential, a WKB approximation can be successfully implemented (Gosselin et al., 2009). The WKB cross section is very close to the more exact usual DWBA quantum approach (Alder et al., 1956) but much computationally heavier.

Cross sections exhibit usually low values in the 10^{-30} cm^2 range which can be at least partially compensated by the huge number of free electrons in high temperature plasmas (above the nuclear transition energy). However, these cross sections also exhibit a non-physical behavior close to the energy threshold where it does not drop to zero as it should do, as can be seen on Fig. 6. This can be explained by an "acceleration" of the incident electron by the unscreened potential as the global neutrality of the atom is not verified. Using a screened potential in the future will allow to get rid of this artifact.

By folding this cross section with the free electron distribution, we easily deduce an electron inelastic scattering excitation rate in plasma as shown on Fig. 7. It is negligible at low temperatures when there are very few free electrons, and these electrons do not have a high enough energy to be above the threshold. This changes when the temperature reaches values around the nuclear transition energy. At high temperatures, the excitation rate does not vary much as the lowering cross section is compensated by the increasing velocity of the electrons.

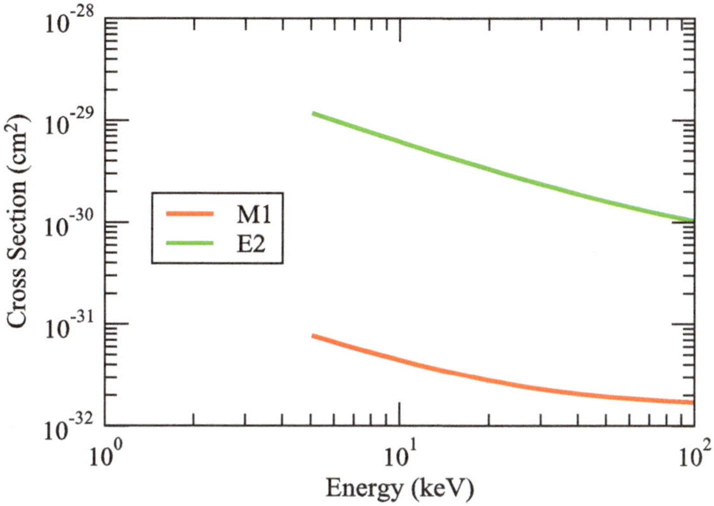

Fig. 6. WKB Inelastic electron scattering cross section of ^{171}Tm.

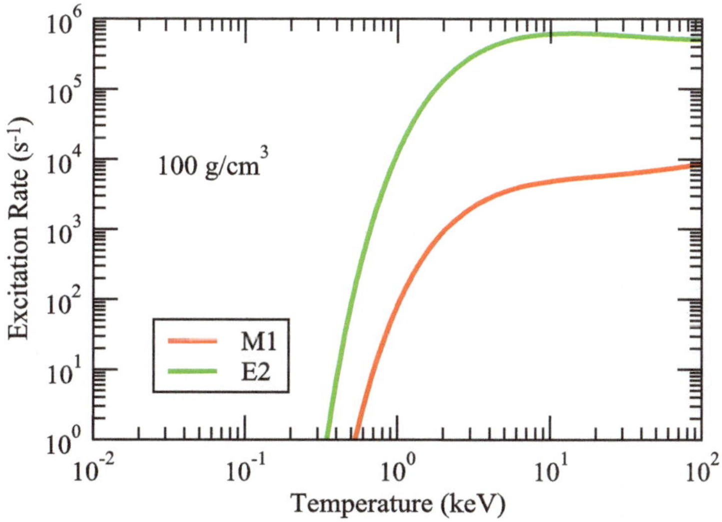

Fig. 7. Inelastic electron scattering excitation rate of ^{171}Tm.

2.5 Lifetime evolution

Fig. 8 summarizes excitation rates of the four electromagnetic processes present in plasma. NEEC dominates at low temperatures, radiative excitation at the hottest temperatures and NEET in between. The electron inelastic scattering never dominates although this situation could change at higher plasma density.

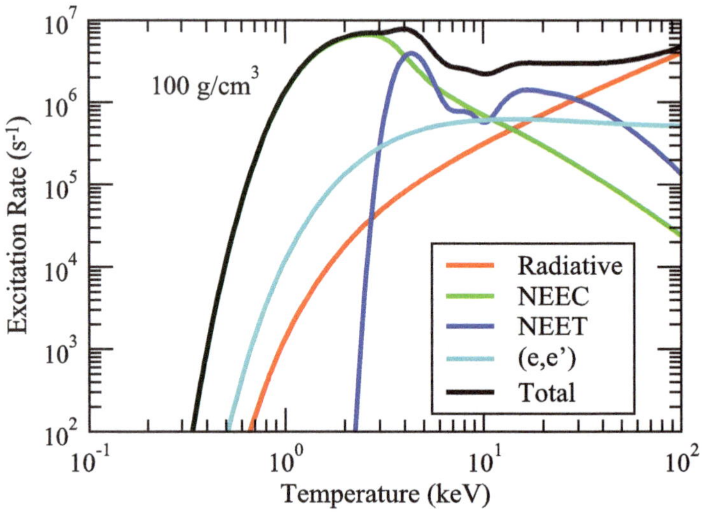

Fig. 8. Nuclear excitation rate in plasma of ^{171}Tm.

Highly variable rates of excitation also means highly variable rates of de-excitation which means that the nuclear lifetime of the excited level now depends on the plasma temperature. Fig. 9 shows these variations which cover nearly two orders of magnitude in the particular example of ^{171}Tm at 100 g/cm^3. Some other nuclei even exhibit larger variations, such as ^{201}Hg (Gosselin et al., 2007) where the lifetime can be increased by a factor of more than 10^4.

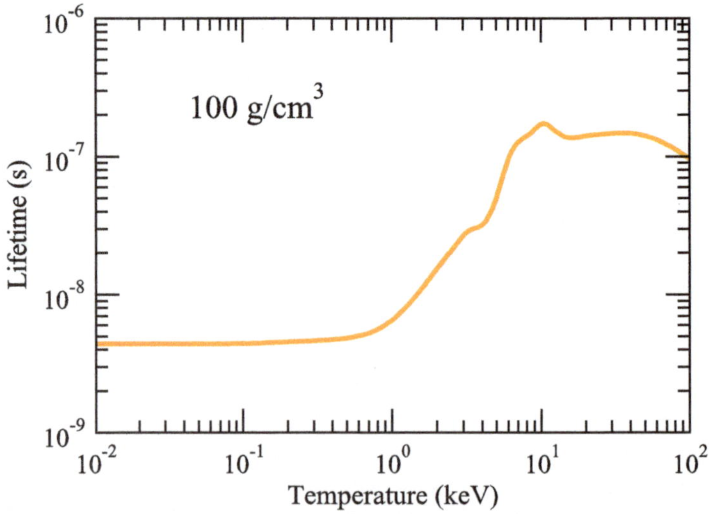

Fig. 9. Lifetime in plasma of ^{171}Tm.

However, this lifetime is an extension of the usual notion of lifetime in the laboratory. This plasma lifetime is the characteristic time required to get from any relative populations of both the ground and excited states to the LTE (for the nuclear levels) relative populations, that is populations whose ratio is given by the Boltzmann factor. For high temperatures and a higher spin for the excited state, the population of the excited state reaches a higher value than that of the ground state.

3. Stellar transition rates

As shown before in this chapter, thermalization between the ground state and the first excited state of ^{171}Tm is achieved on a timescale far below 1 second, i.e. much faster than the timescales of the astrophysical s-process. This result remains valid as long as the levels under study are connected by a direct γ-transition or a γ-cascade. A similar result has been derived explicitly for the ground state band of ^{176}Lu (Gintautas et al., 2009). Thus, because of the prompt thermalization within γ-bands it is obvious that the stellar transition rate of any photon-induced reaction has to be calculated including all contributions of thermally excited states. This holds in particular for (γ,n) and (γ,α) reactions for the astrophysical γ-process, but also for the photodestruction of isomers in the s-process. Because the required photon energies for (γ,n) and (γ,α) reactions is of the order of several MeV, the influence of the surrounding plasma remains small, and we focus on the photodestruction of isomers in the following.

The *K*-isomers are found in heavy deformed nuclei. Transitions between states with large differences in the *K* quantum number are strongly suppressed by selection rules. Thus, there are no direct transitions between low-*K* states and high-*K* states. As a consequence, thermalization has to proceed via higher-lying so-called intermediate states (IMS) with intermediate *K* quantum number which have a decay branching to the low-*K* and to the high-*K* part of the excitation spectrum of the respective nucleus. The transition rates from the low-*K* side and the high-*K* side of the spectrum to the IMS define the timescale for thermalization. Typically, these IMS are located at excitation energies above 500 keV. They decay down to the lowest states with low and high *K* by γ-cascades where energies of the individual γ-transitions are obviously much smaller than the excitation energy, i.e. the energies may be as low as 100 keV or even below. This is the energy region where the plasma effects become important. The full formalism for the calculation of stellar reaction rates is given in an earlier work (Gosselin et al., 2010). The essential result is that the stellar reaction rate λ^* for transitions from the low-*K* to the high-*K* states (and reverse) can be derived by a formula similar to Eq. (2) which leads to

$$\lambda^* \propto \frac{\Gamma_1 \Gamma_2}{\Gamma_1 + \Gamma_2} \tag{12}$$

where Γ_1 and Γ_2 are the total decay widths to the low-*K* and the high-*K* states respectively in the deformed nucleus under stellar conditions (i.e., summed over all levels, each particular transition width $\Gamma_{i \to f}$ being modified by the plasma environment according to the discussion in this work). It should be noted that the smaller of the two widths Γ_1 and Γ_2 essentially defines the stellar reaction rate λ^* for transitions from the low-*K* to the high-*K* states because the larger width cancels out in the above equation.

4. Some selected examples

In the following section we discuss three examples in greater detail. The first two examples are so-called *K*-isomers which are relevant in the astrophysical s-process (^{176}Lu and ^{180}Ta). The last example is the ^{7}Be(p,γ)^{8}B reaction with its low reaction Q-value of 137 keV.

4.1 ^{176}Lu in the astrophysical s-process

^{176}Lu is a so-called s-only nucleus because it is synthesized only in the astrophysical s-process. It is produced either in its high-*K* $J^{\pi};K=7^{-};7$ ground state with a half-life of 38 giga-years (quasi-stable for the s-process) or in its low-lying low-*K* $J^{\pi};K=1^{-};0$ isomer at 123 keV with a half-life of less than 4 hours. The ^{175}Lu(n,γ)^{176}Lu reaction produces most of ^{176}Lu in the low-*K* isomeric state which decays by β^{-} emission to ^{176}Hf. ^{176}Lu can only survive if the isomer is coupled to the high-*K* ground state (Heil et al., 2008). The most relevant IMS for this coupling is located at 839 keV with $J^{\pi};K=5^{-};4$, although other IMS have been suggested very recently (Gintautas et al., 2009; Dracoulis et al., 2010).

The IMS at 839 keV decays predominantly to the high-*K* part of ^{176}Lu; thus, the decay branch to the low-*K* part defines the stellar reaction rate. The decay properties of the IMS at 839 keV are well known from various γ-spectroscopic studies (Doll et al., 1999; Klay et al., 1991; Lesko et al., 1991). The lowest γ-ray energy is 123 keV, and three further γ-rays are observed at higher energies. Because of the relatively high energies the stellar reaction rate is only weakly affected. The dominating effect is NEEC in this case. However, the modification of the stellar transition rate remains far below a factor of two in the astrophysically relevant energy region (Gosselin et al., 2010).

4.2 ^{180}Ta and its uncertain nucleosynthetic origin

^{180}Ta is the rarest nucleus in our solar system (Lodders, 2003), and it is the only nucleus which does not exist in its ground state, but in an isomeric state. The ground state is a low-*K* state with $J^{\pi};K=1^{+};1$ and a short half-life of 8.154 hours. The isomer is located at an excitation energy of 77 keV; it is a high-*K* state with $J^{\pi};K=9^{-};9$. Because of its huge *K* quantum number, its decay to the ground state is highly suppressed, and also the energetically possible β-decays to ^{180}Hf and ^{180}W are largely hindered. The half-life of the isomer is unknown with a lower limit of 7.1 x 10^{15} years (Hult et al, 2006; Wu & Niu, 2003). Despite significant effort, the nucleosynthetic origin of ^{180}Ta is still uncertain.

Various astrophysical sites and corresponding processes have been suggested for the nucleosynthesis of ^{180}Ta. Very recently it has been concluded that a large contribution to the solar abundance can be produced in the neutrino burst during type II supernovae in the so-called ν-process by the ^{180}Hf(ν,e^{-})^{180}Ta reaction (Hayakawa et al., 2010). Alternatively, in the same astrophysical site the classical p- or γ-process may produce some ^{180}Ta by photodestruction of ^{181}Ta in the ^{181}Ta(γ,n)^{180}Ta reaction (Arnould & Goriely 2003; Utsunomiya et al., 2006). Similar conditions for the temperature of several billions Kelvin (or $kT \approx 200 - 300$ keV) occur in type Ia supernovae, and it has been found that some ^{180}Ta can also be made in that site (Travaglio et al., 2011). In addition, some ^{180}Ta may also be produced in the s-process via β-decay of thermally excited ^{179}Hf to ^{179}Ta and subsequent neutron capture in the ^{179}Ta(n,γ)^{180}Ta reaction or via isomeric β-decay of ^{180m}Hf in the

^{179}Hf(n,γ)^{180m}Hf(β^{-})^{180m}Ta reaction chain (Schumann & Käppeler, 1999; Beer & Ward, 1981; Yokoi & Takahashi, 1983; Mohr et al., 2007).

A common problem in all above production scenarios is the survival of ^{180}Ta in its isomeric state. If the production occurs in a high-temperature environment like any supernova explosion, then ^{180}Ta is produced in thermal equilibrium between the low-*K* ground state band and the high-*K* isomeric band. Following the evolution of the isomer-to-ground state ratio, freeze-out is found around 40 keV with a survival probability of ^{180}Ta in its isomeric state of 0.38±0.01 (Hayakawa et al., 2010). Under s-process conditions with its lower temperature, it is not clear how much ^{180}Ta can survive in the isomeric state because of mixing between hotter and cooler areas of the thermally pulsing AGB star (Mohr et al., 2007). The resulting yield of ^{180}Ta depends sensitively on the properties of the lowest intermediate state (IMS) which couples the low-*K* and high-*K* bands.

There is indirect confirmation for the existence of IMS which couple the low-*K* ground state and the high-*K* isomeric state from photoactivation experiments (Belic et al., 2002; Collins et al., 1990). However, no direct γ-transition has been observed up to now. Based on reasonable estimates for transition strengths it has been suggested (Mohr et al., 2007) that the lowest IMS is located at an excitation energy of 594 keV with J=(5). This state decays with a transition energy of 72 keV and a half-life of about 16 ns to the low-*K* side of ^{180}Ta. A weak branch to the high-K side can be expected via a transition to the 7^{+} state at 357 keV. This may lead to thermalization of ^{180}Ta within days at the s-process temperatures around 25 keV (Mohr et al., 2007).

Because of the low transition energy of only 72 keV, a significant modification of the corresponding transition strength can be expected. Enhancements of the radiative strength of up to a factor of 10 for this transition have been calculated (Gosselin et al., 2010) which are mainly based on NEEC. However, unfortunately the influence on the stellar transition rate between low-*K* states and high-*K* states remains very small. The stellar transition rate as given in Eq. (12) is essentially defined by the weak branch of the 594 keV state which has a transition energy of more than 200 keV. As transitions with such high energies are practically not affected by the plasma environment, the stellar transition rate remains almost unchanged.

4.3 ^{8}B, the ^{7}Be(p,γ)^{8}B reaction, and solar neutrino production

The ^{7}Be(p,γ)^{8}B reaction is the key reaction for the production of high-energy neutrinos in our sun. It is one of the very few examples for a capture reaction between light nuclei where low-energy γ-rays play a significant role. The Q-value of this reaction is extremely low (137 keV), and together with the most effective energy of about 18 keV (at temperatures around 15 million Kelvin, typical for the center of our sun) we find a transition energy of about 155 keV in the low-mass (Z=5) nucleus ^{8}B. The γ-energy of 155 keV is still too high to be significantly influenced by the surrounding plasma. Although experimental conditions in the laboratory (either a proton beam and neutral ^{7}Be target or a ^{7}Be beam in arbitrary charge state on a neutral hydrogen target) are quite different from the stellar environment of ^{7}Be, the cross sections from laboratory experiments do not require a plasma correction for the electromagnetic transition strength to calculate the stellar reaction rates, as e.g. summarized in the NACRE compilation (Angulo et al., 1999) or in a compilation of solar fusion cross sections (Adelberger et al., 2011). However, two further effects of the surrounding plasma have to be kept in mind here: (i) The electron capture decay of ^{7}Be is significantly affected under solar conditions because electrons have to be captured from the surrounding plasma instead of the bound *K*-shell electrons for ^{7}Be in a neutral beryllium atom. (ii) Electron screening affects all capture cross sections at very low energies. Because the electronic environment is different in laboratory experiments and under stellar conditions, different corrections have to be applied here.

5. Conclusion

The description of nuclear excitation in hot astrophysical plasma environment requires an accurate knowledge of each individual excitation process. The dominant processes are photo-excitation, where a photon close to the resonant nuclear transition energy is absorbed by the nucleus, NEEC and NEET, where an electron from the continuum (NEEC) or an outer electronic shell (NEET) is captured in a vacancy of the electronic cloud, and inelastic electron scattering, where an incident electron gives a part of its energy to excite the nucleus.

Results for the excitation of the first isomeric state of ^{171}Tm clearly exhibit a strong dependence upon the plasma temperature. The nuclear lifetime under plasma conditions is more than one order of magnitude higher than the laboratory value. All these calculations are made at Local Thermodynamic Equilibrium (LTE), a condition encountered in many astrophysical plasmas.

However, nuclear excitation models in plasma need to be elaborated further. NEET rates require to take into account detailed electronic configuration, as the mean RAAM configuration is more often than not far from a real configuration on which NEET can occur. For high density plasmas, electron inelastic scattering is a major process and screening effects will have to be added to the description.

In most astrophysical scenarios the influence of the surrounding plasma on astrophysical reaction rates will remain relatively small because the γ-transition energies are too high to be significantly affected by the electronic environment. Note that typical γ-transition energies exceed by far 1 MeV whereas plasma effects become most important below 100 keV. Nevertheless, it should be always kept in mind that γ-transitions with energies below about 100 keV may be modified dramatically. As soon as such a transition defines the stellar transition rate, the calculated stellar reaction rate without consideration of plasma effects may be wrong; this error may reach one order of magnitude in special cases, i.e. for very low γ-transition energies.

Such cases with small γ-transition energies (and thus huge influence of the plasma environment) can be found in particular in the astrophysical s-process where the production and destruction of so-called *K*-isomers proceeds by low-energy γ-transitions which connect the low-*K* and high-*K* parts of the excitation spectrum of heavy nuclei via intermediate states. However, for the two most prominent examples (^{176}Lu and ^{180}Ta) it is found that plasma effects remain relatively small for the resulting stellar reaction rates although one particular transition in ^{180}Ta is enhanced by about a factor of 10.

6. Acknowledgments

This work was supported by OTKA (NN83261).

7. References

Abdallah, J. & Sherrill, M.E. (2008). *High Energy Density Physics,* Vol 83, pp 195-245

Adelberger, E. G.; et al. (2011). *Review of Modern Physics,* Vol 83, pp 195-245

Alder, K.; Bohr, A.; Huus, T.; Mottelson, B.; & Winther, A. (1956). *Review of Modern Physics,* Vol 28 pp 432-542

Angulo, C.; et al. (1999). *Nuclear Physics A,* Vol 656, pp 3-187

Arnould, M. & Goriely, S. (2003). *Physics Reports,* Vol 384, pp 1-84

Arnould, M. & Goriely, S. (2012). The r-process of nucleosynthesis: The puzzle is still with us, In: *Nucleosynthesis,* InTech, 978-953-308-32-9, Rijeka, Croatia.

Baglin, C. M. (2002). *Nuclear Data Sheets,* Vol 96, pp 399-610; available online at *http://www.nndc.bnl.gov/ensdf.*

Basunia, M. S. (2006). *Nuclear Data Sheets,* Vol 107, pp 791-1026; available online at *http://www.nndc.bnl.gov/ensdf*

Beer, H. & Ward, R. A. (1981). *Nature,* Vol 291, pp 308-310

Belic, D.; et al. (2002). *Physical Review C,* Vol 65, pp 035801

Bosch, F.; et al. (1996). *Physical Review Letters,* Vol 77, pp 5190-5193

Collins, C. B.; et al. (1990). *Physical Review C,* Vol 42, pp R1813-R1816

Doll, C.; et al. (1999). *Physical Review C,* Vol 59, pp 492-499

Doolen, G. D. (1978). *Physical Review C,* Vol 18, pp 2547-2559

Dracoulis, G. D.; et al. (2010). *Physical Review C,* Vol 81, pp 011301(R)

Farkas, J.; et al. (2009). *Journal of Physics G,* Vol 36, pp 105101

Faussurier, G.; Blancard, C. & Decoster, A. (1997). *Physical Review E,* Vol 56, pp 3474

Gallino, R.; et al. (1998). *Astrophysical Journal,* Vol 497, pp 388-403

Gintautas, V.; Champagne, A. E.; Kondev, F. G. & Longland, R. (2009). *Physical Review C,* Vol 80, pp 015806

Gosselin, G.; Méot, V. & Morel, P. (2007). *Physical Review C,* Vol 76, pp 044611

Gosselin, G. & Morel, P. (2004). *Physical Review C,* Vol 70, pp 064603

Gosselin, G.; Morel, P. & Mohr, P. (2010). *Physical Review C,* Vol 81, pp 055808

Gosselin, G.; Pillet, N.; Méot, V.; Morel, P. & Dzyublik, A. Ya. (2009). *Physical Review C,* Vol 79, pp 014604

Hamilton, W. D. (1975). *The Electromagnetic Interaction in Nuclear Spectroscopy,* Elsevier, ISBN 0 444 10519 O, New York, USA

Hayakawa, T.; Mohr, P.; Kajino, T.; Chiba, S. & Mathews, G. J. (2010). *Physical Review C,* Vol 82, pp 055801

Heil, M.; et al. (2008). *Astrophysical Journal,* Vol 673, pp 434-444

Hult, M.; Gasparro, J.; Marissens, G.; Lindahl, P.; Wäthjen, U.; Johnston, P. & Wagemans, C. (2006). *Physical Review C,* Vol 74, pp 054311

Iliadis, C. (2007). *Nuclear Physics of Stars,* Wiley-VCH, 978-3-527-40602-9, Weinheim, Germany.

Käppeler, F.; Gallino, R.; Bisterzo, S. & Aoki, W. (2011). *Review of Modern Physics,* Vol 83, pp 157-193

Kimball, J. C.; Bittel, D. & Cue, N. (1991). *Physics Letters A,* Vol 152, pp 367-370

Klay, N.; et al. (1991). *Physical Review C,* Vol 44, pp 2801-2838
Kücük, I. (2012). Screening Factors and Thermonuclear Reaction Rates for Low Mass Stars, In: *Nucleosynthesis,* InTech, 978-953-308-32-9, Rijeka, Croatia.
Lesko, K. T.; et al. (1991). *Physical Review C,* Vol 44, pp 2850-2864
Lodders, K. (2003). *Astrophysical Journal,* Vol 591, pp 1220-1247
Marss, R. (2010), *XI'th International Symposium on EBIS and EBIT, 7-10 April 2010,* UCRL-PRES-427008, Livermore, California, USA
Matteucci, F. (2012). Stellar Nucleosynthesis, In: *Nucleosynthesis,* InTech, 978-953-308-32-9, Rijeka, Croatia.
Messiah, A. (1961). *Quantum Mechanics,* North Holland, ISBN 0 471 59766 X, Amsterdam, Holland
Mohr, P.; Käppeler, F. & Gallino, R. (2007). *Physical Review C,* Vol 75, pp 012802(R)
Mohr, P.; Fülöp, Zs. & Utsunomiya, H. (2007b). *European Physical Journal A,* Vol 32, pp 357-369
Mohr, P.; Bisterzo, S.; Gallino, R.; Käppeler, F.; Kneissl, U. & Winckler, N. (2009). *Physical Review C,* Vol 79, pp 045804
Morel, P.; Méot, V.; Gosselin, G.; Gogny, D. & Yunnes, W. (2004). *Physical Review A,* Vol 69, pp 063414
Ohtsuki, T.; Yuki, H.; Muto, M.; Kasagi, J. & Ohno, K. (2004). *Physical Review Letters,* Vol 93, pp 112501
Pumo, M. L. (2012). The s-process nucleosynthesis in massive stars: current status and uncertainties due to convective overshooting, In: *Nucleosynthesis,* InTech, 978-953-308-32-9, Rijeka, Croatia.
Rauscher, T.; Mohr, P.; Dillmann, I. & Plag, R. (2011). *Astrophysical Journal,* Vol 738, pp 143
Rolfs, C. E. & Rodney, W. S. (1988). *Cauldrons in the Cosmos,* The University of Chicago Press, 0-226-72456-5, Chicago, IL, USA.
Rozsnyai, B. F. (1972). *Physical Review A,* Vol 5, pp 1137-1149
Schumann, M. & Käppeler, F. (1999). *Physical Review C,* Vol 60, pp 025802
Takahashi, K. & Yokoi, K. (1987). *Atomic Data Nuclear Data Tables,* Vol 36, pp 375-409
Travaglio, C.; Roepke, F.; Gallino, R. & Hillebrandt, W. (2011). *Astrophysical Journal,* Vol 739, pp 93
Utsunomiya, H.; Mohr, P.; Zilges, A. & Rayet, M. (2006). *Nuclear Physics A,* Vol 777, pp 459-478
Ward, R. A. & Fowler, W. A. (1980). *Astrophysical Journal,* Vol 238, pp 266-286
Wu, S. C. & Niu, H. (2003). *Nuclear Data Sheets,* Vol 100, pp 483-705; available online at *http://www.nndc.bnl.gov/ensdf*
Yokoi, K. & Takahashi, K. (1983). *Nature,* Vol 305, pp 198-200

3

Stellar Nucleosynthesis Nuclear Data Mining

Boris Pritychenko
National Nuclear Data Center, Brookhaven National Laboratory, Upton, NY USA

1. Introduction

In the past 100 years, astronomy, astrophysics and cosmology have evolved from the observational and theoretical fields into more experimental science, when many stellar and planetary processes are recreated in physics laboratories and extensively studied (Boyd et al., 2009). Many astrophysical phenomena have been explained using our understanding of nuclear physics processes, and the whole concept of stellar nucleosynthesis has been introduced. The importance of nuclear reactions as a source of stellar energy was recognized by Arthur Stanley Eddington as early as 1920 (Eddington, 1920). Later, nuclear mechanisms by which hydrogen is fused into helium were proposed by Hans Bethe (Bethe, 1939). However, neither of these contributions explained the origin of elements heavier than helium.

Further developments helped to identify the Big Bang, stellar and explosive nucleosynthesis processes that are responsible for the currently-observed variety of elements and isotopes (Burbidge et al., 1957; Cameron, 1957; Hoyle, 1946; Merrill, 1952). Today, nuclear physics is successfully applied to explain the variety of elements and isotope abundances observed in stellar surfaces, the solar system and cosmic rays via network calculations and comparison with observed values.

A comprehensive analysis of stellar energy production, metallicity and isotope abundances indicates the crucial role of proton-, neutron- and light ion-induced nuclear reactions and α-, β-decay rates. These subatomic processes govern the observables and predict the star life cycle. Calculations of the transition rates between isotopes in a network strongly rely on theoretical and experimental cross section and decay rate values at stellar temperatures. Consequently, the general availability of nuclear data is of paramount importance in stellar nucleosynthesis research.

This chapter will provide a review of theoretical and experimental nuclear reaction and structure data for stellar and explosive nucleosynthesis and modern computation tools and methods. Examples of evaluated and compiled nuclear physics data will be given. Major nuclear databases and their input for nucleosynthesis calculations will be discussed.

2. Nucleosynthesis and its data needs

Nucleosynthesis is an important nuclear astrophysics phenomenon that is responsible for presently observed chemical elements and isotope abundances. It started in the early Universe and presently proceeds in the stars. The Big Bang nucleosynthesis is responsible for a relatively high abundance of the lightest primordial elements in the Universe from ^{1}H to ^{7}Li, and it precedes stars formation and stellar nucleosynthesis. The general consistency between theoretically predicted and observed lightest elements abundances serves as a strong evidence for the Big Bang theory (Kolb & Turner, 1988).

The currently-known variety of nuclei and element abundances is shown in Fig. 1 and Fig. 2. These Figures indicate a large variety of isotopes in Nature (Anders & Grevesse, 1989) and the strong need for additional nucleosynthesis mechanisms beyond the Big Bang theory.

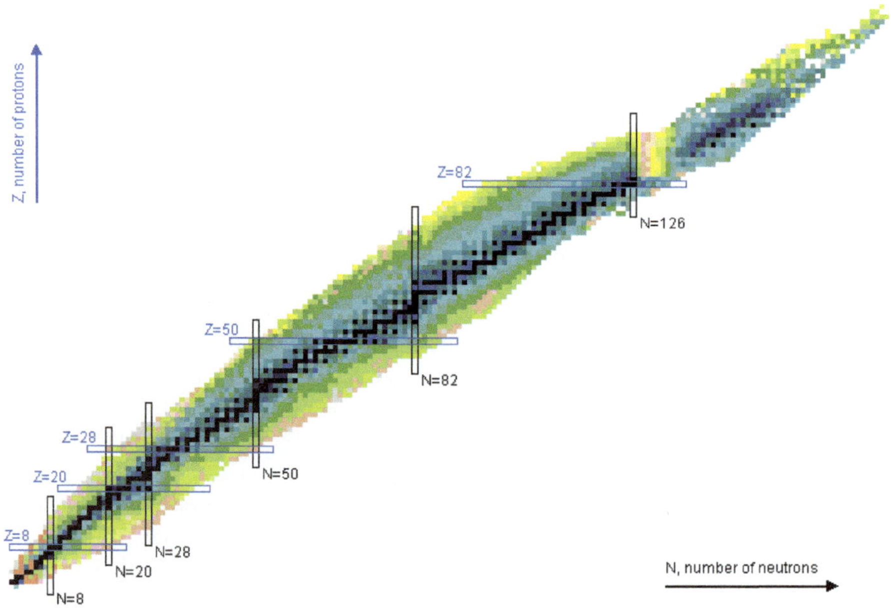

Fig. 1. The chart of nuclides. Stable and long-lived ($>10^{15}$ s) nuclides are shown in black. Courtesy of NuDat Web application (*http://www.nndc.bnl.gov/nudat*).

These additional mechanisms have been pioneered by Eddington (Eddington, 1920) via introduction of a revolutionary concept of element production in the stars. Present nucleosynthesis models explain medium and heavy element abundances using the stellar nucleosynthesis that consists of burning (explosive) stages of stellar evolution, photo disintegration, and neutron and proton capture processes. The model predictions can be verified through the star metallicity studies and comparison of calculated isotopic/elemental abundances with the observed values, as shown in Fig. 2 .

Nowadays, there are many well-established theoretical models of stellar nucleosynthesis (Boyd et al., 2009; Burbidge et al., 1957); however, they still cannot reproduce the observed abundances due to many parameter uncertainties. The flow of the nuclear physics processes in the network calculations is defined by the nuclear masses, reaction, and decay rates, and strongly correlated with the stellar temperature and density. These calculations depend heavily on our understanding of nuclear physics processes in stars, and the availability of high quality nuclear data.

The common sources of stellar nucleosynthesis data include KADONIS, NACRE and REACLIB dedicated nuclear astrophysics libraries (Angulo et al., 1999; Cyburt et al., 2010;

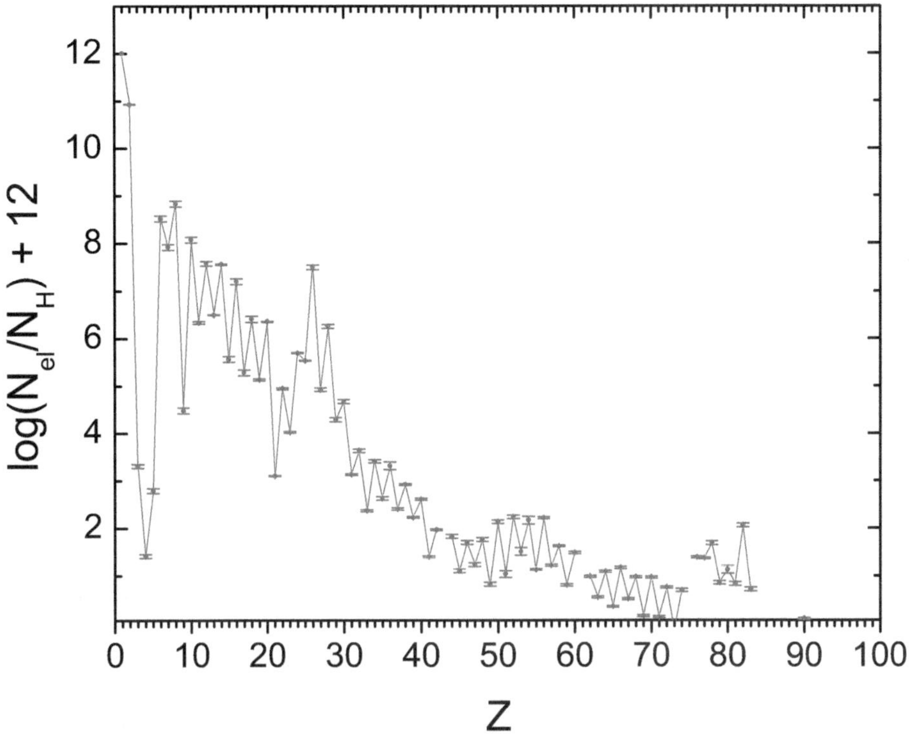

Fig. 2. Solar system elemental abundances; data are taken from (Grevesse & Sauval, 1998).

Dillmann et al., 2006) and general nuclear science and industry databases. Dedicated libraries are optimized for nuclear astrophysics applications, and contain pre-selected data that are often limited to the original scope. In many cases, these sources reflect the present state of nuclear physics, when experimental data are not always available or limited to a single measurement. Such limitation highlights the importance of theoretical calculations that strongly dependent on nuclear models. Another problem arises from the fact that nucleosynthesis processes are strongly affected by the astrophysical site conditions.

To broaden the scope of the traditional nuclear astrophysics calculations, we will investigate applicability of nuclear physics databases for stellar nucleosynthesis data mining. These databases were developed for nuclear science, energy production and national security applications and will provide complementary astrophysics model-independent results.

3. Stellar nucleosynthesis

This section will briefly consider the light elements and concentrate on the active research subject of production of medium and heavy elements beyond iron via slow and rapid neutron capture and associated data needs. Finally, photo disintegration and proton capture nucleosynthesis processes will be reviewed.

3.1 Burning phases of stellar evolution

Fusion reactions are responsible for burning phases of stellar evolution. These reactions produce light, tightly-bound nuclei and release energy. The process of new element creation proceeds before nuclear binding energy reaches maximum value in the Fe-Ni region. Four important cases will be reviewed: pure hydrogen burning, triple alpha process, CNO cycle, and stellar burning. These processes take place in stars with a mass similar to our Sun, as shown in Fig. 3. The data needs for these processes are addressed in the IAEA FENDL (Aldama & Trkov, 2004) and EXFOR (Experimental Nuclear Reaction Data) (NRDC, 2011) databases.

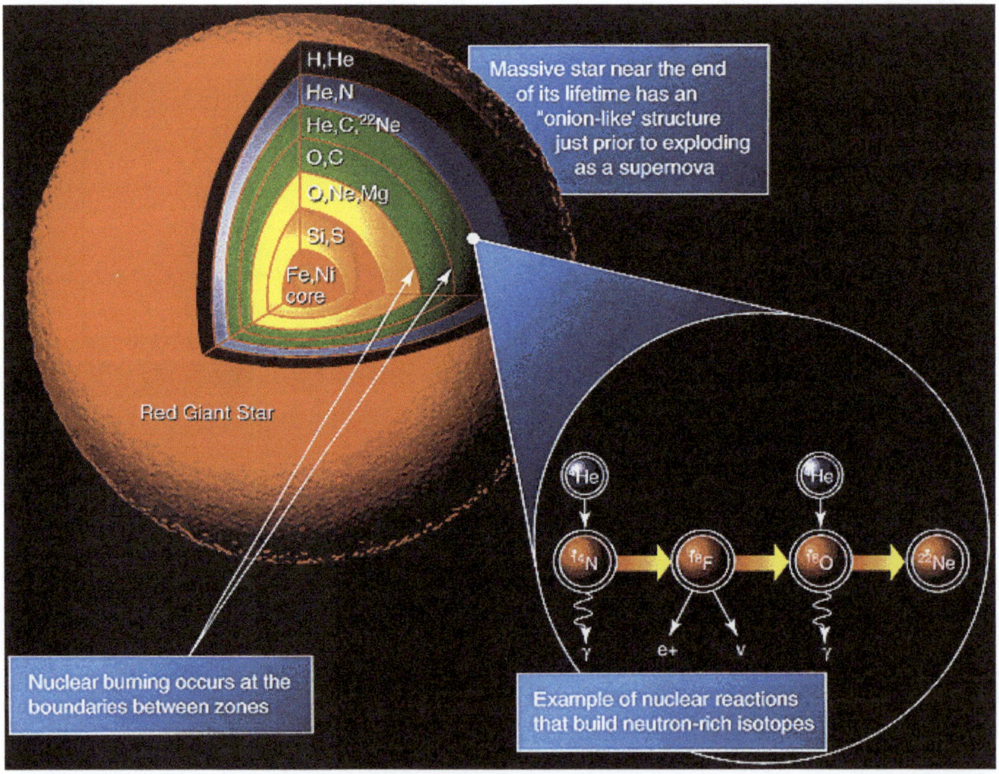

Fig. 3. Cross section of a Red Giant showing nucleosynthesis and elements formed. Courtesy of Wikipedia (*http://en.wikipedia.org/wiki/Stellar_nucleosynthesis*).

3.1.1 Pure hydrogen burning

Hydrogen is the most abundant element in the Universe. The proton-proton chain dominates stellar nucleosynthesis in stars comparable to our Sun

$$p + p \rightarrow d + \beta^{+} + \nu + 0.42MeV \tag{1}$$

Further analysis of the pp-process (Burbidge et al., 1957) indicates extremely low cross sections for E<1 MeV nuclear projectiles and explains the necessity of large target mass and density for sustainable nuclear fusion reaction

$$\sigma \sim 3E^{4.5} \times 10^{-24} b \tag{2}$$

Next, the deuterium produced in the first stage can fuse with another hydrogen

$$d + p \rightarrow^{3} He + \gamma + 5.49MeV \tag{3}$$

Finally, ^{4}He will be produced in the pp I and pp II branches.

3.1.2 Triple alpha process

The hydrogen burning in stars leads to the production of the helium core at a star's center. Further helium burning goes through the 3α process

$$\begin{aligned} \alpha + \alpha &\Leftrightarrow {}^{8}Be - 93.7keV \\ {}^{8}Be + \alpha &\Leftrightarrow {}^{12}C^{*} + 7.367MeV \end{aligned} \tag{4}$$

3.1.3 CNO cycle

Carbon-Nitrogen-Oxygen cycle leads to the production of elements heavier than carbon and consists of major CNO-I (regeneration of carbon and alpha particle)

$$^{12}C \rightarrow^{13} N \rightarrow^{13} C \rightarrow^{14} N \rightarrow^{15} O \rightarrow^{15} N \rightarrow^{12} C \tag{5}$$

and minor CNO-II branches (production of ^{16}O and proton):

$$^{15}N \rightarrow^{16} O \rightarrow^{17} F \rightarrow^{17} O \rightarrow^{14} N \rightarrow^{15} O \rightarrow^{15} N \tag{6}$$

3.1.4 Advanced stages of stellar burning

Further burning includes carbon, neon, oxygen and silicon burning. Here, we have high-temperature and density burning stages when photonuclear processes produce additional α-particles and create a complex network of reactions with light particles. These reactions produce fusion nuclei up to the Fe-Ni peak. The Fe-Ni region nuclei are the most tightly bound in Nature and fusion reactions stop. Further element production proceeds via neutron and proton captures and photo disintegration.

3.2 s-process

The slow-neutron capture (s-process) is responsible for creation of $\sim$50 % of the elements beyond iron. In this region, neutron capture becomes dominant because of the increasing Coulomb barrier and decreasing binding energies. This s-process takes place in the Red Giants and AGB stars, where neutron temperature (kT) varies from 8 to 90 keV. A steady supply of neutrons is available due to H-burning and He-flash reactions

$$\begin{aligned} &^{13}C(\alpha, n)^{16}O \\ &^{22}Ne(\alpha, n)^{25}Mg \end{aligned} \tag{7}$$

Fig. 2 indicates a high abundance of ^{56}Fe nuclei due to termination of the explosive nucleosynthesis. It is natural to assume that iron acts as a seed for the neutron capture reactions that eventually produce medium and heavy elements. The neutron capture time of s-process takes approximately one year. Consequently, the process path lies along the nuclear valley of stability up to the last long-lived nucleus of ^{209}Bi.

In the giant stars, neutron reaction rates define the elemental abundances, and in some cases branching points (created by the competition between neutron capture and β-decay) strongly affect the heavy isotope production rates. Therefore, special attention has to be paid to the branching points at ^{79}Se,^{134}Cs,^{147}Pm, ^{151}Sm, ^{154}Eu, ^{170}Yb and ^{185}W, and the neutron poison (absorption) ^{16,18}O,^{22}Ne(n,γ) reactions. The s-process astrophysical site conditions imply the following data needs

- Reaction rates
  - neutron-induced
  - charged particle
- Half-lives

Convincing proof of s-process existence and its role in Nature could come from the calculation of isotopic abundances and comparison with observed values (Anders & Grevesse, 1989; Grevesse & Sauval, 1998). Present-day s-process nucleosynthesis calculations often are based on the dedicated nuclear astrophysics data tables, such as works of (Bao et al., 2000), and (Rauscher & Thielemann, 2000). These data tables contain quality information on Maxwellian-averaged cross sections ($\langle\sigma_{\gamma}^{Maxw}(kT)\rangle$) and astrophysical reaction rates ($R(T_9)$). However, it is essential to produce complementary neutron-induced reaction data sets for an independent verification and to expand the boundaries of the existing data tables.

Recent releases of ENDF/B-VII evaluated nuclear reaction libraries (Chadwick et al., 2006) and publication of the Atlas of Neutron Resonances reference book (Mughabghab, 2006) created a unique opportunity of applying these data for non-traditional applications, such as s-process nucleosynthesis (Pritychenko et al., 2010). Many neutron cross sections for astrophysical range of energies, including ^{56}Fe(n,γ) reaction as shown in Fig. 4, are available in the ENDF (Evaluated Nuclear Data File) and EXFOR libraries. The feasibility study of the evaluated nuclear data for s-process nucleosynthesis will be presented below.

3.2.1 Calculation of Maxwellian-averaged cross sections and uncertainties

The Maxwellian-averaged cross section can be expressed as

$$\langle\sigma^{Maxw}(kT)\rangle = \frac{2}{\sqrt{\pi}}\frac{(m_2/(m_1+m_2))^2}{(kT)^2}\int_0^{\infty}\sigma(E_n^L)E_n^L e^{-\frac{E_n^L m_2}{kT(m_1+m_2)}}dE_n^L, \tag{8}$$

where k and T are the Boltzmann constant and temperature of the system, respectively, and E is an energy of relative motion of the neutron with respect to the target. Here, E_n^L is a neutron energy in the laboratory system and m_1 and m_2 are masses of a neutron and target nucleus, respectively.

The astrophysical reaction rate for network calculations is defined as

$$R(T_9) = N_A\langle\sigma v\rangle = 10^{-24}\sqrt{(2kT/\mu)}N_A\sigma^{Maxw}(kT), \tag{9}$$

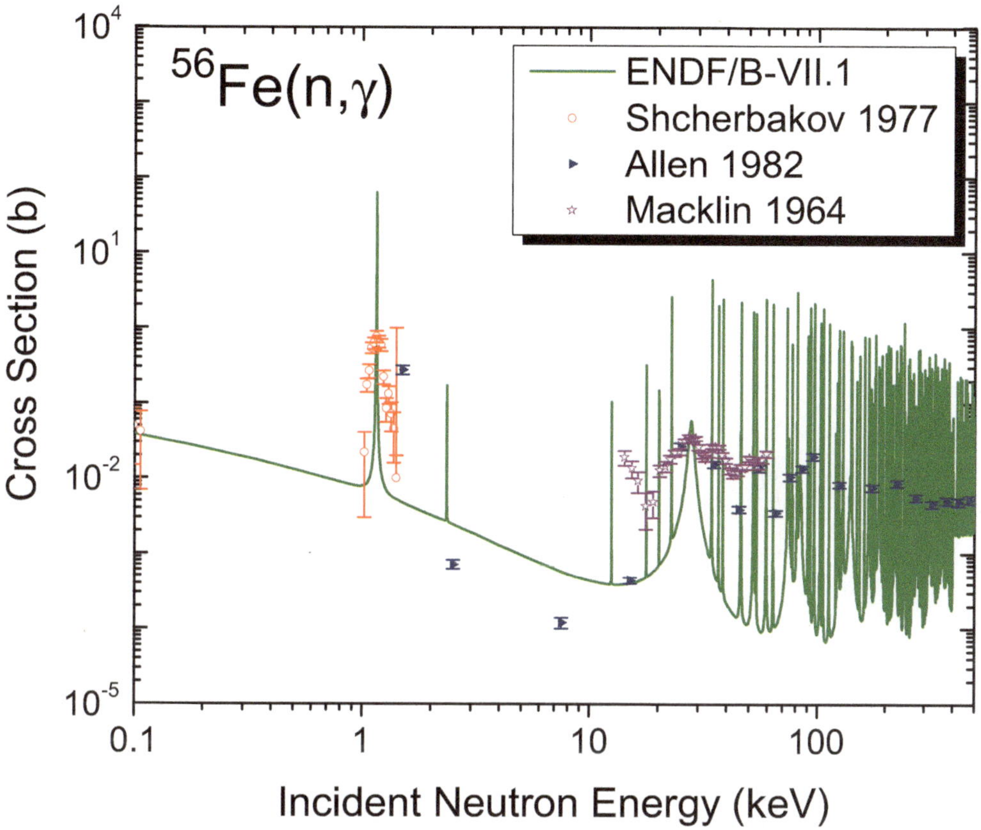

Fig. 4. ENDF/B-VII.1 and EXFOR libraries $^{56}Fe(n,\gamma)$ cross sections (Chadwick et al., 2006; NRDC, 2011) for astrophysical range of energies.

where T_9 is temperature expressed in billions of Kelvin, N_A is an Avogadro number. T_9 is related to the kT in MeV units as follows

$$11.6045 \times T_9 = kT \tag{10}$$

It is commonly known that for the equilibrium s-process-only nuclei product of $\langle \sigma_\gamma^{Maxw}(kT) \rangle$ and solar-system abundances ($N_{(A)}$) is preserved (Rolfs & Rodney, 1988)

$$\sigma_A N_{(A)} = \sigma_{A-1} N_{(A-1)} = constant \tag{11}$$

The stellar equilibrium conditions provide an important test for the s-process nucleosynthesis in mass regions between neutron magic numbers N=50,82,126 (Arlandini et al., 1999). To investigate this phenomenon, we will consider ENDF libraries. These data were never adjusted for nuclear astrophysics models and are essentially model-independent.

ENDF library is a core nuclear reaction database containing evaluated (recommended) cross sections, spectra, angular distributions, fission product yields, thermal neutron scattering, photo-atomic and other data, with emphasis on neutron-induced reactions. ENDF library evaluations cover all neutron reaction channels within 10^{-5} eV - 20 MeV energy range. In many cases, evaluations contain information on neutron cross section covariances. An example of ENDF/B-VII.1library ^{56}Fe(n,γ) neutron cross section covariances (uncertainties) is shown in Fig. 5.

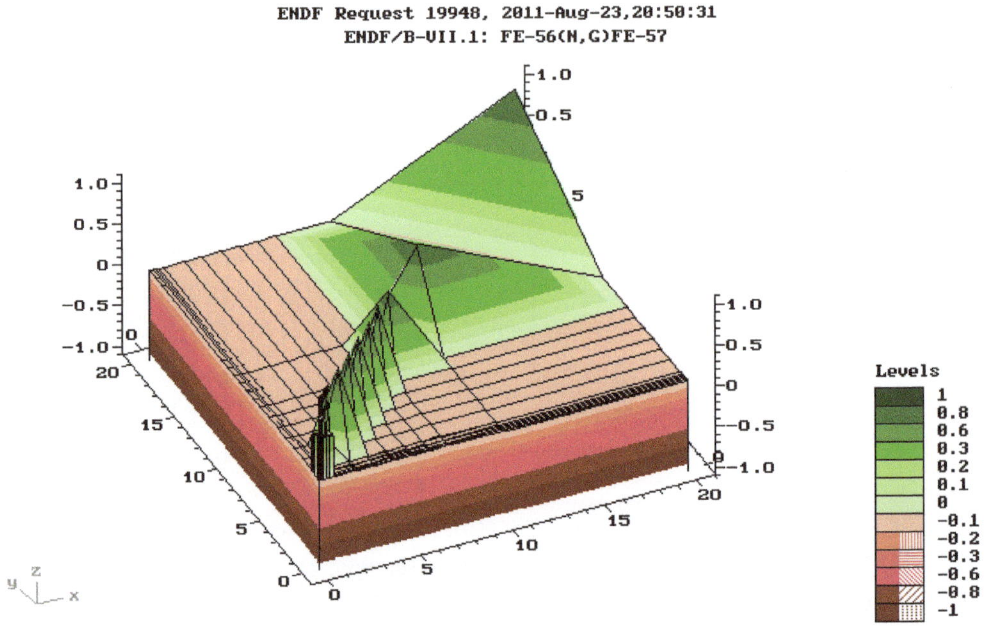

Fig. 5. ENDF/B-VII.1 library ^{56}Fe(n,γ) cross section covariances (Zerkin et al., 2005).

The evaluated neutron libraries are based on theoretical calculations using EMPIRE, TALYS and Atlas collection of nuclear reaction model codes (Herman et al., 2007; Koning et al., 2008; Mughabghab, 2006) that are often adjusted to fit experimental data (NRDC, 2011). The model codes are essential for neutron cross section calculations of short-lived radioactive nuclei where experimental data are not available. The ENDF data files are publicly available from the NNDC (National Nuclear Data Center) Sigma Web Interface: *http://www.nndc.bnl.gov/sigma* (Pritychenko & Sonzogni, 2008).

Previously, Maxwellian-averaged cross sections and astrophysical reaction rates were produced using the Simpson method for the linearized ENDF cross sections (Pritychenko et al., 2010) . A similar effort was completed at the Japanese Atomic Energy Agency (Nakagawa et al., 2005). Fig. 6 shows the NNDC website (Pritychenko et al., 2006) that provides access to nuclear databases and the NucRates Web application (*http://www.nndc.bnl.gov/astro*). The NucRates application was designed for online calculations of $\langle\sigma^{Maxw}(kT)\rangle$ and $R(T_9)$ using all major evaluated nuclear reaction libraries, astrophysical neutron-induced reactions and kT values ranging from 10^{-5} eV to 20 MeV.

The Simpson method allowed quick calculation of integral values; however, the degree of precision was within $\sim$1%. Precision can be improved with the linearized ENDF files because the cross section value is linearly-dependent on energy within a particular bin (Pritychenko,

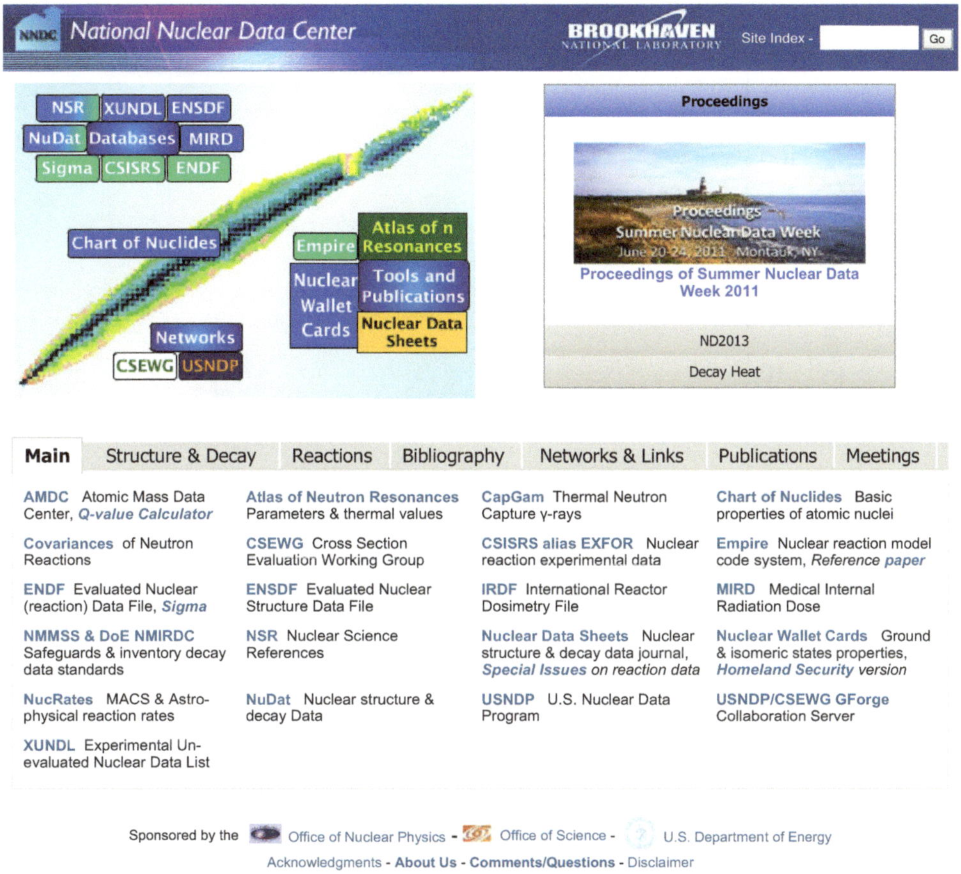

Fig. 6. The NNDC website (*http://www.nndc.bnl.gov*) provides access to nuclear databases and Maxwellian-averaged cross sections and astrophysical reaction rates online calculations.

2010)

$$\sigma(E) = \sigma(E_1) + (E - E_1)\frac{\sigma(E_2) - \sigma(E_1)}{E_2 - E_1}, \tag{12}$$

where $\sigma(E_1), E_1$ and $\sigma(E_2), E_2$ are cross section and energy values for the corresponding energy bin. The last equation is a good approximation of neutron cross section values for a sufficiently dense energy grid. This allowed deduction of $\langle\sigma_\gamma^{Maxw}(kT)\rangle$ definite integrals for separate energy bins using Doppler-broadened cross sections and the Wolfram Mathematica online integrator (Wolram, 2011). Further, summing integrals for all energy bins will produce a precise ENDF value for the Maxwellian-averaged cross section.

The product values of the ENDF/B-VII.1 $\langle\sigma_\gamma^{Maxw}(30keV)\rangle$ times solar abundances (Anders & Grevesse, 1989) are plotted in Fig. 7. They reveal that the ENDF/B-VII.1 library data closely replicate a two-plateau plot (Rolfs & Rodney, 1988). The current result provides a powerful testimony for stellar nucleosynthesis.

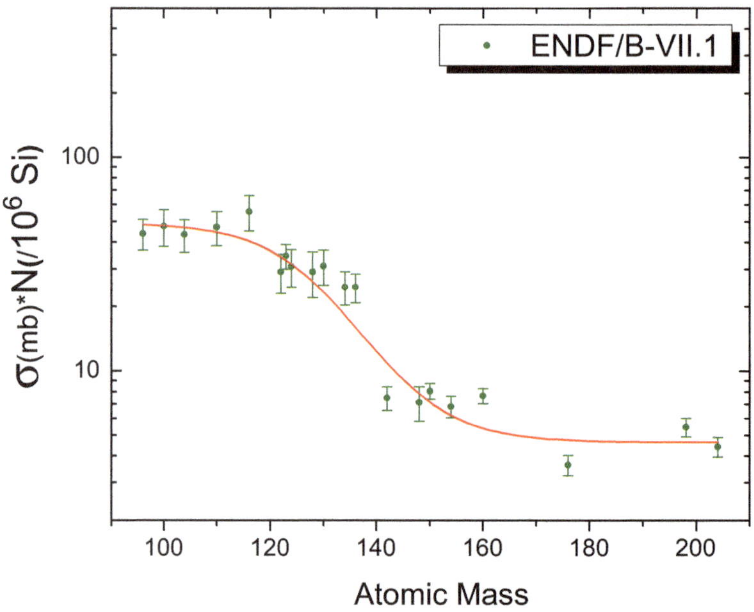

Fig. 7. ENDF/B-VII.1 library product of neutron-capture cross section (at 30 keV in mb) times solar system abundances (relative to Si = 10^6) as a function of atomic mass for nuclei produced only in the *s*-process.

The predictive power of stellar nucleosynthesis calculations depends heavily on the neutron cross section values and their covariances. To understand the unique isotopic signatures from the presolar grains, ~1% cross section uncertainties are necessary (Käppeler, 2011). Unfortunately, present uncertainties are often much higher, as shown in Fig. 8.

The situation gets even more complex after considering s-process branches where β-decay and neutron capture rates are

$$\lambda_n \sim \lambda_\beta \tag{13}$$

Here, the stellar thermal environment may affect β-decay rates and change the process path. Branching is particularly important for unstable isotopes such as ^{134}Cs. This isotope can either decay to ^{134}Ba, if neutron flux is low or capture neutron and produce ^{135}Cs. The ^{134}Cs β-decay lifetime may vary from ~1 y to 30 day over the temperature range of (100-300)$\times 10^6$ K, which further complicates the calculations.

Finally, the calculated values of Maxwellian-averaged cross sections at kT=30 keV, for selected s-process nuclei, derived from the JENDL-4.0, ROSFOND 2010 and ENDF/B-VII.1 libraries (Chadwick et al., 2006; Shibata et al., 2011; Zabrodskaya et al., 2007) have been produced and shown in Table 1. The tabulated results are compared with the KADONIS values (Dillmann et al., 2006). Due to a limited number of ENDF covariance files, Low-Fidelity cross section covariances (Little et al., 2008) were used to calculate uncertainties for ENDF/B-VII.1 data.

The complete sets of ENDF s-process nucleosynthesis data sets are available for download from the NucRates Web application *http://www.nndc.bnl.gov/astro*. These complimentary data

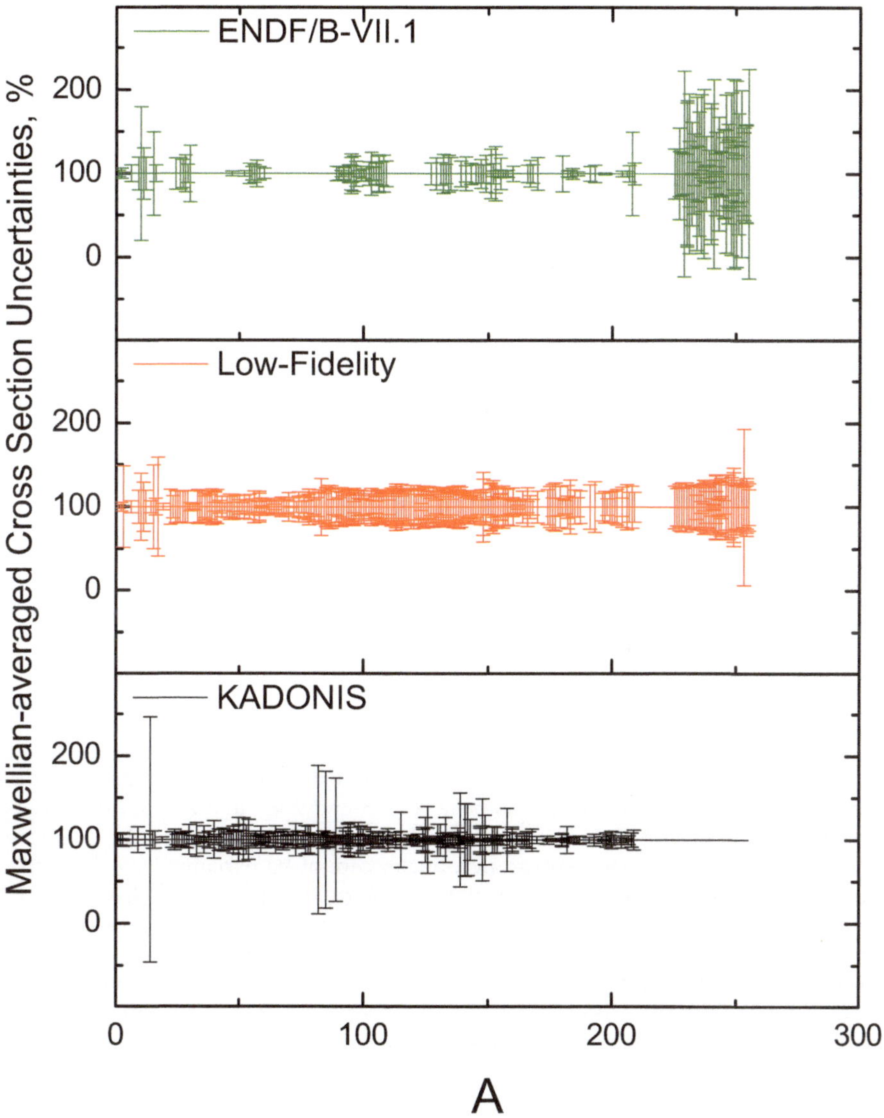

Fig. 8. Maxwellian-averaged cross section uncertainties for ENDF/B-VII.1, Low-Fidelity and KADONIS libraries (Chadwick et al., 2006; Dillmann et al., 2006; Little et al., 2008).

Isotope	JENDL-4.0	ROSFOND 2010	ENDF/B-VII.1	KADONIS
42-Mo- 96	1.052E+2	1.035E+2	1.035E+2±1.700E+1	1.120E+2±8.000E+0
44-Ru-100	2.065E+2	2.062E+2	2.035E+2±3.949E+1	2.060E+2±1.300E+1
46-Pd-104	2.700E+2	2.809E+2	2.809E+2±4.923E+1	2.890E+2±2.900E+1
48-Cd-110	2.260E+2	2.346E+2	2.349E+2±4.263E+1	2.370E+2±2.000E+0
50-Sn-116	9.115E+1	1.002E+2	1.003E+2±1.875E+1	9.160E+1±6.000E-1
52-Te-122	2.644E+2	2.639E+2	2.349E+2±4.882E+1	2.950E+2±3.000E+0
52-Te-123	8.138E+2	8.128E+2	8.063E+2±1.063E+2	8.320E+2±8.000E+0
52-Te-124	1.474E+2	1.473E+2	1.351E+2±2.697E+1	1.550E+2±2.000E+0
54-Xe-128	2.582E+2	2.826E+2	2.826E+2±6.823E+1	2.625E+2±3.700E+0
54-Xe-130	1.333E+2	1.518E+2	1.518E+2±2.835E+1	1.320E+2±2.100E+0
56-Ba-134	2.301E+2	2.270E+2	2.270E+2±4.038E+1	1.760E+2±5.600E+0
56-Ba-136	7.071E+1	7.001E+1	7.001E+1±1.087E+1	6.120E+1±2.000E+0
60-Nd-142	3.557E+1	3.701E+1	3.343E+1±4.251E+1	3.500E+1±7.000E-1
62-Sm-148	2.361E+2	2.444E+2	2.449E+2±4.507E+1	2.410E+2±2.000E+0
62-Sm-150	4.217E+2	4.079E+2	4.227E+2±3.607E+2	4.220E+2±4.000E+0
64-Gd-154	9.926E+2	1.010E+3	9.511E+2±1.096E+2	1.028E+3±1.200E+1
66-Dy-160	8.702E+2	8.293E+2	8.328E+2±6.769E+1	8.900E+2±1.200E+1
72-Hf-176	5.930E+2	4.529E+2	4.531E+2±4.896E+1	6.260E+2±1.100E+1
80-Hg-198	1.612E+2	1.612E+2	1.613E+2±1.635E+1	1.730E+2±1.500E+1
82-Pb-204	8.355E+1	7.242E+1	7.242E+1±7.624E+0	8.100E+1±2.300E+0

Table 1. Evaluated nuclear reaction and KADONIS libraries Maxwellian-averaged neutron capture cross sections in mb at kT=30 keV for s-process nuclei.

sets demonstrate a strong correlation between nuclear astrophysics and nuclear industry data needs, the large nuclear astrophysics potential of ENDF libraries, and a perspective beneficial relationship between both fields.

3.3 r-process

The detailed analysis of stable and long-lived nuclei indicates the large number of isotopes that lie outside of the s-process path peaks near A=138 and 208. In addition, the large gap between s-process nucleus ^{209}Bi and ^{232}Th,^{235,238}U effectively terminates the s-process at ^{210}Po. Production of the actinide neutron-rich nuclei cannot be explained by the s-process nucleosynthesis and requires introduction of rapid neutron capture or r-process. In this case, neutron capture timescale has to be less than typical β-decay lifetimes of $\sim$ms for neutron-rich nuclides. It implies neutron fluxes 10^{10}-10^{11} higher than those of the s-process. Such conditions can be found in ν-driven core-collapse supernova and neutron stars. From here, one may conclude that r-process temperature depends on the site and may lie within a (0.5-10) $\times 10^9$ K range. However, it is still not clear where r-process takes place and how it proceeds.

Among many unknowns of the r-process is process path. The path is defined by the nuclear masses and β-decay half-lives. The 2003 & 2011 experimental Atomic Mass Evaluations (Audi et al., 2003; Audi & Meng, 2011) do not cover nuclei far from stability near the r-process expected path. To resolve this problem, theoretical mass calculations based on the FRDM and other models have been performed with 25% uncertainties (Aprahamian, 2011). This calculation helped to identify the list of critical nuclei along the r-process path and demonstrated strong connections between nuclear astrophysics and nuclear structure. Further progress will require mass and half-life measurements of unstable nuclei. The r-process data needs can be summarized as follows

- nuclear masses
- β-decay half-lives
- n-capture rates
- neutrino interaction rates
- fission probabilities
- fission products distribution

Recent r-process estimates (Cowan & Thielemann, 2004) demonstrate sharp abundance peaks for the A=130,195 (N=82,126) nuclei with large N/Z ratios and another broad peak at A=160. Further analysis of r-process abundances shows (Boyd, 2008)

$$dN_{Z,A}/dt = \lambda_{Z-1}N_{Z-1} - \lambda_Z N_Z \quad (14)$$

For neutron closed shells, nuclei most likely will experience β-decay rather than absorb another neutron

$$\lambda_Z N_Z = N_Z/\tau_Z \quad (15)$$

Thus near neutron closed shells, the relationship between abundances and β-decay lifetimes is

$$dN_{Z,A}/dt = N_{Z-1,A-1}/\tau_{Z-1} - N_{Z,A}/\tau_Z \quad (16)$$

Finally, for equilibrium conditions one can deduce r-process analog of the equation 11

$$N_{Z-1}/\tau_{Z-1} = N_Z/\tau_Z \quad (17)$$

From the last formula, one can conclude that closed shell nuclei with the largest half-lives will have the largest abundances. In order to obtain the complete picture, all half-lives and decay modes have to be determined. A list of properties relevant to r-process A$\sim$130 nuclei is shown in Table 2. It includes β-decay half-lives and emission probabilities for delayed neutrons. The delayed neutrons provide an additional neutron source for the r-process and may shift the location of the abundance peak. The tabulated data were taken from the Evaluated Nuclear Structure Data File (ENSDF) database (Burrows, 1990) and the relation between nuclear lifetime and half-life is

$$\tau = T_{1/2}/0.693 \quad (18)$$

The regularly updated ENDF and ENSDF database evaluations could provide valuable for r-process data in the actinide region. The extremely neutron-rich superheavy fission nuclei play an important role in element production. Presently, these nuclei can be studied only with theoretical model calculations. These calculations could be calibrated using the existing actinide data for Maxwellian-averaged neutron cross sections, half-lives, spontaneous fission and delayed neutrons probabilities.

An example of nuclear reaction and structure data sets for the Z=90-110 region is shown in Table 3. The tabulated values demonstrate an increasingly complex nature of nuclear decay for Z>95 nuclei where spontaneous fission and β-decay play an important role. Spontaneous fission fragments are of interest to the r-process studies. Complimentary information on fission fragments distribution can be obtained from the Sigma Web interface *http://www.nndc.bnl.gov/sigma* (Pritychenko & Sonzogni, 2008).

Isotope	$T_{1/2}$, msec	β^--decay, %	β-n Emission, %
^{127}Ag	109±25	100	?
^{128}Ag	58±5	100	?
^{129}Ag	46±7	100	?
^{130}Ag	≈50	?	?
^{130}Cd	162±7	100	3.5±1.0
^{131}Cd	68±3	100	3.5±1.0
^{132}Cd	97±10	100	60±15
^{131}In	280±30	100	≤2.0±0.3
^{132}In	207±6	100	6.3±0.9
^{133}In	165±3	100	85±10
^{134}In	140±4	100	65
^{135}In	92±10	100	>0
^{130}Sn	223200±4200	100	?
^{131}Sn	56000±5000	100	?
^{132}Sn	39700±800	100	?
^{133}Sn	1460±30	100	0.0294±24
^{134}Sn	1050±11	100	17±13
^{135}Sn	530±20	100	21±3
^{136}Sn	250±30	100	30±5
^{137}Sn	190±60	100	58±15
^{138}Sn	>0.000408	?	?
^{139}Sn	?	?	?
^{140}Sn	?	?	?
^{136}Sb	923±14	100	16.3±3.2
^{137}Sb	450±50	100	49±10
^{138}Sb	≥0.0003	?	?
^{139}Sb	>0.00015	?	?
^{137}Te	2490±50	100	2.99±16
^{138}Te	1400±400	100	6.3±2.1
^{139}Te	>0.00015	?	?

Table 2. Properties of neutron-rich nuclides relevant to the A=130 r-process peak. All data are taken from the ENSDF database (*http://www.nndc.bnl.gov/ensdf*) (Burrows, 1990). The ? symbol was used where data were not available.

3.4 p-process

A detailed analysis of the Fig. 1 data indicates between 29 and 35 proton-rich nuclei that cannot be produced in the s- or r-processes. A significant fraction of these nuclei originate from the γ-process (Boyd, 2008; Woosley & Howard, 1978). This process could take place in Type-II supernovae at (2-3)$\times T_9$. It begins with (γ,n) reactions that synthesize proton-rich heavy nuclei that are followed by charged-particle emitting reactions. Such process includes an extensive reaction network consisting of approximately 20,000 reactions and 2,000 nuclei. Due to lack of experimental data, p-process network calculations are often based on theoretical model predictions. This situation can be improved via addition of the known experimental reaction cross sections.

Isotope	$\sigma(n,F)$, mb	$T_{1/2}$, y	SF, %	α-decay, %	β-decay, %
^{232}Th	1.672E1	1.40E10	1.1E-9	100	?
^{235}U	1.376E3	7.04E8	7E-9	100	?
^{238}U	7.839E1	4.468E9	5.45E-5	100	?
^{237}Np	9.953E2	2.144E6	$\leq$2E-10	100	?
^{239}Pu	1.885E3	2.411E4	3.1E-10	100	?
^{241}Am	7.767E2	432.6	3.6E-10	100	?
^{250}Cm	3.192E2	8.3E3	74	18	8
^{250}Bk	1.209E3	3.212 h	?	?	100
^{252}Cf	2.133E3	2.645	3.092	96.908	?
^{255}Es	3.108E2	39.8 d	0.0041	8	92
^{255}Fm	2.776E3	20.07 h	2.4E-5	100	?

Table 3. Nuclear reaction and structure properties of several actinides. Reaction cross sections were calculated from ENDF/B-VII.1 library at kT=400 keV and decay data were taken from the ENSDF database (*http://www.nndc.bnl.gov/ensdf*). The ? symbol was used where data were not available.

Reaction	EXFOR	NSR
p,γ	396	2162
p,n	666	2571
p,α	337	1031
α,γ	144	522
α,n	343	1321
α,p	166	848

Table 4. Total number of p-process reaction entries in EXFOR (*http://www-nds.iaea.org/exfor*) and NSR (*http://www.nndc.bnl.gov/nsr*) databases as of August 2011.

These reactions have been compiled in the EXFOR database since 70ies, and the database content is shown in Fig. 9. For historic reasons, it is relatively complete for neutron-, proton- and alpha-induced reaction compilations, and has a limited number of compilations for heavy-ion and photonuclear reactions. The IAEA EXFOR Web interface: *http://www-nds.iaea.org/exfor* (Zerkin et al., 2005) allows user-friendly nuclear astrophysics data search using multiple parameters, such as target, nuclear reaction, cross section, and energy range. The interface is currently used for a p-process nuclei data mining operation at ATOMKI (Szücs et al., 2010). EXFOR is the best source of experimental nuclear reaction data; however, it is not complete, and it takes ~1-2 y before article compilation is completed.

To overcome this problem the Nuclear Science References (NSR) database (*http://www.nndc.bnl.gov/nsr*) (Pritychenko et al., 2011) is recommended. Table 4 shows EXFOR and NSR database content for the ATOMKI project scope of (p,γ), (p,n), (p,α), (α,γ), (α,n) and (α,p) reactions. The tabulated data indicate a factor of 3-5 difference between two databases. This is mostly due to the fact that multiple article can be combined into a single EXFOR entry and gaps in the EXFOR coverage.

Another important tool for p-process nucleosynthesis studies is the nuclear reaction reciprocity theorem. It allows extracting a reaction cross section if an inverse reaction is

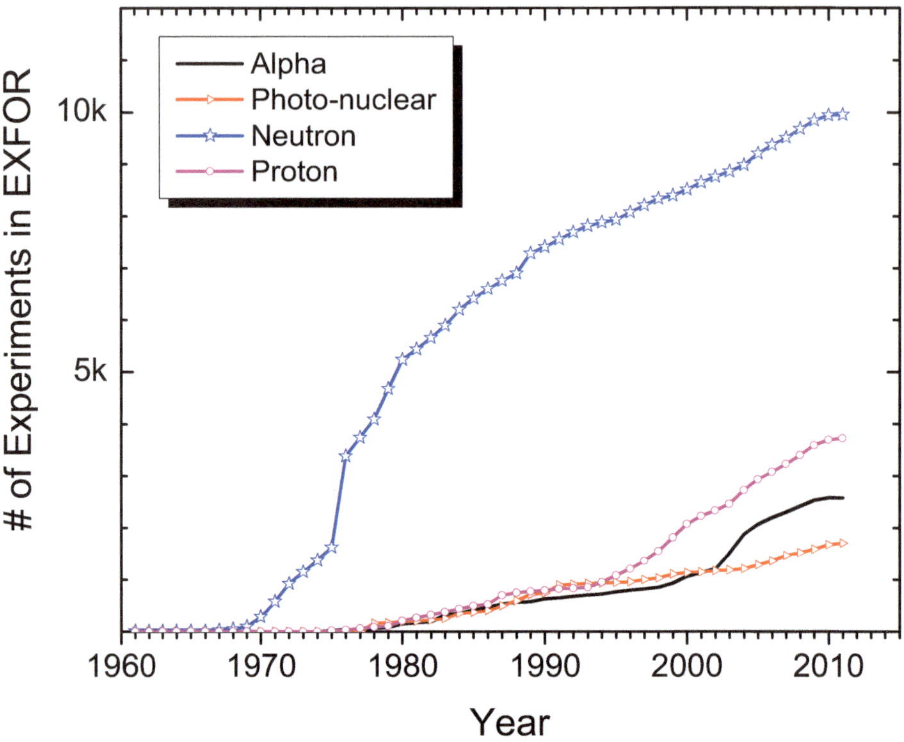

Fig. 9. Time and content evolution of the EXFOR database. Initially database scope was limited to neutron-induced reaction cross sections, later scope expansion included charge particle and photo-nuclear reactions.

known. For 1 + 2 → 3 + 4 and 3 + 4 → 1 + 2 processes the cross section ratio is

$$\frac{\sigma_{34}}{\sigma_{12}} = \frac{m_3 m_4 E_{34}(2J_3+1)(2J_4+1)(1+\delta_{12})}{m_1 m_2 E_{12}(2J_1+1)(2J_1+1)(1+\delta_{34})}, \tag{19}$$

where E_{12}, E_{34} are kinetic energies in the c.m. system, J is angular momentum, and δ_{12}=δ_{34}=0.

3.5 rp-process

The rp-process (rapid proton capture process) consists of consecutive proton captures onto seed nuclei to produce heavier elements. It occurs in a number of astrophysical sites including X-ray bursts, novae, and supernovae. The rp-process requires a high-temperature environment ($\sim 10^9$ K) to overcome Coulomb barrier for charged particles. This process contributes to observed abundances of light and medium proton-rich nuclei and compliments the p-process. The rp-process data needs include

- nuclear masses
- proton capture rates

4. Conclusion

Finally, a review of stellar nucleosynthesis and its data needs has been presented. Several nuclear astrophysics opportunities and the corresponding computation tools and methods have been identified. Complimentary sets of nuclear data for s-process nucleosynthesis have been produced. These, nuclear astrophysics model-independent data sets are based on the latest evaluated nuclear libraries and low-fidelity covariances data. This analysis indicates that the nucleosynthesis processes and respective abundances are strongly affected by both nuclear reaction cross sections and nuclear structure effects. Several nuclear structure data sets that include β-decay half-lives, spontaneous fission, and delayed neutron emission probabilities have been considered.

Present results demonstrate a wide range of uses for nuclear reaction cross sections and structure data in stellar nucleosynthesis. They provide additional benchmarks and build a bridge between nuclear astrophysics and nuclear industry applications. Further work will include extensive data analysis, neutron physics, and network calculations.

We are grateful to M. Herman (BNL) for the constant support of this project, to S. Goriely (Universite Libre de Bruxells) and R. Reifarth (Goethe University) for productive discussions, and to V. Unferth (Viterbo University) for a careful reading of the manuscript. This work was sponsored in part by the Office of Nuclear Physics, Office of Science of the U.S. Department of Energy under Contract No. DE-AC02-98CH10886 with Brookhaven Science Associates, LLC.

5. References

Aldama, D.L. & Trkov, A. (2004). FENDL-2.1 Update of an evaluated nuclear data library for fusion applications. *International Atomic Energy Agency*, INDC(NDS)-467, Distribution FE, December 2004.

Anders, E. & Grevesse, N. (1989). Abundances of the elements - Meteoritic and solar. *Geochimica et Cosmochimica Acta*, Vol. 53, January 1989, pp. 197-214.

Angulo, C.; Arnould, M.; Rayet, M.; Descouvemont, P.; Baye, D.; Leclercq-Willain, C.; Coc, A.; Barhoumi, S.; Aguer, P.; Rolfs, C.; Kunz, R.; Hammer, J.W.; Mayer, A.; Paradellis, T.; Kossionides, S.; Chronidou, C.; Spyrou, K.; Degl'Innocenti, S.; Fiorentini, G.; Ricci, B.; Zavatarelli, S.; Providencia, C.; Wolters, H.; Soares, J.; Grama, C.; Rahighi, J.; Shotter, A. & Lamehi Rachti, M. (1999). A compilation of charged-particle induced thermonuclear reaction rates. *Nuclear Physics A*, Vol. 656, No.1, August 1999, pp. 3-183.

Aprahamian, A. (2011). R-process mass sensitivities. *Proc. XIV International Symposium on Capture γ-ray Spectroscopy*, August 28-September 2, 2011, University of Guelph, Guelph, Ontario, Canada.

Arlandini, C.; Käppeler, F. ; Wisshak, K.; Gallino, R.; Lugaro, M.; Busso, M. & Straniero, O. (1999). Neutron Capture in Low-Mass Asymptotic Giant Branch Stars: Cross Sections and Abundance Signatures. *The Astrophysical Journal*, Vol. 525, November 1999, pp. 886-900.

Audi, G.; Wapstra, A.H. & Thibault, C. (2003). The AME2003 atomic mass Evaluation (II). Tables, graphs, and references. *Nuclear Physics A*, Vol. 729, Issue 1, December 2003, pp. 337-676.

Audi, G. & Meng, W. (2011). Atomic Mass Evaluation 2011. *Private Communication*.

Bao, Z.Y.; Beer, H.; Käppeler, F.; Voss, F.; Wisshak, K.; Rauscher, T. (2000). Neutron Cross Sections for Nucleosynthesis Studies. *Atomic Data and Nuclear Data Tables*, Vol. 76, Issue 1, September 2000, pp. 70-154.

Bethe, H.A. (1939). Energy Production in Stars. *Physical Review*, Vol. 55, Issue5, March 1939, pp. 434-456.

Boyd, R.N. (2008). *An introduction to nuclear astrophysics*, The University of Chicago Press, 2008.

Boyd, R.N; Bernstein, L. & Brune, C. (2009). Studying Nuclear Astrophysics at NIF. *Physics Today*, August 14, 2009.

Burbidge, E.M.; Burbidge, G.R.; Fowler, W.A. & Hoyle, F. (1957). Synthesis of the Elements in Stars. *Review of Modern Physics*, Vol. 29, No.4, October 1957, pp. 547-654.

Burrows, T.W. (1990). The evaluated nuclear structure data file: Philosophy, content, and uses. *Nuclear Instruments and Methods in Physics Research Section A*, Vol. 286, Issue 3, January 1990, pp. 595-600.

Cameron, A.G.W. (1957). Stellar evolution, nuclear astrophysics, and nucleogenesis. *Chalk River Report*, CLR-41, Chalk River, Ontario, June 1957, pp. 1-197.

Chadwick, M.B.; Obložinský, P.; Herman, M.; Greene, N.M.; McKnight, R.D.; Smith, D.L.; Young, P.G.; MacFarlane, R.E.; Hale, G.M.; Frankle, S.C.; Kahler, A.C.; Kawano, T.; Little, R.C.; Madland, D.G.; Moller, P.; Mosteller, R.D.; Page, P.R.; Talou, P.; Trellue, H.; White, M.C.; Wilson, W.B.; Arcilla, R.; Dunford, C.L.; Mughabghab, S.F.; Pritychenko, B.; Rochman, D.; Sonzogni, A.A.; Lubitz, C.R.; Trumbull, T.H.; Weinman, J.P.; Brown, D.A.; Cullen, D.E., Heinrichs, D.P.; McNabb, D.P.; Derrien, H.; Dunn, M.E.; Larson, N.M.; Leal, L.C.; Carlson, A.D.; Block, R.C.; Briggs, J.B.; Cheng, E.T.; Huria, H.C.; Zerkle, M.L.; Kozier, K.S.; Courcelle, A.; Pronyaev, V. & van der Marck, S.C. (2006). ENDF/B-VII.0: Next Generation Evaluated Nuclear Data Library for Nuclear Science and Technology. *Nuclear Data Sheets*, Vol. 107, Issue 12, December 2006, pp. 2931-3060.

Cowan, J. J. & Thielemann, F.-K. (2004) R-process Nucleosynthesis in Supernovae. *Physics Today* Vol. 57, October 2004, pp. 47-53.

Cyburt, R.H.; Amthor, A.M.; Ferguson, R.; Meisel, Z.; Smith, K.; Warren, S.; Heger, A.; Hoffman, R.D., Rauscher, T.; Sakharuk, A.; Schatz, H.; Thielemann, F.L. & Wiescher, M. (2010). The JINA REACLIB Database: Its Recent Updates and Impact on Type-I X-ray Bursts. *The Astrophysical Journal Supplement Series*, Vol. 189, July 2010, pp. 240-252.

Dillmann, I.; Heil, M.; Käppeler, F.; Plag, R.; Rauscher , T. & Thielemann, F.-K. (2006). KADoNiS - The Karlsruhe Astrophysical Database of Nucleosynthesis in Stars. *Proceedings of Capture Gamma-ray Spectroscopy and Related Topics: 12th International Symposium*, Vol. 819, pp. 123-127, ISBN: 0-7354-0313-9, Notre Dame, Indiana (USA), 4-9 September 2005, AIP Conference Proceedings.

Eddington, A.S. (1920). The Internal Constitution of the Stars. *Nature*, Vol. 106, September 1920, pp. 14-20.

Grevesse, N. & Sauval, A. J. (1998). Standard Solar Composition. *Space Science Reviews*, v. 85, Issue 1/2, 1998, pp. 161-174.

Herman, M.; Capote, R.; Carlson, B.V.; Obložinský, P.; Sin, M.; Trkov, A.; Wienke, H. & Zerkin, V. (2007). EMPIRE: Nuclear Reaction Model Code System for Data Evaluation. *Nuclear Data Sheets*, Vol. 108, Issue 12, December 2007, pp. 2655-2715.

Hoyle, F. (1946). The synthesis of the elements from hydrogen. *Monthly Notices of the Royal Astronomical Society* Vol. 106, 1946, pp. 343-383.

Käppeler, F. (2011). Reaction cross sections for s, r, and p process. *Progress in Particle and Nuclear Physics*, Vol. 66, Issue 2, April 2011, pp. 390-399.

Käppeler, F.; Gallino, R.; Bisterzo, S. & Wako Aoki. (2011). The s-process: Nuclear physics, stellar models, and observations. *Review of Modern Physics*, Vol. 83, January-March 2011, pp. 157-193.

Kolb, E. & Turner, M. (1988). *The Early Universe*, Addison-Wesley, 1988.

Koning, A.J.; Hilaire, S, & Duijvestijn, M.C. (2008). TALYS-1.0. *Proceedings of International Conference on Nuclear Data for Science and Technology*, Vol. 1, pp. 211-214, ISBN:978-2-7598-0090-2, Nice, France, April 22-27, 2007, EDP Sciences, 2008.

Little, R.C.; Kawano, T.; Hale, G.D.; Pigni, M.T.; Herman, M.; Oblozinsky, P.; Williams, M.L.; Dunn, M.E.; Arbanas, G.; Wiarda, D.; McKnight, R.D.; McKamy, J.N. & Felty, J.R. (2008). Low-fidelity Covariance Project. *Nuclear Data Sheets*, Vol. 109, Issue 12, December 2008, pp. 2828-2833.

Merrill, S. P. W. (1952). Spectroscopic Observations of Stars of Class S. *The Astrophysical Journal*, Vol. 116, February 1952, pp. 21-26.

Mughabghab, S.F. (2006). *Atlas of Neutron Resonances, Resonance Parameters and Thermal Cross Sections. Z=1-100*, Elsevier, 2006.

Nakagawa, T.; Chiba, S; Hayakawa, T. & Kajino, T. (2005). Maxwellian-averaged neutron-induced reaction cross sections and astrophysical reaction rates for kT = 1 keV to 1 MeV calculated from microscopic neutron cross section library JENDL-3.3. *Atomic Data and Nuclear Data Tables*, Vol. 91, Issue 2, November 2005, pp. 77-186.

NRDC, International Network of Nuclear Reaction Data Centres. (2011). *Compilation of experimental nuclear reaction data (EXFOR/CSISRS)*, (Available from *http://www-nds.iaea.org/exfor/*, *http://www.nndc.bnl.gov/exfor/*).

Pritychenko, B.; Sonzogni, A.A.; Winchell, D.F.; Zerkin, V.V; Arcilla, R.; Burrows, T.W.; Dunford, C.L.; Herman, M.W.; McLane, V.; Obložinský, P.; Sanborn, Y. & Tuli, J.K. (2006). Nuclear Reaction and Structure Data Services of the National Nuclear Data Center. *Annals of Nuclear Energy*, Vol. 33, Issue 4, March 2006, pp. 390-399.

Pritychenko, B & Sonzogni, A.A. (2008). Sigma: Web Retrieval Interface for Nuclear Reaction Data. *Nuclear Data Sheets*, Vol. 109, Issue 12, December 2008, pp. 2822 - 2827.

Pritychenko, B.; Mughabghab, S.F. & Sonzogni, A.A. (2010). Calculations of Maxwellian-averaged Cross Sections and Astrophysical Reaction Rates Using the ENDF/B-VII.0, JEFF-3.1, JENDL-3.3 and ENDF/B-VI.8 Evaluated Nuclear Reaction Data Libraries. *Atomic Data and Nuclear Data Tables*, Vol. 96, Issue 6, November 2010, pp. 645-748.

Pritychenko, B. (2010). Complete calculation of evaluated Maxwellian-averaged cross sections and their uncertainties for s-process nucleosynthesis. *Proceedings of 11th Symposium on Nuclei in the Cosmos, NIC XI*, PoS(NIC-XI)197, Heidelberg, Germany, July 19-23, 2010, Proceedings of Science, 2010.

Pritychenko, B.; Běták, E.; Kellett, M.A.; Singh, B. & Totans, J. (2011). The Nuclear Science References (NSR) database and Web Retrieval System. *Nuclear Instruments and Methods in Physics Research Section A*, Vol. 640, Issue 1, 1 June 2011, pp. 213-218.

Rauscher T. & Thielemann F. (2000). Astrophysical Reaction Rates from Statistical Model Calculations. *Atomic Data and Nuclear Data Tables*, Vol.75, Issue 1, January 2000, 1.

Rolfs, C.E. & Rodney, W.S. (1988). *Cauldrons in the Cosmos*, The University of Chicago Press, 1988.

Shibata, K.; Iwamoto, O.; Nakagawa, T.; Iwamoto, N; Ichihara, A.; Kunieda, S.; Chiba, S.; Furutaka, K.; Otuka, N.; Ohsawa, T.; Murata, T.; Matsunobu, H.; Zukeran, A.; Kamada, S. & Katakura, J.-i. (2011). JENDL-4.0: A New Library for Nuclear Science and Engineering. *Journal of Nuclear Science and Technology*, Vol. 48, No. 1, January 2011, pp. 1-30.

Szücs, T.; Dillmann, I.; Plag, R. & Fülöp, Zs. (2010). The new p-process database of KADoNiS. *Proceedings of 11th Symposium on Nuclei in the Cosmos, NIC XI* PoS(NIC-XI)247, Heidelberg, Germany, July 19-23, 2010, Proceedings of Science, 2010.

Wolfram Corporation. (2011). *Wolfram Mathematica Online Integrator*, Available from ⟨http://integrals.wolfram.com⟩.

Woosley, S. E. & Howard, W. M. (1978). The p-process in supernovae. *The Astrophysical Journal Supplement*, Vol. 36, 1978, pp. 285-304.

Zabrodskaya, S.V.; Ignatyuk, A.V.; Koscheev, V.N.; Manokhin, V.N.; Nikolaev, M.N. & Pronyaev, V.G. (2007). ROSFOND - Russian National Library of Neutron Data. *Voprosy Atomnoj Nauki i Techniki, Seriya: Nuclear Constants* Vol. 1-2, 2007, p. 3-21 .

Zerkin, V.V.; McLane, V.; Herman, M.W. & Dunford, C.L. (2005). EXFOR-CINDA-ENDF: Migration of Databases to Give Higher-Quality Nuclear Data Services. *Proceedings of International Conference on Nuclear Data for Science and Technology*, Vol. 769, pp. 586-589, ISBN: 0-7354-0254-X, Santa Fe, New Mexico (USA), 26 September-1 October 2004, AIP Conference Proceedings.

4

The r-Process of Nucleosynthesis: The Puzzle Is Still with Us

Marcel Arnould and Stephane Goriely
Université Libre de Bruxelles
Belgium

1. Introduction

Through the increasing number and quality of astronomical observations, our picture of the composition of the various constituents of the Universe is getting quickly more and more complete, and concomitantly more and more complex. Despite this spectacular progress, the solar system (hereafter SoS) continues to provide a body of abundance data whose quantity, quality and coherence remain unmatched. This concerns especially the heavy elements (defined here as those with atomic numbers in excess of the value $Z = 26$ corresponding to iron), and in particular their isotopic compositions, which are the prime fingerprints of astrophysical nuclear processes. Except in a few cases, these isotopic patterns indeed remain out of reach even of the most-advanced stellar spectroscopic techniques available today. No wonder then that, from the early days of its development, the theory of nucleosynthesis has been deeply rooted in the SoS composition, especially in the heavy element domain.

Since the early days of the development of the theory of nucleosynthesis, it has been proved operationally rewarding to introduce three categories of heavy nuclides referred to as s-, p-, and r-nuclides. This splitting corresponds to the 'topology' of the chart of the nuclides, which exhibits three categories of stable heavy nuclides: those located at the bottom of the valley of nuclear stability, called the s-nuclides, and those situated on the neutron-deficient or neutron-rich side of the valley, named the p- or r-nuclides, respectively. Three different mechanisms are called for to account for the production of these three types of stable nuclides. They are naturally referred to as the s-, r-, and p-processes. An extensive survey of the p-process can be found in Arnould & Goriely (2003), while some aspects of the s-process are the subject of another chapter of this book.

The main aim of this chapter is to provide a rather brief summary of some selected astrophysics and nuclear physics aspects of the r-process already reviewed in detail by Arnould et al. (2007), but that we consider to be worth emphasizing again for one reason or another. Some update of this review is also proposed, based on papers published after September 2006.

2. The bulk solar system abundances of the r-nuclides: too often neglected severe intricacies

The bulk SoS material has long been recognized to be made of a well-mixed blend of many nucleosynthesis contributions over the approximate 10 Gy that have elapsed between the formations of the Galaxy and of the SoS. Since the nineteen fifties, much effort has been devoted to the highly complex derivation of a meaningful set of elemental abundances representative of this material at the SoS formation some 4.6 Gy ago. This research has led to refinements over the years, but not to drastic changes (see (Lodders, 2010) for the most recent compilation).

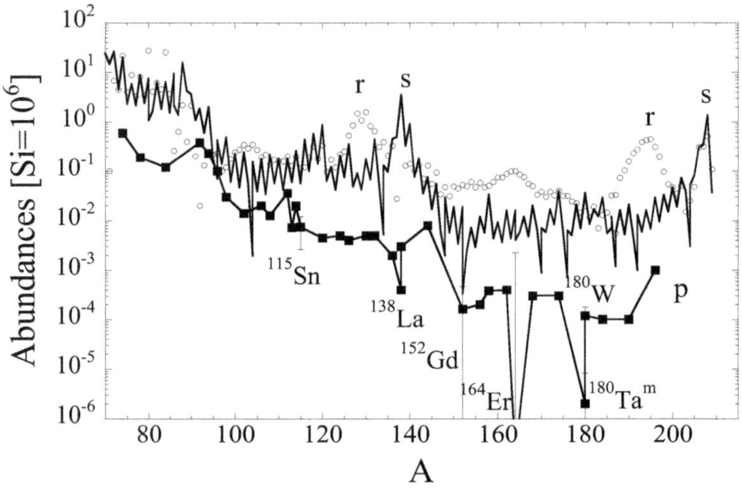

Fig. 1. Decomposition of the solar abundances of heavy nuclides into s-process (*solid line*), r-process (*open circles*) and p-process (*squares*) contributions. The uncertainties on the abundances of some p-nuclides due to a possible s-process contamination are represented by vertical bars (from (Arnould & Goriely, 2003)). See Figs. 2 - 4 for the uncertainties in the s- and r-nuclide data

As mentioned above, it is very useful to split the abundance distribution of the nuclides heavier than iron into three separate distributions giving the image of the SoS content of the p-, s- and r-nuclides. A rough representation of this splitting is displayed in Fig. 1. In its details, the procedure of decomposition is by far not as obvious as it might be thought from the very definition of the different types of the nuclides heavier than iron, and is to some extent dependent on the models for the synthesis of these nuclides. These models predict in particular that the stable nuclides located on the neutron-rich(deficient) side of the valley of nuclear stability are produced, to a first good approximation, by the r-(p-)process only. These stable nuclides are naturally called 'r-only' and 'p-only' nuclides, and their abundances are deduced directly from the SoS abundances. The situation is much more intricate for the nuclides situated at the bottom of the valley of nuclear stability. Some of them are produced solely by the s-process, the typical flow of which is located very close to the valley (see the review of the s-process in this volume). They are referred to as 's-only' nuclides, and are encountered only when a stable r- or p-isobar exists, which 'shields' the s-isobar from the r- and p-processes. As a result, only even-Z heavy elements possess an s-only isotope. In the other cases, the nuclides at the bottom of the valley of nuclear stability may be produced by both the s- and the r-processes in proportions that vary from nuclide to nuclide. They are referred to as 'sr' nuclides, and the relative contributions of the two processes to their abundances are the most intricate to evaluate[1].

[1] The abundances of some nuclides may also result result from a contribution of the s- and p-processes, they are referred to as 'sp' nuclides

Schematic SoS distributions of the p-, s- and r-nuclides are displayed in Fig. 1. It appears that about half of the heavy nuclei in the SoS material come from the s-process, and the other half from the r-process, whereas the p-process is responsible for the production of about 0.01 to 0.001 of the abundances of the s- and r-isobars. Some elements have their abundances dominated by an s- or r-nuclide. They are naturally referred to as s- or r-elements. Clearly, p-elements do not exist. These general characteristics are not drastically sensitive to the particular model used to obtain this splitting. In contrast, most details of the decomposition between the s- and r-contributions do, and many uncertainties remain. *It is highly unfortunate that they are classically put under the rug, especially in view of the sometimes very detailed and far-reaching considerations that have the s-r SoS splitting as an essential starting point.*

The canonical s-process model has been widely used to split the s- and r-contributions to the SoS material. In short, this model assumes that stellar material composed only of iron nuclei is subject to neutron densities and temperatures that remain constant over the whole period of the neutron irradiation. In addition, the SoS s-abundance pattern is viewed as originating from a superposition of two exponential distributions of the time-integrated neutron exposure $\tau_{\rm n} = \int_0^t N_{\rm n} v_T {\rm d}t$ (where v_T is the most probable relative neutron-nucleus velocity at temperature T, and N_n is the neutron density). These distributions are traditionally held responsible for the so-called weak ($70 \lesssim A \lesssim 90$) and main ($A \gtrsim 90$) components of the s-process. A third exponential distribution is sometimes added in order to account for the $204 < A \leq 209$ s-nuclides. Through an adequate fitting of the parameters of the two or three τ-distributions, the superposition of the two or three resulting abundance components reproduces quite successfully the abundance distribution of the s-only nuclides in the SoS, from which it is concluded that the s-contribution to the sr-nuclides can be predicted reliably. It has to be stressed that this result is rooted only in the nuclear properties of the species involved in the s-process, and does not rely at all on specific astrophysics scenarios.

In spite of the claimed success of the canonical model, some of its basic assumptions deserve questioning. This concerns in particular a presumed exponential form for the distribution of the neutron exposures τ, which has been introduced in view of its mathematical ease in abundance calculations. In addition, the canonical model makes it difficult to evaluate uncertainties of nuclear or observational nature. As a result, the concomitant uncertainties in the SoS r-abundances are traditionally not evaluated.

For about ten years, it has become fashionable to predict the s- and r-contributions from s-process calculations performed in the framework of some models for stars of various masses and initial compositions (e.g. Arlandini et al., 1999; Simmerer et al., 2004). *We strongly suggest not to use this procedure for predicting the SoS s- and r-components.* Large uncertainties indeed remain in the s-abundances predicted from all model stars. In addition, the SoS s-nuclide abundances result from a long evolution of the galactic composition that cannot be mimicked reliably enough.

A third approach has been proposed by Goriely (1997), and is referred to as the multi-event s-process model (MES). It cures to a large extent many of the shortcomings of the canonical model, and is in fact the only attempt so far to evaluate on a quantitative basis the uncertainties in the derived SoS s- and r-abundances. This makes quite incomprehensible the fact that it has remained largely unnoticed for more than ten years. In view of the importance to evaluate the uncertainties affecting the SoS distribution of the abundances of the r-nuclides, we review the MES in some detail.

The MES relies on a superposition of a given number of canonical events, each of them being defined by a neutron irradiation on the ^{56}Fe seed nuclei during a time $t_{\rm irr}$ at a constant temperature T and a constant neutron density $N_{\rm n}$. In contrast to the canonical model, no hypothesis is made concerning any particular distribution of the neutron exposures. Only a set of canonical events that are considered as astrophysically plausible is selected a priori. We adopt here about 500 s-process canonical events covering ranges of astrophysical conditions that are identified as relevant by the canonical model, that is $1.5 \times 10^8 \le T \le 4 \times 10^8$ K, $7.5 \le \log N_{\rm n}[{\rm cm}^{-3}] \le 10$, and 40 chosen $t_{\rm irr}$-values, corresponding to evenly distributed values of $n_{\rm cap}$ in the $5 \le n_{\rm cap} \le 150$ range, where

$$n_{\rm cap} = \sum_{Z,A} A\, N_{Z,A}(t = t_{\rm irr}) - \sum_{Z,A} A\, N_{Z,A}(t = 0) \tag{1}$$

is the number of neutrons captured per seed nucleus (^{56}Fe) on the timescale $t_{\rm irr}$, the summation extending over all the nuclides involved in the s-process. For each of the selected canonical events, the abundances $N_{Z,A}$ are obtained by solving a reaction network including 640 nuclear species between Cr and Po. Based on these calculated abundances, an iterative inversion procedure described in Goriely (1997) allows to identify a combination of events from the considered set that provides the best fit to the solar abundances of a selected ensemble of nuclides. This set includes 35 nuclides comprising the s-only nuclides, complemented with ^{86}Kr and ^{96}Zr (largely produced by the s-process in the canonical model), ^{152}Gd and ^{164}Er (not produced in the p-process, but able to be produced in solar abundances in the s-process (Arnould & Goriely, 2003)), and ^{208}Pb (possibly produced by the strong s-process component in the canonical model).

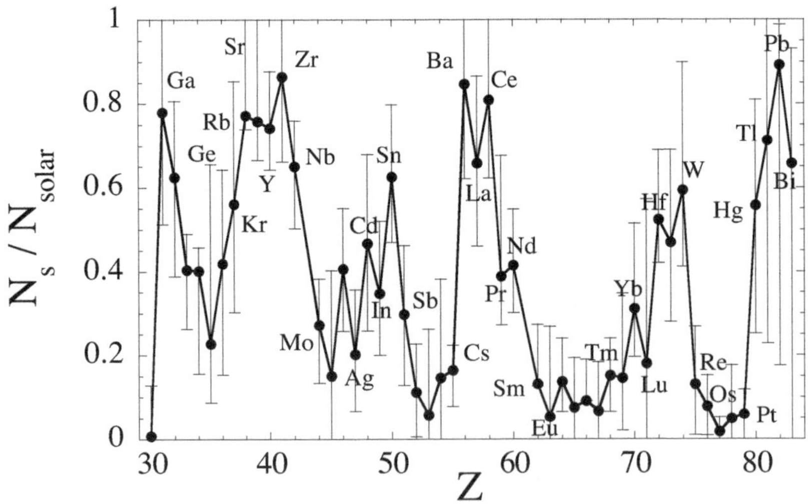

Fig. 2. The s-process contribution to the elemental SoS abundances $N_{\rm solar}$ (Palme & Beer, 1993) of the elements with $Z \ge 30$ predicted from the MES procedure. Uncertainties are represented by vertical bars

On grounds of the solar abundances of Palme & Beer (1993), it is demonstrated in Goriely (1997) that the MES provides an excellent fit to the abundances of the 35 nuclides included

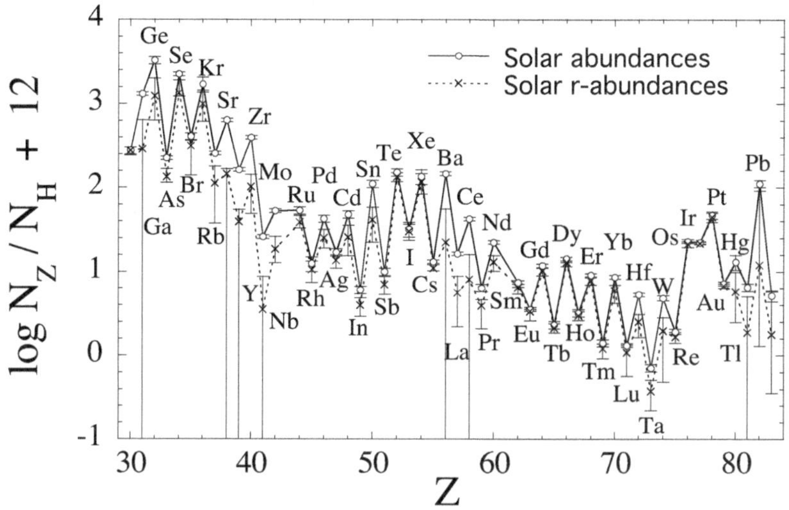

Fig. 3. SoS r-residuals and their uncertainties for the $Z \geq 30$ elements based on the s-abundances of Fig. 2. The abundances N_{H} and N_{Z} refer to hydrogen and to the element with atomic number Z

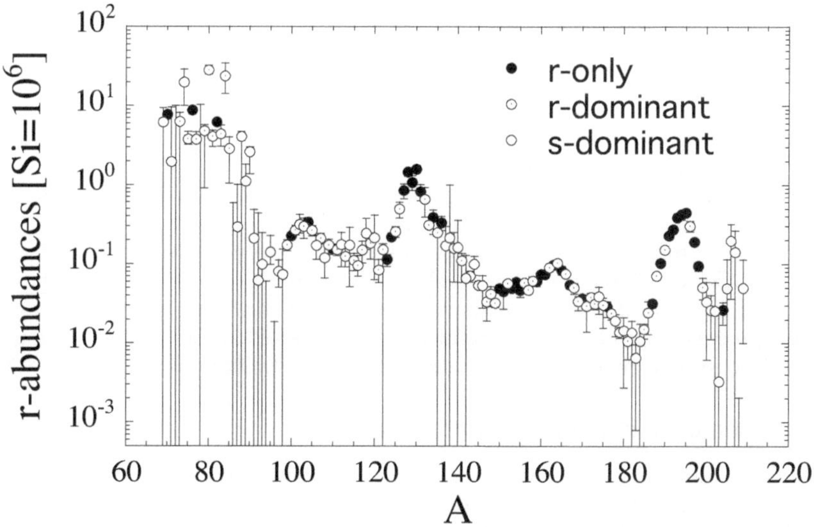

Fig. 4. SoS isotopic r-residuals corresponding to the elemental abundances of Fig. 3. Different symbols identify different relative levels of r-process contribution. The s-dominant nuclides are those predicted by MES to have more than 50% of their abundances produced by the s-process. The s-process contribution varies between 10 and 50% in the case of the r-dominant species, and does not exceed 10% for the r-only nuclides

in the considered set of species, and in fact performs to a quite-similar overall quality as that of the exponential canonical model predictions of Palme & Beer (1993) with a distribution of neutron irradiations that agrees qualitatively with the exponential distributions assumed in the canonical model, even though some deviations are noticed with respect to the canonical weak and strong components.[2]. An even better fit than in the canonical framework is obtained for the s-only nuclides (see Goriely, 1999, for details). *The MES model is therefore expected to provide a decomposition of the solar abundances into their s- and r-components that is likely to be more reliable than the one derived from the canonical approach without any a priori assumption regarding the distribution of neutron exposures.*

Compared with the canonical approach, the MES model has the additional major advantage of allowing a systematic study of the various uncertainties affecting the abundances derived from the parametric s-process model, and consequently the residual r-nuclide abundances. The uncertainties in these residuals have been studied in detail by Goriely (1999) from due consideration of the uncertainties in (i) the observed SoS abundances as given by Palme & Beer (1993) (see footnote[1]), (ii) the experimental and theoretical radiative neutron-capture rates involved in the s-process network, and in (iii) the relevant β-decay and electron-capture rates. Total uncertainties resulting from a combination of (i) to (iii) have finally been evaluated. The r-residuals obtained from such a study for the elements with $Z \geq 30$ are displayed in Fig. 3. Under the classical adoption of terrestrial isotopic compositions of the considered elements, the corresponding SoS isotopic r-residuals and their uncertainties have been calculated by Goriely (1999) and are shown in Fig. 4. They are presented in tabular form in Goriely (1999) and Arnould et al. (2007). Different situations can be identified concerning the uncertainties affecting the r-residuals. Many sr-nuclides are predicted to have a small s-process component only. The r-process contribution to these species, referred to as r-dominant, is clearly quite insensitive to the s-process uncertainties. The situation is just the opposite in the case of s-dominant nuclides.

Some r-process residuals suffer from remarkably large uncertainties, which quite clearly cannot be ignored when discussing the r-process and the virtues of one or another model for this process. This concerns in particular the elements Rb, Sr, Y, Zr, Ba, La, Ce and Pb. Some of them, and in particular Ba or La, are often used as tracers of the levels of s- or r-processing during the galactic history (see Sect. 9). Lead has also a special status in the studies of the s-process (e.g. Goriely, 2005, for references), as well as of the r-process (see Sect. 9). It could well be of pure s-nature if a strong s-process component can indeed develop in some stars, but a pure r-process origin cannot be excluded. These uncertainties largely blur any picture one might try to draw from observations and from simplistic theoretical considerations.

3. Isotopic anomalies in the SoS composition

The bulk SoS composition has been of focal interest since the very beginning of the development of the theory of nucleosynthesis. Further astrophysical interest and excitement have developed with the discovery of the fact that a minute fraction of the SoS material has an isotopic composition deviating from that of the bulk. Such 'isotopic anomalies' are observed in quite a large suite of elements ranging from C to Nd (including the rare gases), and are now known to be carried by high-temperature inclusions of primitive meteorites, as well as by various types of meteoritic grains. The inclusions are formed from SoS material out of

[2] A MES calculation with the revised solar abundances of Lodders (2010) has not been done, but is expected not to give significantly different results from those reported here

equilibrium with the rest of the solar nebula. The grains are considered to be of circumstellar origin, and to have survived the process of incorporation into the SoS.

Isotopic anomalies contradict the canonical model of an homogeneous and gaseous protosolar nebula, and provide new clues to many astrophysical problems, like the physics and chemistry of interstellar dust grains, the formation and growth of grains in the vicinity of objects with active nucleosynthesis, the circumstances under which stars (and in particular SoS-type structures) can form, as well as the early history of the Sun (in the so-called 'T-Tauri' phase) and of the SoS solid bodies. Last but not least, they raise the question of their nucleosynthesis origin and offer the exciting perspective of complementing the spectroscopic data for chemically peculiar stars in the confrontation between abundance observations and nucleosynthesis models for a very limited number of stellar sources, even possibly a single one. This situation is in marked contrast with the one encountered when trying to understand the bulk SoS composition, which results from the mixture of a large variety of nucleosynthesis events, and consequently requires the modelling of the chemical evolution of the Galaxy.

Among the identified anomalies, several concern the p-, s- and r-nuclides. Those attributed to the p- and r-processes are discussed in some detail by Arnould & Goriely (2003) and Arnould et al. (2007), and are not reviewed here. As a very brief summary, let us just say that various blends of p-, s- and r-nuclides that differ more or less markedly from the bulk SoS mixture depicted in Sect. (2) are identified in a variety of meteorites at various scales, including bulk samples, refractory inclusions or grains viewed from their many highly anomalous isotopic signatures as grains of circumstellar origins. This is generally interpreted in terms of the decoupling between the three mechanisms producing these nuclides, and of the non-uniform mixing of their products. One of the surprises of main relevance in the discussion of the r-process is that those grains that are generally interpreted in terms of supernova (SN) condensates do not carry the unambiguous signature of the r-process that would be expected if indeed SNe are the provileged r-process providers (see Sect. 7).

4. The r-nuclide content of the Galaxy and its evolution

In practice, the question of the evolution of the galactic content of the nuclides heavier than iron concerns the s- and r-nuclides only. It is traditionally assumed indeed that the p-nuclides are just as rare in all galactic locations as in the SoS. In such conditions, the p-nuclide abundances outside the SoS are out of reach of spectroscopic studies. On the other hand, the s-process is discussed elsewhere in this volume, so that we focus here only on the evolution of the r-nuclide galactic content.

A substantial observational work has been conducted in recent years on this subject. This effort largely relies on the abundance evolution of Eu, classified as an r-process element on grounds of SoS abundance analyses. The main conclusions derived from this observational work up to 2007 have been discussed in some detail by Arnould et al. (2007). Some new observations have confirmed previous ones, or have brought some interesting new pieces of information which, however, are not considered to bring a real breakthrough in our global understanding of the r-process. Let us just give a very limited flavour of the conclusions that can be drawn from the myriad of available observations:

(i) the Eu data are classically used to support the idea that the r-process has contributed very early to the heavy element content of the Galaxy. However, the observed Eu abundance scatter introduces some confusion when one tries to establish a clear trend of the Eu enrichment with metallicity. It is also difficult to identify the value of [Fe/H] at which the signature of the s-process becomes identifiable. This conclusion relies in particular on La abundances derived from observation, La being classically considered as an s-element in the SoS (even if a non-negligible r-process contribution cannot be excluded). Recent observations also indicate that star to star variations in the r-process content of metal-poor globular clusters may be a common, although not ubiquitous, phenomenon (Roederer, 2011);

(ii) a most useful information on the relative evolution of the s- and r-process efficiencies in the Galaxy would be provided by the knowledge of the isotopic composition of the neutron-capture elements. Such data are unfortunately very scarce, and concern the Ba isotopic composition in a limited sample of stars. They are still under some debate, but raise the possibility that Ba may be mainly of s-process venue in some metal-poor stars (Gallagher et al. (2010), who refer to this possibility as an 'inconvenient truth');

(iii) much excitement has been raised by the observation that the patterns of abundances of heavy neutron-capture elements between Ba and Pb in r-process-rich metal-poor stars are remarkably similar to the SoS one. This claimed 'convergence' or 'universality' has to be taken with some care, however, as it largely relies on the assumption that the decomposition between s- and r-process contributions in metal-poor stars is identical to the SoS one, which has yet to be demonstrated. An interpretation of this universality is proposed in Sect. 6.1;

(iv) no universality appears to hold for $Z \lesssim 58$ and for $Z \gtrsim 76$, where this concerns in particular the Pb-peak elements and the actinides. This situation has far-reaching consequences, particularly regarding the possibility of building galactic chronologies on the actinide content of very metal-poor stars (see Sect. 9);

(v) the different behaviours of the abundance patterns of the elements below and above Ba have laid the ground for speculations on possible different sites of the r-process. It is discussed by Arnould et al. (2007) that *the interest of these speculations is rather limited at the present stage of desperate search for a single suitable site for the r-process* (see Sects. 7.2 and 8);

(vi) the knowledge of the r-process content of the galactic cosmic rays (GCRs) is of substantial interest, as they are made of much younger material than the SoS (i.e., about 20 to 30 Myr old). In spite of substantial progress made in the measurement of the GCR composition, much remains to be done. At this time, there is no clear proof that the r-process(es) that has (have) contributed to this recent sample of galactic material is (are) of a similar nature as those responsible for the bulk SoS. The identification of actinides in the GCRs has been made possible recently. An accurate measurement of their abundances would be of great value to discriminate between various GCR sources that have been proposed. As in the stellar case, information on the isotopic composition of the GCRs would be of prime interest in helping to evaluate the fractional contribution of freshly synthesized r-process material to this sample of young galactic material.

5. Nuclear physics input for the r-process

It is easy to conceive that the nuclear physics that enters the r-process modelling depends to a more or less large extent on the astrophysics conditions that are suitable for the development of that process. Even if the proper site(s) of the r-process has (have) not been reliably identified yet, it appears reasonable to say from existing studies that a huge body of nuclear data are potentially needed for the purpose of r-process predictions. This includes the static properties (like masses, matter and charge distributions, single-particle spectra, or pairing characteristics) of thousands of nuclides from hydrogen to the superheavy region located between the valley of β-stability and the neutron drip line. Their decay characteristics (α-decay, spontaneous fission, β-decay, β-delayed neutron emission or fission), and reactivity (nucleon or α-particle captures, photoreactions, or neutrino captures) are needed as well.

A major effort has been devoted in recent years to the measurement of nuclear data of relevance to the r-process. Still, a large body of information remains, and will remain in a foreseeable future, out of reach of experimental capabilities. This is of course the direct consequence of the huge number of nuclear species that may be involved in one r-process or another, along with the fact that nuclei very far from the valley of stability are likely to enter the process. Theory has thus mandatorily to complement the laboratory measurements (see e.g. Dillmann & Litvinov (2011) and Habs et al. (2011) for a non-exhaustive list of recent laboratory efforts). Predictions based as much as possible on global universal and microscopic models have to be favoured, as discussed in Arnould et al. (2007) where the reader is referred to for many details on the relevant nuclear physics.

6. The astrophysics of the r-process: parametrized site-free scenarios

6.1 Canonical and 'multi-event r-process (MER)' high-temperature models

Since the early days of the theory of nucleosynthesis, it has been proposed that the r-process results from the availability of neutron concentrations that are so high that neutron captures (especially of the radiative type) are faster than β-decays, at least for a substantial number of neutron-rich nuclides that are not located too far from the valley of nuclear stability. This is in marked contrast to the situation envisioned for the s-process. Such conditions clearly provide a natural way to transform any pre-existing material into very neutron-rich species, viewed as the progenitors of the r-nuclides. A classical additional hypothesis has been that the otherwise unspecified stellar location is hot enough to allow (γ,n) photodisintegrations to counteract to a more or less large extent the action of the inverse radiative neutron captures. Finally, it is supposed that a cooling of the material that allows the 'freezing-out' of the highly temperature-sensitive photodisintegrations occurs concomitantly with a decrease of the neutron density to values that are low enough to freeze the neutron captures.[3] The hypothesized high neutron fluence and temperature have been the framework adopted by the vast majority of studies of the r-process. In many cases, the consequences of such an assumption have been scrutinized only from a purely nuclear physics point of view, just considering that one astrophysical site or the other, and in particular the inner regions of massive star SN explosions, could be the required neutron provider.

The simplest and most widely used form of this high-temperature r-process scenario is commonly referred to as the canonical r-process model (see e.g. Arnould et al., 2007, for details). It assumes constant thermodynamic conditions (temperatures, densities, neutron concentrations) during a given irradiation time. Over this period, pre-existing iron-peak material is driven by neutron captures into a location of the neutron-rich region determined by the neutron fluence and by the reverse photodintegrations. When the irradiation stops, all neutron captures and photodisintegrations are abruptly frozen (this implies that the capture of neutrons produced in β-delayed processes is neglected). Even if this canonical model does not make reference to any specific astrophysics scenario, but builds on nuclear properties only, it has helped greatly paving the way to more sophisticated approaches of the r-process.

[3] Let us recall that a transformation is said to be 'frozen' if its typical mean lifetime gets longer than a typical evolutionary timescale of the considered astrophysical site

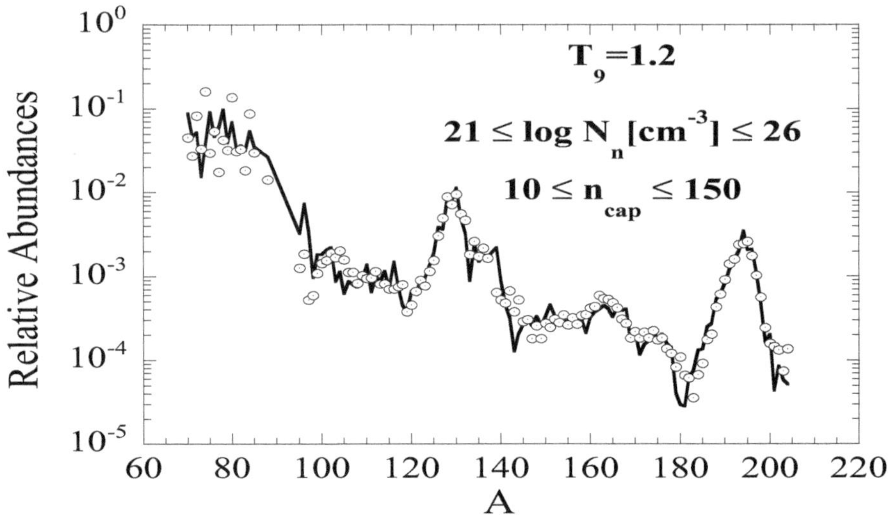

Fig. 5. Comparison between the SoS r-abundances (Fig. 4; the uncertainties are not shown) and a MER fit obtained assuming a constant temperature $T_9 = 1.2$ (in billion K) and a superposition of events characterized by free neutron concentrations N_n and numbers of neutrons captured by iron seed n_{cap} located in the indicated ranges. Within these prescribed limits, the characteristic of the involved events and their relative contributions to the displayed yields are derived from an iterative inversion procedure (see Arnould et al., 2007, for details and references)

A parametric approach of the r-process referred to as the 'multi-event r-process (MER)' has been developed recently (see Arnould et al., 2007, for details). It drops some of the basic assumptions of the canonical model, but keeps the simplification of constant thermodynamic conditions and neutron concentrations during the irradiation time, after which all nuclear reactions are frozen. It is an especially efficient tool to carry out systematic studies of the impact of uncertainties of nuclear physics nature on yield predictions, which, in this approach, necessitate the solution of huge nuclear reaction networks involving thousands of nuclear species, and the availability of an even larger body of input nuclear data (reaction and β-decay rates, fission probabilities,...). Figure 5 provides an example of fit to the SoS r-abundance distribution obtained with the use of MER.

6.2 Universality, you have said universality?

As a special application, MER has been used to interpret the high similarity of the SoS r-abundances between Ba and Os and those observed in r-process-rich metal-poor stars. This situation is generally interpreted as the signature of a 'universality' of the r-process (see Sect. 4). As reviewed by Arnould et al. (2007), the main conclusions drawn from the MER results are that

(1) *the pattern of abundances in the Ba to Os range is mainly governed by nuclear physics properties* (and in particular by the fact that even Z elements have more stable isotopes that can be fed by the r-process). If this is indeed true, a possible universality in this Z range does not tell

much about specific astrophysical conditions. It is quite remarkable that this result is largely ignored in the literature on the r-process;

(2) the convergence of abundances in the above mentioned range does not provide any demonstration of any sort of a more global universality involving lighter and heavier elements. With time, these reservations have received mounting support from observation, as noted in Sect. 4.

6.3 Dynamical high-temperature r-process approaches (DYR)

In associating the r-process with SN explosions, several attempts to go beyond the canonical and MER models have been made by taking into account some evolution of the characteristics of the sites of the r-process during its development. These models are coined 'dynamical' (DYR) in the following in order to remind of the time variations of the thermodynamic state of the r-process environment (see Arnould et al., 2007, for references). These models do not rely on any specific explosion scenario. They just assume that a material that is initially hot enough for allowing a nuclear statistical equilibrium (NSE) to be achieved expands and cools in a prescribed way on some selected timescale. This evolution is fully parameterized.

With the requirement of charge and mass conservation, and if the relevant nuclear binding energies are known, the initial NSE composition is determined from the application of the nuclear Saha equation (e.g. Sect. 7-2 of Clayton, 1968, for a general presentation) for an initial temperature and density (or, equivalently, entropy), and electron fraction (net electron number per baryon) Y_e. These three quantities are of course free parameters in a site-free r-process approach. The evolution of the abundances during expansion and cooling of the material from the NSE state is derived by solving an appropriate nuclear reaction network. The freeze-out of the charged-particle induced reactions might be followed by an r-process.

With temperature, density and Y_e as free parameters, many choices of initial NSE compositions may clearly be made, involving a dominance of light or heavy nuclides, as illustrated in Fig. 6. However, in view of its relevance to the SN models, an initial NSE at temperatures of the order of 10^{10} K is generally considered. It favours the recombination of essentially all the available protons into α-particles (the region noted NSE [n,α] in Fig. 6). The evolution of this initial composition to the stage of charged-particle induced reaction freeze-out has been analyzed in detail by Meyer et al. (1998)), and we just summarize here some of its most important features that are of relevance to a possible subsequent r-process:

(1) at some point in the course of the expansion and cooling of the initially α-rich material, full NSE breaks down as the result of the slowness of a fraction of the charged-particle reactions relative to the expansion timescale. The formation of quasi-equilibrium (QSE) clusters results. In this state, the intra-QSE composition still follows the NSE Saha equation, but the relative inter-cluster abundances do not, and depend on the kinetics of the nuclear flows into and out of the QSE clusters. To be more specific, the QSE phase is dominated in its early stages by a light cluster made of neutrons, α-particles and traces of protons, and by a cluster made of ^{12}C and heavier species. The population of the latter is determined mainly by the $\alpha + \alpha + n$ reaction, followed by $^9\mathrm{Be}(\alpha,n)^{12}\mathrm{C}(n,\gamma)^{13}\mathrm{C}(\alpha,n)^{16}\mathrm{O}$, as first noticed by Delano & Cameron (1971);

(2) as the temperature decreases further, the QSE clusters fragment more and more into smaller clusters until total breakdown of the QSE approximation, at which point the abundances of all nuclides have to be calculated from a full nuclear reaction network. In the

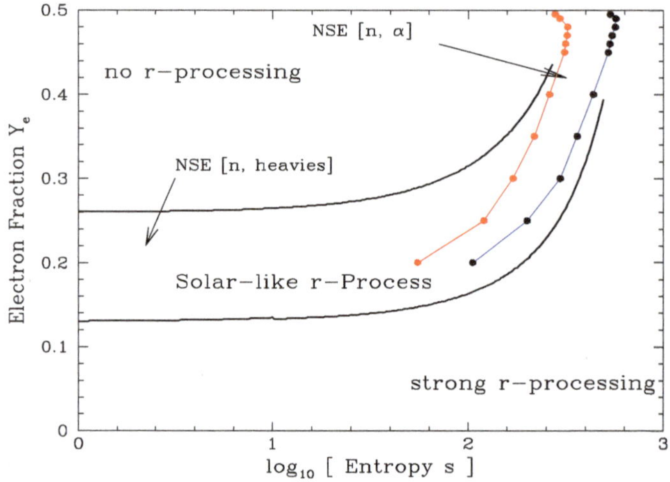

Fig. 6. The likelihood of a DYR r-process for given combinations of the electron fraction Y_e and the entropy per baryon s. A SoS-like r-process is expected for a suitable superposition of conditions between the black lines. The results inferred from an initial NSE phase at low s are smoothly connected to those of various nuclear network calculations for high s values. In the latter cases, the assumed expansion timescales imply that the charged-particle induced reaction freeze-out is reached after dynamical timescales τ_{dyn} in excess of about 50 - 100 ms. The two dotted lines represent the contours of successful r-processing for $\tau_{dyn} = 50$ ms (left line) and 100 ms (right line) (see Hoffman et al., 1997, for details)

relevant α-particle-rich environment, the reaction flows are dominated by (α, γ) and (α, n) reactions with the addition of radiative neutron captures. Nuclei as heavy as Fe or even beyond may result. For a low enough temperature, all charged-particle-induced reactions freeze-out, only neutron captures being still possible. This freeze-out is made even more efficient if the temperature decrease is accompanied with a drop of the density ρ, which is especially efficient in bringing the operation of the ρ^2-dependent $\alpha + \alpha + n$ reaction to an end. In the following, the neutron-rich α-rich process summarized above will be referred to as the α-process for simplicity, and for keeping the terminology introduced by Woosley & Hoffman (1992).

The composition of the material at the time of freeze-out depends on the initial Y_e, on the entropy s (see Meyer et al., 1998, for a detailed discussion), as well as on the dynamical timescale τ_{dyn}. The heavy nuclei synthesized at that moment may have on average neutron numbers close to the $N = 50$ closed shell, and an average mass number around $A = 100$. These nuclei can be envisioned to be the seeds for a subsequent r-process, in replacement of the iron peak assumed in the canonical and MER models. For a robust r-process to develop, favourable conditions have to be fulfilled at the time of the α-process freeze-out. In particular, the ratio at that time of the neutron concentration to the abundance of heavy neutron-rich seeds has to be high enough for allowing even the heaviest r-nuclides to be produced. As an example, $A = 200$ nuclei can be produced if an average of 100 neutrons are available per $A = 100$ nuclei that could emerge from the α-process. The availability of a large enough number of neutrons per seed can be obtained under different circumstances: (i) at high enough entropies (high enough temperatures and low enough densities), even in absence of a large neutron excess, as it is the case if Y_e is close to 0.5 (Woosley & Hoffman, 1992), (ii) at lower entropies if Y_e is low enough, and/or (iii) if the temperature decrease is fast enough for avoiding a too prolific production of heavy seeds. Figure 6 sketches in a semi-quantitative way the conclusions of the discussion above concerning the likelihood of development of a successful r-process in terms of entropy and Y_e.

6.4 A high-density r-process scenario (HIDER)

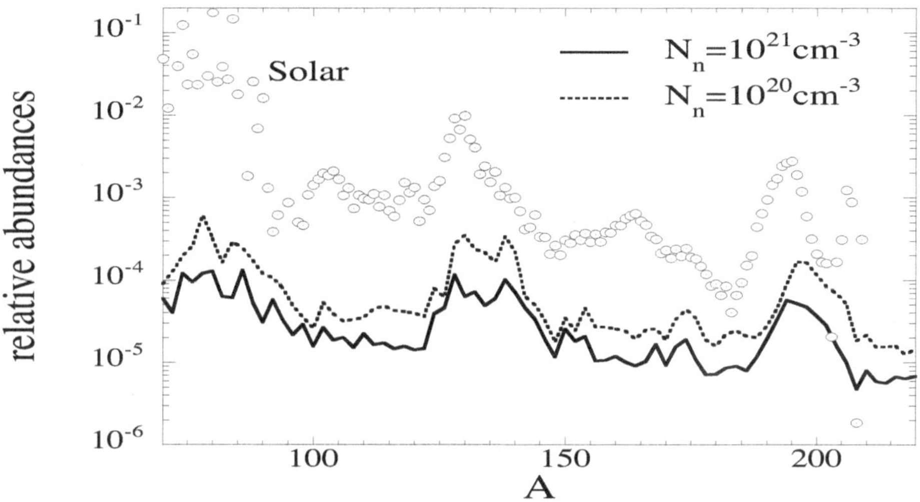

Fig. 7. Abundance distributions predicted by the steady flow HIDER for $N_n = 10^{20}$ and 10^{21} cm^{-3}. The details of the adopted nuclear physics can be found in Arnould et al. (2007). The SoS abundances are shown for illustrative purpose

Early in the development of the theory of nucleosynthesis, an alternative to the high-T r-process canonical model has been proposed (Tsuruta & Cameron, 1965). It relies on the fact that very high densities (say $\rho > 10^{10}$ gcm^{-3}) can lead material deep into the neutron-rich side of the valley of nuclear stability as a result of the operation of endothermic free electron captures. This so-called 'neutronization' of the material is possible even at the $T = 0$ limit. The astrophysical plausibility of this scenario in accounting for the production of the r-nuclides has long been questioned, and has remained largely unexplored until the study of the composition of the outer and inner crusts of neutron stars (Baym et al., 1971) and of the decompression of cold neutronized matter resulting from tidal effects of a black hole on a neutron star companion (Lattimer et al., 1977). The decompression of cold neutron star matter has been studied further recently (Sect. 8).

In view of the renewed interest for a high-density r-process, a simple steady flow model, referred to in the following as HIDER, may be developed. Irrespective of the specific details of a given astrophysical scenario, it allows to follow in a very simple and approximate way the evolution of the composition of an initial cold (say $T = 0$) highly neutronized matter under the combined effect of β-decays and of the captures of free neutrons that are an important initial component of the considered material. These are the only two types of transformations that have to be considered if fissions are disregarded, and if any heating of the material resulting from the β-decay energy deposition is neglected, so that photodisintegrations can be ignored.

The predictions of HIDER under the additional assumption of a steady flow ($\mathrm{d}N(A)/\mathrm{d}t = 0$, $N(A)$ being the total abundance of all the isobars with mass number A; see Arnould et al. (2007) for details) are illustrated in Fig. 7. This model does roughly as well in reproducing the three SoS abundance peaks as a steady state high-T canonical model for comparable neutron densities. In other words, a high-T environment is not a necessary condition to account either for the location, or for the width of the observed SoS r-abundance peaks.

7. Supernovae: a high-temperature site for the r-process, really?

7.1 The current status of supernova simulations: a brief review

As recalled above, the r-process has for long been associated with SN explosions. Over the years, many attempts have been conducted to better characterize the SN site where an r-process can develop. One has to acknowledge that these efforts have not been really successful up to now. Even simulating the explosion itself represents a real nightmare, as we briefly review below (see Arnould et al., 2007, for additional details and references).

At the end of their nuclear evolution, stars between about 10 and 100 $M_\odot$ develop a core made of nuclides of the iron group ('iron core') at temperatures in excess of about 4×10^9 K. As these nuclides have the highest binding energy per nucleon, further nuclear energy cannot be released at this stage, so that the iron core contracts and heats up. This triggers endothermic photodisintegrations of the iron-group nuclides down to α-particles, and even nucleons. The corresponding energy deficit is accompanied with a pressure decrease which can be responsible of the acceleration of the contraction into a collapse of the core. Endothermic electron captures can make things even worse. To a first approximation, this gravitational instability sets in near the classical Chandrasekhar mass limit for cold white dwarfs, $M_{\mathrm{Ch}} = 5.83\, Y_{\mathrm{e}}^2$ (as above, Y_{e} is the electron mole fraction).

The gravitational collapse of the iron core does not stop before the central densities exceed the nuclear matter density $\rho_0 \approx 2.5 \times 10^{14}\,\mathrm{g\,cm^{-3}}$ by about a factor of two. At this point, the innermost ($M \lesssim 0.5\, M_\odot$) material forms an incompressible, hot and still lepton-rich 'proto-neutron' star (PNS) whose collapse is stopped abruptly. A shock wave powered by the gravitational binding energy released in the collapse propagates supersonically into the infalling outer layers. For many years there has been hope that this shock could be sufficiently strong for ejecting explosively most of the material outside the core, producing a so-called 'prompt core collapse supernova' (PCCSN) with a typical kinetic energy of 1–2 $\times 10^{51}$ ergs, as observed. The problem is that the shock is formed roughly half-way inside the iron core, and looses a substantial fraction of its energy in the endothermic photodisintegrations of the iron-group nuclei located in the outermost portion of the core. The shock energy loss is aggravated further by the escape of the neutrinos produced by electron captures on the abundant free protons in the shock-heated material. Detailed one-dimensional hydrodynamic simulations conclude that the initially outgoing shock wave transforms within a few milliseconds after bounce into an accretion shock. The matter behind the shock continues to accrete on the PNS. The bottom line is that no recent simulation is able to predict a successful PCCSN for a Fe-core progenitor star ($M \gtrsim 10 M_\odot$).

Even so, some hope to get a CCSN of a non-prompt type has been expressed if there is a way to 'rejuvenate' the shock efficiently enough to obtain an explosive ejection of the material outside the PNS. This rejuvenation remains a matter of intensive research. Neutrinos might well play a pivotal role in this matter. They are produced in profusion from the internal energy reservoir of the PNS that cools and deleptonizes hundreds of milliseconds after bounce, and their total energy might amount to several 10^{53} ergs, that is about 100 times the typical explosion energy of a Type II supernova (SNII). The deposition of a few percent of this energy would thus be sufficient to unbind the stellar mantle and envelope, and provoke a 'delayed' CCSN (DCCSN) (these qualitative statements assume that a black hole is not formed instead of a PNS; see below). Very many attempts to evaluate the precise level of neutrino energy deposition have been conducted over the last decades, based on more or less controversial simplifications of the treatment of the neutrino transport (e.g. Liebendorfer et al., 2005, for a recent re-analysis of the problem, which is made even more complex by the due consideration of neutrino flavour mixing). In fact, theoretical investigations and numerical simulations performed with increasing sophistication over the past 20 years have not been able to come up with a clearly successful DCCSN for a Fe-core progenitor. This conclusion is apparently robust to changes in the highly complex physical ingredients (like the neutrino interactions, or the equation of state), and in the numerical techniques (e.g. Liebendorfer et al., 2005). In fact, the neutrino energy deposition should have to be significantly enhanced over the calculated one in order to trigger an explosion.

This adverse situation may not mark the end of any hope to get a DCCSN, however. In the case of single stars considered here, one might just have to limit the considerations to stars in the approximate 9 to 10 $M_\odot$ range. These stars possibly develop O-Ne cores instead of iron cores at the termination of their hydrostatic evolution. Efficient endothermic electron captures could trigger the collapse of that core, which could eventually transform into a so-called electron-capture SN that may be of the SNII or Type Ia (SNIa) type, depending upon the extent of the pre-explosion wind mass losses (binary systems might offer additional opportunities of obtaining electron-capture SNe, as mentioned below). Just as in the case of more massive stars, a PCCSN is not obtained. However, a successful DCCSN is predicted (Kitaura et al., 2006). The neutrino heating is efficient enough for rejuvenating the shock wave about 150 ms after bounce, and mass shells start being ablated from the PNS surface about 50 ms later, leading to a so-called 'neutrino-driven wind'.[4] No information is provided by the current simulations on the conditions at times much later than a second after bounce. Note that the predicted successful delayed electron capture SN is characterized by a low final explosion energy (of the order of 10^{50} ergs, which is roughly ten times lower than typical SN values), and by just a small amount of ejected material (only about 0.015 $M_\odot$). These features might suggest a possible connection with some subluminous SNII events and with the Crab nebula.

Note that the outcome of a failed CCSN is the transformation of the PNS into a black hole through the fallback onto the neutron star of the material that cannot be shock ejected. A black hole is even expected to form 'directly' instead by fallback in $M \gtrsim 40M_\odot$ non-rotating stars, at least under the assumption of no strong mass losses. In fact, this assumption is likely to be invalid for a large fraction at least of the not too low metallicity $M \gtrsim 40M_\odot$ stars which transform through strong steady mass losses into Wolf-Rayet stars that might eventually experience a DCCSN of the SNIb/c type.

A major effort has been put recently in the development of simulations of explosions that go beyond the one-dimensional approximation. This is motivated not only by the difficulty of obtaining successful CCSNe in one-dimensional simulations, as briefly reviewed above,

[4] Unless otherwise stated, neutrino-driven winds refer to transonic as well as subsonic winds. In case one has to be more specific, a subsonic wind is referred to as a breeze

but also by the mounting observational evidence that SN explosions deviate from spherical symmetry, not talking about the possible connection between the so-called soft long-duration gamma-ray bursts and grossly asymmetric explosions accompanied with narrow jets of relativistic particles, referred to as JetSNe. The multi-dimensional extension of the simulations opens the potentiality to treat in a proper way different effects that may turn out to be essential in the CCSN or JetSNe process. They include fluid instabilities, rotation and magnetic fields, or, possibly, acoustic power triggering (see Arnould et al., 2007, for more details and references). These effects come of course on top of the neutrino transport already built into the one-dimensional models. The exact properties of this transport remain a subject of active research. In particular, the role of collective neutrino oscillations on the SN explosion has been studied recently by Duan & Friedland (2011). They consider that this phenomenon still needs a more complete physical understanding, and still has a rich potential of surprises.

7.2 Neutrino-driven winds and the r-process

The neutrino-driven winds that may accompany successful DCCSNe are certainly interesting from a purely hydrodynamical point of view, even if they may turn out not to be the key triggering agents of DCCSNe. In addition, their nucleosynthesis has been scrutinized in detail, especially following the excitement raised by the hope that they could provide a natural site for an α-process and for a subsequent dynamical r-process (Sect. 6.3, and e.g. Takahashi et al. (1994) for early calculations). This hope has gained support from a one-dimensional DCCSN simulation of an iron-core progenitor predicting that entropies as high as about 400 could be attained in the wind more than 10 seconds after bounce (Woosley et al., 1994). Such a high entropy allows the development of a robust r-process for a large variety of values of the neutron excess or Y_e and dynamical timescale τ_{dyn} (see Sect. 6.3 and Fig. 6). However, another one-dimensional iron-core DCCSN model has predicted about five times lower entropies, so that the development of an extended r-process is severely endangered (Takahashi et al., 1994). The subsequent studies have confirmed that this r-process scenario could only be recovered at the expense of some twists that are difficult to justify in general (e.g. Qian & Woosley, 1996) and that large late time entropies (Woosley et al., 1994) were unrealistic due to some problems with the equation of state (Roberts et al., 2010). As a consequence, recent studies conclude that *the simplest model of the neutrino-driven wind can most likely not produce the r-process* (Roberts et al., 2010).

In such an unsatisfactory state of affairs, the best one can do is to try to understand better the physics of neutrino-driven winds through the development of (semi-)analytical models some aspects of which may be inspired by (failed) explosion simulations, and to try to delineate on such grounds favourable conditions for the development of the r-process. These analytical models confirm that the wind nucleosynthesis depends on Y_e, entropy s, and τ_{dyn}, as in the α-process discussed in Sect. 6.3. The wind mass-loss rate $\dot{M}$ is influential as well. Ultimately, the quantities acting upon the synthesis in the neutrino-driven DCCSN model depend crucially on the details of the interaction of neutrinos with the innermost SN layers, as well as on the mechanisms that might aid to get a successful DCCSN, and whose relative importance remains to be quantified in detail.

Several wind models of analytical nature exist. They differ in their level of physical sophistication and in their way to parametrize the wind characteristics. In all cases, the wind is assumed to be spherically symmetric, which appears to be a reasonable first approximation even in two-dimensional simulations, at least late enough after core bounce. In addition, the wind is generally treated as a stationary flow, meaning no explicit time dependence of any physical quantity at a given radial position. Newtonian and post-Newtonian descriptions of a spherically symmetric stationary neutrino-driven (supersonic) wind or (subsonic) breeze emerging from the surface of a PNS have been developed. The reader is referred to Arnould et al. (2007) for the presentation of a Newtonian, adiabatic and steady-state model for the wind and breeze regimes, and for a general-relativistic steady-state wind solution.

A comment is in order here concerning the relevance of the wind or breeze regimes. Is one of the two favoured by the DCCSN physics? This question is far from being just academic, as it is likely that its answer may have some impact on the predicted development of the r-process. It is quite intricate as well. One difficulty arises as the neutrino-driven material is likely not to flow unperturbed to infinity in a variety of DCCSN situations. The wind may in particular interact with matter and radiation in that portion of the star through which the SN shock has already passed. This interaction is likely to depend, among other things, on the pre-SN structure. As an example, it is clearly more limited as the mass of the outer layers decreases when going from massive SNII progenitors to SNIb/c events whose progenitors (Wolf-Rayet stars) have lost their extended H-rich envelope prior to the explosion. The interaction of the material ablated from the PNS and the outer SN layers has several important consequences. It may give rise to a reverse shock responsible for the fallback of a more or less large amount of material onto the PNS, and whose properties (location and strength) alter more or less deeply the characteristics of the neutrino-ejected material. For large enough energies of the reverse shock, the wind may indeed transform into a breeze.

Figure 8 displays the evolution of temperature and density for a particular breeze solution discussed in Arnould et al. (2007). Some snapshots of the corresponding progressive build-up of heavy nuclei by the α-process, followed by the production of r-nuclides are shown in Fig. 9. The r-nuclide abundance distributions calculated for a breeze solution that is slightly different from the one displayed in Fig. 8 (even if the PNS mass and entropy are the same; see Arnould et al., 2007, for details) and three different mass loss rates (leading to three different cooling timescales) are shown in Fig. 10. The influence on the r-abundances of different values of other breeze parameters is discussed by Arnould et al. (2007).

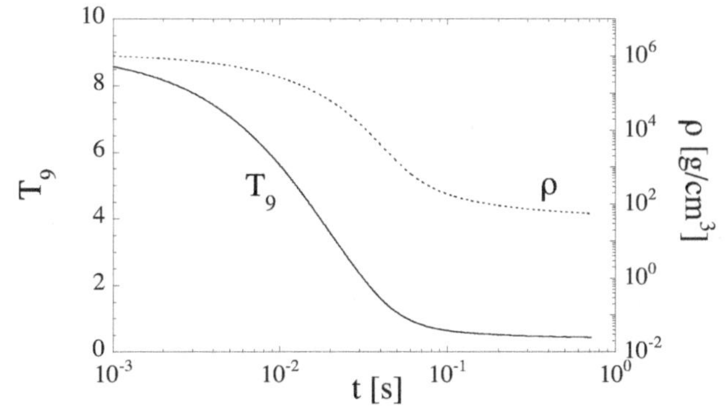

Fig. 8. Evolution of density ρ and of temperature T_9 (in 10^9K) calculated in the Newtonian approximation for a breeze starting about 14 km away from a $1.5M_\odot$ PNS with a temperature $T_9 = 9$. The displayed solution depends on other parameters (including entropy and mass loss rate) that are discussed in Arnould et al. (2007)

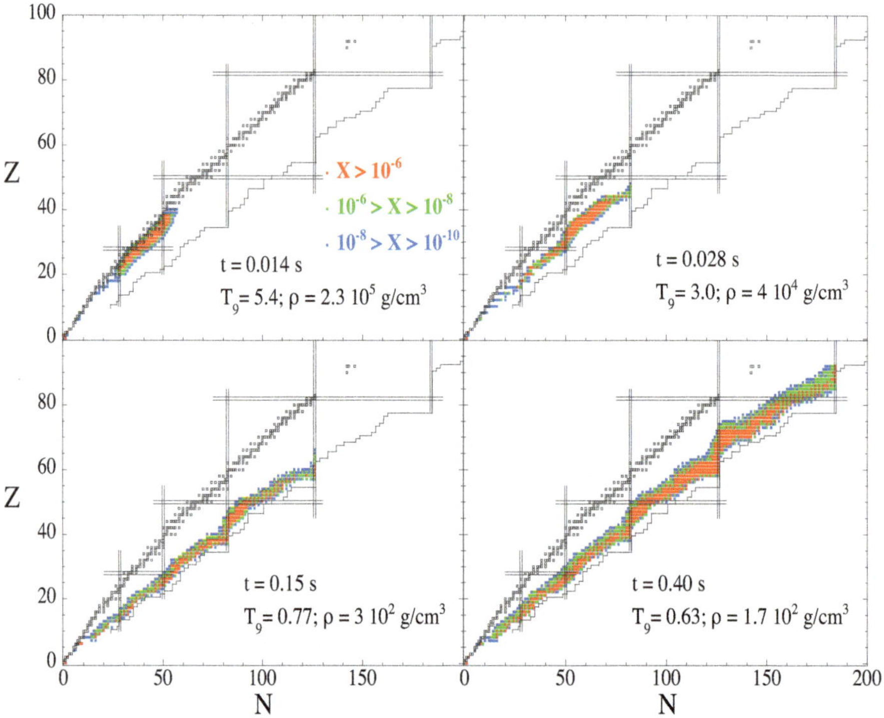

Fig. 9. Snapshots of the nuclear flows calculated for the breeze solution of Fig. 8. The first three panels ($t \leq 0.15$ s) describe the progressive build-up of heavy nuclei by the α-process. These act as the seeds for the r-process that develops after the α-process freeze-out, as shown in the last panel

The influence of the neutrino luminosity on the r-process is also reviewed by Arnould et al. (2007). The neutrino interaction is detrimental to the r-process as a result of the reduction of the number of neutrons available per seed nucleus. In addition, the abundance distribution is reshaped by the neutrino interactions. Such effects strongly depend on the adopted neutrino luminosities and temperatures, which remain rather uncertain. The impact of neutrinos on the r-process has been further explored recently (Duan & Friedland, 2011; Duan et al., 2011). They demonstrate that neutrino flavour transformations, and more specifically collective oscillations, might play a role in the SN explosion and might decrease further the r-process yields.

Let us finally report on some recent work concerning

(1) the possible development of an r-process in electron-capture SNe studied in 1D (Wanajo et al., 2009) and 2D (Wanajo et al., 2011) simulations. These models show that some neutron-rich material can be expelled. In particular, the 2D calculations predict the ejection of lumps of material enriched with some r-nuclides up to about Zr that are reminiscent of the abundance patterns in some r-nuclide deficient halo stars. However, the possibility of development of

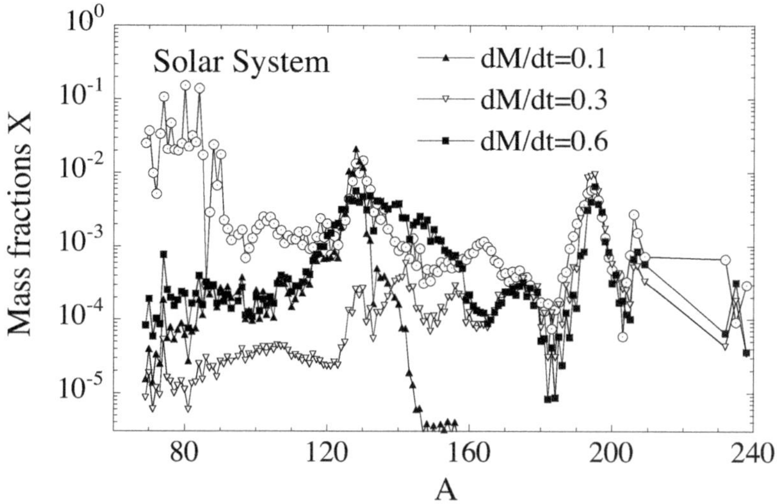

Fig. 10. Distribution of the r-nuclide abundances derived for a breeze solution obtained with a PNS mass of 1.5 $M_\odot$, an initial electron fraction $Y_e = 0.48$, and different values of the mass loss rate dM/dt (in units of $10^{-5} M_\odot/s$). The open dots give the SoS r-nuclide abundances normalized to $\sum_i X_i = 1$. See Arnould et al. (2007) for details, particularly on the selected breeze solution

a limited r-process in electron-capture SNe remains to be confirmed by more reliable pre-SN models and by 3D simulations with high enough resolution;

(2) a further examination of the possibility of r-nuclide production in the He-shell of massive stars. As reviewed by Arnould et al. (2007), a neutron-capture episode could be encountered in explosive He-burning as a result of the neutrons produced by (α,n) reactions on pre-existing ^{22}Ne or Mg isotopes. This neutron supply is, however, found to be so weak that it could only lead to a limited redistribution of pre-explosion heavy nuclide abundances. It has been speculated that some meteoritic r-nuclide anomalies could be generated this way. The (α,n) neutron production could be augmented by neutral (Epstein et al., 1988) or charged-current (Banerjee et al., 2011) neutrino reactions on ^{4}He, the necessary neutrinos streaming out of the PNS at the centre of the exploding star. It is concluded by Banerjee et al. (2011) that r-nuclide abundance peaks at $A = 130$ and 195 could be expected in very metal-poor ($Z \lesssim 10^{-3} Z_\odot$) stars.

8. Compact objects: a site for the high-density r-process scenario?

As reminded in Sect. 6.4, the decompression of the crust of cold neutron stars (NSs) has long been envisioned as a possible site for the development of a high-density r-process (HIDER). Recently, special attention has been paid to the coalescence of two NSs or of a NS and a BH. This interest follows the confirmation by hydrodynamic simulations that a non-negligible amount of matter can be ejected in such events (Janka et al., 1999; Oechslin et al., 2007; Rosswog et al., 2004), and that this material should be enriched in r-nuclides (Freiburghaus et al., 1999; Goriely, 2005). The ejection of initially cold, decompressed NS matter might also occur in other astrophysical scenarios like giant flares in soft-gamma repeaters, the explosion of a NS eroded below its minimum mass, or the equatorial shedding of material from very rapidly rotating supramassive or ultramassive NSs (see Arnould et al., 2007, for more details).

Little effort has so far been devoted to determining the composition of the matter that undergoes a decompression from initially cold NS crust conditions. The first detailed calculation was performed by Meyer (1989) in a systematic parametric study, but only the decompression down to densities around the neutron drip density $\rho_{\rm drip} \simeq 4.2 \times 10^{11} {\rm g/cm}^3$ was followed. Most of the subsequent investigations of the ejecta from coalescing NSs (Arnould et al., 2007; Freiburghaus et al., 1999; Goriely, 2005; Metzger et al., 2010; Roberts et al., 2011) were parametrized in one way or another, which makes their results and conclusions open to questions. In Goriely (2005) and Arnould et al. (2007) the thermodynamic profiles were constructed from a simple decompression model (see Goriely et al., 2011), but the neutron enrichment (or equivalently the electron fraction Y_e) was consistently taken from β-equilibrium assumed to have been achieved at the initial density prior to the decompression. It was found, as shown in Fig. 11, that the final composition of the material ejected from the inner crust depends on the initial density, at least for the outer parts of the inner crust at $\rho_{\rm drip} \leq \rho \leq 10^{12} {\rm g/cm}^3$. For the deeper inner crust layers ($\rho > 10^{12}$ ${\rm g/cm}^3$), large neutron-to-seed ratios drive the nuclear flow into the very heavy-mass region, leading to multiple fission recycling. As a consequence, the resulting abundance distribution becomes independent of the initial conditions, especially of the initial density. It was found to be in close agreement with the solar distribution for $A > 140$ nuclei, as shown in Fig. 12 (Arnould et al., 2007; Goriely, 2005).

Different approaches were taken by Freiburghaus et al. (1999); Metzger et al. (2010) and Roberts et al. (2011). In their calculations, while the density evolution of the mass elements was adopted from hydrodynamical simulations, both the initial neutron enrichment and the temperature history were taken as free parameters. In particular, Y_e was chosen in order to obtain, after decompression, an r-abundance distribution as close as possible to the solar distribution. This led to values of $Y_e = 0.1$ (Freiburghaus et al., 1999) or 0.2 (Roberts et al., 2011), corresponding to relatively near-surface layers of the inner crust and to nuclear flows that are not subject to multiple fission cycles.

Recent nucleosynthesis calculations (Goriely et al., 2011) have performed in the framework of the hydrodynamic simulation of a double 1.35 $M_\odot$ and of a 1.20–1.50 $M_\odot$ NS binary, where about 3×10^{-3} and 6×10^{-3} $M_\odot$, respectively, of the system is found to become gravitationally unbound. In contrast to previous studies, detailed information about the density, Y_e, and entropy evolution of the ejecta is extracted from the hydrodynamical simulations and included in the network calculations. The ejected mass elements (referred to as 'particles') originate essentially from the inner crust, and more precisely from two different layers at densities below and above roughly $0.3 \times \rho_S$ (where ρ_S is the nuclear saturation density), as shown in the density histogram of Fig. 13. The low-density set of surface particles is ejected through tidal forces without being much affected by the coalescence. When the NSs come in contact for the first time, some particles deeper in the inner crust are shocked and ejected perpendicularly to the orbital plane. These particles have a higher initial density ($\rho \gtrsim 0.4\, \rho_S$) than the crust particles ejected by tidal forces. Both types of expelled particles have a low initial electron fraction ranging between 0.015 and 0.050 (see Fig. 13, where $Y_e = 0.015$ in about 45% of the mass).

As far as the temperature history is concerned, most of the particles are heated during the ejection process to temperatures above 1 MeV. At these temperatures, a particle has a NSE composition at its own density and Y_e. At the time the drip density is reached,

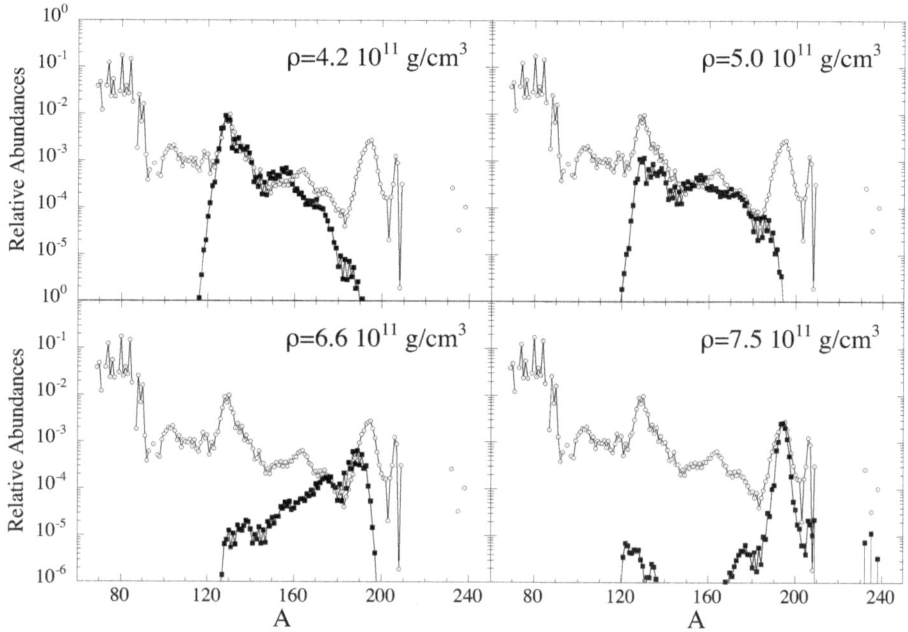

Fig. 11. Composition of the ejected NS inner crust material with four different initial densities lower than 10^{12} g/cm^3 (black squares). The abundances are normalized to the SoS r-abundance distribution shown by open dots (from Arnould et al., 2007)

most of the ejected matter has cooled below 1 MeV (see Fig. 13, right panel) and the NSE has frozen out. As soon as the particle temperature has dropped below 10^{10} K, the particle composition, initially at NSE, must be followed by a full network calculation at the temperatures determined on the basis of the laws of thermodynamics allowing for possible nuclear heating through β-decays, fission and α-decay processes, as described in Meyer (1989).

In this specific r-process scenario, the neutron density is initially so high ($N_n \simeq 10^{33-35}\text{cm}^{-3}$) that the nuclear flow follows for the first hundreds of ms a path licking the neutron drip line. For $Z \geq 103$, fission becomes efficient and recycling takes place two to three times before the neutrons are totally exhausted, as shown in Fig. 14 by the oscillating behaviour of the time evolution of the charge $\langle Z \rangle$ and mass number $\langle A \rangle$ mass-averaged over all the ejected particles. After several hundreds of ms, the density has dropped by a few orders of magnitude and the neutron density experiences a dramatic fall-off as a result of the neutron exhaution by captures. During this period of time, the nuclear flow around the $N = 126$ region follows an isotonic chain. When the neutron density reaches some $N_n = 10^{20}$ cm^{-3}, the timescale of neutron captures by the most abundant $N = 126$ nuclei becomes longer than a few seconds, and the nuclear flow is dominated by β-decays back to the stability line (as well as by fissions and α-decays for the heaviest species). During the decompression, the average temperature remains rather low (around $T_9 = 0.3 - 0.5$; Fig. 14), so that photoreactions do not play a major role.

The final mass-averaged composition of the ejected particles is shown in Fig. 15. The $A = 195$ abundance peak related to the $N = 126$ shell closure is produced in SoS distribution, and

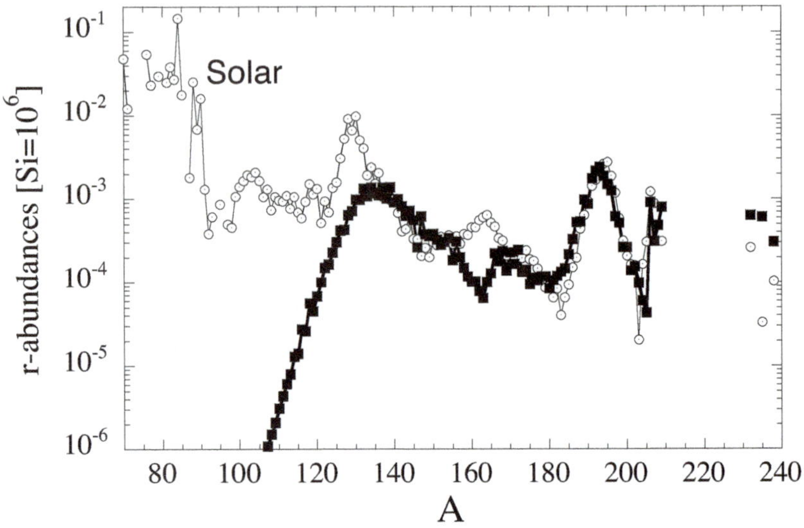

Fig. 12. Final r-abundance distribution (black squares) for a clump of material with initial density $\rho = 10^{14}\text{g/cm}^3$ expanding on a timescale $\tau_{exp} = 6.5$ ms. The SoS r-abundance distribution is also shown (from Goriely, 2005)

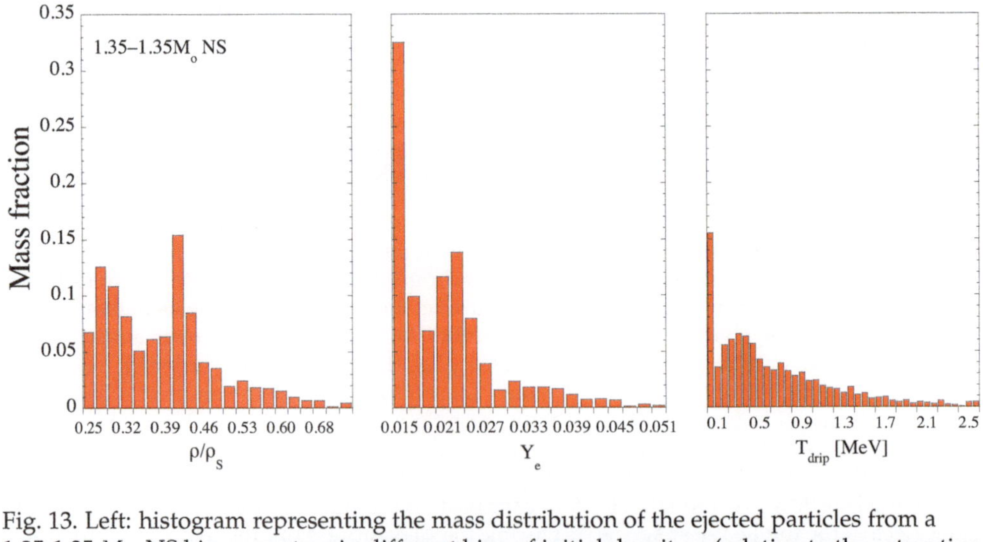

Fig. 13. Left: histogram representing the mass distribution of the ejected particles from a 1.35-1.35 $M_{\odot}$ NS binary system in different bins of initial density ρ (relative to the saturation density $\rho_S \simeq 2.6 \times 10^{14}\text{g/cm}^3$); Middle: same as the left panel for the electron fraction Y_e; Right: Same as the left panel for the temperature reached at the drip density

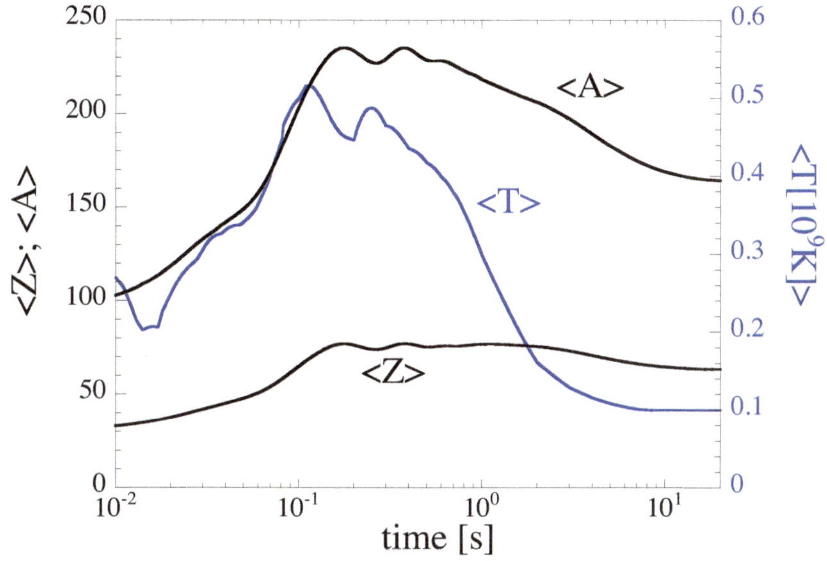

Fig. 14. Time variation of the charge number $\langle Z \rangle$, mass number $\langle A \rangle$ and temperature $\langle T \rangle$ mass-averaged over all the ejected particles

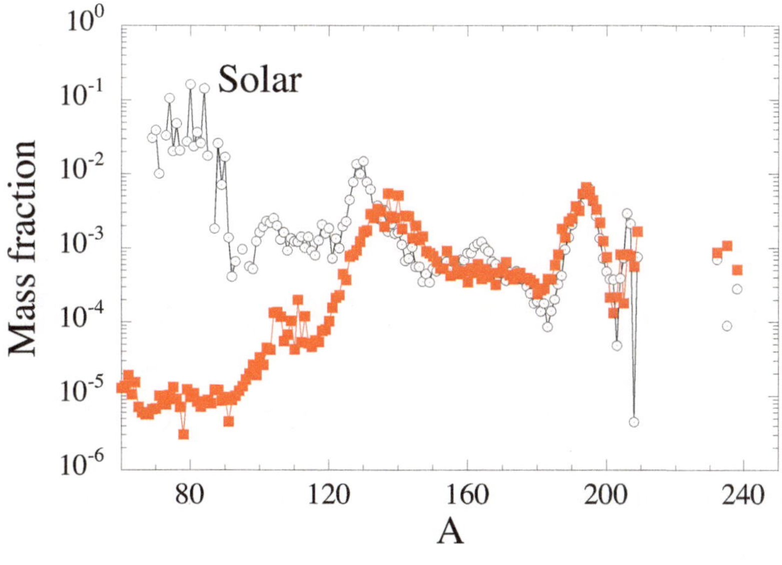

Fig. 15. Final abundance distribution of the matter ejected from the 1.35–1.35 $M_{\odot}$ NS merger as a function of the atomic mass. The distribution is compared with the solar r-abundance distribution (open dots)

is found to be almost insensitive to all input parameters, such as the initial abundances, the expansion timescales or the adopted nuclear models. In contrast, the second peak around $A = 140$ originates exclusively from the fission recycling, which is found to take place in the $A \simeq 280$–290 region at the time all the neutrons have been captured. These nuclei are found to fission symmetrically, as seen in Fig. 15 where the $A \simeq 140$ peak corresponds to the mass-symmetric fragment distribution. It should be emphasized that significant uncertainties still affect the prediction of fission probabilities and fragment distributions, so that the exact strength and location of the $A \simeq 140$ fission peak depend on the adopted nuclear model.

Although the outer crust ($\rho \leq \rho_{drip} \simeq 4.2 \times 10^{11}$g/cm^3) is far less massive than the inner crust, the ejection of the inner crust cannot take place without leading at the same time to the ejection of at least some outer crust material. The whole outer crust typically amounts to 10^{-5} to 10^{-4} $M_\odot$, depending on the NS mass and radius (Pearson et al., 2011), so that its contribution, even for complete ejection, remains negligible compared to the 10^{-3} to 10^{-2} $M_\odot$ of the inner crust material ejected during a NS–NS merger event. Still, the composition of the outer crust initially heated to NSE at a temperature of about 10^{10} K can be estimated before and after the decompression that follows its ejection (Goriely et al., 2011). Note that the amount of unbound outer crust and the dynamics of its ejection cannot be reliably calculated from the hydrodynamical model. Details about the adopted decompression model can be found in Goriely et al. (2011).

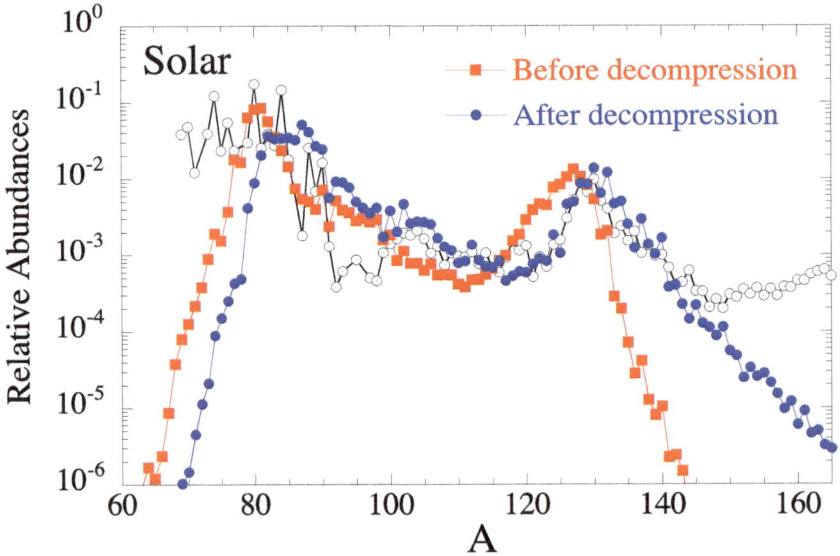

Fig. 16. Abundance distribution before and after decompression of a layer initially in NSE at $T_9 = 8$ and $\rho_0 = 3.4 \times 10^{11}$g/cm^3 with initial pressure $P_0 = 4 \times 10^{-4}$ MeV/fm^3. The distributions are compared with the solar r-abundance distributions (open dots). See Goriely et al. (2011) for details

Before decompression and at temperatures corresponding to 8–10 $\times 10^9$ K, the Coulomb effect due to the high densities in the crust leads to an overall content in neutron-rich nuclei of the outer crust close to the SoS r-abundance distribution, as shown in Fig. 16. Such a distribution differs, however, from the SoS one due to a shift of the second peak to lower mass numbers. During the decompression, the free neutrons (initially liberated as a result of the high temperatures) are re-captured, which leads to a final distribution of stable neutron-rich nuclei with a mass distribution of $80 \leq A \leq 140$ nuclei in excellent agreement with the SoS one, provided that the outer crust is initially at a temperature around 8×10^9 K, and that all the layers of the outer crust are ejected. The decompression of the outer NS crust provides suitable conditions for a robust r-processing of the r-nuclides with $A \leq 140$. The overall abundance distribution depends on the fraction of the ejected crust, but also on the initial temperature at which the NSE has been frozen in. Figure 17 shows the abundance distributions obtained for different initial temperatures between 7 and 10×10^9 K, assuming that the whole outer crust is ejected. The final composition should carry the imprint of the temperature at which the NSE is frozen prior to the ejection. Temperatures typically around 8–9 $\times 10^9$ K correspond to those at which NSE can be dynamically achieved in cooling events Goriely et al. (2011).

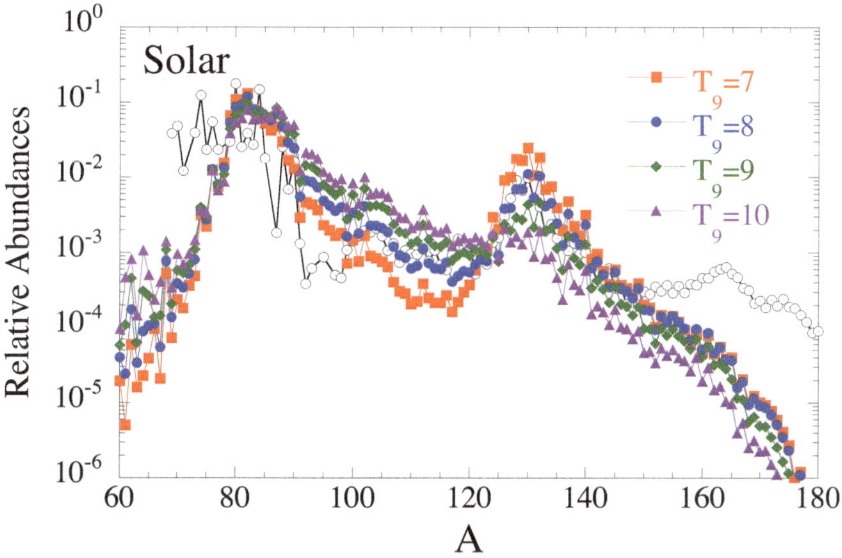

Fig. 17. Final abundance distributions of the outer crust material after decompression, if initially in NSE at different temperatures ranging between $T_9 = 7$ and 10. The distributions are compared with the SoS r-abundance distributions (open dots). See Goriely et al. (2011) for details

It could be seen as purely fortuitous that temperatures around $T_9 \simeq 8$ give rise to r-abundance distributions in agreement with the solar distribution. In fact, the temperatures considered here typically correspond to those at which NSE can be dynamically achieved in cooling events. As shown in Goriely et al. (2011), for matter with initial values of $Y_e = 0.33 - 0.40$, it takes about 1 to 20 ms to reach a NSE at $T_9 = 8$ and $\rho = 3 \times 10^{11}$ g/cm^{-3}, while at $T_9 \simeq 9$, around 0.2 ms are required for the most neutron-rich conditions ($Y_e = 0.33$). Such ms timescales are characteristic of dynamical scenarios of interest here for the potential mass ejection (e.g. in the bursts of soft gamma-repeaters, during NS mergers, ...), so that prior to the ejection the NSE should be achieved at temperatures typically above $T_9 \simeq 8 - 9$. It can be expected to be frozen during the ejection.

9. The evolution of the r-nuclide content of the Galaxy and nucleo-cosmochronology: inquietudes in a minefield

A quite natural astrophysicists' dream is to understand the wealth of data on the evolution of the r-nuclide content of the Galaxy that are accumulating from very many spectroscopic observations (Sect. 4). These observations clearly demonstrate a huge complexity that will probably keep rising as new observations become available. The best one can thus imagine at this stage is to explain broad trends which may be identified through the analysis of the r-nuclide abundance information. In this exercise, one has always to keep in mind that, at best, more or less reliable *elemental* abundances are derived from spectral analyses that often rely on approximate classically used techniques. This necessitates to disentangle the s- and r-process contributions to a given elemental abundance. It is generally done by assuming that these two nucleosynthesis contributions are at a largely metallicity-independent relative level, and thus do not differ widely from the SoS case. This assumption cannot yet be ascertained in any quantitative way, and is in fact not expected to hold for the s-process.

Various galactic chemical evolution toy models have been constructed, which often focus on the evolution of the abundances of two representative elements, Ba (a s-process element in the SoS) and Eu (a r-process element in the SoS). They adopt different schematic descriptions of the galactic halo and disk, and different prescriptions for the physical input quantities to these models. In particular, fully ad-hoc assumptions are made or free parameters are chosen concerning the r-process yields from stars of different masses and metallicities.

Some predictions for [Eu/Fe] have been made recently by Wanajo & Ishimaru (2006). They are based on a homogeneous one zone model in which it is assumed that stars in prescribed mass ranges produce an artificially selected amount of r-nuclides through the neutrino wind or prompt explosion mechanisms (Sect. 7.2). As expected, the predicted [Eu/Fe] ratio is very sensitive to the selected stellar mass ranges. This result might be optimistically considered as providing a way to constrain the site(s) of the r-process from observation. Reality is most likely less rosy, as very many uncertainties and severe approximations drastically blur the picture.

The assumption of the homogeneity of the interstellar medium at all times is dropped by Argast et al. (2004). The resulting inhomogeneous model might increase the plausibility of the predictions especially at early times in the galactic history. With the granularity of the nucleosynthesis events duly considered, one might hope to better account for the large observed scatter of the r-nuclide abundances at very low metallicities. In addition, the model of Argast et al. (2004) takes into account the r-process contribution from NS mergers (Sect. 8) on top of the one from SNe in selected mass ranges. The many other simplifications generally made in other chemical evolution models are also adopted by Argast et al. (2004). This concerns in particular the r-process yields from SNe, as well as from NS mergers, that are just taken to be SoS-like. From their predicted [Eu/Fe] ratio, Argast et al. (2004) conclude that the scenario assuming the predominance of SNII events in the 20 to 50 $M_{\odot}$ range allows the best fit to the observations. This result is obtained for total masses of r-nuclides per SN varying from about $10^{-4} M_{\odot}$ down to about $10^{-7} M_{\odot}$ when going from 20 to 50 $M_{\odot}$ stars. Again, this conclusion has to be taken with great care in view of the many uncertainties and approximations involved in the chemical evolution model. Within the same model, it is also claimed that NS mergers are ruled out as the major source of r-nuclides in the Galaxy. This conclusion relies on a very approximate and highly uncertain time-dependent frequency of the events. In order to cope at best with observational constraints, coalescence timescales and amount of r-nuclides ejected per merger are adopted by Argast et al. (2004) to vary from about 0.1 to $10^{-4}M_{\odot}$, depending upon other parameters of the NS merging model.

All in all, we consider that the galactic chemical evolution models devised up to now are by far too schematic and uncertain to provide a reliable tool to account for the observed evolution of the r-nuclide content of the Galaxy, or for constraining the possible sites of the r-process. A priority should clearly be at least to identify with a reasonable confidence a single site for the r-process before dwelling on the grand project of constructing models for the evolution of the r-nuclide content of a whole galaxy, and of the SoS in particular.

As far as nucleo-cosmochronology is concerned, the actinides produced by the r-process enter in particular attempts to estimate the age of the Galaxy through their present SoS content, or through their abundances evaluated at the surface of very metal-poor stars. These attempts face some severe problems related to the nuclear physics and astrophysics uncertainties that affect the predictions of the actinide production. The situation is worsened further by the especially large uncertainties in the contribution of the r-process to the solar-system Pb and Bi content (Arnould et al., 2007). *Concerning the* ^{232}Th - ^{238}U *and* ^{235}U - ^{238}U *pairs classically used to date the Galaxy from their present meteoritic abundances, the opinion has been expressed more than 20 years ago by Yokoi et al. (1983) that they have just limited chronometric virtues.* This is in marked contrast to a widely-spread and repeated claim following Fowler & Hoyle (1960), and based on simple analytic models for the evolution of Th and U in the Galaxy. In addition, *the chronometric predictions based on the observations of Th and U in very metal-poor stars have to be considered with great care* (see Arnould et al., 2007, for a brief review and references). In order for them to be reliable, it is not only required that the production of the actinides by the r-process is well known, but also, and very decisively, that the production of Th with respect to U and to the Pt peak is universal. Observation demonstrates now that this is not the case. In spite of this, age evaluations based on this technique continue to appear in the literature.

10. By way of a very brief conclusion

The bottom line of this brief review of the r-process is that unanswered questions are by far more numerous than solved problems when one is dealing with this nucleosynthesis mechanism. They concern especially the astrophysics of the process, as no single site has been identified decisively yet. It also raises many nuclear physics questions. In such conditions, the modelling of the evolution of the r-nuclide content of the Galaxy and actinide-based chronometric evaluations cannot be put on solid grounds yet. This is in fact a very pleasing situation, as hope for many exciting discoveries is still ahead of us!

11. References

Argast, D., *et al.* (2004), *Astron. Astrophys.,* 416, 997

Arlandini, C, *et al.* (1999). *Astrophys. J.*, 525, 886

Arnould, M., & Goriely, S. (2003). *Phys. Rep.*, 384, 1

Arnould, M., Goriely, S., & Takahashi, K.(2003), *Phys. Rep.*, 384, 1

Banerjee, P., Haxton, W.C., & Qian, Y.-Z. (2011), *Phys. Rev. Lett.*, 106, 201104

Baym, G., Bethe, H.A., & Pethick, C.J. (1971) *Nucl. Phys. A*, 175, 225

Clayton, D.D. (1968), *Principles of Stellar Evolution and Nucleosynthesis* (Chicago: Univ. Chicago Press)

Delano, M.D. & Cameron, A.G.W. (1971), *Astophys. Space Sci.*, 10, 203

Dillmann, I., & Litvinov, Yu.A. (2011), *Prog. Part. Nucl. Phys.*, 66, 358

Duan, H., & Friedland, A. (2011) *Phys. Rev. Lett.*, 106, 091101
Duan, H., *et al.* (2011) *J. Phys. G: Nucl. Part. Phys.* 38, 035201
Epstein, R.I., Colgate, S.A., & Haxton, W.C. (1988), *Phys. Rev. Lett.*, 61, 2038
Fowler, W.A., & Hoyle, F. (1960), *Ann. Phys.* 10, 280
Freiburghaus, C., Rosswog, S., & Thielemann, F.-K. (1999), *Astrophys. J.*, 525, L121
Gallagher, A.J., *et al.* (2010), *Astron. Astrophys.*, 523, A24
Goriely, S. (1997), *Astron. Astrophys.*, 327, 845
Goriely, S. (1999), *Astron. Astrophys.*, 342, 881
Goriely, S., & Siess, L. (2005) in *From Lithium to Uranium: Elemental tracers of early cosmic evolution* eds: V. Hill et al., Proc. of IAU symposium Nr 228, (Cambridge: Cambridge University Press), p. 451
Goriely, S., *et al.* (2011), *Astron. Astrophys.*, 531, A78
Goriely, S., Bauswein, A., & Janka, H.-T. (2011), *Astrophys. J.*, 738, L32
Habs, D., *et al.* (2011), *Appl. Phys. B* 103, 471
Hoffman, R.D., *et al.* (1997), *Astrophys. J.*, 482, 951
Janka, H.T., *et al.* (1999), *Astrophys. J.*, 527, L39
Kitaura, F.S., *et al.* (2006), *Astron. Astrophys.*, 450, 345
Lattimer, J.M., *et al.* (1977), *Astrophys. J.*, 213, 225
Liebendörfer, M. *et al.* (2005), *Astrophys. J.*, 620, 840
Lodders, K. (2010). In: *Principles and Perspectives in Cosmochemistry*, A. Goswami & B.E. Reddy , (Ed.), 379-417, Springer-Verlag Berlin Heidelberg
Metzger, B.D., *et al.* (2010), *Mon. Not. Roy. Astron. Soc.*, 406, 2650
Meyer, B.S. 1989, *Astrophys. J.*, 343, 254
Meyer, B.S., *et al.* (1998), *Astrophys. J.*, 498, 808
Oechslin, R., Janka, H.-T., & Marek, A. (2007), *Astron. Astrophys.*, 467, 395
Palme, H., & Beer, H. (1993), in *Landolt Börnstein*, New Series, Group VI, Astron. & Astrophys., Vol. 3, Subvol. a, (Berlin: Springer), p. 196
Pearson, J.M., Goriely, S., & Chamel, N. (2011), *Phys. Rev. C*, 83, 065810
Roberts, L. F., Woosley, S. E.; & Hoffman, R. D. (2010) *Astrophys. J.*, 722, 954
Roberts, L.F., *et al.* 2011, *Astrophys. J.*, 736, L21
Roederer, I.U. (2011) *Astrophys. J. Let.* 732, L17
Rosswog, S., Speith, R., & Wynn, G.A. (2004), *Mon. Not. Roy. Astron. Soc.*, 351, 1121
Qian, Y.-Z., & Woosley, S.E. (1996), *Astrophys. J.*, 471, 331
Simmerer, J., *et al.* (2004), *Astrophys. J.*, 617, 1091
Takahashi, K., *et al.* (1994), *Astron. Astrophys.*, 286, 857
Tsuruta, S., & Cameron, A.G.W. (1965), *Can. J. Phys.* 43, 2056
Wanajo, S., & Ishimaru, Y. (2006). *Nucl. Phys.* A77, 676c
Wanajo, S., *et al.* (2009), *Astrophys. J.*, 695, 208
Wanajo, S., *et al.* (2011), *Astrophys. J.*, 726, L15
Woosley, S.E., & Hoffman, R.D. (1992), *Astrophys. J.*, 395, 202
Woosley, S.E., *et al.* (1994), *Astrophys. J.*, 433, 229
Yokoi, K., *et al.* (1983), *Astron. Astrophys.*, 117, 65

5

The s-Process Nucleosynthesis in Massive Stars: Current Status and Uncertainties due to Convective Overshooting

M. L. Pumo
Università degli Studi di Padova,
Dip. di Fisica e Astronomia "G. Galilei", Padova
INAF-Osservatorio Astronomico di Padova, Padova
Bonino-Pulejo Foundation, Messina
INAF-Osservatorio Astrofisico di Catania, Catania
Italy

1. Introduction

From time out of mind, men wish to know more about the world surrounding them and, in particular, to understand the origin of the matter. However, only in the late 1950s, thanks to the pioneering work of Burbidge et al. (1957, the famous B^2FH) and the independent analysis of Cameron (1957), the basic principles of explaining the origin of the elements were laid down in the theory of nucleosynthesis, giving rise to a dedicated, interdisciplinary field of research referred to as "nuclear astrophysics".

Since the early days of its development, nuclear astrophysics has improved at an impressive pace. Nowadays, it is a peculiar mixing of knowledge that blends the progresses in experimental and theoretical nuclear physics, ground-based and space observational astronomy, cosmochemistry, and theoretical astrophysics.

Different works have been devoted to review the achievements reached by the nuclear astrophysics, either providing an overall picture of this field of research (e.g. Arnould & Takahashi, 1999; Jordi & Iliadis, 2011; Rauscher & Patkós, 2011; Wallerstein et al., 1997), or focusing on various its subfields such as the Big-Bang nucleosynthesis and primordial abundances (e.g. Iocco et al., 2009; Steigman , 2007; Tytler et al., 2000), the nucleosynthesis mechanisms of specific types of trans-iron elements (e.g. Arnould & Goriely, 2003; Arnould et al., 2007; Käppeler, 1999; Käppeler et al., 2011; Meyer, 1994), the hydrostatic and explosive stellar nucleosynthesis (e.g. Busso et al., 1999; Chiosi, 2007; Jordi & Hernanz, 2007; Woosley et al., 2002, and references therein), the nucleosynthesis by spallation (e.g. Reeves, 1994; Vangioni-Flamet al., 2000), the experimental techniques and theoretical methods used to investigate nuclear processes of astrophysical interest (e.g. Angulo, 2009; Baur et al., 2003; Bertulani & Gade, 2010; Costantini, 2009; Spitaleri, 2010; Thielemann et al., 2001; Typel & Baur, 2003).

This chapter is concerned with one of the aforementioned subfields of the nuclear astrophysics. In particular, it is devoted to the so-called s-process, which is a nucleosynthesis mechanism responsible for the production of about half of all the trans-iron elements. The chapter starts out in Sect. 2.1 with some basic considerations on the solar system composition, considered as "standard of reference" dataset for cosmic abundances. A brief description of the nucleosynthesis mechanisms responsible for the production of trans-iron nuclei follows in Sect. 2.2. Afterward (see Sect. 3), the s-process nucleosynthesis mechanism is reviewed and its different components are discussed. A specific attention is paid to the so-called "weak component" occurring in massive stars ($M_{ZAMS} \gtrsim 13M_{\odot}$), showing its sensitivity to stellar mass and metallicity (see Sect. 4.1). Moreover the uncertainties affecting the efficiency of this component are described (see Sect. 4.2), placing particular emphasis upon uncertainties due to convective overshooting (see Sect. 5). Prospects of improvements in modeling the s-process weak component and their possible consequences for some open astrophysical questions are also briefly discussed (see Sect. 6).

2. General considerations on the origin of the elements

2.1 Cosmic abundances

Whatever nucleosynthesis model is built, a comparison between the model predictions to the observed cosmic abundances is needed.

Even if the possibility of defining a truly "standard" set of observed cosmic abundances has to be considered with caution (see e.g. Grevesse et al., 1996), the composition of the material from which the solar system formed $\sim$ 5.6 Gy ago (referred hereafter also as solar composition) is usually considered as the "standard of reference" dataset (e.g. Asplund et al., 2009, and references therein). Such a choice is essentially due to the fact that the solar composition is the only comprehensive sample with a well defined isotopic abundance distribution (e.g. Arnett, 1996; Käppeler, 1999), since it can be derived using different sources of information that include, among others, the Earth, the Moon, other solar system planets, the Sun, meteorites, and material from the interplanetary medium. The methods employed to gather abundances information combine spectral analysis (e.g. via spectroscopy of the solar photosphere), laboratory measurements of matter samples (e.g. via mass spectrometry of meteorites, Lunar glasses, and material from the Earth's crust and carried by space probes from the interplanetary medium), and particle detection from space-based experiments (e.g. via analysis of solar wind and solar energetic particles).

Specifically, the elemental solar composition (displayed in right panel of Fig. 1) is largely grounded on abundance analyses of a peculiar class of uncommon meteorites — the so-called CI chondrites — which are believed to reflect the composition of the "primitive" solar system. However information derived from the Sun (i.e. using the solar photospheric spectrum, the impulse flare spectra, and the analysis of solar wind and solar energetic particles) have to be used for determining the abundances of H, C, N, O and the noble gases He, Ne, and Ar; while analyses based on theoretical considerations are required for the "heavy" noble gases Kr and Xe. The nuclear solar composition (displayed in left panel of Fig. 1) is determined considering the terrestrial isotopic compositions as the most representative ones for all the elements of the primitive solar system with the exception of the H and all the noble gases, for which other sources of information (e.g. Jupiter's atmosphere, solar wind and lunar samples) are used (for details see e.g. Lodders, 2003, but also the other widely used compilations of the solar composition by Anders & Grevesse, 1989; Grevesse & Noels, 1993; Grevesse & Sauval, 1998; Lodders at al., 2009, and Asplund et al., 2009).

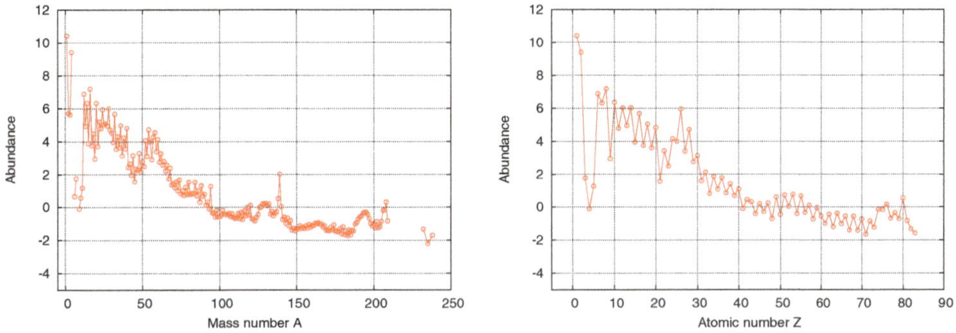

Fig. 1. Nuclear (left panel) and elemental (right panel) solar compositions from the compilation of Lodders (2003). In both panels, the abundances are normalized to 10^6 Si atoms and the y axes are logarithmic. In left panel, only the most abundant isobar is reported for a given mass number A.

Looking at Fig. 1, it is easy to notice that the solar distribution is clearly dominated by the H and He abundances, followed by a "deep" (with respect to the neighboring nuclides) minimum of the abundances for the elements Li, Be, and B. Subsequently, the main feature of the distribution is the presence of a series of peaks at the locations of the "α-elements[1]", superimposed on an exponentially decreasing curve from the A $\simeq$ 12-16 mass region down to the Sc, followed by a pronounced peak centered around ^{56}Fe. From this peak on, the distribution becomes fairly flat with a variety of superimposed peaks which correspond to nuclides having magic numbers[2] of neutrons.

As already pointed out by Burbidge et al. (1957), these features are a "reflection" of the different nucleosynthesis processes responsible for the production of the various isotopes. Even without going into details because it is out of the purpose of this chapter (for details the interested reader is referred to the reviews/books of Arnett, 1996; Arnould & Takahashi, 1999; Clayton, 1983; Jordi & Iliadis, 2011; Käppeler, 1999; Rolfs & Rodney, 1988, and references therein), we remind that the nuclides with $A < 12$ are produced by Big-Bang nucleosynthesis and via spallation mechanisms. The minimum at the elements Li, Be, and B primarily reflects the difficult to synthesize such rare and fragile nuclides due to the stability gaps at $A = 5$ and 8. Various charged-particle induced reactions, which occur inside the stars during quiescent evolutionary phases and are accountable for the stellar energy production, are instead responsible for the nucleosynthesis of all the elements from the A $\simeq$ 12-16 mass region up to the region of the iron peak. The exponentially drop in the elemental abundances

[1] Alpha-elements are so-called since their most abundant isotopes are integer multiples of the mass of the α particle. The most abundant are ^{16}O and ^{12}C, followed by ^{20}Ne, ^{24}Mg, ^{28}Si, ^{32}S, ^{36}Ar, and ^{40}Ca.

[2] In the nuclear shell model, a magic number is a number of nucleons (either protons or neutrons) such that they can be arranged into complete shells within a nucleus. The seven most widely recognised magic numbers are: 2, 8, 20, 28, 50, 82, and 126. Nuclei consisting of such a magic number of nucleons have a particularly tightly bound configuration (see e.g. Povh et al., 2008, for details).

up to the Sc is just related to the increase of the Coulomb barrier with increasing the nuclear charged of the particles involved in the different charged-particle induced reactions, that increasingly hampers such reactions. Exceptions to this trend are the peaks at the locations of the α-elements and the iron peak. The greater stability of these nuclei (compared to the one of the neighbouring nuclides) leads to a their more abundant production and, consequently, gives rise to the above mentioned local maxima in the solar distribution. Finally, neutron capture chains and an additional mechanism characterized by photo-disintegrations of preexisting nuclei are called for in order to synthesize the trans-iron elements, as better described in the following Sect. 2.2. Indeed, if the trans-iron elements had been synthesized by charged-particle induced reactions, the peaks present in the trans-iron region of the solar distribution would not be explained and the abundances of the trans-iron elements would exhibit a much higher decrease (with the mass number A) than the one actually observed in the solar distribution.

2.2 Nucleosynthesis mechanisms for the production of trans-iron nuclei

The stable trans-iron nuclides can be classified into three categories according to their position on the chart of nuclides (see Fig. 2): those located at the bottom of the valley of nuclear stability, called the s-nuclei, and those situated on the neutron-deficient or neutron-rich side of the valley, named the p- or r-nuclei, respectively.

As pointed out above, the charged-particle induced reactions are not able to account for the production of these three types of trans-iron nuclides, which are instead primarily produced by three different mechanisms, naturally referred to as the s-, r-, and p-processes.

The first two processes (namely, s- and r-processes) can take place through neutron captures and subsequent β-decays in astrophysical environments where at least one neutrons source is working efficiently (e.g. Käppeler, 1999). As a result of each (n, γ) capture reaction, a generic nucleus (Z, A) is transformed into the heavier isotope (Z, A+1). If this isotope is stable against β-decay, an additional neutron capture can take place, leading to the isotope (Z, A+2). Otherwise, if the produced isotope is unstable, it can either decay into the isobar (Z+1, A+1) or capture another neutron (e.g. Clayton, 1983). The question whether this unstable isotope decays or captures a neutron depends on the values of τ_β and $\tau_{n\gamma}$, which are the β-decay lifetime and the average time between two successive neutron captures for such unstable isotope, respectively. If the relation $\tau_\beta \gg \tau_{n\gamma}$ is valid for the majority of the unstable nuclides involved in the nucleosynthesis process, the sequence of neutron captures and β-decays is called s-process ("s" stands just for "slow" neutron capture); otherwise, if neutron capture proceeds on a rapid time scale compared to the β-decay lifetimes (i.e. relation $\tau_\beta \ll \tau_{n\gamma}$ is valid for the majority of the unstable nuclides), the sequence of reactions is named r-process, where "r" stands for "rapid" neutron capture (e.g. Rolfs & Rodney, 1988).

Fig. 3 reveals that the s-process involves the addition of neutrons to seed nuclei, which are nuclear species of the iron peak region (mainly ^{56}Fe). In particular, this nucleosynthesis process produces nuclei from the A $\simeq$ 60 mass region up to ^{209}Bi, closely following the valley of nuclear stability. Indeed, since the neutron capture is slow compared to the β-decay rates, the neutron capture chains move through the stable isotopes of a given element until an unstable isotope is reached. At this point, a β-decay occurs and the neutron capture chains

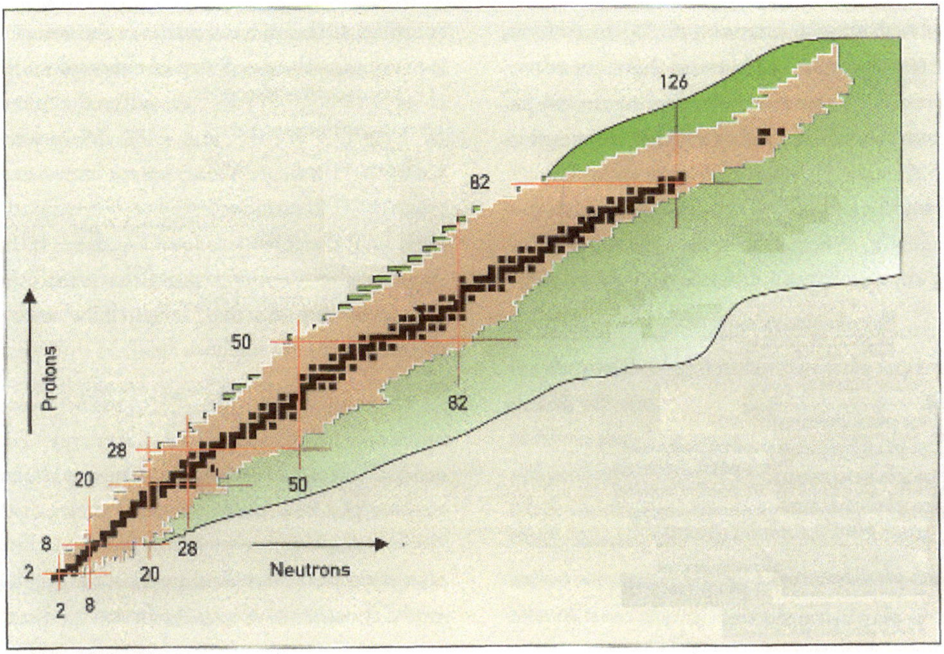

Fig. 2. Chart of nuclides where bound nuclear systems are reported as a function of the neutron number N (x axis) and the proton number Z (y axis). The brown squares represent stable plus very long-lived (lifetime $> 10^5$ yr) nuclei that define the so-called valley of nuclear stability. The red lines show the magic numbers for the valley of nuclear stability. The pink region designates the zone of unstable nuclei that has been explored in laboratory; while the green region indicates the so-called "terra incognita" (Latin expression for "unknown land") occupied with unstable nuclei that remain to be explored. The explorable nuclear landscape is bounded by the proton drip line to the upper left and neutron drip line to the bottom right. Both drip lines are marked by thin lines. (Figure adapted from http://www.pas.rochester.edu/~cline/Research/sciencehome.htm)

resume in the element having the nuclear charge increased by one unit (e.g. Clayton, 1983, for details).

As with the s-process, also the r-process has the nuclides of the iron peak region as seeds. However, the path of the r-process moves along the extreme neutron-rich side of the valley of nuclear stability (see Fig. 3), where the values of the neutron binding energies approach zero (so-called neutron drip line; see also caption of Fig. 2). Indeed, since the neutron capture proceeds on very rapid time scale compared to the β-decay lifetimes, a generic nucleus (Z, A) is transformed into the heavier neutron-rich isotope (Z, A+i) by a series of (n, γ) capture reactions. Only when it is reached a point where the (n, γ) capture reaction and its inverse (γ, n) reaction are in equilibrium, capture reactions stop and a β-decay can occur, transforming the nucleus (Z, A+i) into its isobaric neighbour (Z+1, A+i). A this point, the chain of capture reactions restarts and the nucleus (Z+1, A+i) absorbs neutrons until the balance

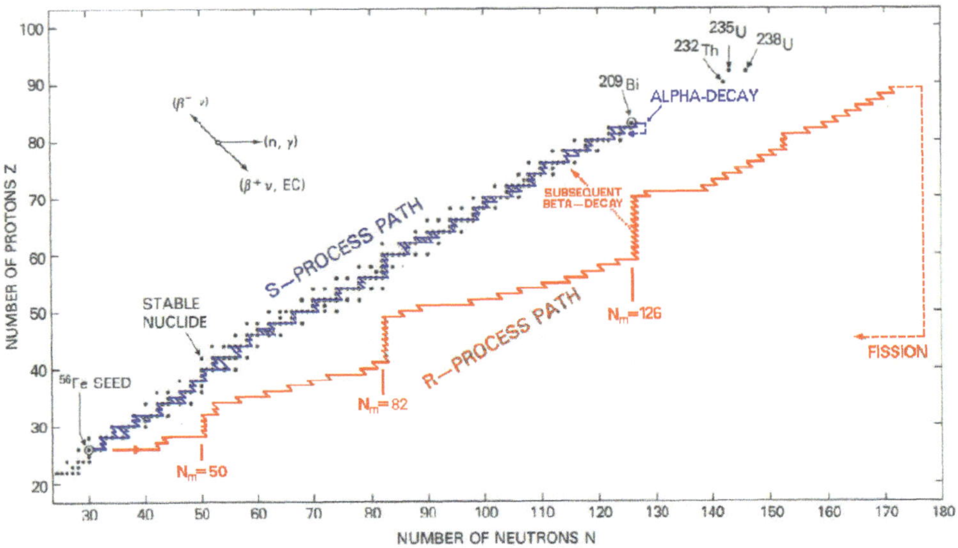

Fig. 3. Paths of the s- and r-processes in the N-Z plan. Both processes start with the nuclides of the iron peak region (mainly ^{56}Fe) as seeds. The s-process path (blue line) follows the valley of nuclear stability and stops at the ^{209}Bi, after which one enters the region of α-instability. The r-process path (red line) moves far to the neutron-rich side of the valley of nuclear stability, bypasses nuclei with natural α-radioactivity (which stops the s-process path), and terminates by β-delayed fission and neutron-induced fission at A $\simeq$ 270 (e.g. Thielemann et al., 2001). The chains of β-decays from the path of the r-process towards the valley of nuclear stability, which occur after the r-process neutron irradiation, are schematized by a dashed line. The sequences of waiting points at nuclei with magic neutron numbers N_m = 50, 82, and 126 are also highlighted. (Figure adapted from Rolfs & Rodney, 1988)

between capture reactions and photodisintegrations is again reached for its isotopic neighbour (Z+1, A+i+k). This recurring sequence of "neutron captures plus β-decay" reactions where it is necessary to "wait" a β-decay prior of restarting another series of neutron captures leads to a so-called waiting point on the N-Z plan. In particular, if the r-process path reaches a nucleus with a magic neutron number N_m, the next heavier isotope with N_m+1 neutrons has a relatively small neutron binding energy and it is relatively easy to reach the equilibrium between the (n, γ) capture reaction and its inverse (γ, n). Consequently, the sequence of neutron capture events stops and, after one neutron capture and a subsequent β-decay, the (Z, N_m) nucleus is transformed into its isobaric neighbour (Z+1, N_m), which is again a nucleus with the same magic number N_m of neutrons. Therefore, it is expected the occurrence of a series of waiting points at the same magic neutron number N_m, that increases the Z value by one unit at a time and moves the resulting nuclei nearer and nearer to the valley of nuclear stability. Only when these nuclei are enough close to the valley of nuclear stability that the neutron binding energy becomes sufficiently large to break through the neutron magic bottleneck at N_m, the chain of capture reactions restarts (e.g. see in Fig. 3 the sequences of waiting points at nuclei with N_m = 50, 82, and 126). After the synthesizing event (when the neutron irradiation stops), the very neutron-rich unstable nuclides undergo chains of β-decay, which end at the most neutron-rich stable isobar for each value of A. In this way, it is possible to produce either neutron-rich stable nuclei bypassed by the s-process[3] or nuclides lying also on the s-process path (see e.g. Arnould & Takahashi, 1999; Arnould et al., 2007; Clayton, 1983; Rolfs & Rodney, 1988, for details).

The p-process is instead responsible for the production of proton-rich isotopic species (named p-nuclei), which are skipped by the s-process and r-process paths because they are "shielded" from formation both via β-decay by the presence of other stable isobars and via neutron capture by the lack of other less massive stable isotopes (neutron capture will obviously never bring a more massive stable isotope to a lighter one). The general consensus is that the p-nuclei are synthesized through "photo-erosion" of neutrons, α particles and protons involving heavy ($A \gtrsim 75$) isotopes previously formed via the s- and/or r-processes. A variety of astrophysical sites (e.g. the Ne-O-rich layer of massive stars during their pre-supernova phase or their explosion as core-collapse supernova, the C-rich zones of Chandrasekhar-mass white dwarfs exploding as Type Ia supernovae, and the exploding sub-Chandrasekhar mass white dwarfs on which He-rich material has been accreted from a companion star) have been identified as possible active p-nuclei contributors, but a complete and self-consistent model for the origin of the p-isotopes remain to be soundly based (for details see e.g. Arnould & Goriely, 2003, and references therein).

3. The s-process

3.1 An analytical approach: the σN-curve and the different components of s-process

As mentioned above, the path of the s-process follows the valley of nuclear stability, synthesizing about half of all the trans-iron elements. The time dependence of the abundance of these synthesized nuclei can be described on first approximation[4] using an analytical approach. It is based on the assumption that the relation $\tau_\beta \gg \tau_{n\gamma}$ is valid for all the unstable nuclides involved in the nucleosynthesis process and, consequently, that a generic unstable nucleus (Z, A+1) immediately decays into its isobar (Z+1, A+1), which is usually stable. Although this assumption does not hold in every instance (cf. Sect. 3.2), it enable us to describe the s-process in a relatively simple way and with no general damage to the theoretical description of this nucleosynthesis process (e.g. Clayton, 1983; Rolfs & Rodney, 1988, for details). Indeed, according to such assumption, one can neglect the abundances of unstable species and assume that, fixed the mass number A, there is only one stable nuclide with a given Z (i.e. only one isotope for each element) involved in the nucleosynthesis process. As a consequence, the time dependence of the abundance N_A[5] of the nuclei synthesized by

[3] Such neutron-rich stable nuclei are referred to as "r-only" products. Similarly, nuclides produced by the s-process and bypassed by the r-process are referred to as "s-only" nuclei.

[4] A dedicated numerical modelling is required for a more detailed description (see e.g. the numerical approaches described in Pignatari et al., 2008; Prantzos et al., 1987; Pumo et al., 2010, and references therein).

[5] The abundance is labeled only by the mass number A and not additionally by the nuclear charge Z because the latter quantity is uniquely defined within the aforementioned assumption that there is only one stable nuclide with a given Z involved in the nucleosynthesis process.

the s-process can be written as

$$\frac{dN_A(t)}{dt} = N_n(t)N_{A-1}(t) < \sigma v >_{A-1} - N_n(t)N_A(t) < \sigma v >_A, \quad (1)$$

where $N_n(t)$ is the neutron density at time t, while $< \sigma v >_{A-1}$ and $< \sigma v >_A$ are the reaction rate of the capture reaction involving the isotope with mass number $A-1$ and A, respectively. The first term on the right side of the above equation describes the production of an isotope with mass number A by neutron capture of its lighter "neighbour" with mass number $A-1$, and the second one represents the destruction of the isotope with mass number A, again due to a neutron capture reaction. All the terms in the equation are time-dependent because of their dependence upon the temperature that, in turn, depends on the time.

Relation 1 defines a set of coupled differential equations, which can be solved analytically under the additional assumption that the temperature is essentially constant during the event of neutron irradiation. Indeed, with this further hypothesis, it is possible to replace $< \sigma v >_{A-1}$ with $\sigma_{A-1} v_T$, where σ_{A-1} is the Maxwellian-averaged neutron-capture cross section for the isotope with mass number $A-1$ and v_T is the thermal velocity given by the relation $v_T \simeq \left(\frac{2kT}{M_n}\right)^{1/2}$, in which k and T have their standard meaning of Boltzmann constant and temperature, and M_n is the neutron mass (see e.g. Rolfs & Rodney, 1988, for details). Similarly $< \sigma v >_A$ can be replaced with $\sigma_A v_T$ and, consequently, relation 1 becomes

$$\frac{dN_A(t)}{dt} = v_T N_n(t) \left[N_{A-1}(t)\sigma_{A-1} - N_A(t)\sigma_A\right]. \quad (2)$$

Moreover it is possible to introduce a new variable, the so-called time-integrated neutron flux τ, which is a measurement of the total neutron irradiation per unit of area and is defined as $\tau = v_T \int_0^t N_n(t)dt$. Replacing the time t with this new variable τ, the relation 2 reduces to

$$\frac{dN_A(t)}{d\tau} = \sigma_{A-1}N_{A-1} - \sigma_A N_A, \quad (3)$$

where the rate of change of the N_A is now with respect to τ.

The process described by the set of coupled equations 3 has the property to be "self-regulating" (see also Clayton, 1983; Rolfs & Rodney, 1988), that is the tendency to reach a state of equilibrium where $\frac{dN_A(t)}{d\tau} \simeq 0 \rightarrow \sigma_{A-1}N_{A-1} \simeq \sigma_A N_A \simeq$ constant. For a true equilibrium condition in the s-process, the σN value should be strictly constant over the whole mass region from ^{56}Fe up to ^{209}Bi. However, looking at Fig. 4, it is easy to notice that such a condition is only satisfied locally in mass regions between magic neutron numbers and, for that reason, the condition is called the *local condition approximation*.

The existence of a local equilibrium indicates that the observed distribution of $\sigma_A N_A$ can not be generated by a uniform exposure of iron-peak nuclei to a single neutron flux, but it is necessary a superposition of different exposures to various neutron fluxes. As a consequence, the quantity $\sigma_A N_A$ can be written as (see e.g. Clayton, 1983, for details)

$$\sigma_A N_A = \int_0^\infty \rho(\tau)\sigma N(\tau)\, d\tau,$$

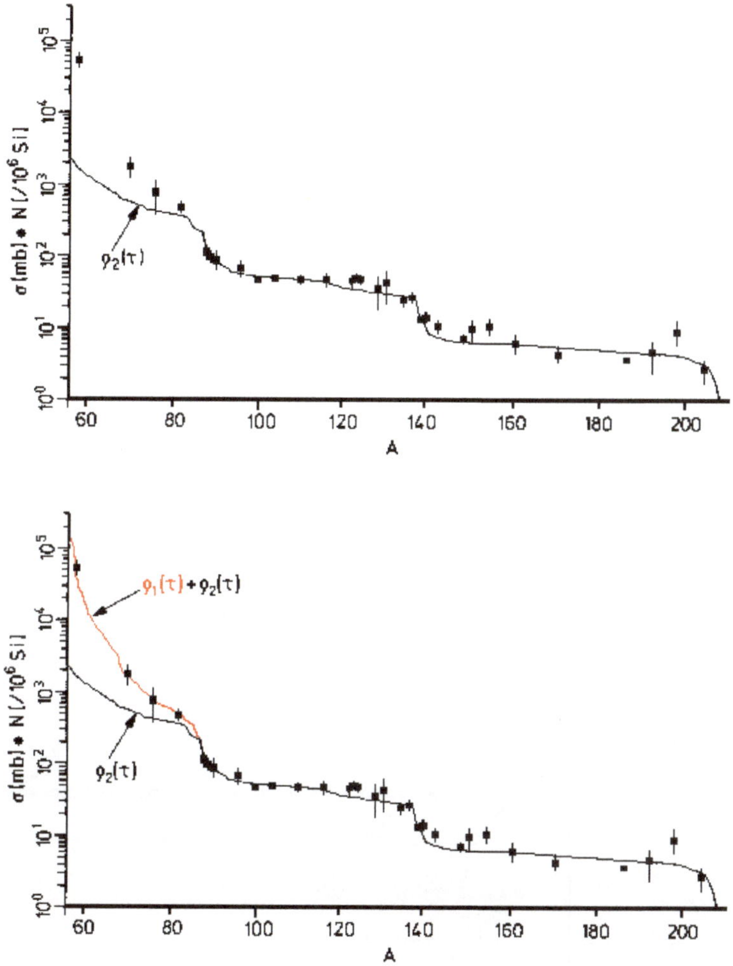

Fig. 4. Observed distribution of the product $\sigma_A N_A$ as a function of the mass number A for the s-only nuclei. The neutron-capture cross section σ is measured at 30 keV in millibarn, and the abundances are the solar ones (normalized to 10^6 Si atoms). The solid lines represent theoretical calculations for the single exponential distribution $\rho_2(\tau)$ (both panels) and for the sum of the exponential distributions $\rho_1(\tau) + \rho_2(\tau)$ (bottom panel). (Figures adapted from Rolfs & Rodney, 1988)

where $\rho(\tau)$ represents the continuous distribution of multiple neutron exposures and is defined by the following relation

$$\rho(\tau) \propto \frac{1}{\tau_0} exp\left(-\frac{\tau}{\tau_0}\right),$$

in which τ_0 is a parameter representing the mean neutron exposure.

It is found (see e.g. Clayton, 1983; Käppeler, 1999; Käppeler et al., 2011; Rolfs & Rodney, 1988) that at least two exponential neutron exposures, $\rho_1(\tau)$ and $\rho_2(\tau)$, are necessary to roughly reproduce the $\sigma_A N_A$ curve reported in Fig. 4. The first one, called weak component, has a relatively low mean neutron exposure ($\tau_{0,1} \simeq 0.06$ neutrons millibarn^{-1}, with 1 neutron millibarn$^{-1} \equiv 10^{27}$ neutrons cm^{-2}) that allows for the synthesis of s-species in the $60 \lesssim A \lesssim 90$ mass range (see also the red line in bottom panel of Fig. 4). The second one, called main component, has a higher mean neutron exposure ($\tau_{0,2} \simeq 0.25$ neutrons millibarn^{-1}) and allows for the synthesis of the remaining s-nuclei.

3.2 The s-process branchings

As above mentioned, the analytical approach discussed in the previous Sect. 3.1 is based on the assumption that the relation $\tau_\beta \gg \tau_{n\gamma}$ is valid for all the unstable nuclides involved in the nucleosynthesis process. However, the s-process path can actually encounter unstable nuclei for which the decay rate becomes comparable to the neutron capture rate (i.e. $\tau_\beta \simeq \tau_{n\gamma}$). This leads to a splitting of the path, which is called s-process branching. As a consequence, a fraction of the s-process flow proceeds through neutron capture, while the other one goes through the β-decay (see Fig. 5).

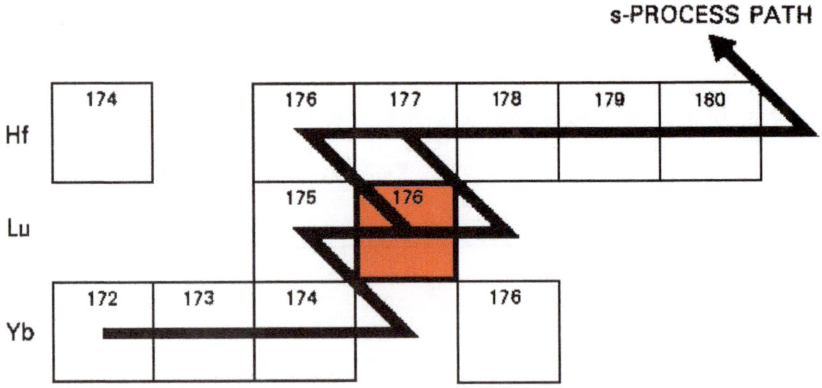

Fig. 5. A section of the chart of nuclides showing the s-process branching at the isotope ^{176}Lu (highlighted in red). (Figure adapted from Rolfs & Rodney, 1988)

A comparison of abundances of nuclei reached at a branching starting at a nucleus with mass number A can provide information on the physical conditions of the environment in which the s-process takes place, through the analysis of the so-called branching ratio R defined as

$$R = \frac{1}{\tau_\beta N_n < \sigma v >_A}.$$

Indeed, the above mentioned ratio can be also deduced directly from the observations, considering that R can be also written in terms of the observed values of σN as $R = \frac{(\sigma N)_{Z+1}}{(\sigma N)_{A+1}}$ or $R = \frac{(\sigma N)_{A-1}}{(\sigma N)_{A+1}} - 1$. As a consequence, if τ_β is temperature-independent and $< \sigma v >_A$ is known, the observed ratio R can provide information about the s-process neutron density N_n. Instead, if N_n is known, the ratio R can become a sensitive s-process "thermometer" or "barometer", every time that the s-process branching involves reactions which critically depend on temperature or density (see e.g. Rolfs & Rodney, 1988, for details).

3.3 Actual sites of the s-process

To have a more complete picture of the s-process it is necessary to individuate the astrophysical sites where the s-process can take place, keeping in mind that more than one s-process component is required in order to explain the observed solar distribution of s-nuclei abundances.

Current views on the subject suggest that the main and weak components of s-process correspond, in terms of stellar environments, to two distinct categories of stars in different evolutionary phases (e.g. Käppeler, 1999).

In particular, the main component is associated with low-mass stars ($M_{ZAMS} \sim 1.5 - 3M_\odot$) during their thermally pulsing asymptotic giant branch (TP-AGB) phase, when the H- and He-burning shells surrounding the degenerate stellar core are alternately activated. The predominant neutron source is the ^{13}C(α,n)^{16}O reaction, but the ^{22}Ne(α,n)^{25}Mg reaction can be also marginally activated, leading to a variation of some abundance ratios of nuclides belonging to s-process branchings (for details on main component the interested reader is referred to the reviews of Busso et al., 1999; Käppeler, 1999; Käppeler et al., 2011, and references therein).

The weak component occurs in massive stars ($M_{ZAMS} \gtrsim 13M_\odot$) primarily during their core He-burning phase, and the most important neutron source is the ^{22}Ne(α,n)^{25}Mg reaction. This reaction is efficiently activated only at the end of the core He-burning phase, when the temperature is $\gtrsim 2,5 \times 10^8$ K. The available ^{22}Ne is produced at the beginning of the core He-burning phase via the reaction sequence ^{14}N(α,γ)^{18}F(e^+ ν)^{18}O(α,γ)^{22}Ne, where ^{14}N derives from the CNO cycle activated during the previous H-burning phase (for details see e.g. Woosley et al., 2002, and references therein).

In addition to these two components, other kinds of stars, such as massive AGB ($M_{ZAMS} \sim 4 - 7M_\odot$) and super-AGB stars ending their life as NeO white dwarfs ($M_{ZAMS} \sim 7.5 - 10M_\odot$; for details see e.g. Fig. 1 in Pumo et al. 2009b, but also Pumo & Siess 2007 or Pumo 2007 and references therein), could also contribute to the nucleosynthesis of s-species, but this hypothesis still needs further investigation (Pumo et al., 2009a, and references therein). Moreover, some studies (Busso et al., 1999; Gallino et al., 1998; Goriely & Siess, 2004; Lugaro et al., 2003) suggest the existence of a "strong" component, which occurs in low-metallicity stars of low-intermediate mass during the TP-AGB phase, and which is supposed to be responsible for the synthesis of "massive" (around ^{208}Pb) s-species. Furthermore, Travaglio et al. (2004) propose the existence of an additional component referred to as lighter element primary s-process (LEPP), but its nature is still unclear and under debate (e.g. Pignatari et al., 2010; Tur et al., 2009, and references therein).

4. Production of s-nuclei in massive stars

4.1 Sensitivity to stellar mass and metallicity

There is a wide consensus about the main characteristics of weak component of the s-process and, in particular, about its sensitivity to stellar mass and metallicity (e.g. Käppeler, 1999).

As for the dependence on the stellar mass, quantitative studies (see e.g. Prantzos et al. 1990; Käppeler et al. 1994; Rayet & Hashimoto 2000; The et al. 2000, 2007) show that the s-process weak component efficiency[6] decreases with decreasing initial stellar mass, and that the shape of the distribution of the overproduction factors as a function of the mass number essentially does not depend on the initial stellar mass value. This behavior is connected to the fact that the reaction $^{22}Ne(\alpha,\ n)^{25}Mg$ becomes efficient only for $T \gtrsim 2,5 \times 10^8\ K$, so the production of s-nuclei is more and more efficient when the initial stellar mass is increased, because more massive models burn helium at a "time averaged" higher temperature; however the ratio of the overproduction factor F_i of a given s-only nucleus i to the average overproduction factor F_0 (see footnote 6 for details on F_i and F_0) remains fairly constant irrespective of the stellar mass, so the shape of the distribution of the overproduction factors does not change when the initial stellar mass is increased (see also the behavior of the s-only nuclei distribution in Fig. 6).

---

[6] Usually the s-process efficiency is analyzed in terms of the following efficiency indicators (see e.g. Prantzos et al. 1987; The et al. 2000; Costa et al. 2006; Pumo et al. 2006, 2010):

- the average overproduction factor F_0 for the 6 s-only nuclei ^{70}Ge, ^{76}Se, ^{80}Kr, ^{82}Kr, ^{86}Sr and ^{87}Sr, given by

$$F_0 = \frac{1}{N_s} \sum_i F_i \text{ with } F_i = \frac{X_i}{X_{i,ini}},\ N_s = 6$$

  where F_i is the overproduction factor, X_i is the mass fraction (averaged over the convective He-burning core) of s-only nucleus i at the end of s-process, $X_{i,ini}$ is the initial mass fraction of the same nucleus, and N_s is the number of the s-only nuclei within the mass range $60 \leq A \leq 87$;

- the maximum mass number A_{max} for which the species in the $60 \leq A \leq A_{max}$ mass range are overproduced by at least a factor of about 10;

- the number of neutrons captured per initial ^{56}Fe seed nucleus n_c defined as

$$n_c = \sum_{A=57}^{209} (A-56) \frac{[Y_A - Y_A(0)]}{Y_{56}(0)}$$

  where $Y_{56}(0)$ is the initial number fraction of ^{56}Fe, Y_A is the final number fraction of the nucleus with mass number A, and $Y_A(0)$ is the initial one.

In addition to the previous parameters, the maximum convection zone mass extension and the duration of the nucleosynthesis event are also used for characterizing the efficiency of the s-process weak component during the core He-burning.

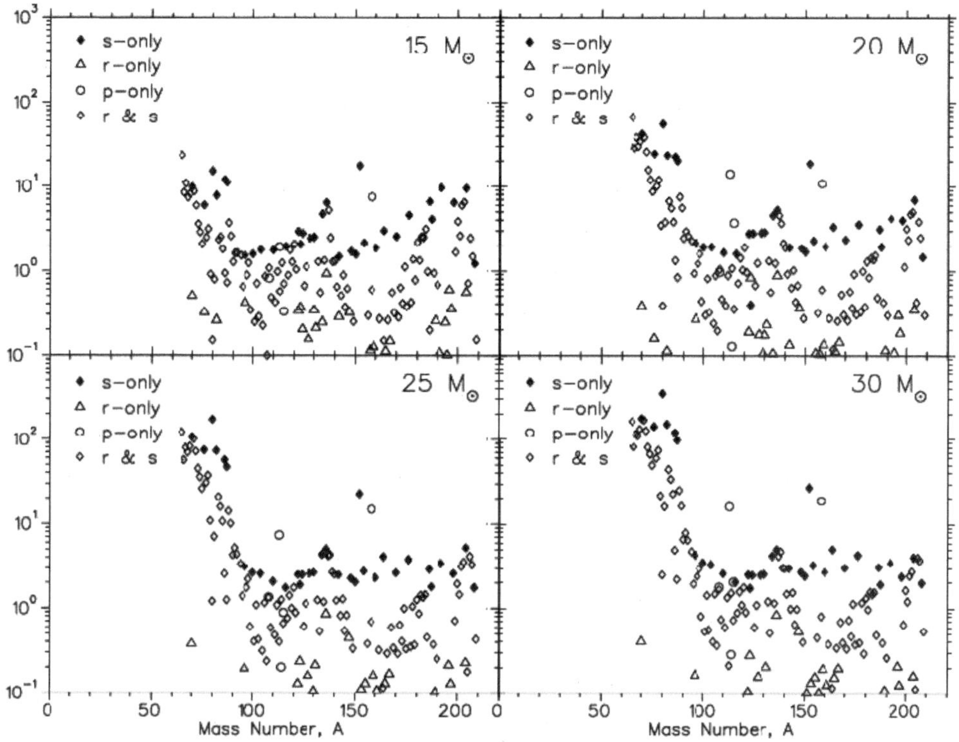

Fig. 6. Overproduction factor at the end of the core He-burning as function of the mass number A for stellar models of different initial mass (namely, $M_{ZAMS} = 15, 20, 25$, and $30 M_\odot$). All the stable trans-iron nuclei of the network used to simulate the s-process are reported. The primary nucleosynthesis production process of each nucleus is indicated by the symbol type. (Figure adapted from The et al., 2007)

As for the effect of metallicity, the s-process weak component efficiency depends on the so-called *source/seed* ratio[7] (see e.g. Prantzos et al., 1990; Rayet & Hashimoto, 2000). If the *source/seed* ratio is constant with the metallicity Z, the efficiency is expected to increase when increasing the Z value, because the effect of the ^{16}O primary poison becomes less important when the abundances of the *source* nuclei increase with Z. For a non-constant *source/seed* ratio that increases when decreasing Z, the efficiency (measured in terms of the number of neutrons captured per initial ^{56}Fe seed nucleus n_c; see also footnote 6) has a non-linear behavior with Z, which reflects the interplay between two opposite factors: from one hand, the aforementioned role of the ^{16}O primary poison, which tends to decrease n_c with decreasing Z, because its

[7] Considering that the ^{22}Ne is the main neutron provider during the core He-burning phase and neglecting all the heavier ^{56}Fe nuclei, one obtains:

$$\text{source/seed} \simeq {}^{22}\text{Ne}/{}^{56}\text{Fe}.$$

This last quantity approximately corresponds to the $^{14}N/^{56}Fe$ ratio at the end of the core H-burning phase which, in turn, is roughly equal to the $O/^{56}Fe$ ratio at the ZAMS (see Prantzos et al., 1990; Rayet & Hashimoto, 2000, for details).

abundance remains the same independently of Z, so its relative importance increases as Z decreases; on the other hand the effect of the increased *source/seed* ratio, which tends to increase n_c with decreasing Z, because the number of available neutrons per nucleus *seed* increases as Z decreases.

4.2 Sources of uncertainties

Although the general features of the s-process weak component seem to be well established, there are still some open questions linked to the nuclear physics, the stellar evolution modeling, and the possible contribution to the s-nucleosynthesis from post-He-burning stellar evolutionary phases (see e.g. Costa et al., 2006; Pumo et al., 2006; Woosley et al., 2002).

The uncertainties due to nuclear physics are linked both with the reaction rates of reactions affecting the stellar structure evolution (as, for example, the triple-alpha, the $^{12}C(\alpha,\gamma)^{16}O$ and the $^{12}C + ^{12}C$ reactions) and with reaction rates on which the so-called "neutron economy" (i.e. the balance between neutron emission and captures) is based. Many works (see e.g. Bennett et al., 2010; Hoffman et al., 2001; Käppeler et al., 1994; Pignatari et al., 2010; Rayet & Hashimoto, 2000; The et al., 2000; 2007; Tur et al., 2007; 2009) have been devoted to analyze these uncertainties, showing that such uncertainties still affect significantly the s-process weak component efficiency.

The contribution to the synthesis of s-nuclei during the post-core-He-burning evolutionary phases was also explored by many authors (see e.g. Arcoragi et al. 1991; Raiteri et al. 1993; The et al. 2000, 2007; Hoffman et al. 2001; Rauscher et al. 2002; Limongi & Chieffi 2003; Tur et al. 2007, 2009), including in some cases the explosive burning (see e.g. Hoffman et al. 2001; Rauscher et al. 2002; Limongi & Chieffi 2003; Tur et al. 2007, 2009). All these studies have shown that a significant production of s-nuclei in massive stars can continue during the post-core-He-burning evolutionary phases and that the abundance of s-nuclei ejected in the interstellar medium after the core-collapse supernova events can be substantially modified by the explosive nucleosynthesis.

As for the impact of uncertainties owing to stellar evolution modeling, the determination of the size of the convective core and, more in general, of the mixing regions represents one of the major source of uncertainties still affecting the s-process weak component efficiency, as described in detail in the next Sect. 5.

5. Role of the convective overshooting

The determination of the mixing regions and, in particular, of the size of the convective core can directly affect the efficiency of the s-process nucleosynthesis by influencing the chemical and local temperature stratification (see e.g. Canuto, 1997; Deng & Xiong, 2008; Molawi & Forestini, 1994), by determining the amount of stellar material which experiences neutron irradiation (see e.g. Langer et al., 1989), and by giving rise to a variation of the s-process lifetime (see e.g. Costa et al., 2006; Pumo et al., 2010).

The convective core's extension of a star with a given initial mass and metallicity is determined in turn by a series of physical parameters such as the choice of the convective instability criterion (Schwarzschild's or Ledoux's criteria), the extra mixing processes induced by axial rotation and convective overshooting (see e.g. Chiosi et al., 1992; Woosley et al., 2002).

A series of studies have been devoted to analyze the effects of these physical parameters on the evolution of massive stars (see e.g. Meynet & Maeder 1997, 2000; Heger et al. 2000; Woosley et al. 2002; Hirschi et al. 2004; Limongi & Chieffi 2006; El Eid et al. 2009) and to examine the corresponding impact on the s-process weak component (see e.g. Costa et al., 2006; Langer et al., 1989; Pignatari et al., 2008; Pumo et al., 2006).

As far as the convective overshooting is concerned, one finds that this extra mixing process leads to an increase of the convective core mass (see also Fig. 7) and to a variation of the chemical and temperature stratification that, in turn, tend to enhance the s-process weak component efficiency, by giving rise to an increase of the amount of material that experiences neutron irradiation and to a variation of the s-process lifetime (see e.g. Costa et al., 2006, for more details).

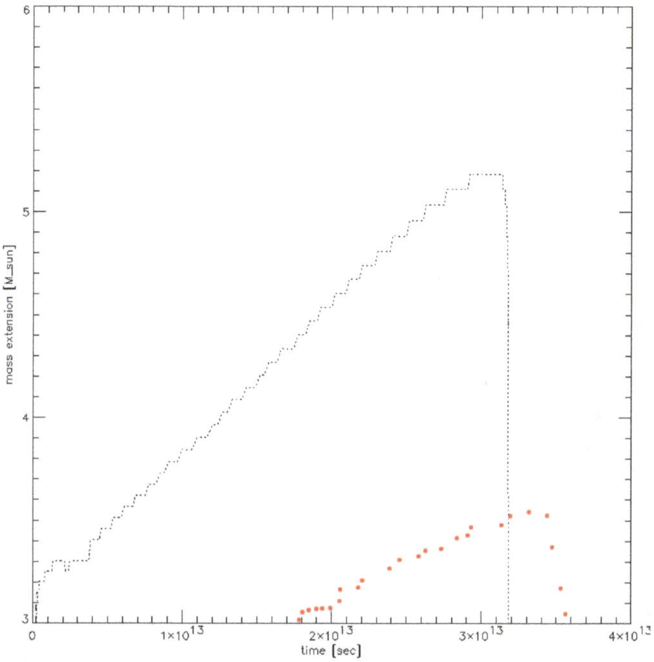

Fig. 7. Mass extension of the convection zone (in unit of $M_{\odot}$) during the core He-burning s-process as a function of time (measured from the core He-burning ignition) for a Z=0.005, M=20$M_{\odot}$ model with (dotted line) and without (red dots) convective overshooting. Details on the input physics used to calculate the models can be found in Pumo et al. (2010).

Recent works (Costa et al., 2006; Pumo et al., 2006; 2010) have been devoted to perform a comprehensive and quantitative study on the impact of the convective overshooting on the s-process, using a diffusive approach to model the convective overshooting (for details see e.g. Freytag et al., 1996; Herwig et al., 1997, and reference therein). The results show that models with overshooting give a higher s-process efficiency compared with "no-overshooting" models, with enhancements for the main s-process indicators F_0 and n_c until a factor $\sim$ 6 and $\sim$ 3, respectively.

6. Final remarks

The results reported in Sect. 5 clearly show the high level of uncertainty (up to a factor ~ 6) in the modeling of the weak s-process component due to the current lack of a self-consistent theory describing mixing processes inside the stars, indicating that a detailed scrutiny of the impact of the stellar evolution modelling uncertainties on this component remain to be performed prior to giving a final conclusion on the the s-process weak component efficiency.

In particular, prior to giving a final conclusion on the possible contribution of post-He burning phases to the s-process yields from a quantitative point of view, some additional investigation taking into account stellar evolution uncertainties in addition to the nuclear physics ones should be performed. Moreover, this additional investigation may shed light on different open questions (see also Pumo et al., 2010) linked, for example, to the effective existence of the LEPP process and to the model for the p-process taking place in the core-collapse supernovae O-Ne layers (see also Costa et al., 2006, and references therein), because the relevant s-nuclei are p-process seeds (cf. Sect.2.2 and see also e.g. Arnould & Goriely, 2003, and references therein).

7. Acknowledgments

M.L.P. acknowledges financial support from the Bonino-Pulejo Foundation and from the PRIN-INAF 2009 "Supernovae Variety and Nucleosynthesis Yields".

8. References

Arcoragi, J. -P.; Langer, N., Arnould, M. (1991). Neutron capture nucleosynthesis and the evolution of 15 and 30 solar mass mass-losing stars. II - The core carbon-burning phase, , In: *Astronomy and Astrophysics*, 249, 134

Anders, E.; Grevesse, N. (1989). Abundances of the elements - Meteoritic and solar, In: *Geochim. Cosmochim. Acta*, 53, 197

Angulo, C. (2009). Experimental Tools for Nuclear Astrophysics, In: *Lect. Notes Phys.*, 764, 253

Arnett, D. (1996). *Supernovae and Nucleosynthesis*, Princeton Univ. Press, ISBN 0-691-01147-8, Princeton, New Jersey

Arnould, M.; Goriely, S. (2003). The p-process of stellar nucleosynthesis: astrophysics and nuclear physics status, In: *Phys. Rept.*, 384, 1

Arnould, M.; Takahashi, K. (1999). Nuclear astrophysics, In: *Rep. Prog. Phys.*, 62 , 395 (arXiv:astro-ph/9812243)

Arnould, M.; Goriely, S.; Takahashi, K. (2007). The r-process of stellar nucleosynthesis: Astrophysics and nuclear physics achievements and mysteries, In: *Phys. Rept.*, 450, 97

Asplund, M.; Grevesse, N.; Sauval, A. J.; Scott, P. (2009). The Chemical Composition of the Sun, In: *Annual Rev. of Astronomy and Astrophysics*, 47, 481

Baur, G.; Hencken, K.; Trautmann, D. (2003). Electromagnetic dissociation as a tool for nuclear structure and astrophysics, In: *Progress in Particle and Nuclear Physics*, 51, 487

Bennett, M. E.; Hirschi, R.; Pignatari, M.; Diehl, S.; Fryer, C.; Herwig, F.; Hungerford, A.; Magkotsios, G.; Rockefeller, G.; Timmes, F.; Wiescher, M.; Young, P. (2010). The effect of 12C + 12C rate uncertainties on s-process yields, In: *Journal of Physics: Conference Series*, 202, 012023

Bertulani, C. A.; Gade, A. (2010). Nuclear astrophysics with radioactive beam, In: *Physics Reports*, 485, 195

Burbidge, E. M.; Burbidge, G. R.; Fowler, W. A.; Hoyle, F. (1957). Synthesis of the Elements in Stars, In: *Rev. Mod. Phys.*, 29, 547

Busso, M.; Gallino, R.; Wasserburg, G. J. (1999). Nucleosynthesis in Asymptotic Giant Branch Stars: Relevance for Galactic Enrichment and Solar System Formation, In: *Annual Rev. of Astronomy and Astrophysics*, 37, 239

Cameron, A. G. W. (1957). Stellar Evolution, Nuclear Astrophysics, and Nucleogenesis, In: *Report CRL-41*, A.E.C.L. Chalk River, Canada

Canuto, V. M. (1997). Overshooting in Stars: Five Old Fallacies and a New Model, In: *Astrophysical Journal*, 601, L71

Chiosi, C. (2007). Stellar Nucleosynthesis and Chemical Evolution of Galaxies, In: *EAS Publications Series*, 27, 25

Chiosi, C.; Bertelli, G.; Bressan, A. (1992). New developments in understanding the HR diagram, In: *Annual Rev. of Astronomy and Astrophysics*, 30, 235

Clayton, D. D. (1983). *Principles of Stellar Evolution and Nucleosynthesis*, Reprint. Originally pubblished: New-York: McGraw-Hill [1968], ISBN 0-226-10953-4, Chicago and London

Costa, V.; Pumo, M. L.; Bonanno, A.; Zappalà, R. A. (2006). The s-process weak component: uncertainties due to convective overshooting, In: *Astronomy and Astrophysics*, 447, 641

Costantini, H.; Formicola, A.; Imbriani, G.; Junker, M.; Rolfs, C.; Strieder, F. (2009). LUNA: a laboratory for underground nuclear astrophysics. In: *Reports on Progress in Physics*, 72, 086301

Deng, L.; Xiong, D. R. (2008). How to define the boundaries of a convective zone, and how extended is overshooting?, In: *Monthly Notices of the Royal Astronomical Society*, 386, 1979

El Eid, M. F.; The, L. -S.; Meyer, B. S. (2009). Massive Stars: Input Physics and Stellar Models, In: *Space Sci. Rev.*, 147, 1

Freytag, B.; Ludwig, H.-G.; Steffen, M. (1996). Hydrodynamical models of stellar convection. The role of overshoot in DA white dwarfs, A-type stars, and the Sun, In: *Astronomy and Astrophysics*, 313, 497

Gallino, R.; Arlandini, C.; Busso, M.; Lugaro, M.; Travaglio, C.; Straniero, O.; Chieffi, A.; Limongi, M. (1998). Evolution and Nucleosynthesis in Low-Mass Asymptotic Giant Branch Stars. II. Neutron Capture and the s-Process, In: *Astrophysical Journal*, 497, 388

Goriely, S.; Siess, L. (2004). S-process in hot AGB stars: A complex interplay between diffusive mixing and nuclear burning, In: *Astronomy and Astrophysics*, 421, L25

Grevesse, N.; Noels, A. (1993). Cosmic abundances of the elements, In: *In Origin and Evolution of the Elements*, ed. Prantzos, Vangioni-Flam, Cassé, p 15. Cambridge, UK: Cambridge Univ. Press.

Grevesse, N.; Sauval, A. J. (1998). Standard Solar Composition, In: *Space Science Reviews*, 85, 161

Grevesse, N.; Noels, A.; Sauval, A. J. (1996). Standard Abundances, In: *Astronomical Society of the Pacific Conference Series*, 99, 117

Heger, A.; Langer, N.; Woosley, S. E. (2000). Presupernova Evolution of Rotating Massive Stars. I. Numerical Method and Evolution of the Internal Stellar Structure, In: *Astrophysical Journal*, 528, 368

Herwig, F.; Blöcker, T.; Schoenberner, D.; El Eid, M. (1997). Stellar evolution of low and intermediate-mass stars. IV. Hydrodynamically-based overshoot and nucleosynthesis in AGB stars, In: *Astronomy and Astrophysics*, 324, 81

Hirschi, R.; Meynet, G.; Maeder, A. (2004). Stellar evolution with rotation. XII. Pre-supernova models, In: *Astronomy and Astrophysics*, 425, 649

Hoffman, R. D.; Woosley, S. E.; Weaver, T. A. (2001). Nucleosynthesis below A=100 in Massive Stars, In: *Astrophysical Journal*, 549, 1085

Iocco, F; Mangano, G.; Miele, G.; Pisanti, O.; Serpico, P. D. (2009). Primordial Nucleosynthesis: from precision cosmology to fundamental physics, In: *Phys. Rept.*, 472, 1

Jordi, J.; Hernanz, M. (2007). Nucleosynthesis in classical nova explosions, In: *Journal of Physics G: Nuclear and Particle Physics*, 34, R431

Jordi, J.; Iliadis, C. (2011). Nuclear astrophysics: the unfinished quest for the origin of the elements, In: *Reports on Progress in Physics*, Accepted for publication (arXiv:1107.2234v1)

Käppeler, F. (1999). The origin of the heavy elements: The s process, In: *Progress in Particle and Nuclear Physics*, 43, 419

Käppeler, F.; Wiescher, M.; Giesen, U.; Goerres, J.; Baraffe, I.; El Eid, M.; Raiteri, C. M.; Busso, M.; Gallino, R.; Limongi, M.; Chieffi, A. (1994). Reaction rates for O-18(alpha, gamma)Ne-22, Ne-22(alpha, gamma)Mg-26, and Ne-22(alpha, n)Mg-25 in stellar helium burning and s-process nucleosynthesis in massive stars, In: *Astrophysical Journal*, 437, 396

Käppeler, F.; Gallino, R.; Bisterzo, S.; Aoki, W. (2011). The s process: Nuclear physics, stellar models, and observations, In: *Review of Modern Physics*, 83, 157

Langer, N.; Arcoragi, J.-P.; Arnould, M. (1989). Neutron capture nucleosynthesis and the evolution of 15 and 30 solar-mass stars. I - The core helium burning phase, In: *Astronomy and Astrophysics*, 210, 187

Limongi, M.; Chieffi, A. (2003). The Nucleosynthesis of ^{26}Al and ^{60}Fe in Solar Metallicity Stars Extending in Mass from 11 to 120 Msolar: The Hydrostatic and Explosive Contributions, In: *Astrophysical Journal*, 592, 404

Limongi, M.; Chieffi, A. (2006). Evolution, Explosion, and Nucleosynthesis of Core-Collapse Supernovae, In: *Astrophysical Journal*, 647, 483

Lodders, K. (2003). Solar System Abundances and Condensation Temperatures of the Elements, In: *Astrophysical Journal*, 591, 1220

Lodders, K.; Palme., H; Gail, H-P. (2009). Abundances of the Elements in the Solar System, In: *Landolt-Börnstein, New Series in Astronomy and Astrophysics*, 4, 44

Lugaro, M.; Herwig, F.; Lattanzio, J. C.; Gallino, R.; Straniero, O. (2003). s-Process Nucleosynthesis in Asymptotic Giant Branch Stars: A Test for Stellar Evolution, In: *Astrophysical Journal*, 586, 1305

Meyer, B. S. (1994). The r-, s-, and p-Processes in Nucleosynthesis, In: *Annual Rev. of Astronomy and Astrophysics*, 32, 153

Meynet, G.; Maeder, A. (1997). Stellar evolution with rotation. I. The computational method and the inhibiting effect of the μ-gradient, In: *Astronomy and Astrophysics*, 321, 465

Meynet, G.; Maeder, A. (2000). Stellar evolution with rotation. V. Changes in all the outputs of massive star models, In: *Astronomy and Astrophysics*, 361, 101

Mowlavi, N.; Forestini, M. (1994). Semi-convection and overshooting in intermediate mass and massive stars, In: *Astronomy and Astrophysics*, 282, 843

Pignatari, M.; Gallino, R.; Meynet, G.; Hirschi, R.; Herwig, F.; Wiescher, M. (2008). The s-Process in Massive Stars at Low Metallicity: The Effect of Primary ^{14}N from Fast Rotating Stars, In: *Astrophysical Journal*, 687, L95

Pignatari, M.; Gallino, R.; Heil, M.; Wiescher, M.; Käppeler, F.; Herwig, F.; Bisterzo, S. (2010). The Weak s-Process in Massive Stars and its Dependence on the Neutron Capture Cross Sections, In: *Astrophysical Journal*, 710, 1557

Povh, B.; Rith, K.; Scholz, C.; Zetsche, F. (2008). *Particles and Nuclei - An Introduction to Physical Concepts*, 6th ed., Springer, ISBN 978-3-540-79367-0

Prantzos, N.; Arnould, M.; Arcoragi, J.-P. (1987). Neutron capture nucleosynthesis during core helium burning in massive stars, In: *Astrophysical Journal*, 315, 209

Prantzos, N.; Hashimoto, M.; Nomoto, K. (1990). The s-process in massive stars - Yields as a function of stellar mass and metallicity, In: *Astronomy and Astrophysics*, 234, 211

Pumo, M. L. (2007). Evolutionary properties of $\sim$ 7-13 $M_\odot$ stars and the associated nucleosynthesis, In: *Mem. S.A.It.*, 78, 689

Pumo, M. L.; Siess, L. (2007). Post-He-Burning Phases and Final Fate of Super-AGB Stars, In: *Astronomical Society of the Pacific*, 378, 133

Pumo, M. L.; Costa, V.; Bonanno, A.; Belvedere, G.; Pizzone, R. G.; Tumino, A.; Spitaleri, C.; Cherubini, S.; La Cognata, M.; Lamia, L.; Musumarra, A.; Romano, S.; Zappalà, R. A. (2006). The role of overshooting treatment in the s-process weak component efficiency, In: *Mem. S.A.It. Suppl.*, 9, 263

Pumo, M. L.; Ventura, P.; D'Antona, F.; Zappalà, R. A. (2009a). The "double final fate" of super-AGB stars and its possible consequences for some astrophysical issues, In: *Mem. S.A.It.*, 80, 161

Pumo, M. L.; Turatto, M.; Botticella, M. T.; Pastorello, A.; Valenti, S.; Zampieri, L.; Benetti, S.; Cappellaro, E.; Patat, F. (2009b). EC-SNe from Super-Asymptotic Giant Branch Progenitors: Theoretical Models Versus Observations, In: *Astrophysical Journal*, 705, L138

Pumo, M. L.; Contino, G.; Bonanno, A.; Zappalà, R. A. (2010). Convective overshooting and production of s-nuclei in massive stars during their core He-burning phase, In: *Astronomy and Astrophysics*, 524, A45

Raiteri, C. M.; Gallino, R.; Busso, M.; Neuberger, D.; Käppeler, F. (1993). The Weak s-Component and Nucleosynthesis in Massive Stars, In: *Astrophysical Journal*, 419, 207

Rauscher, T.; Patkós, A. (2011). Origin of the Chemical Elements, In: *Handbook of Nuclear Chemistry*, ISBN 978-1-4419-0719-6. Springer Science+Business Media B.V., p. 611 (arXiv:1011.5627)

Rauscher, T.; Heger, A.; Hoffman, R. D.; Woosley, S. E. (2002). Nucleosynthesis in Massive Stars with Improved Nuclear and Stellar Physics, In: *Astrophysical Journal*, 576, 323

Rayet, M.; Hashimoto, M. (2000). The s-process efficiency in massive stars, In: *Astronomy and Astrophysics*, 354, 740

Reeves, H. (1994). On the origin of the light elements (Z<6), In: *Reviews of Modern Physics*, 66, 193

Rolfs, C. E.; Rodney, W. S. (1988). *Cauldrons in the Cosmos*, Univ. Chicago Press, ISBN 0-226-72457-3, Chicago and London

Spitaleri, C. (2010). The Trojan Horse Method in Nuclear Astrophysics, In: *AIP Conf. Proc.*, 1304, 192

Steigman, G. (2007). Primordial Nucleosynthesis in the Precision Cosmology Era, In: *Annu. Rev. Nucl. Part. Sci.*, 57, 463

The, L.-S.; El Eid, M. F.; Meyer, B. S. (2000). A New Study of s-Process Nucleosynthesis in Massive Stars, In: *Astrophysical Journal*, 533, 998

The, L.-S.; El Eid, M. F.; Meyer, B. S. (2007). s-Process Nucleosynthesis in Advanced Burning Phases of Massive Stars, In: *Astrophysical Journal*, 655, 1058

Thielemann, F.-K.; Brachwitz, F.; Freiburghaus, C.; Kolbe, E.; Martinez-Pinedo, G.; Rauscher, T.; Rembges, F.; Hix, W. R.; Liebendörfer, M.; Mezzacappa, A.; Kratz, K.-L.; Pfeiffer, B.; Langanke, K.; Nomoto, K.; Rosswog, S.; Schatz, H.; Wiescher, W. (2001). Element synthesis in stars, In: *Progress in Particle and Nuclear Physics*, 46, 5

Travaglio, C.; Gallino, R.; Arnone, E.; Cowan, J.; Jordan, F.; Sneden, C. (2004). Galactic Evolution of Sr, Y, And Zr: A Multiplicity of Nucleosynthetic Processes, In: *Astrophysical Journal*, 601, 864

Tur, C.; Heger, A.; Austin, S. M. (2007) On the Sensitivity of Massive Star Nucleosynthesis and Evolution to Solar Abundances and to Uncertainties in Helium-Burning Reaction Rates, In: *Astrophysical Journal*, 671, 821

Tur, C.; Heger, A.; Austin, S. M. (2009). Dependence of s-Process Nucleosynthesis in Massive Stars on Triple-Alpha and $^{12}C(\alpha,\gamma)^{16}O$ Reaction Rate Uncertainties, In: *Astrophysical Journal*, 702, 1068

Typel, S.; Baur, G. (2003). Theory of the Trojan-Horse method, In: *Annals of Physics*, 305, 228

Tytler, D.; O'Meara, J. M.; Suzuki, N.; Lubin, D. (2000). Review of Big Bang Nucleosynthesis and Primordial Abundances, In: *Physica Scripta*, 85, 12

Vangioni-Flam, E.; Cassé, M.; Audouze, J. (2000). Lithium-beryllium-boron: origin and evolution, In: *Phys. Rept.*, 333, 365

Wallerstein, G.; Iben, I.; Parker, P.; Boesgaard, A. M; Hale, G. M.; Champagne, A. E.; Barnes, C. A.; Käppeler, F.; Smith, V. V.; Hoffman, R. D.; Timmes, F. X.; Sneden, C.; Boyd, R. N.; Meyer, B. S.; Lambert, D. L. (1997). Synthesis of the elements in stars: forty years of progress, In: *Reviews of Modern Physics*, 69, 995

Woosley, S. E.; Heger, A.; Weaver, T. A. (2002). The evolution and explosion of massive stars, In: *Reviews of Modern Physics*, 74, 1015

6

Energy Generation Mechanisms in Stellar Interiors

İbrahim Küçük
Erciyes University, Science Faculty, Department of Astronomy and Space Sciences, Kayseri, Turkey

1. Introduction

It is well known that in the calculation of thermonuclear reaction rates in dense and cool plasma, the screening effects must be taken into account. The internal structure and evolution of stars can be obtained with the calculations of equation of state of gas structure, opacity and thermonuclear reaction rates. For high density it is necessary to calculate the equation of state by inserting deviations from ideal gas under high density and temprature. For low mass high density stars the effect of degenerate non-ideal interactions must be taken into account in equation of state. The rate of fusion nuclear reactions in stellar matter is important for the evolution of the star. In dense ionized matter, the rate of nuclear reactions is enhanced by screening effects (Alastuey and Jancovici 1978). One measure of nonideality in plasmas is the so-called coupling parameter Γ. In a plasma where particles have average distance $<r>$ from each other, we can define Γ as the ratio of average potential binding energy over mean kinetic energy kT,

$$\Gamma = \left(\frac{e^2 / <r>}{kT} \right) \tag{1}$$

Plasmas with Γ >>1 are strongly coupled, and those with Γ <<1 are weakly coupled (Basu et al. 1999).

2. Thermonuclear reaction rates

The reaction rate, r_{12}, between two nuclei,1 and 2, is given by,

$$r_{12} = \frac{N_1 N_2 \langle \sigma \upsilon \rangle}{(1 + \delta_{12})} cm^{-3} / sn \tag{2}$$

where N_1 and N_2 are the number densities of 1 and 2. For a gas of mass density, ρ, the number density, N_i , of the nucleide, *i*, is often expressed in terms of its mass fraction, X_i, by the relation

$$N_i = \rho N_A \frac{X_i}{A_i}\ cm^{-3} \tag{3}$$

where N_A is the Avogadro's number and A_i is the atomic mass of i in atomic mass units. In a stellar environment, the reaction rate per particle pair is calculated as

$$\langle \sigma \upsilon \rangle = \frac{(8/\pi)^{1/2}}{M^{1/2}(kT)^{3/2}} \int \sigma\, E \exp(-E/kT)\, dE \tag{4}$$

where σ is the energy-dependent reaction cross section.

At low energies, nonresonant charged particle interactions are dominated by the coulomb-barrier penetration factors. It is therefore convenient to factor out this energy dependence and express the cross section, $\sigma(E)$,by

$$\sigma = \frac{S(E)}{E} \exp[-(E_G/E)^{1/2}] \tag{5}$$

where the kinetic energy, E, of the center-of-mass system, the cross section factor S(E), the Gamow energy, E_G , is given by

$$E_G = (2\pi\alpha\ Z_0 Z_1)^2 (Mc^2/2) = \left[9.8948 Z_1 Z_2 A^{1/2} \right]^2 keV \tag{6}$$

(Lang 1999). Far from a nuclear resonance, the cross section factor, S(E), is a slowly varying function of E, and can be conveniently expressed as the first three terms of a Maclaurin Series in the center-of-momentum energy E. Thus,

$$S(E) = S(0)\left[1 + \frac{S'(0)}{S(0)} E + \frac{1}{2}\frac{S''(0)}{S(0)} E^2 \right] \tag{7}$$

where the prime indicates differentiation with respect to E. The values of S and associated derivates are quoted at zero energy. Substitution of Equations 5 and 7 into Equation 4 yields,

$$\langle \sigma \upsilon \rangle = \frac{(8/\pi)^{1/2}}{M^{1/2}(kT)^{3/2}} \int S(E) \exp\left(-E_G{}^{1/2}/E^{1/2} - E/kT\right) dE = \left(\frac{2}{M}\right)^{1/2} \frac{\Delta E_0}{(kT)^{3/2}} S_{eff} \exp(-\tau) \tag{8}$$

(Fowler, Caughlan, and Zimmerman 1967) where,

$$\Delta E_0 = 4(E_0 kT/3)^{1/2} \tag{9}$$

$$E_0 = \left[\pi\alpha\ Z_0 Z_1 kT (Mc^2/2)^{1/2} \right]^{2/3} = 1.2204 \left[Z_1^2 Z_2^2 A T_6^2 \right]^{1/3} keV \tag{10}$$

$$\tau = 3E_0/kT = 3\left[\pi\alpha\ Z_0 Z_1 (Mc^2/2kT)^{1/2} \right]^{2/3} = 42.487 \left[Z_1^2 Z_2^2 A \right]^{1/3} T_6^{-1/3} \tag{11}$$

$$S_{eff} = S(0)\left[1+\frac{5}{12\tau}+\frac{S'(0)}{S(0)}\left(E_0+\frac{35}{36}kT\right)+\frac{1}{2}\frac{S''(0)}{S(0)}\left(E_0{}^2+\frac{89}{36}E_0kT\right)\right] \text{ keV-barn} \quad (12)$$

Putting these equations into Equation 8 yields,

$$\langle\sigma\upsilon\rangle = \left\{1.3006\times10^{-15}\left(Z_0Z_1/A\right)^{1/3}S_{eff}\right\}T_6{}^{-2/3}\exp(-\tau)\ cm^3s^{-1} \quad (13)$$

This is the obtained relation of cross section ($T_6 = T/10^6$). The mean lifetime, $\tau_2(1)$, of nucleus 1 for destruction by nucleus 2 is given by the relation,

$$\lambda_2(1) = \frac{1}{\tau_2(1)} = N_2\langle\sigma\upsilon\rangle = \rho\, N_A\frac{X_2}{A_2}\langle\sigma\upsilon\rangle\quad sn^{-1} \quad (14)$$

where $\lambda_2(1)$, is the decay rate of 1 for interaction with 2 (Fowler, Caughlan, and Zimmerman 1967). Putting equations (12), (13) and (5) into equation (2) yields for unscreened reaction rate

$$r_{unscreened} = KX_1X_2g\rho\exp(-\tau)\,T_6^{-2/3} \quad (15)$$

where g is:

$$g = \left[1+\frac{5}{12\tau}+\frac{S'(0)}{S(0)}\left(E_0+\frac{35}{kT}\right)+\frac{1}{2}\frac{S''(0)}{S(0)}\left(E_0+\frac{89}{36}E_0kT\right)\right] \quad (16)$$

and K is given as,

$$K = 7.83x10^8\left(\frac{N_A}{A_1A_2}\right)\left(\frac{Z_1Z_2}{A}\right)^{1/3}S(0). \quad (17)$$

3. The electron screening effect and enhancement factors

At the high temperatures in stellar interiors all the atoms are ionized and the gas density ρ is high. The average distance between nucleus and neighbouring electrons is small. Each nucleus is then completely screened by a spherically symmetric negative charge cloud. The radius of this charge cloud is of the same order as the interparticle distance or larger, depending on the ratio of coulomb repulsion between neighbouring charges to the mean thermal energy. Hence, when two nuclei approach each other in a collision each of them carries its screening affects the interaction energy between the nuclei. The nucleus is surrounded in its immediate vicinity by electrons only the nuclei staying outside a sphere containing nearly electrons which effectively screen the nucleus (Salpeter 1954; hereafter S54). As explained by S54, the rate of a fusion of two nuclei charges Z_1 and Z_2 is increased by factor

$$f = \exp\left[-\frac{U(0)}{kT}\right] \quad (18)$$

$$= \exp \Lambda$$

where $\Lambda = U(0)/kT$ is the natural strength screening parameter.

$$\left(-\frac{U(0)}{kT}\right)_{ws} = 0.188 Z_1 Z_2 \frac{\rho^{\frac{1}{2}}}{T_6^{\frac{3}{2}}} \zeta \tag{19}$$

This is the so-called standard assumption or weak screening which was orginally presented by S54 (Dzitko et al. 1995). In Table 1 the results of screening strength parameter and enhancement factors calculated by using Küçük, Kızıloğlu and Civelek 1998 stellar evolutionary program are listed. The input physics is as follows: OPAL opacity has been used and the ratio of mixing length to the scale height is taken to be α=1.50. The accepted chemical composition is X=0.699, and Z=0.019 (Küçük, Kızıloğlu and Civelek 1998).

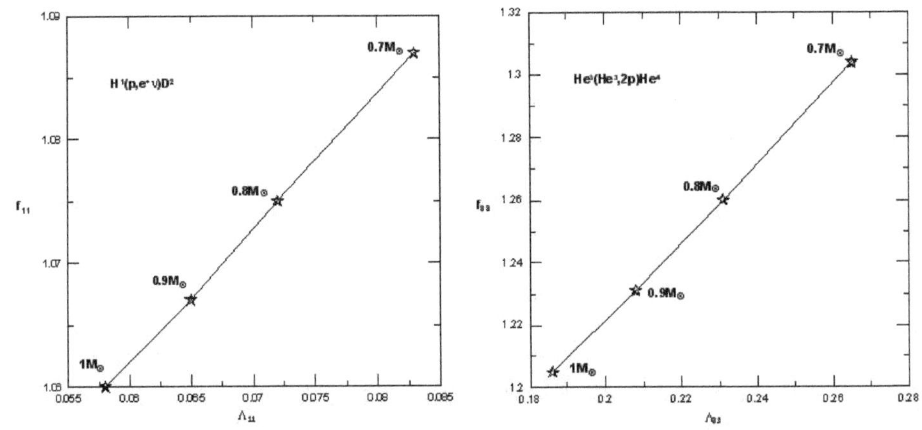

Fig. 1. Variation of the screening factors (S54) for $H^1(p,e^+\upsilon)D^2$ reaction with the screening strength parameter Λ (left). Variation of the screening factors (S54) for $He^3(He^3,2p)He^4$ reaction with the screening strength parameter (right)

Mass	$T_c \times 10^7$ (K)	ρ_c (gr/cm^3)	$\Lambda_{1,1}$	$\Lambda_{3,3}$	$\Lambda_{3,4}$	$f_{1,1}$	$f_{3,3}$	$f_{3,4}$
0.7M☉	10.26	76.99	0.083	0.265	0.265	1.087	1.304	1.304
0.8M☉	10.85	68.59	0.072	0.231	0.231	1.075	1.260	1.260
0.9M☉	12.17	78.03	0.065	0.208	0.208	1.067	1.231	1.231
1.0M☉	13.02	76.79	0.058	0.186	0.186	1.060	1.205	1.205

Table 1. Calculated screening strength parameter and enhancement factors.

Table 2 gives enhancement factors calculated with the various prescriptions for the main reactions involved in hydrogen burning: Küçük and Çalışkan 2010 (hereafter KÇ2010), S54, Salpeter and Van Horn 1969 (hereafter SVH), Graboske et al. 1973 (hereafter GDGC), Mitler 1977 hereafterML) and Tsytovich 2000 (hereafter TS).

	$f_{1,1}$	$f_{3,3}$	$f_{3,4}$
KÇ2010	1.046	1.166	1.166
S54	1.050	1.215	1.215
SVH	1.045	1.186	1.186
GDGC	1.050	1.115	1.115
ML	1.045	1.176	1.176
TS	0.950	0.830	0.827

Table 2. Enhancement factors.(T_c~15.54x10^6 K, ρ_c~160.8 gr/cm^3, X_c=0.35 and Y_c=0.62) (Dzitko, H., et al. 1995).

$$r_{3,3} = 2.671 \times 10^5 \, , E_{3,3} = 5.503 \, .$$

Thermonuclear reaction rates in stars, are calculated by multiplying the screened reaction rate enhancement factor, $f_{i,j}$ (i,j=1,2,3,...), with unscreened reaction.

$$r_{screened} = f_{i,j} \times r_{unscreened} \tag{20}$$

Then energy generation rate in stars given by,

$$\varepsilon_{nuc} = Q \times r_{screened} \tag{21}$$

In this study, various data (Weiss et al., 2001, Morel et.al. 1999, Bahcall 1989) are used for the calculation of thermonuclear reaction rates. The results for some reaction rates are as follows:

For $H^1\left(p,e^{+}\upsilon\right)D^2$ reaction,

$$r_{1,1} = 1.15 \times 10^{11} X_1^2 f_{1,1} g_{1,1} \rho \exp\left(-\frac{33.81}{T_6^{1/3}}\right) T_6^{-2/3}$$

$$g_{1,1} = 1 + 0.0123 T_6^{1/3} + 0.00114 T_6^{2/3} + 9.8 \times 10^{-4} T_6 \tag{22}$$

$$Q_{1,1} = 1.442 \quad MeV \, ,$$

for $D^2(p,\gamma)He^3$ reaction,

$$r_{2,1} = 5.305 \times 10^{28} X_1 X_2 f_{2,1} g_{2,1} \rho \exp\left(-\frac{37.21}{T_6^{1/3}}\right) T_6^{-2/3}$$

$$g_{2,1} = 1 + 0.0112 T_6^{1/3} + 0.299 T_6^{2/3} + 0.00234 T_6$$

$$Q_{2,1} = 5.494 \quad MeV \tag{23}$$

for $D^2(d,n)He^3$ reaction,

$$r_{2,2} = 3.154 \times 10^{33} X_2^2 f_{2,2} g_{2,2} \rho \exp\left(-\frac{42.58}{T_6^{1/3}}\right) T_6^{-2/3}$$

$$g_{2,2} = 1$$

$$Q_{2,2} = 3.269 \quad MeV \tag{24}$$

for $He^3(He^3,2p)He^4$ reaction,

$$r_{3,3} = 1.859 \times 10^{35} X_3^2 f_{3,3} g_{3,3} \rho \exp\left(-\frac{122.76}{T_6^{1/3}}\right) T_6^{-2/3}$$

$$g_{3,3} = 1 + 3.39 \times 10^{-3} T_6^{1/3}$$

$$Q_{3,3} = 12.860 \quad MeV \tag{25}$$

for $He^3(\alpha,\gamma)Be^7$ reaction

$$r_{3,4} = 2.795 \times 10^{31} X_3 X_4 f_{3,4} g_{3,4} \rho \exp\left(-\frac{128.26}{T_6^{1/3}}\right) T_6^{-2/3}$$

$$g_{3,4} = 1 + 3.25 \times 10^{-3} T_6^{1/3} - 3.547 \times 10^{-3} T_6^{2/3} - 8.07 \times 10^{-5} T_6$$

$$Q_{3,4} = 1.588 \quad MeV \tag{26}$$

for $Be^7(p,\gamma)B^8$ reaction

$$r_{7,1} = 2.32 \times 10^{30} X_7 X_1 f_{7,1} g_{7,1} \rho \exp\left(-\frac{102.62}{T_6^{1/3}}\right) T_6^{-2/3}$$

$$g_{7,1} = 1 + 4.06 \times 10^{-3} T_6^{1/3}$$

$$Q_{7,1} = 0.137 \quad MeV \tag{27}$$

for $C^{12}(p,\gamma)N^{13}$ reaction

$$r_{1,12} = 1.089 \times 10^{33} X_1 X_{12} f_{1,12} g_{1,12} \rho \exp\left(-\frac{136.9}{T_6^{1/3}}\right) T_6^{-2/3}$$

$$g_{1,12} = 1 + 3.04 \times 10^{-3} T_6^{1/3} + 6.41 \times 10^{-3} T_6^{2/3} + 1.36 \times 10^{-4} T_6$$

$$Q_{1,12} = 1.944 \quad MeV \tag{28}$$

for $N^{14}(p,\gamma)O^{15}$ reaction

$$r_{1,14} = 2.088 \times 10^{32} X_1 X_{14} f_{1,14} g_{1,14} \rho \exp\left(-\frac{152.28}{T_6^{1/3}}\right) T_6^{-2/3}$$

$$g_{1,14} = 1 + 2.74 \times 10^{-3} T_6^{1/3} - 8.08 \times 10^{-3} T_6^{2/3} - 1.547 \times 10^{-4} T_6$$

$$Q_{1,14} = 7.297 \quad MeV \tag{29}$$

for $O^{16}(p,\gamma)F^{17}$ reaction

$$r_{1,16} = 5.54 \times 10^{33} X_1 X_{16} f_{1,16} g_{1,16} \rho \exp\left(-\frac{166.92}{T_6^{1/3}}\right) T_6^{-2/3}$$

$$g_{1,16} = 1 + 2.491 \times 10^{-3} T_6^{1/3} - 1.18 \times 10^{-2} T_6^{2/3} - 2.07 \times 10^{-4} T_6$$

$$Q_{1,16} = 0.60 \quad MeV \tag{30}$$

The star's thermonuclear reaction rate and energy generation can be obtained from these relations. For example; $H^1(p,e^+\upsilon)D^2$ and $He^3(He^3,2p)He^4$ reactions occured at the center, when $0.7M_\odot$ model reaches its ZAMS. The reaction rates and energy generations are found to be $r_{1,1} = 2.154 \times 10^5$, $E_{1,1} = 2.305$ and $r_{3,3} = 2.013 \times 10^5$, $E_{3,3} = 4.147$, respectively. Similiarly, for $0.8M_\odot$ the results are $r_{1,1} = 3.262 \times 10^5$, $E_{1,1} = 3.490$ and $r_{3,3} = 2.671 \times 10^5$, $E_{3,3} = 5.503$.

4. Conclusion

During the lifetime of a star its composition, the relative abundances of different elements and isotopes, change basically because of nuclear reactions in the deep interior. In the evolution of 0.7, 0.8, 0.9 and $1M_\odot$ stars, the hydrogen burning begins and gets equlibrium with increasing central pressure and temperature. Equlibrium time increases when going lower masses. These are $\sim 2 \times 10^{10}$ and $\sim 5 \times 10^9$ years for $0.7M_\odot$, and $1M_\odot$, respectively. It is well known that in the calculation of nuclear reaction rates in dense, relatively cool stellar plasma, the screening of the coulomb interaction between the reacting nuclei by the surrounding ions and electrons must be taken into account. When screening factors calculated with different methods are considered one can see the differences. When the energies released due to nuclear reaction rates are calculated for ^1H - ^1H and ^3He - ^3He reactions and for each mass, respectively, it is seen that, the released energy increases with increasing mass. As a conclusion reaction rates and electron screening factors which are added to the evolutionary programs in details, causes some shifts.

5. References

Adelberger, E. G., and 29 coauthors. 1998, RvMP, Volume 70, Issue 4, pp. 1265-1291.
Alastuey, A. and Jancovici, B. 1978, ApJ, 226, 1034-1040.
Bahcall, J. N. 1989, *Neutrino Astrophysics* (Cambridge University Press).
Basu, S., et al. 1999, ApJ, 518, 985-993.
Caughlan, G. R. and Fowler, W. A. 1988, Atomic Data Nucl. Data Tables, Vol 40, 283-34.
Dzitko, H., Turck-Chieze, S., Delbourgo-Salvador, P., Lagrange, C. 1995, ApJ, 447, 428-442.
Fowler, W. A., Caughlan, G. R. and Zimmerman, B. A. 1967, A&A, 5, 525.
Graboske, H. C., Dewitt, H. E., Grossman, A. S., Cooper, M. S. 1973, *ApJ*, 181, 457–474.
Küçük, 'İ., Kızıloğlu, N., Civelek, R. 1998, *Ap&SS*, 3, 279.
Küçük, I. and Çalışkan Ş., 2010, *JAA*, 31, 135-145
Lang, R. K. 1999, *Astrophysical Formulae Volume I Radiation, Gas Processes and High Energy Astrophysics*, Astronomy and Astrophysics Library (Printed in Germany).
Mitler, H. E. 1977, *ApJ*, 212, 513–532.
Morel, P., Pichon, B., Provost, J., Berthomieu, G. 1999, *A&A*, 350, 275–285.
Salpeter, E. E. 1954, *Australian J. Phys.*, 7, 373.
Salpeter, E. E., van Horn, H. M. 1969, *ApJ*, 155, 183.
Siess, L., Dufour, E., Forestini, M. 2000, *A&A*, 358, 593–599.
Siess, L. 2007, *A&A*, 476, 893.
Tsytovich, V. N. 2000, *A&AL*, 356, L57–L61.
Weiss, A., Flaskamp, M., Tsytovich, V. N. 2001, *A&A*, 371, 1123–1127.

7

The Lane-Emden-Fowler Equation and Its Generalizations – Lie Symmetry Analysis

Chaudry Masood Khalique
International Institute for Symmetry Analysis and Mathematical Modelling, Department of Mathematical Sciences, North-West University, Mmabatho Republic of South Africa

1. Introduction

In the study of stellar structure the Lane-Emden equation (1; 2)

$$\frac{d^2y}{dx^2} + \frac{2}{x}\frac{dy}{dx} + y^r = 0, \tag{1}$$

where r is a constant, models the thermal behaviour of a spherical cloud of gas acting under the mutual attraction of its molecules and subject to the classical laws of thermodynamics. This equation was proposed by Lane (1) (see also (3)) and studied in detail by Emden (2). Fowler (4; 5) considered a generalization of Eq. (1), called Emden-Fowler equation (6), where the last term is replaced by $x^{\nu-1}y^r$.

The Lane-Emden equation (1) also models the equilibria of nonrotating fluids in which internal pressure balances self-gravity. When spherically symmetric solutions of Eq. (1) appeared in (7), they got the attention of astrophysicists. In the latter half of the twentieth century, some interesting applications of the isothermal solution (singular isothermal sphere) and its nonsingular modifications were used in the structures of collisionless systems such as globular clusters and early-type galaxies (8; 9).

The work of Emden (2) also got the attention of physicists outside the field of astrophysics who investigated the generalized polytropic forms of the Lane-Emden equation (1) for specific polytropic indices r. Some singular solutions for $r = 3$ were produced by Fowler (4; 5) and the Emden-Fowler equation in the literature was established, while the works of Thomas (10) and Fermi (11) resulted in the Thomas-Fermi equation, used in atomic theory. Both of these equations, even today, are being investigated by physicists and mathematicians. Other applications of Eq. (1) can be found in the works of Meerson et al (12), Gnutzmann and Ritschel (13), and Bahcall (14; 15).

Many methods, including numerical and perturbation, have been used to solve Eq. (1). The reader is referred to the works of Horedt (16; 17), Bender (18) and Lema (19; 20), Roxbough and Stocken (21), Adomian et al (22), Shawagfeh (23), Burt (24), Wazwaz (25) and Liao (26) for a sample. Exact solutions of Eq. (1) for $r = 0, 1$ and 5 have been obtained (see for example Chandrasekhar (7), Davis (27), Datta (28) and Wrubel (29)). Usually, for $r = 5$, only a one-parameter family of solutions is presented. A more general form of (1), in which the coefficient of y' is considered an arbitrary function of x, was investigated for first integrals by Leach (30).

Many problems in mathematical physics and astrophysics can be formulated by the generalized Lane-Emden equation

$$\frac{d^2y}{dx^2} + \frac{n}{x}\frac{dy}{dx} + f(y) = 0, \tag{2}$$

where n is a real constant and $f(y)$ is an arbitrary function of y. For $n = 2$ the approximate analytical solutions to the Eq. (2) were studied by Wazwaz (25) and Dehghan and Shakeri (31).

Another form of $f(y)$ is given by

$$f(y) = (y^2 - C)^{3/2}. \tag{3}$$

Inserting (3) into Eq. (1) gives us the "white-dwarf" equation introduced by Chandrasekhar (7) in his study of the gravitational potential of degenerate white-dwarf stars. In fact, when $C = 0$ this equation reduces to Lane-Emden equation with index $r = 3$.

Another nonlinear form of $f(y)$ is the exponential function

$$f(y) = e^y. \tag{4}$$

Substituting (4) into Eq. (1) results in a model that describes isothermal gas spheres where the temperature remains constant.

Equation (1) with

$$f(y) = e^{-y}$$

gives a model that appears in the theory of thermionic currents when one seeks to determine the density and electric force of an electron gas in the neighbourhood of a hot body in thermal equilibrium was thoroughly investigated by Richardson (32).

Furthermore, the Eq. (1) appears in eight additional cases for the function $f(y)$. The interested reader is referred to Davis (27) for more detail.

The equation

$$\frac{d^2y}{dx^2} + \frac{2}{x}\frac{dy}{dx} + e^{\beta y} = 0, \tag{5}$$

where β is a constant, has also been studied by Emden (2). In a recent work (33) an approximate implicit solution has been obtained for Eq. (5) with $\beta = 1$.

Furthermore, more general Emden-type equations were considered in the works (34–38). See also the review paper by Wong (39), which contains more than 140 references on the topic.

The so-called generalized Lane-Emden equation of the first kind

$$x\frac{d^2y}{dx^2} + \alpha\frac{dy}{dx} + \beta x^{\nu} y^n = 0, \tag{6}$$

and generalized Lane-Emden equation of the second kind

$$x\frac{d^2y}{dx^2} + \alpha\frac{dy}{dx} + \beta x^{\nu} e^{ny} = 0, \tag{7}$$

where α, β, ν and n are constants, have been recently studied in (40; 41). In Goenner (41), the author uncovered symmetries of Eq. (6) to explain integrability of (6) for certain values of the parameters considered in Goenner and Havas (40). Recently, the integrability of the generalized Lane-Emden equations of the first and second kinds has been discussed in Muatjetjeja and Khalique (42).

In this chapter, firstly, a generalized Lane-Emden-Fowler type equation

$$x\frac{d^2y}{dx^2}+n\frac{dy}{dx}+x^{\nu}f(y)=0, \tag{8}$$

where n and ν are real constants and $f(y)$ is an arbitrary function of y will be studied. We perform the Lie and Noether symmetry analysis of this problem. It should be noted that Eq. (8) for the power function $F(y)=y^r$ is related to the Emden-Fowler equation $y''+p(X)y^r=0$ by means of the transformation on the independent variable $X=x^{1-n}$, $n\neq 1$ and $X=\ln x$, $n=1$.

Secondly, we consider a generalized coupled Lane-Emden system, which occurs in the modelling of several physical phenomena such as pattern formation, population evolution and chemical reactions. We perform Noether symmetry classification of this system and compute the Noether operators corresponding to the standard Lagrangian. In addition the first integrals for the Lane-Emden system will be constructed with respect to Noether operators.

2. Lie point symmetry classification of (8)

We start by determining the equivalence transformations of Eq. (8). We recall (43) that an equivalence transformation

$$\bar{x}=\bar{x}(x,y),\ \bar{y}=\bar{y}(x,y)$$

is a nondegenerate change of variables such that the family of Eqs. (8) remains invariant, i.e., Eq. (8) becomes

$$\bar{x}\frac{d^2\bar{y}}{d\bar{x}^2}+n\frac{d\bar{y}}{d\bar{x}}+\bar{x}^{\nu}\bar{f}(\bar{y})=0$$

with $\bar{f}$ depending on $\bar{y}$. Equivalence transformations are essential for simplifying the determining equation and for obtaining disjoint classes.

For Eq. (8) the equivalence transformations are

$$\begin{aligned}\bar{x}&=e^{a_2}x,\\ \bar{y}&=e^{a_3}y+a_1,\\ \bar{f}&=e^{a_3-(1+\nu)a_2}f,\end{aligned} \tag{9}$$

where a_1, a_2 and a_3 are constants. For details of computations see (44).

If X, given by

$$X=\xi(x,y)\frac{\partial}{\partial x}+\eta(x,y)\frac{\partial}{\partial y}$$

is an admitted generator of a symmetry group of Eq. (8), then

$$X^{[2]}\left(x\frac{d^2y}{dx^2}+n\frac{dy}{dx}+x^{\nu}f(y)\right)\bigg|_{(8)}=0, \tag{10}$$

where $X^{[2]}$ is the second prolongation of X, gives the determining equations for the symmetry. This gives rise to

$$\xi=b(x),$$
$$\eta=c(x)y+d(x),$$

$$x^{\nu-1}(cy+d)f'(y)+\left[x^{\nu-1}(2b'-c)+(\nu-1)x^{\nu-2}b\right]f(y)$$
$$+\left(\frac{n}{x}c'+c''\right)y+\left(\frac{n}{x}d'+d''\right)=0. \tag{11}$$

If f is an arbitrary function, the above system yields $\xi=0$, $\eta=0$, meaning that the principal Lie algebra of Eq. (8) is trivial.

The function f depends upon y only. Thus Eq. (11) only holds if its coefficients identically vanish or they are proportional to a function $\alpha=\alpha(x)$, i.e.,

$$c=r\alpha,\ \ d=q\alpha,\ \ 2b'-c+(\nu-1)x^{-1}b=p\alpha,$$
$$c''x^{-\nu+1}+nx^{-\nu}c'=h\alpha,\ \ d''x^{-\nu+1}+nx^{-\nu}d'=g\alpha, \tag{12}$$

where r, q, p, h and g are constants. Thus Eq. (11) becomes

$$(ru+q)F'(u)+pF(u)+hu+g=0, \tag{13}$$

which is our classifying relation. This relation is invariant under the equivalence transformations (9) if

$$\bar{r}=r,\ \bar{q}=(ra_1+q)e^{-a_3},\ \bar{p}=p,$$
$$\bar{h}=he^{(\nu+1)a_2},\ \bar{g}=e^{-a_3+(\nu+1)a_2}(ha_1+g). \tag{14}$$

The relations in (14) are used to find the non-equivalent forms of f and this leads to the following eight cases.

Case 1. $n\neq(1-\nu)/2$, $f(y)$ arbitrary but not of the form contained in Cases 3, 4, 5 and 6.

No Lie point symmetry exits in this case.

Case 2. $n=(1-\nu)/2$, $f(y)$ arbitrary but not of the form contained in Cases 4, 5 and 6.

We obtain one Lie point symmetry

$$X=x^{(1-\nu)/2}\frac{\partial}{\partial x} \tag{15}$$

for the corresponding Eq. (8).

Case 3. $f(y)$ is linear in y.

This case is well known and the corresponding Eq. (8) has $sl(3, \Re)$ symmetry algebra. (See, for example, (45)).

Case 4. $f(y) = K - \delta y^2/2$, where $\delta = \pm 1$ and K is a constant.

Here we have six subcases:

4.1. $n = 2\nu + 3$, $K = 0$. The corresponding Eq. (8) admits a single Lie point symmetry

$$X_1 = x\frac{\partial}{\partial x} - (\nu + 1)y\frac{\partial}{\partial y}. \tag{16}$$

Note that this is subsumed in Case 5.1 below.

4.2. $n = 12\nu + 13$, $K = 0$. Here the corresponding Eq. (8) admits the same symmetry as in Case 4.1.

4.3. $n = (\nu + 4)/3$, $K = 0$. In this subcase the corresponding Eq. (8) admits a two-dimensional symmetry Lie algebra which is spanned by the operators (16) and

$$X_2 = x^{(2-\nu)/3}\frac{\partial}{\partial x} - \frac{\nu + 1}{3}x^{-(\nu+1)/3}y\frac{\partial}{\partial y}.$$

Note that this is contained in Case 5.2 below.

4.4. $n = 7\nu + 8$, $K = 0$. The corresponding Eq. (8) admits the symmetry operator (16) and in addition, the symmetry operator

$$X_2 = x^{\nu+2}\frac{\partial}{\partial x} - \left[3(\nu + 1)x^{\nu+1}y + \frac{24(\nu + 1)^3}{\delta}\right]\frac{\partial}{\partial y}.$$

4.5. $n = (7\nu + 13)/6$, $K = 0$. In this subcase the corresponding Eq. (8) admits two Lie point symmetries, namely, the symmetry given by (16) and the symmetry

$$X_2 = x^{(5-\nu)/6}\frac{\partial}{\partial x} - \left[\frac{2}{3}(\nu + 1)x^{-(\nu+1)/6}y - \frac{(\nu + 1)^3}{9\delta}x^{-7(\nu+1)/6}\right]\frac{\partial}{\partial y}.$$

4.6. $n = (1 - \nu)/2$, $K = 0$. The corresponding Eq. (8) admits two Lie point symmetries and they are (15) and (16).

Case 5. $f(y) = -\delta_1/\sigma - y\delta_2/(\sigma + 1) + Ky^{-\sigma}$, where $\delta_1, \delta_2 = 0, \pm 1$, $\sigma \neq -1, 0$ and K is a constant.

Three subcases arise:

5.1. $n = \dfrac{\sigma - 2\nu - 1}{\sigma + 1}$, $\sigma \neq 3$, $\delta_1, \delta_2 = 0$. In this subcase we have one Lie point symmetry generator

$$X_1 = x\frac{\partial}{\partial x} + \frac{\nu + 1}{\sigma + 1}y\frac{\partial}{\partial y} \tag{17}$$

admitted by the corresponding Eq. (8).

5.2. $n = \dfrac{\sigma - \nu - 2}{\sigma - 1}$, $\sigma \neq 3$, $\delta_1, \delta_2 = 0$. Here the corresponding Eq. (8) admits a two-dimensional symmetry Lie algebra spanned by the operators (17) and

$$X_2 = x^{\frac{\sigma+\nu}{\sigma-1}} \frac{\partial}{\partial x} + \frac{\nu+1}{\sigma-1} x^{\frac{\nu+1}{\sigma-1}} y \frac{\partial}{\partial y}. \tag{18}$$

5.3. $n = \dfrac{1-\nu}{2}$, (This subcase corresponds to $\sigma = 3$), $\delta_1, \delta_2 = 0$. The corresponding Eq. (8) in this case admits three Lie point symmetry generators and these are given by (17), (18) with $\sigma = 3$ and (15).

Case 6. $f(y) = Ke^{-\delta_1 y} + \delta_2 y + \delta_3$, where $\delta_1 = \pm 1, \delta_2, \delta_3 = 0, \pm 1$ and K is a constant.

We have three subcases.

6.1. For all values of $n \neq 1, (1-\nu)/2, \delta_2, \delta_3 = 0$ one Lie point symmetry generator

$$X_1 = x\frac{\partial}{\partial x} + \frac{\nu+1}{\delta_1}\frac{\partial}{\partial y} \tag{19}$$

is admitted by the corresponding Eq. (8).

6.2. $n = 1, \delta_2, \delta_3 = 0$. In this subcase the corresponding Eq. (8) admits the Lie point symmetry (19) and in addition the Lie point symmetry

$$X_2 = x \ln x \frac{\partial}{\partial x} + \frac{1}{\delta_1}[2 + (\nu+1)\ln x]\frac{\partial}{\partial y}.$$

6.3. $n = (1-\nu)/2, \delta_2, \delta_3 = 0$. The corresponding Eq. (8) admits two Lie point symmetries. These symmetries are given by (15) and (19).

Case 7. $f(y) = -\delta_1 \ln y - \delta_2 y + K$, where $\delta_1, \delta_2 = 0, \pm 1$ and K is a constant.

This reduces to Case 2.

Case 8. $f(y) = -\delta_1 y \ln y + Ky + \delta_2$, where $\delta_1, \delta_2 = 0, \pm 1$ and K is a constant.

This also reduces to Case 2.

2.1 Integration of (8) for different fs

The main purpose for calculating symmetries is to use them to solve or reduce the order of differential equations. Here we use the symmetries calculated above to integrate Eq. (8) for three functions f. Other cases can be dealt in a similar manner. We recall that for any two-dimensional Lie algebra with symmetries G_1 and G_2 satisfying the Lie bracket relationship $[G_1, G_2] = \lambda G_1$, for some constant λ, the usual reduction of order is through the normal subgroup G_1 (46). We first consider Case 4.4. The corresponding Eq. (8) admits the two symmetries

$$X_1 = x\frac{\partial}{\partial x} - (\nu+1)y\frac{\partial}{\partial y}, \; X_2 = x^{\nu+2}\frac{\partial}{\partial x} - \left[3(\nu+1)x^{\nu+1}y + \frac{24(\nu+1)^3}{\delta}\right]\frac{\partial}{\partial y}.$$

Since $[X_1, X_2] = (\nu+1)X_2$, we may use X_2 to reduce the corresponding Eq. (8) to quadratures. The invariants of X_2 are found from

$$\frac{dx}{x^{\nu+2}} = \frac{dy}{-[3(\nu+1)x^{\nu+1}y + 24(\nu+1)^3/\delta]} = \frac{dy'}{-[3(\nu+1)^2x^{\nu}y + (4\nu+5)x^{\nu+1}y']}$$

and are

$$t = x^{3\nu+3}y + \frac{12}{\delta}(\nu+1)^2x^{2\nu+2}, \; s = x^{4\nu+5}y' + 3(\nu+1)x^{4\nu+4}y + \frac{24}{\delta}(\nu+1)^3x^{3\nu+3}.$$

This leads to the first-order equation

$$\frac{ds}{dt} = \frac{\delta t^2}{2s}$$

which can be immediately integrated to give

$$s^2 = \frac{\delta}{3}t^3 + C_1,$$

where C_1 is an arbitrary constant of integration. Reverting to the x and y variables we obtain a first-order differential equation whose solution can be written as

$$y = x^{-3(\nu+1)}t - \frac{12}{\delta}(\nu+1)^2x^{-(\nu+1)},$$

where t is given by

$$\int \frac{dt}{\pm\sqrt{C_1 + \delta t^3/3}} = -\frac{1}{\nu+1}x^{-(\nu+1)} + C_2,$$

in which C_1 and C_2 are integration constants. Hence we have quadrature of Eq. (8) for given f.

We now consider Case 5.2. The two symmetries admitted by the corresponding Eq. (8) are

$$X_1 = x\frac{\partial}{\partial x} + \frac{\nu+1}{\sigma+1}y\frac{\partial}{\partial y}, \; X_2 = x^{\frac{\sigma+\nu}{\sigma-1}}\frac{\partial}{\partial x} + \frac{\nu+1}{\sigma-1}x^{\frac{\nu+1}{\sigma-1}}y\frac{\partial}{\partial y}$$

with $[X_1, X_2] = \dfrac{(\nu+1)}{\sigma-1}X_2$. Following the above procedure we find that the solution of the corresponding Eq. (8) is

$$y = tx^{(\nu+1)/(\sigma-1)},$$

where t is defined by

$$\int \frac{dt}{\pm\sqrt{C_1 + 2Kt^{1-\sigma}/(1-\sigma)}} = \frac{1-\sigma}{1+\nu}x^{(1+\nu)/(1-\sigma)} + C_2,$$

in which C_1 and C_2 are arbitrary constants of integration.

Finally for Case 6.2 the corresponding Eq. (8) admits the two symmetries

$$X_1 = x\frac{\partial}{\partial x} + \frac{\nu+1}{\delta_1}\frac{\partial}{\partial y}, \; X_2 = x\ln x\frac{\partial}{\partial x} + \frac{1}{\delta_1}[2+(\nu+1)\ln x]\frac{\partial}{\partial y}$$

with $[X_1, X_2] = X_1$. In this case the solution of the corresponding Eq. (8) is

$$y = \frac{1}{\delta_1}\ln\left(\frac{x^{\nu+1}}{t}\right),$$

where t is given by

$$\int \frac{dt}{\pm\sqrt{2\delta_1 K t^3 + t^2[(\nu+1)^2 - 2\delta_1 C_1]}} = \ln x + C_2,$$

in which C_1 and C_2 are arbitrary integration constants.

3. Noether classification and integration of (8) for different fs

In this section we perform a Nother point symmetry classification of Eq. (8) with respect to the standard Lagrangian. We then obtain first integrals of the various cases, which admit Noether point symmetries and reduce the corresponding equations to quadratures.

It can easily be verified that the standard Lagrangian of Eq. (8) is

$$L = \frac{1}{2}x^n y'^2 - x^{n+\nu-1}\int f(y)dy. \tag{20}$$

The determining equation (see (47)) for the Noether point symmetries corresponding to L in (20) is

$$X^{[1]}(L) + LD(\xi) = D(B), \tag{21}$$

where X given by

$$X = \xi(x,y)\frac{\partial}{\partial x} + \eta(x,y)\frac{\partial}{\partial y} \tag{22}$$

is the generator of Noether symmetry and $B(x,y)$ is the gauge term and D is the total differentiation operator defined by (48)

$$D = \frac{\partial}{\partial x} + y'\frac{\partial}{\partial y} + y''\frac{\partial}{\partial y'} + \cdots. \tag{23}$$

The solution of Eq. (21) results in

$$\xi = a(x),$$

$$\eta = \frac{1}{2}[a' - nx^{-1}a]y + b(x), \tag{24}$$

$$B = \frac{1}{4}x^n\left[a'' - n\left(\frac{a}{x}\right)'\right]y^2 + b'x^n y + c(x), \tag{25}$$

$$[-(n+\nu-1)x^{n+\nu-2}a - a'x^{n+\nu-1}]\int f(y)dy + [-\frac{1}{2}x^{n+\nu-1}a'y$$
$$+\frac{1}{2}nx^{n+\nu-2}ay - x^{n+\nu-1}b]f(y) = \frac{1}{4}a'''x^n y^2 + \frac{1}{2}nx^{n-2}a'y^2$$
$$-\frac{1}{2}nx^{n-3}ay^2 - \frac{1}{4}n^2x^{n-1}\left(\frac{a}{x}\right)' y^2 + b''x^n y + b'nx^{n-1}y + c'(x). \tag{26}$$

The analysis of Eq. (26) leads to the following eight cases:

Case 1. $n \neq \frac{1-\nu}{2}$, $f(y)$ arbitrary but not of the form contained in cases 3, 4, 5 and 6.

We find that $\xi = 0$, $\eta = 0$, $B =$ constant and we conclude that there is no Noether point symmetry.

Noether point symmetries exist in the following cases.

Case 2. $n = \frac{1-\nu}{2}$, $f(y)$ arbitrary.

We obtain $\xi = x^{\frac{1-\nu}{2}}$, $\eta = 0$ and $B =$ constant. Therefore we have a single Noether symmetry generator $X = x^{\frac{1-\nu}{2}}\partial/\partial x$. For this case the integration is trivial even without a Noether symmetry. The Noetherian first integral (47) is

$$I = \frac{1}{2}x^{1-\nu}y'^2 + \int f(y)dy$$

from which, setting $I = C$, one gets quadrature.

Case 3. $f(y)$ is linear in y.

We have five Noether point symmetries associated with the standard Lagrangian for the corresponding differential equation (8) and $sl(3,\Re)$ symmetry algebra. This case is well-known, see, e.g., (45).

Case 4. $f = \alpha y^2 + \beta y + \gamma$, $\alpha \neq 0$

There are four subcases. They are as follows:

4.1. If $n = 2\nu + 3$, $\beta = 0$ and $\gamma = 0$, we obtain $\xi = x$, $\eta = -(\nu+1)y$ and $B =$ constant. This is contained in Case 5.1 below.

4.2. If $n = 2\nu + 3$, $\nu \neq -1$, $\beta^2 = 4\alpha\gamma$, we get $\xi = x$, $\eta = -(\nu+1)(y + \beta/2\alpha)$ and $B = \frac{\beta\gamma}{6\alpha}x^{3\nu+3}$. We have

$$X = x\frac{\partial}{\partial x} - (\nu+1)(y+\beta/2\alpha)\frac{\partial}{\partial y}.$$

In this case the Noetherian first integral (47) is

$$I = -\frac{1}{2}x^{2\nu+4}y'^2 - \frac{1}{3}\alpha x^{3\nu+3}y^3 - \frac{1}{2}\beta x^{3\nu+3}y^2 - \gamma x^{3\nu+3}y - (\nu+1)x^{2\nu+3}yy'$$
$$-(\nu+1)\frac{\beta}{2\alpha}x^{2\nu+3}y' - \frac{\beta\gamma}{6\alpha}x^{3\nu+3}.$$

Thus the reduced equation is

$$\frac{1}{2}x^{2\nu+4}y'^2+\frac{1}{3}\alpha x^{3\nu+3}y^3+\frac{1}{2}\beta x^{3\nu+3}y^2+\gamma x^{3\nu+3}y+(\nu+1)x^{2\nu+3}yy' +(\nu+1)\frac{\beta}{2\alpha}x^{2\nu+3}y'+\frac{\beta\gamma}{6\alpha}x^{3\nu+3}=C, \tag{27}$$

where C is an arbitrary constant. We now solve Eq. (27). For this purpose we use an invariant of X (see (49)) as the new dependent variable. This invariant is obtained by solving the Lagrange's system associated with X, viz.,

$$\frac{dx}{x}=\frac{dy}{-(\nu+1)(y+\beta/2\alpha)},$$

and is

$$u=x^{\nu+1}y+\frac{\beta}{2\alpha}x^{\nu+1}.$$

In terms of u Eq. (27) becomes

$$C=\frac{1}{2}(\nu+1)^2u^2-\frac{1}{2}x^2u'^2-\frac{1}{3}\alpha u^3,$$

which is a first-order variables separable ordinary differential equation. Separating the variables we obtain

$$\frac{du}{\pm\sqrt{(\nu+1)^2u^2-(2/3)\alpha u^3-2C}}=\frac{dx}{x}.$$

Hence we have quadrature or double reduction of our Eq. (8) for the given f.

4.3. If $n=(\nu+4)/3$, $n\neq(1-\nu)/2,-1$, $\beta=0$ and $\gamma=0$, we find $\xi=x^{(2-\nu)/3}$, $\eta=-\frac{\nu+1}{3}x^{-(\nu+1)/3}y$ and $B=\frac{(\nu+1)^2}{18}y^2+k$, k a constant. This is subsumed in Case 5.2 below.

4.4. If $n=(1-\nu)/2$, $n\neq(\nu+4)/3$, β and γ are arbitrary, we obtain $\xi=x^{\frac{1-\nu}{2}}$, $\eta=0$. This reduces to Case 2.

Case 5. $f=\alpha y^r, \alpha\neq 0, r\neq 0,1$.
Here we have two subcases.

5.1. If $n=\frac{r+2\nu+1}{r-1}$, we obtain $\xi=x$, $\eta=\frac{\nu+1}{1-r}y$ and $B=$ constant. The solution of Eq. (8) for the above n and f is given by

$$y=ux^{\frac{\nu+1}{1-r}}, \tag{28}$$

where u satisfies

$$\int\frac{du}{\pm\sqrt{(\nu+1)^2(1-r)^{-2}u^2-2\alpha(1+r)^{-1}u^{1+r}-2C_1}}=\ln xC_2, \tag{29}$$

in which, C_1 and C_2 are arbitrary constants of integration.

We note that when $r=5$ and $\nu=1$, we get $n=2$. This gives us the Lane-Emden equation $y''+(2/x)y'+y^5=0$. Its general solution is given by Eq. (29) and we recover the solution given in (50). Only a one-parameter family of solutions is known in the other literature, namely,

$y = [3a/(x^2+3a^2)]^{1/2}$, $a =$ constant (see, e.g., (27) or (51)). Here we have determined a two-parameter family of solutions. Another almost unknown exact solution of $y'' + (2/x)y' + y^5 = 0$, which is worth mentioning here, is given by

$$xy^2 = \left[1 + 3\cot^2\left(\frac{1}{2}\ln\frac{x}{c}\right)\right]^{-1}, \tag{30}$$

where c is an arbitrary constant.

5.2. If $n = \dfrac{r+\nu+2}{r+1}$, with $r \neq -1$, we have $\xi = x^{\frac{r-\nu}{r+1}}$, $\eta = -\left(\dfrac{\nu+1}{r+1}\right)x^{-\frac{\nu+1}{r+1}}y$ and $B = \dfrac{(\nu+1)^2}{2(r+1)^2}y^2 + k$, where k is a constant.

In this case the solution of the corresponding Eq. (8) is

$$y = ux^{-\frac{\nu+1}{r+1}}, \tag{31}$$

where u is given by

$$\int \frac{du}{\pm\sqrt{C_1 - 2\alpha(r+1)^{-1}u^{r+1}}} = \left(\frac{r+1}{\nu+1}\right)x^{\frac{\nu+1}{r+1}} + C_2, \tag{32}$$

in which, C_1 and C_2 are arbitrary constants.

5.3. If $n = \dfrac{1-\nu}{2}$, we obtain $\xi = x^{\frac{1-\nu}{2}}$, $\eta = 0$ and $B =$ constant. This reduces to Case 2.

Case 6. $f = \alpha\exp(\beta y) + \gamma y + \delta$, $\alpha \neq 0, \beta \neq 0$.
Here again we have two subcases.

6.1. If $n = \dfrac{1-\nu}{2}$, we obtain $\xi = x^{\frac{1-\nu}{2}}$, $\eta = 0$ and $B = k$, k a constant. This reduces to Case 2.

6.2. If $n = 1$, $\nu \neq -1$, $\gamma = 0$ and $\delta = 0$, we deduce that $\xi = x$, $\eta = -(\nu+1)/\beta$ and $B = k$, k a constant.

The solution of the corresponding Eq. (8) for this case is

$$y = \frac{\nu+1}{\beta}\ln\left(\frac{u}{x}\right), \tag{33}$$

where u is defined by

$$\int \frac{du}{\pm u\sqrt{1 - 2\alpha\beta(\nu+1)^{-2}u^{\nu+1} + 2C_1\beta^2(\nu+1)^{-2}}} = \ln xC_2, \tag{34}$$

in which, C_1 and C_2 are integration constants.

Case 7. $f = \alpha\ln y + \gamma y + \delta$, $\alpha \neq 0$.
If $n = \dfrac{1-\nu}{2}$, we obtain $\xi = x^{\frac{1-\nu}{2}}$, $\eta = 0$ and $B = k$, k a constant. This reduces to Case 2.

Case 8. $f = \alpha y\ln y + \gamma y + \delta$, $\alpha \neq 0$.
If $n = \dfrac{1-\nu}{2}$, we obtain $\xi = x^{\frac{1-\nu}{2}}$, $\eta = 0$ and $B = k$, k a constant. This reduces to Case 2.

4. Systems of Lane-Emden-Fowler equations

The modelling of several physical phenomena such as pattern formation, population evolution, chemical reactions, and so on (see, for example (52)), gives rise to the systems of Lane-Emden equations, and have attracted much attention in recent years. Several authors have proved existence and uniqueness results for the Lane-Emden systems (53; 54) and other related systems (see, for example (55–57) and references therein). Here we consider the following generalized coupled Lane-Emden system (58)

$$\frac{d^2u}{dt^2} + \frac{n}{t}\frac{du}{dt} + f(v) = 0, \tag{35}$$

$$\frac{d^2v}{dt^2} + \frac{n}{t}\frac{dv}{dt} + g(u) = 0, \tag{36}$$

where n is real constant and $f(v)$ and $g(u)$ are arbitrary functions of v and u, respectively. Note that system (35)-(36) is a natural extension of the well-known Lane-Emden equation. We will classify the Noether operators and construct first integrals for this coupled Lane-Emden system.

It can readily be verified that the natural Lagrangian of system (35)-(36) is

$$L = t^n \dot{u}\dot{v} - t^n \int f(v)dv - t^n \int g(u)du. \tag{37}$$

The determining equation (see (58)) for the Noether point symmetries corresponding to L in (37) is

$$X^{[1]}(L) + LD(\tau) = D(B), \tag{38}$$

where X is given by

$$X = \tau(t,u,v)\frac{\partial}{\partial t} + \xi(t,u,v)\frac{\partial}{\partial u} + \eta(t,u,v)\frac{\partial}{\partial v}, \tag{39}$$

with first extension (59)

$$X^{[1]} = X + (\dot{\xi} - \dot{u}\dot{\tau})\frac{\partial}{\partial \dot{u}} + (\dot{\eta} - \dot{v}\dot{\tau})\frac{\partial}{\partial \dot{v}}, \tag{40}$$

where $\dot{\tau}, \dot{\xi}$ and $\dot{\eta}$ denote total time derivatives of τ, ξ and η respectively. Proceeding as in Section 3, (see details of computations in (58)) we obtain the following seven cases:

Case 1. $n \neq 0$, $f(u)$ and $g(v)$ arbitrary but not of the form contained in cases 3, 4, 5 and 6.

We find that $\tau = 0$, $\xi = 0$, $\eta = 0$, $B =$ constant and we conclude that there is no Noether point symmetry.

Noether point symmetries exist in the following cases.

Case 2. $n = 0$, $f(u)$ and $g(v)$ arbitrary.

We obtain $\tau = 1$, $\xi = 0$, $\eta = 0$ and $B =$ constant. Therefore we have a single Noether symmetry generator

$$X_1 = \frac{\partial}{\partial t} \tag{41}$$

with the Noetherian integral given by

$$I = \dot{u}\dot{v} + \int f(u)du + \int g(v)dv.$$

Case 3. $f(v)$ and $g(u)$ constants. We have eight Noether point symmetries associated with the standard Lagrangian for the corresponding system (35)-(36) and this case is well-known.

Case 4. $f = \alpha v + \beta,\ \ g = \gamma u + \lambda,$ where α, β, γ and λ are constants, with $\alpha \neq 0$ and $\gamma \neq 0$.

There are three subcases, namely

4.1. For all values of $n \neq 0, 2$, we obtain $\tau = 0,\ \xi = a(t),\ \eta = l(t)$ and $B = t^n \dot{l} u + t^n \dot{a} v - \lambda \int t^n a\, dt - \beta \int t^n l dt + C_1, C_1$ a constant. Therefore we obtain Noether point symmetry

$$X_1 = a(t)\frac{\partial}{\partial u} + l(t)\frac{\partial}{\partial v}, \tag{42}$$

where $a(t)$ and $l(t)$ satisfy the second-order coupled Lane-Emden system

$$\ddot{l} + \frac{n}{t}\dot{l} + \gamma a = 0,\ \ddot{a} + \frac{n}{t}\dot{a} + \alpha l = 0. \tag{43}$$

The first integral in this case is given by

$$I_1 = t^n \dot{l} u + t^n \dot{a} v - \lambda \int t^n a\, dt - \beta \int t^n l dt - a t^n \dot{v} - l t^n \dot{u}.$$

4.2. $n = 2$. In this subcase the Noether symmetries are X_1 given by the operator (42) and

$$X_2 = \frac{\partial}{\partial t} - u t^{-1} \frac{\partial}{\partial u} - v t^{-1} \frac{\partial}{\partial v}. \tag{44}$$

The value of B for the operator X_2 is given by $B = uv$.

The associated first integral for X_2 is given by

$$I_2 = uv + \frac{\alpha}{2} t^2 v^2 + \frac{\gamma}{2} t^2 u^2 + u t \dot{v} + v t \dot{u} + t^2 \dot{u}\dot{v}.$$

In this subcase, we note that the first integral corresponding to X_1 is subsumed in Case 4.1 above with $\beta,\ \lambda = 0$.

4.3. $n = 0$. Here the Noether operators are X_1 given by the operator (42) and

$$X_2 = \frac{\partial}{\partial t}, \text{ with } B = C_2, C_2 \text{ a constant.} \tag{45}$$

This reduces to Case 2.

We note also that the first integral associated with X_1 is contained in Case 4.1 above where $a(t)$ and $l(t)$ satisfy the coupled system

$$\ddot{l} + \gamma a = 0,\ \ddot{a} + \alpha l = 0. \tag{46}$$

Case 5. $f = \alpha v^r, g = \beta u^m, m \neq -1$ and $r \neq -1$ where α, β are constants, with $\alpha \neq 0$ and $\beta \neq 0$.

There are three subcases,viz.,

5.1. If $n = \dfrac{2m+2r+mr+3}{rm-1}$, $rm \neq 1, m \neq -1, m \neq 1$ and $r \neq -1$, we obtain $\tau = t$, $\xi = -\dfrac{(1+n)}{m+1}u$, $\eta = -\dfrac{(1+n)}{r+1}v$ and $B = $ constant.

Thus we obtain a single Noether point symmetry

$$X = t\frac{\partial}{\partial t} - \frac{(1+n)}{m+1}u\frac{\partial}{\partial u} - \frac{(1+n)}{r+1}v\frac{\partial}{\partial v} \tag{47}$$

with the associated first integral

$$I = \beta t^{n+1}\frac{u^{m+1}}{m+1} + \alpha t^{n+1}\frac{v^{r+1}}{r+1} + \frac{(n+1)}{m+1}t^n u\dot{v} + \frac{(n+1)}{r+1}t^n v\dot{u} + t^{n+1}\dot{u}\dot{v}.$$

We now consider the case when $m = -1$ and $r = -1$, in Case 5. Here we have two subcases

Case 5.2. $n = 0, (m = -1, r = -1)$.

This case provides us with two Noether symmetries namely,

$$X_1 = u\frac{\partial}{\partial u} - v\frac{\partial}{\partial v} \text{ and } X_2 = \frac{\partial}{\partial t} \text{ with } B = 0 \text{ for both cases.} \tag{48}$$

We obtain the Noetherian first integrals corresponding to X_1 and X_2 as

$$I_1 = \dot{u}v - u\dot{v},\ I_2 = \dot{u}\dot{v} + \ln u + \ln v,$$

respectively.

Case 5.3. $n = -1\ (m = -1, r = -1)$.

Here we obtain two Noether symmetry operators, viz.,

$$\begin{aligned} X_1 &= u\frac{\partial}{\partial u} - v\frac{\partial}{\partial v} \text{ with } B = 0 \text{ and } X_2 = t\frac{\partial}{\partial t} + 2u\frac{\partial}{\partial u} \\ &\text{with } B = -2\ln t \end{aligned} \tag{49}$$

and first integrals associated with X_1 and X_2 are given by

$$I_1 = \dot{u}vt^{-1} - u\dot{v}t^{-1}, I_2 = -2\ln t + \ln u + \ln v - 2u\dot{v}t^{-1} + \dot{u}\dot{v},$$

respectively.

Case 6. $f = \alpha\exp(\beta v) + \lambda,\ g = \delta\exp(\gamma u) + \sigma,\ \ \alpha, \beta, \lambda, \gamma, \delta,$ and σ are constants, with $\alpha \neq 0, \beta \neq 0, \delta \neq 0, \gamma \neq 0$.

There are two subcases. They are

6.1. If $n = 1$, $\lambda = 0$ and $\sigma = 0$, we obtain $\tau = t$, $\xi = -\dfrac{2}{\gamma}$, $\eta = -\dfrac{2}{\beta}$ and $B = C_3$, C_3 a constant. Therefore we have a single Noether point symmetry

$$X_1 = t\frac{\partial}{\partial t} - \frac{2}{\gamma}\frac{\partial}{\partial u} - \frac{2}{\beta}\frac{\partial}{\partial v} \tag{50}$$

and this results in the first integral

$$I = t^2\dot{u}\dot{v} + \frac{\alpha t^2}{\beta}\exp(\beta v) + \frac{\delta t^2}{\gamma}\exp(\gamma u) + \frac{2}{\gamma}t\dot{v} + \frac{2}{\beta}t\dot{u}.$$

6.2. If $n = 0$, $\lambda = 0$ and $\sigma = 0$, we deduce that $\tau = 1$, $\xi = 0$, $\eta = 0$ and $B = C_4$, C_4 a constant. The Noether operator is given by

$$X_1 = \frac{\partial}{\partial t}. \tag{51}$$

This reduces to Case 2.

Case 7. $f = \alpha \ln v + \beta$, $g = \gamma \ln u + \lambda$, where α, β, γ and λ are constants with $\alpha \neq 0, \gamma \neq 0$. If $n = 0$, we obtain $\tau = 1$, $\xi = 0$, $\eta = 0$ and $B = C_5$, C_5 a constant. This reduces to Case 2.

5. Concluding remarks

In this Chapter we gave a brief history of the Lane-Emden-Fowler equation and its applications in various fields. Several methods have been employed by scientists to solve the Lane-Emden-Fowler equation. Various generalizations of the Lane-Emden-Fowler equations were given which can be found in the literature. Also we gave the extension of the Lane-Emden equation to the System of Lane-Emden equations. We presented the complete Lie symmetry group classification of a generalized Lane-Emden-Fowler equation and performed the Lie and Noether symmetry analysis of this problem. It should be noted that Lie symmetry method is the most powerful tool to solve nonlinear differential equations. Finally, we classified a generalized coupled Lane-Emden system with respect to the standard first-order Lagrangian according to its Noether point symmetries and obtained first integrals for the corresponding Noether operators.

6. Acknowledgement

The author thanks Dr Ben Muatjetjeja for fruitful discussions.

7. References

[1] Lane JH, On the theoretical temperature of the sun under the hypothesis of a gaseous mass maintaining its volume by its internal heat and depending on the laws of gases known to terrestrial experiment, *The American Journal of Science and Arts*, 2nd series 50 (1870), 57-74.

[2] Emden R, *Gaskugeln, Anwendungen der mechanischen Warmen-theorie auf Kosmologie und meteorologische Probleme*, Leipzig, Teubner, 1907.

[3] Thomson W, Collected Papers, Vol. 5, p. 266. Cambrige University Press, 1991.

[4] Fowler RH, The form near infinity of real, continuous solutions of a certain differential equation of the second order, *Quart. J. Math. (Oxford),* 45 (1914), 289-350.

[5] Fowler RH, Further studies of Emden's and similar differential equations, *Quart. J. Math. (Oxford),* 2 (1931), 259-288.

[6] Mellin CM, Mahomed FM and Leach PGL, Solution of generalized Emden-Fowler equations with two symmetries, *Int. J. Nonlinear Mech.,* 29 (1994), 529-538.

[7] Chandrasekhar S, *An Introduction to the Study of Stellar Structure,* Dover Publications Inc., New York, 1957.

[8] Binney J and Tremaine S, *Galactic Dynamics,* Princeton University Press, Princeton, 1987.

[9] Rix HW, de Zeeuw PT, Cretton N, van der Marel RP and Carollo, C. M. *ApJ,* 488 (1997), 702

[10] Thomas LH, *Proc. Cambridge Phil. Soc.,* 23, (1927), 542.

[11] Fermi, E. *Rend. Accad. Naz. Lincei,* 6, (1927) 602.

[12] Meerson E, Megged E and Tajima T, On the Quasi-hydrostatic Flowsof Radiatively Cooling Self-gravitating Gas Clouds, *Ap. J.,* 457 (1996), 321.

[13] Gnutzmann S and Ritschel U, Analytic solution of Emden-Fowler equation and critical adsorption in spherical geometry, *Z. Phys. B: Condens Matter.,* 96 (1995), 391.

[14] Bahcall NA, The Galaxy Distribution in the Cluster Abell 2199, *Ap. J.,* 186 (1973), 1179.

[15] Bahcall NA, Core radii and central densities of 15 rich clusters of galaxies, *Ap. J.,* 198 (1975), 249.

[16] Horedt GP, Seven-digit tables of Lane-Emden functions, *Astron. Astrophys.,* 126 (1986), 357-408.

[17] Horedt GP, Approximate analytical solutions of the Lane-Emden equation in N-dimensional space, *Astron. Astrophys.,* 172 (1987), 359-367.

[18] Bender CM, Milton KA, Pinsky SS and Simmons Jr LM, *et al.* A new perturbative approach to nonlinear problems, *J. Math. Phys.,* 30 (1989), 1447-1455.

[19] Lima PM, Numerical methods and asymptotic error expansions for the Emden-Fowler equations, *J. Comput. Appl. Math..,* 70 (1996), 245-266.

[20] Lima PM, Numerical solution of a singular boundary-value problem in non-newtonian fluid mechanics, *Appl. Num. Math.,* 30 (1999), 93-111.

[21] Roxburgh IW and Stockman LM, Power series solutions of the polytrope equations, *Monthly Not. Roy. Astron. Soc.,* 303 (1999), 466-470.

[22] Adomian G, Rach R and Shawagfen NT, On the analytic solution of the Lane-Emden equation, *Found. Phys. Lett.,* 8 (1995), 161-181.

[23] Shawagfeh NT, Nonperturbative approximate solution for Lane-Emden equation, *J. Math. Phys.,* 34 (1993), 4364-4369.

[24] Burt PB, Nonperturbative solution of nonlinear field equations, *Nuov. Cim.,* 100B (1987), 43-52.

[25] Wazwaz AM, A new algorithm for solving differential equations of Lane-Emden type *Appl. Math. Comput.,* 118 (2001), 287-310.

[26] Liao S, A new analytic algorithm of Lane-Emden type equations, *Appl. Math. Comput.,* 142 (2003), 1-16.

[27] Davis HT, *Introduction to Nonlinear Differential and Integral Equations,* Dover Publications Inc., New York, 1962.

[28] Datta BK, Analytic solution to the Lane-Emden equation, *Nuov. Cim.* 111B (1996), 1385-1388.

[29] Wrubel MH, *Stellar Interiors. In Encyclopedia of Physics,* S. Flugge, Ed. Springer Verlag, Berlin, 1958, p 53.
[30] Leach PGL, First integrals for the modified Emden equation $\ddot{q} + \alpha(t)\dot{q} + q^n = 0$, *J. Math. Phys.,* 26 (1985), 2510-2514.
[31] Dehghan M and Shakeri F, Approximate solution of a differential equation arising in astrophysics using the variational iteration method, *New Astronomy,* 13 (2008), 53-59.
[32] Richardson OW, *The Emission of Electricity from Hot Bodies.* 2nd edition, London, 1921.
[33] Momoniat E and Harley C, Approximate implicit solution of a Lane-Emden equation, *New Astronomy,* 11 (2006), 520-526.
[34] Bozhkov Y and Martins ACG, Lie point symmetries of the Lane-Emden systems, *J. Math. Anal. Appl.,* 294 (2004), 334-344.
[35] Bozhkov Y and Martins ACG, Lie point symmetries and exact solutions of quasilinear differential equations with critial exponents, *Nonlin. Anal.,* 57 (2004), 773-793.
[36] Govinder KS and Leach PGL, Integrability analysis of the Emden-Fowler equation, *J. Nonlin. Math. Phys.,* 14 3 (2007), 435-453.
[37] Kara AH and Mahomed FM, Equivalent Langrangians and solutions of some classes of nonlinear equations $\ddot{q} + p(t)\dot{q} + r(t)q = \mu\dot{q}^2q^{-1} + f(t)q^n$, *Int. J. Nonlin. Mech,* 27 (1992), 919-927.
[38] Kara AH and Mahomed FM, A note on the solutions of the Emden-Fowler equation, *Int. J. Nonlin. Mech.,* 28 (1993), 379-384.
[39] Wong JSW, On the generalized Emden-Fowler equation, *SIAM Review,* 17 (1975), 339-360.
[40] Goenner H and Havas P, Exact solutions of the generalized Lane-Emden equation, *J. Math. Phys.,* 41 (2000), 7029-7042.
[41] Goenner H, Symmetry transformations for the generalized Lane-Emden equation, *Gen. Rel. Grav.,* 33 (2001), 833-841.
[42] Muatjetjeja B and Khalique CM, Exact solutions of the generalized Lane-Enden equations of the first and second kind, *Pramana,* 77 (2011) 545-554.
[43] Ibragimov NH, (ed) *CRC Handbook of Lie Group Analysis of Differential Equations,* Vols. 1, 2 and 3, CRC Press, Boca Raton, 1994-1996.
[44] Khalique CM, Mahomed FM and Ntsime BP, Group classification of the generalized Emden-Fowler-type equation, *Nonlinear Analysis: Real World Applications,* 10 (2009) 3387-3395.
[45] Lie S, *Differential Equations,* Chelsea, New York, (in German), 1967.
[46] Olver PJ, *Applications of Lie groups to differential equations,* Springer-Verlag, New York, 1993.
[47] Khalique CM and Ntsime BP, Exact solutions of the Lane-Emden-type equation, *New Astronomy,* 13 (2008) 476-480.
[48] Ibragimov NH, *Elementary Lie Group Analysis and Ordinary Differential Equations,* Wiley, Chichester, 1999.
[49] Adam AA and Mahomed FM, Integration of ordinary differential equations vianonlocal symmetries, *Nonlinear Dynamics,* 30 (2002), 267-275.
[50] Khalique CM, Mahomed FM and Muatjetjeja B, Lagrangian formulation of a generalized Lane-Emden equation and double reduction, *J. Nonlinear Math. Phys.,* 15 (2008) 152-161.
[51] Dresner L, *Similarity solutions of nonlinear partial differential equations.* Pitman Advanced Publishing Program, London, 1983.
[52] Zou H, A priori estimates for a semilinear elliptic system without variational structure and their applications. *Mathematische Annalen,* 323 (2002) 713-735.

[53] Serrin J and Zou H, Non-existence of positive solutions of the Lane-Emden system. *Differential Integral Equations* , 9 (1996) 635-653.

[54] Serrin J and Zou H, Existence of positive solutions of Lane-Emden systems. *Atti del Sem. Mat. Fis. Univ. Modena,* 46 (Suppl) (1998) 369-380.

[55] Qi Y, The existence of ground states to a weakly coupled elliptic system. *Nonlinear Analysis,* 48 (2002) 905-925.

[56] Dalmasso R, Existence and uniqueness of solutions for a semilinear elliptic system. *International of Journal of Mathematics and Mathematical Sciences,* 10 (2005) 1507-1523.

[57] Q. Dai and Tisdell CC, Non-degeneracy of positive solutions to homogeneous second order differential systems and its applications. *Acta Math. Sci.,* 29B (2009) 437-448.

[58] Muatjetjeja B and Khalique CM, Lagrangian Approach to a Generalized Coupled Lane-Emden System: Symmetries and First Integrals, *Commun. Nonlin. Sci. Numer. Simulat.,* 15 (2010) 1166-1171.

[59] Gorringe VM and Leach PGL, Lie point symmetries for systems of second order linear ordinary differential equations. *Queastiones Mathematicae,* 11 (1998) 95-117.

Part 2

High Energy Astrophysics

8

The Missing Matter Problem: From the Dark Matter Search to Alternative Hypotheses

S. Capozziello, L. Consiglio, M. De Laurentis, G. De Rosa and C. Di Donato

Dipartimento di Scienze Fisiche, Université di Napoli "Federico II", Compl. Univ. di Monte S. Angelo, Edificio G, Napoli

INFN Sezione di Napoli, Compl. Univ. di Monte S. Angelo, Edificio G, Napoli

Italy

1. Introduction

The nature of dark matter (DM) is one of the greatest today challenges in cosmology and particle physics: on one side it would have a significant impact on the large scale structure in the Universe and, on the other hand, it should lead to the empirical evidence of new unknown particles. The difficulties arise from the fact that DM could be composed by multiple components behaving in different ways depending on the scale. Besides, DM dynamics should be connected to that of dark energy (DE), the other unknown, unclustered form of energy that recent observations pointed out in the last decade. This further ingredient should constitute almost 75 % of cosmic matter-energy budget, turning out that the so-called *"dark side"* problem is dramatic and urgent to be solved.

The DM problem was finally set in the 1970s after several evidences accumulated in the past decades. In 1933 Fritz Zwicky measured the velocity dispersion of galaxies in the Coma cluster and found out that it was about a factor ten larger than expected from the estimated total mass of the cluster (Zwicky, 1933). It is interesting that Zwicky defined such a shortcoming as *the missing matter problem*. These preliminary observations have been confirmed in the 1970s, when data collected on the galactic rotational curves of spiral galaxies, proved the presence of large amounts of mass on scales much larger than the optical size of galactic disks.

Later on, the evidence of DM at various cosmological and astrophysical scales has been established by a wide number of observations, especially the very precise measurements of the cosmic microwave background radiation in the Wilkinson Microwave Anisotropy Probe (WMAP) experiment (Dunkley et al., 2009). Data from weak (Refregier, 2003) and strong (Tyson, 1998) lensing, hot gas in clusters (Lewis et al., 2003), the Bullet Cluster (Clowe et al., 2006), Big Bang nucleosynthesis (BBN) (Fields & Sarkar, 2008), further constraints from large scale structure (Allen, 2003), distant supernovae (Perlmutter et al., 1999; Riess et al., 1998), and the cosmic microwave background (CMB) (Komatsu et al., 2011) also support the evidence for non-luminous matter.

The ensemble of these data provides a strong evidence for a non luminous and non absorbing matter interacting only gravitationally, which is five times more prevalent than ordinary matter and accounts for about a quarter of the Universe. More precisely, current data constrain the energy densities of the Universe in a baryonic component, mostly known ($\Omega_B \lesssim 0.0456 \pm 0.0016$), a DM (CDM) component ($\Omega_{CDM} \simeq 0.227 \pm 0.014$) still unknown, and a DE component ($\Omega_\Lambda \simeq 0.728 \pm 0.015$ with a great uncertainty on the generating mechanisms). The luminous matter in the Universe is less than 1% of the total composition of the Universe. The current most precise estimation of the density of non-baryonic DM Ω_{DM} is obtained combining the measurements of the CMB anisotropy and of the spatial distribution of the galaxies and has found to be $\Omega_{DM}h^2 = 0.110 \pm 0.006$. The "local" DM present in the Galactic disk has an average density of (Kamionkowski & Kinkhabwala, 1998): $\rho_{DM}^{local} \simeq 0.3\frac{GeV}{cm^3}$.

An alternative, intriguing approach is to consider DM, as well as DE, as the manifestation of the break-down of General Relativity (GR) on large scales. A large literature is devoted to the possible modifications of the laws of gravitation, as Extended Theories of Gravity (ETGs), in particular $f(R)$-theories, by introducing into the Lagrangian , physically motivated higher-order curvature invariants and non-minimally coupled scalar fields (Capozziello & De Laurentis, 2011). The interest on such an approach, in early epoch cosmology, is due to the fact that it can "naturally" reproduce inflationary behaviours able to overcome the shortcomings of Standard Cosmological Model and, in late epoch cosmology, it seems capable of matching with several observations overcoming DM and DE problems related to the issue of detection.

In this review paper, we illustrate the experimental status of art to detect DM particles with direct and indirect methods, trying to interpretate and discuss the results obtained so far (Secs. 2,3,4). A possible alternative approach to the DM problem is presented in terms of gravitational effects at astrophysical and cosmological scales (Sec.5). Specifically, corrected gravitational potentials offer the possibility to fit galaxy rotation curves and galaxy clusters haloes without DM. The same approach allows to fit the apparent accelerated Hubble fluid without invoking DE. Conclusions are drawn in Sec. 6.

2. A survey of dark matter candidates

Candidates for non-baryonic DM must satisfy several conditions: *i*) they should be neutral, (otherwise they would interact electromagnetically); *ii*) they should not have color charge (otherwise they could form anomalous nuclear states); *iii*) they should be stable on cosmological time scales (otherwise they would have decayed by now); *iv*) they should interact very weakly with ordinary matter (otherwise they would not be dark), and, finally, *v*) they should have a suitable relic density. If we consider the Standard Model (SM) of Particles, neutrinos seem to be the prominent DM candidate as they interact only weakly and have an extremely low mass (the exact value of neutrinos' mass has yet to be measured, but it is clear, from neutrino oscillation measurements, that their mass is non-zero). There are many sources of neutrinos in the Universe, nevertheless they can account only a small fraction of DM. First of all, they are "hot" since they move at relativistic velocities. Thus they should have had a strong effect on the instabilities that generated the primordial cosmological objects during the earliest Universe, in the sense that galaxy clusters would have formed before galaxies and stars. This is in contrast with most theories and measurements, which are, instead, supported by a model based on cold DM, consisting of particles which move at non-relativistic energies. Thus small-scale perturbations are not suppressed, allowing an early start of structure formation.

Secondly, we know that neutrino mass is rather small. The most recent upper bound on electron neutrino mass is $m_{\nu_e} < 2$ eV (95% c.l.) while the experimental limits on the muon and tau neutrino are even weaker. So neutrinos cannot explain the gravitational effects DM is responsible for. However, extremely, high energetic neutrinos are of interest for DM search as they are among the secondary particles created in the annihilation of other DM candidates.

The most popular hypothesis is that non-baryonic DM consists of some neutral massive weakly interacting particles (WIMPs), which were created in the hot early Universe, decoupled early from ordinary matter in order to seed structure formation, and survived until today. They are supposed (from theoretical considerations and from the fact that they have not been detected yet) to be heavy respect to SM particles, with mass roughly between 10 GeV and few TeV. They interact via weak force and gravity only and are stable with a lifetime at least equal to the age of the Universe. Their relic density can be correctly calculated assuming that WIMPs were in thermal equilibrium with the SM particles in the early Universe. At the early stages, Universe was dense and hot. As the temperature T of the Universe cools, the density of more massive DM particles, with mass greater than T, become Boltzmann exponentially suppressed. When the expansion rate of the Universe, H, exceeds the particle annihilation/creation rate, the WIMPs drop out of thermal equilibrium, and the number density becomes "frozen". Presently the relic density of these particles is approximately given by

$$\frac{\Omega_{DM}h^2}{0.110} \approx \frac{3 \times 10^{-26} cm^3/sec}{\langle \sigma_A v \rangle_{ann}}, \tag{1}$$

where σ_A is the total annihilation cross section, v is the relative velocity of WIMPs and the term in brackets is an average over the thermal distribution of WIMPs velocities. From a cosmological point of view, it is worth noticing that the proper value of $\Omega_{DM}h^2$ density comes out from an annihilation cross section on the electroweak scale. This coincidence, obtained numerically, represents the main reason for believing that WIMPs give the largest contribution to the matter density in the Universe.

The freeze out happens at temperature $T_F \simeq m_{DM}/20$ almost independently of the properties of the WIMPs. This means they are already non-relativistic at the decoupling.

If we search beyond SM, well-motivated scenarios suggest good candidates. As neutralinos and Kaluza-Klein particles on which we are going to focus our discussion.

2.1 WIMPs in supersymmetric extensions of the standard model

This class contains a large amount of DM candidates, which are not predicted in the realm of the SM. In supersymmetric (SUSY) theories, each SM particle has a new yet-undiscovered partner whose spin differs by 1/2 with respect to the supersymmetric partner. In comparison to the mass eigenvalues in the SM, SUSY particles occur in linear combinations. The lightest possible combination (LSP) is the neutralino χ formed by the Bino $\tilde{B}$, Wino $\tilde{W}$ and two Higgsino $\tilde{H}$ states, with a mass range $m_\chi \sim 10$ GeV-TeV. The main ingredient necessary to provide a natural WIMP candidate in the SUSY models is the R-parity conservation defining the R-parity as:

$$R = (-1)^{3(B-L)+2S},$$

where B is the baryon number, L the lepton number and S the spin of the particle. $R = +1$ for ordinary particles and $R = -1$ for SUSY particles. This means that SUSY particles can only be created or annihilated in pairs. By imposing the R-parity conservation, a selection rule on the SUSY particle decays prevents the LSP to decay to an ordinary particle guaranteeing the stability in terms of cosmological abundances. However, this straightforward mechanism has to be experimentally probed. Several indications at Large Hadron Collider (LHC), CERN, seem to exclude minimal SUSY models, so the search of these candidates is, up to now, completely open.

2.2 Extra dimensions and Kaluza-Klein dark matter

An alternative possibility for new weak-scale physics is to search for universal extra dimensions (UED). The motivation to consider UED models is that they provide an interesting and qualitatively different alternative to supersymmetry. The idea of the existence of extra spatial dimensions was introduced, for the first time, by Kaluza and Klein in the 1920's. Such theories attempt to unify the two fundamental forces of gravitation and electromagnetism at the weak scales. Some models assume that all fields of SM propagate in "universal" extradimension. As a consequence, observers in the four-dimensional world see a tower of Kaluza Klein (KK) states for each SM particle (Kolb & Turner, 1990). The SM particles make up the first level of this tower and are referred to as the zero-th KK mode.

Momentum conservation in the extra dimension leads, in the four dimensional world, to a conservation of KK number (N_{KK}), where the KK number of a particle is given by its mode number. All SM particles have $N_{KK} = 0$ while the next most massive set of states have $N_{KK} = 1$. Such KK excitations appear as particles with masses near the TeV scale. Due to conservation of momentum in the higher dimensions, a symmetry called KK parity can arise which can, in some cases, make the lightest KK particle (LKP) stable, in a way which is analogous to R-parity of SUSY models, making it possible for the LKP to be a viable DM candidate. Since DM particles are expected to be neutral, non-baryonic and without color, the first mode KK partners of the neutral gauge bosons or neutrinos are likely choices for the lightest Kaluza-Klein particle (LKP). The identity of the LKP depends on the mass spectrum of the first KK level. The LKP is, most naturally, the first KK excitation of the B_1, the level 1 partner of the hypercharge gauge boson. It has been found that the appropriate relic density is predicted when the mass is moderately heavy, between 600 and 1200 GeV, somewhat heavier than the range favoured by supersymmetry. This range of LKP mass depends on the details of the co-annihilations of LKPs with heavier KK particles. Another difference between DM particles in universal extra dimensions and supersymmetry is that, unlike of LSP neutralino, the bosonic nature of the LKP means there is no chirality suppression of the annihilation signal in fermions. The annihilation rate of the LKP is therefore roughly proportional to the hypercharge of the final state, leading to a large rate in leptons, including neutrinos. The annihilation and co-annihilation cross-sections are determined by SM couplings and the mass spectrum of the first KK level. In contrast to supersymmetry, particles in UED have the same spin of the SM partners. As a result, the couplings become large and non-perturbative at energies far below the Planck scale. The detection of these kind of particles is also expected at LHC.

3. Methods of detection and experimental status

There are three main strategies to search for DM particles. Assuming that DM consists of WIMPs i.e. particles which froze out from thermal equilibrium when they were no more relativistic, one would expect that due to the gravitational interaction, they should have clustered with ordinary matter to form an almost spherical halo around the galaxies. Measurements of rotation curves of a large number of spiral galaxies have suggested the existence of a dark halo more extended than the visible disk. Despite the measurement in case of Milky Way is more difficult and suffers of larger uncertainties, the presence of a halo has been confirmed (Merrifield, 1992). Thus, if DM exists, it should be present also in the Milky Way and a consistent flux of WIMPs is expected to cross the Earth surface. The WIMPs of the dark halo, although with a very low cross section, interact with ordinary matter inducing atomic recoils. The measurement of these rare recoils through their energy deposit, carefully discriminated from the background, is performed by direct techniques.

Another method for DM detection is carried out by indirect techniques, which aims to observe the products of WIMPs annihilation process, as gamma rays, neutrinos, anti-matter cosmic rays occurring mainly inside astrophysical objects.

A third possibility is at particle colliders: here DM may be produced through the SM SM $\rightarrow$ XX process where SM denotes a standard model particle and X represents the WIMP. Such events are, in general, undetectable but are typically accompanied by related production mechanisms, such as SM SM $\rightarrow$ XX + "SM" , where "SM" indicates one or more standard model particles. These events are instead observable and should provide signatures of DM at colliders as LHC.

3.1 WIMPs direct detection

When a WIMP of a certain mass m_X scatters elastically a nucleus of mass m_N, the nuclear recoil occurs at an angle θ with respect to the WIMP initial velocity, with $\cos\theta$ uniformly distributed between -1 and 1 for the isotropic scattering that occurs with zero-momentum transfer. If the WIMP initial energy in the lab frame is $E_i = M_X v^2/2$, the the nucleus recoils (in the lab frame) with energy $E_R = E_i r \frac{(1-\cos\theta)}{2}$ where $r \equiv \frac{4\mu^2}{M_X M_A} = \frac{4 M_X M_A}{(M_X + M_A)^2}$. Note that $r \leq 1$, with $r = 1$ only if $M_X = M_A$. For this isotropic scattering, the recoil energy is therefore uniformly distributed between $0 - E_i r$.

Since typically the mean WIMP velocity is $v \simeq 220 km/s$ (Bernabei et al., 1998b), we can easily estimate that the maximum energy transfer from a WIMP to an electron, initially at rest, is at most in the eV range, while the energy transfer to an atomic nucleus would typically be in the range of tens of keV. This requires to use low threshold detectors, which are sensitive to individual energy deposits of this order of magnitude. In order to compute the WIMP-nucleon interaction rate, one needs the cross-section and the local density of WIMP. Details of this calculations can be found in e.g. (Lewin & Smith, 1996; **?**). We report here only the final expression

$$\frac{dN}{dE_R} = \frac{\sigma_0 \rho_{DM}^{local}}{2\mu^2 m_X} F^2(q) \int_{v_1}^{v_2} \frac{f(v)}{v} dv$$

where ρ_{DM}^{local} is the local WIMP density, μ is the reduced mass of the WIMP nucleus system and $f(v)$ accounts the velocity distribution of WIMPs in the galactic halo. The lower and upper limit of the integral represent respectively the minimum WIMP velocity to produce a recoil of energy E_r, and the maximum WIMP velocity set by the escape velocity of the halo model. $F(q)$ is a dimensionless factor form and σ_0 is the WIMP-nucleus interaction cross-section. This cross section depends strongly on the form of the interactions of the DM particles. It may happen that the WIMP-nucleus cross-section is not sensitive to the spin of the nucleus; in such a case we refer to spin-independent interactions; otherwise, WIMPs couple dominantly to the nucleus spin and this is the case of a spin-dependent interaction. The WIMP-nucleus cross section, may be written in terms of a spin-independent (mostly scalar) and a spin-dependent (mostly axial vector) component. For the former, the interaction will be coherent across the nucleons in the nucleus, while the latter term will only be present for nucleons with nuclear spin 29 or 82 or 83). In most cases, the coherent term is dominant since it is proportional to A^2 where A is the atomic number of nucleus, which favours heavy nuclei. Nevertheless, for very heavy nuclei, as the recoil energy increases, account must also be taken of the nuclear form factor, which may suppress the differential scattering rate significantly as shown in Fig.1. The curves are obtained assuming the spin independent coupling dominant, a standard halo model and choosing a WIMP mass of 100 GeV (Baudis, 2007).

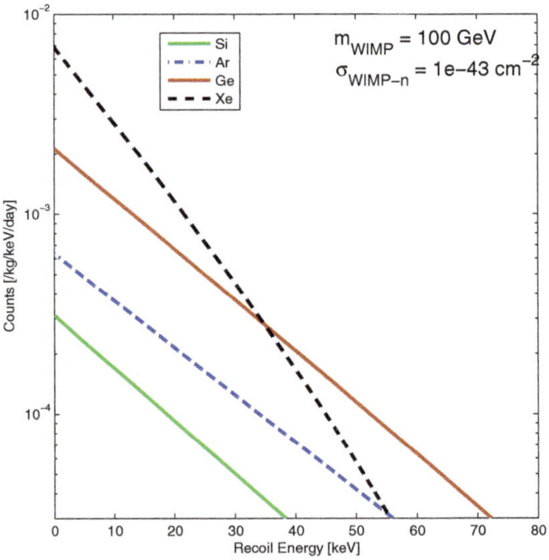

Fig. 1. Differential WIMP recoil spectrum for a WIMP mass of 100 GeV and a WIMP-nucleon cross section $\sigma = 10^{-43}$ cm^2. The spectrum was calculated for Si (light solid), Ar (light dot-dashed), Ge (dark solid), Xe (dark dashed) (Baudis, 2007).

The differential WIMP event rate versus the recoil energy expected for single isotope targets of ^{131}Xe (similar for ^{129}I), ^{73}Ge, ^{28}Si and ^{40}Ar is shown. It can be noticed that for a given cross section for WIMP-nucleon interactions, the interaction rate decreases for smaller nuclei because of a combination of smaller coherence enhancement ($\sim A^2$) and the less effective transfer of recoil energy to a target that is lighter than the WIMP. It is also evident that the recoil spectrum for the heavier Xe nucleus is suppressed significantly by the loss of coherence

Laboratory	Depth (m.w.e.)	Experiment
Soudan, US	2000	CDMS/SuperCDMS, CoGeNT
Yangyang, Korea	2000	KIMS
Kamioka, Japan	2700	XMASS,SuperKamiomkande
Bulby, UK	3200	ZEPLIN
LNGS, Italy	3500	DAMA, CRESST, WARP, XENON100
Modane, France	4800	EDELWEISS
SNOLab, Canada	6000	PICASSO, DEAP/CLEAN

Table 1. Underground laboratories housing DM experiments. The approximate effective shielding depth is mesured in meters of water equivalent (mwe).

for higher q^2 scattering events (form factor suppression). A low analysis threshold is therefore important to maximise the effective search sensitivity of a given detector mass. The influence of threshold energy is even greater for lower-mass WIMPs, where the recoil spectrum slope becomes steeper because of the reduction in typical kinetic energy of the WIMPs. The interaction rate can be calculated within the Minimal SUSY models, but the predicted WIMP-nucleon interaction cross section spans many orders of magnitude. Typical values for the spin-independent cross section are between 10^{-6} pb and 10^{-11} pb (de Austri et al., 2006).

In the recent years, an inelastic dark matter (iDM) model has been proposed as a modification of the elastic WIMP model. iDM model assumes that WIMPs scatter off baryonic matter by simultaneously transitioning to an excited state at an energy δ above the ground state ($\chi N \rightarrow \chi^* N$), while elastic scattering is forbidden or highly suppressed. This fact introduces a minimum velocity for WIMPs to scatter in a detector with a deposited energy E_{nr}.

3.1.1 A review of experiments

The nature of WIMPs makes their detection a hard task. A typical WIMP mass ranges between 10 GeV and few TeV according to the model chosen and the signal, to be detected in terms of deposit energy of the nuclear recoil, is of the order of tens keV. Within the MSSM models, the cross section interaction rates induce at most 1 evt day$^{-1}kg^{-1}$ in the detector, so large target masses and low background detectors are required. Typical background from environmental radioactivity and cosmic radiation are much higher with respect to these low expected rates and experimental purposes need underground laboratories to shield cosmic ray induced backgrounds, and for the selection of extremely radio-pure materials.

Worldwide, a large number of underground laboratories exists, many of them house present and/or future DM experiments (see Tab. 1).

Three different detection principles are at the basis of most particle detectors:

- **ionizing effect of a particle interaction;**
- **scintillation light from electronic excitation;**
- **thermal signal from lattice vibrations.**

A great number of experiments performing one or a combination of the mentioned techniques have been realised throughout the years, mainly divided into two categories: single and double modality. Concerning the former, such experiments must work in ultra-low background conditions because they are not able to perform background rejection. Each event, which falls in the acceptance window of the detector is accepted as a DM candidate with its own deposit energy measurement. For instance CREEST I and CoGeNT belongs to this class of experiments. CRESST I was the first phase of the CRESST experiment and consisted of a cryogenic bolometer based on saphire crystals as target material operating below 10 mK. This technique provides a threshold energy on nuclear recoils of the order of 0.5-0.6 keV, much lower than the minimum thresholds set by other experimental methods. Anyway the reduced target mass (262 g) limit its sensitivity. The CoGeNT experiment uses a single, 440g, high-purity germanium crystal cooled to liquid nitrogen temperatures in its measurements. The detector has the advantage of a very low energy threshold (<0.5 keV) which allows it to search for nuclear recoil events due to DM particles of relatively low mass ($> GeV/c^2$). In addition to a low-background configuration, the detector is able to distinguish and reject background events from the surface by measuring the risetime of the detector's signals. The CoGeNT detector is sensitive only to the ionisation charge from nuclear recoils and sets limits on the mass and interaction cross-section of DM particles by excluding any candidate mass and cross-section pair that would result in a signal above the background of the detector.

The CoGeNT collaboration has recently announced their results on 15 months of data, including the measurement of the spectrum of nuclear recoil candidate events, and the time variation of those events (Aalseth et al., 2011a). These results appear consistent with the signal anticipated from a relatively light DM particle scattering elastically with nuclei. The observed spectrum and rate is consistent with originating from DM particles with a mass in the range of 4.5-12 GeV and an elastic scattering cross section with nucleons of approximately $\sim 10^{-40} cm^2$ (Hooper & Kelso, 2011). In early 2010, the CoGeNT collaboration reported the observation of $\sim$100 events above expected backgrounds over a period of 56 days, with ionisation energies in the range of approximately 0.4 to 1.0 keV (Aalseth et al., 2011b). Concerning the double modality, all the experiments able perform energy measurements and background discrimination of the electron recoils, belong to this category. The most important results come from CDMS and EDELWEISS. The Cryogenic DM Search (CDMS in the Soudan Underground Laboratory) has developed a ionisation/phonon technique which allows a high efficient event by event discrimination between electron and nuclear recoils, based on the simultaneous record of inciting and phonon signal in the Ge and Si detectors packed in towers, operating at 40 mK. The recoil energy threshold is 10 keV. Recently, results have been presented from a re-analysis of the entire five-tower data set acquired with an exposure of 969 kg-days (Ahmed et al., 2011) with a recoil energy extended to 150 keV and an increased sensitivity of the experiment to the inelastic DM (iDM) model. Three DM candidates were found between 25 keV and 150 keV where the probability to observe three or more background events in this energy range is 11%. The CDMS program goes on with a future installation of the SuperCDMS setup at the new deep SNOLab laboratory, with 100 kg of detectors. EDELWEISS experiment (in the Modane Underground Laboratory) is conceptually based on the same idea as CDMS, with 300g germanium monocrystals as target. The typical recoil energy ranges from a few keV to few tens of keV. This recoil is measured at the same time with a heat channel, and a ionisation channel. The energy threshold reaches 10 keV. The EDELWEISS-II experiment has carried out a direct WIMP search with an array of ten 400 g Inter-Digit detectors, achieving an effective exposure of 322 $kg\ days$. The best sensitivity achieved in the elastic spin-independent WIMP-nucleon cross-section is $5 \times 10^{?8} pb$ for a WIMP mass of 80 GeV . The results are detailed in (Armengaud, 2010).

Exploiting the noble gas properties at low temperatures, both scintillation and ionisation signal can be detected by noble liquid detectors. Argon and Xenon at below 88K and 165K respectively behave as dense liquids with good scintillation yield of about 40×10^3 photons/MeV and good electron mobility. The particle interaction in the liquids ionises the medium and produces excited states of the gas atoms, which generate a luminescence signal. There are two excited states both for Ar and Xe (a singlet and a triplet) which differ in lifetime, quite enough to be measured by using a pulse shape discrimination analysis. The noble liquid detectors operate in a dual phase mode: the noble element (in the form of liquid and gas state) is saved in a vessel equipped with an array of photomultiplier. The interaction within the liquid phase produces a first direct scintillation signal, called S1, while, the produced ionisation electrons are drifted toward the liquid-gas interface, by means of a strong electric field applied in the volume. A second light signal S2 is emitted due to the difference of amplitude of the electric fields in both phases. By measuring the relative timings of the different signals, the position of the primary interaction may be reconstructed. Furthermore, since the ionisation yield is smaller for nuclear recoils, the S2/S1 ratio is used to distinguish nuclear from electron recoils.

The XENON10 experiment uses this technology, with PMT arrays both in the gas and liquid phase. The nuclear recoil energy threshold is 4.5 keV and the effective event rate in region of interest (4.5-29.6 keV) after all fiducial cuts is around 2×10^{-3} /keV/kg/day (Angle et al., 2008).

Xenon10 Collaboration has reported results of a search for light DM particle ($\lesssim$ 10 GeV) with a sensitivity threshold of 1.4 keV. Considering spin-independent dark matter- nucleon scattering, a cross sections $\sigma_n > 3.5 \times 10^{-42} cm^2$ is excluded, for a dark matter particle mass m_χ= 8 GeV (Angleet al., 2011).

XENON100 is one of the successor projects of XENON10. The experimental setup has been enlarged and much care has been taken to reduce the background. The XENON100 experiment has recently completed a DM run with 100.9 live-days of data, taken from January to June 2010. A total of three events have been found in the fiducial volume analyzed, compatible with the background prediction of (1.8 ± 0.6) events (Aprile et al., 2011).

The WARP collaboration has built a dual-phase prototype liquid Argon with a fiducial mass of 1.8 kg which achieved a good discrimination of the ^{39}Ar background and reached a background rate of 2×10^{-3} /keV/kg/day after discrimination in the 20-40 keVee [1] range. The WARP collaboration has also produced a large (order of 100 kg) detector at Gran Sasso with a massive active liquid argon shield, but unforeseen technical difficulties seem to prohibit a timely start of the experiment. In the dual phase detectors the major part of the discrimination power comes from the pulse shape. On this basis, a single phase liquid Argon project has been proposed and two detectors, DEAP 3600 and MiniClean with fiducial masses of 1000 kg and 100 kg respectively are under construction at SNOLAB. The pulse-shape based discrimination is expected to provide a sufficient electron recoil background reduction to completely suppress the intrinsic radioactivity, but it is also being considered to fill the detector with argon extracted from underground sources which are depleted in ^{39}Ar by a factor of 20 or more.

[1] keV electron-equivalent.

In the last years a new experiment, called DarkSide (A. Wright, n.d.) has been proposed by a Chinese, Italian, Russian and US Collaboration. It consists of a large depleted argon TPCs for direct DM searches at LNGS. The innovations proposed make DarkSide a detector of unprecedent background-free performance. The mechanical design is in progress. The installation will start in summer 2012. The detector will reach, in 3 years background free operation, a cross-section sensitivity of 1.5×10^{-45} cm^2 for WIMP-nucleon scattering, competitive with the projected sensitivity of other present experiments.

Other projects are based on superheated liquids, used in particle detectors early on in the form of bubble chambers. The PICASSO project is based on this technology, but instead of a monolithic bubble chamber, the detectors consist of tiny droplets immersed in a gel matrix. In appropriate temperature and pressure conditions, the interacting particle creates a bubble in the target volume. The operation conditions can be tuned in a way that the detector is sensitive only for nuclear recoils while is essentially blind to gamma rays and cosmic muon induced events with small energy deposit. The advantage of this technology are the relatively low costs and simple detector production, while a clear disadvantage is that there is no energy information available. PICASSO has operated several of these detectors with a total volume of 4.5 l, containing of order of 70 g of the main WIMP target fluorine each. With this target the experiment is mainly sensitive to spin-dependent WIMP-nucleon interactions. PICASSO reported results on the limits obtained on the spin-dependent cross section for WIMP scattering on ^{19}F, setting a limit for WIMP interactions on protons of 0.16 pb (90% CL).

The largest operating experiment for DM direct search is DAMA (Bernabei et al., 2004) which is based on about 10 kg of scintillating radiopure NaI(Tl) crystals readout by photomultiplier's in a well shielded and controlled environment. Although the scintillation light is detected, a signal shape discrimination can be performed at a reasonable level so the energy evaluation can be coupled to a discrimination method. The large mass and low background allow to investigate the presence of DM exploiting the model independent annual modulation signature.

Since NaI does not provide a strong event-by-event discrimination of electron recoil background, DAMA follows a unique strategy to still get a handle on the background: the Sun orbits our galaxy with a velocity of $\sim$ 220 km/s, while the earth rotates around the sun with $\sim$ 30 km/s. Assuming the DM halo around our galaxy has no net angular momentum, the relative velocity of WIMPs respect to the detector changes over the course of the year. For a given energy threshold this would lead to an annual modulation of the interaction rate with a known phase. DM particles from the Galactic halo are hence expected to show an annual modulation of the event rate induced by the Earth's motion around the Sun (Drukier et al., 1986). DAMA reported an effect of annual modulation of the count rate at low energies, which attributes to a WIMP signal at 6.3σ. This annual fluctuation in the background rates occurs near threshold, in the 2-6 keV region where the pulse shape discrimination start to fail. The variation is well fitted by the cosine function expected for the WIMP signal with a period $T = 0.999 \pm 0.002$ year, i.e. very close to one year, with a phase $t_0 = 146 \pm 7$ days, which is very close to the signal modulation expected for WIMPs (152.5 days or 2nd of June), and with an amplitude of 0.0131 ev/kg/keV/day. The observed modulation amplitude is (0.0200 $\pm$ 0.0032) cev/kg/keV/day with a phase of t_0 = (140$\pm$22) days and a period of (1.00$\pm$0.01) year. In the meanwhile, the DAMA collaboration has upgraded the detector to 250 kg NaI(Tl) called

DAMA/LIBRA which started the operations in 2003. The combination of all DAMA/NaI and DAMA/LIBRA data shows a statistically compelling (8σ) modulation signal at low energies (2-4 keV), compatible with a WIMP interpretation. The DAMA group has not found any modulation effect in the region of energies above 90 keV. If the effect is caused by variation of the Compton background in the $2 \div 6$ keV energy bin it must reveal itself also for the gammas with energies >90 keV. However, other experiments have excluded the region of the allowed parameters M_W and σ_p derived by DAMA from the modulation effect.

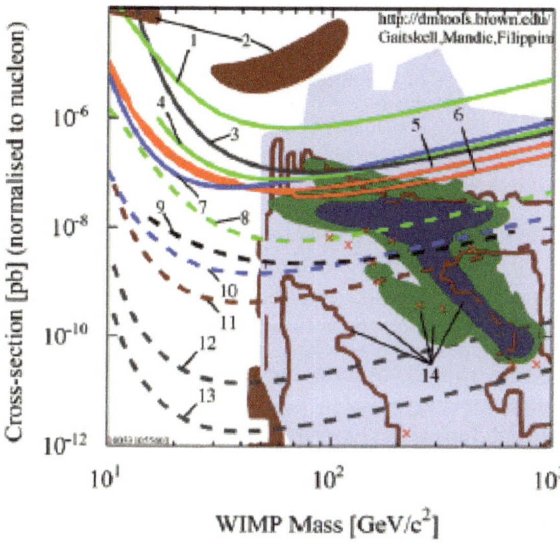

Fig. 2. Experimental results and SUSY predictions for WIMP nucleon cross sections versus WIMP mass 1-ZEPLIN-II, 2-DAMA/LIBRA, 3-EDELWEISS, 4-ZEPLIN-III 1st phase, 5-CDMS, 6-CDMS, 7-Xenon10, 8-ZEPLIN-III 2D phase, 9-XMASS , 10-Xenon100 , 11-LUX, 12-Xenon1t , LZS , MAXG2LXe, 13-LZD, MAXG3LXe and 14-SUSY predictions.

The experiments on WIMP search over the last years have put the limits on the spin-independent cross-section already below 10^{-7} pb. The ultimate goal of DM search experiments is to reach sensitivities down to 10^{-12} pb. This will allow to probe the whole M_W and σ_p parameter space of SUSY predictions. This goal can be achieved only with the use of next-generation detectors of the ton scale and ultimately of the multi-ton scale.

3.2 Indirect detection

After freeze out, DM pair annihilation becomes largely suppressed. However, even if its impact on the DM relic density is negligible, DM annihilation continues and may be observable. Thus, DM particle are expected to either annihilate or decay into SM particles like γ-rays, neutrinos and matter-antimatter particles in the final state and the detection of annihilation products is usually referred to as indirect DM detection.

Indirect searches look for the excesses of annihilation products in the diffuse background or from point sources. In particular, concerning charged products, the search is devoted to the antimatter component due to the large amount of primary matter component in the cosmic rays.

There are several methods for DM indirect detection. Their relative sensitivities are highly dependent on what WIMP candidate is being considered, and the systematic uncertainties and difficulties in determining backgrounds also vary greatly from one method to another.

High WIMP density regions, e.g. the centers of galaxies, but also the core of astrophysical objects such as Sun and Earth, are expected to accumulated large amount of WIMPs. When WIMPs pass through the Sun or the Earth or the galactic center, they may scatter and be slowed below escape velocity. In this way, they may become gravitationally bound in these gravitational wells. Over the age of the solar system, their densities and annihilation rates are greatly enhanced.

Although γ-rays, positrons and anti-protons produced in these annihilations do not escape the Sun or the Earth, neutrinos would be able in this because of their high penetration ability.

These neutrinos can travel to the surface of the Earth, where they may convert to charged leptons through $\nu q \rightarrow lq'$ and the charged leptons may be detected. The resulting neutrino flux could be detectable as a localised emission with earth-based neutrino telescopes by exploiting the Cherenkov light produced by the charged lepton.

Searches for neutrinos are unique among indirect searches since they are, under certain assumptions, probes of scattering cross sections, not annihilation cross sections, and so are competitive with the direct detection searches.

In the Sun both spin-independent as well as spin-dependent scattering can lead to the capture of DM. Among nuclei with a net spin, hydrogen is the only one present in the Sun in significant proportions. Other trace elements in the Sun generally have no net spin, and even when they do, there is no A^2 enhancement in the cross section coherent term because of their low density fraction. Therefore, the only relevant quantity for spin-dependent capture is the WIMP-proton cross section. Spin-independent capture, on the other hand, receives contributions from several elements. In fact, the cross section for spin-independent WIMP-nucleus scattering is strongly enhanced by large A. Therefore, even though the heavier elements in the Sun are rarer, this enhancement makes their contribution to the capture rate significant. In particular, oxygen plays the most important role in spin-independent capture of WIMPs in the Sun.

Neutrino telescopes could probe direct search observation. In fact, the signals observed at the direct detection experiments DAMA, CoGeNT and CRESST could be explained by light WIMPs with sizeable spin-independent cross sections with nucleons. The capture and subsequent annihilation of such particles in the Sun would induce neutrino signals in the GeV range which may be observed at Super-Kamiokande(Superkamiokande website, n.d.).

Actually, detection of these neutrinos in the range $10GeV < E_\nu < 1TeV$ by large neutrino detectors such as SuperKamiokande , AMANDA (AMANDA website, n.d.) and IceCube (IceCube website, n.d.) so far provides only upper limits on the high energy neutrino flux from the center of the Sun or the Earth (see (Kundu & Bhattacharjee, 2011)).

Future neutrino searches at Super-Kamiokande may have lower thresholds and so provide leading bounds on low mass WIMPs. In this way, Super-Kamiokande may test the DAMA and CoGeNT signal regions at high σ_{SI} and $m_X \sim 1$ - 10 GeV (Feng et al., 2009; Hooper et al., 2009; Kumar et al., 2009).

Moreover installing DeepCore, also IceCube significantly lowers its energy threshold and enhances the ability of detecting neutrinos from light WIMP annihilation.

Unlike other indirect DM searches this method does not depend strongly on our galaxy's DM halo profile or on the distribution of DM substructure.

The event rate depends on the DM density in the Sun, which in turn is dictated by the cross section of WIMPs with nucleons. This is constrained by direct detection experiments. The neutrino flux depends on the WIMP density, which is determined by the competing processes of capture, that is from the scattering cross section, and annihilation. Moreover the differential neutrino flux depends also on the way in which neutrinos are produced. Assuming that WIMPs annihilate to $b\bar{b}$ or W^+W^- which decay to neutrinos, as in many neutralino models, the neutrino signal is completely determined by the scattering cross section. For the majority of particle physics models considered (e.g., supersymmetry or KK models), the WIMP capture and annihilation rates reach or nearly reach equilibrium in the Sun. This is often not the case for the Earth. First, the Earth is less massive than the Sun and, therefore, provides fewer targets for WIMP scattering and a less deep gravitational well for capture. Secondly, in the Earth spin-independent interactions may occur. For these reasons, it is unlikely that the Earth will provide any observable neutrino signals from WIMP annihilations in any planned experiments.

If the annihilation occurs in free space, other types of radiation can be detected. The DM particles annihilating within the halo, can arise production of SM particles associated with the emission of γ ray with energies of the order of the DM particle mass. Thus, the γ-ray emission associated with such annihilation provides a chance for DM particles detection. The most striking gamma ray signal would be mono-energetic photons from $\chi\chi \rightarrow \gamma\gamma$, but since WIMPs cannot be charged, these processes are typically loop-induced or otherwise highly suppressed. More commonly, gamma rays are produced when WIMPs annihilate to other particles, which then radiate photons, leading to a smooth distribution of gamma ray energies. If gammas arise together with other particles, we only would expect some enhancement of the spectrum below the WIMP mass. The main difficulty in the extraction of information about DM from the annihilation photons is the presence of large and uncertain gamma-ray backgrounds. On the other hand, photons point back to their source, providing a powerful diagnostic. Possible targets for gamma ray searches are the center of the Galaxy, where signal rates are high but backgrounds are also high and potentially hard to estimate, and dwarf galaxies, where signal rates are lower, but backgrounds are also expected to be low. Moreover, for most particle DM candidates, γ-rays would be accompanied by neutrinos, thus neutrino experiments could help to confirm whether the γ flux originates from DM annihilations or from other astrophysical sources.

Another difficulty for DM search with γ arises from the fact that the DM gamma ray emission is expected to be a function of the DM density profile, which is not experimentally known. The most adopted density profile models are the Navarro-Frenk-White profile for which:

$$\rho(r) = \frac{\rho_0}{\frac{r}{R_s}(1 + \frac{r}{R_s})^2} \tag{2}$$

where ρ_0 and R_s, are parameters which vary from halo to halo, and the Einasto profile for which:

$$\rho(r) \propto exp(-Ar^{\alpha}) \tag{3}$$

The parameter α controls the degree of curvature of the profile.

The expected energy integrated γ-ray flux from a region of volume V at a distance D with DM density ρ is:

$$\phi_{\gamma} = N_{\gamma} \frac{\langle \sigma v \rangle}{m^2} \frac{1}{4\pi D^2} \int_V dV \frac{\rho^2}{2}, \tag{4}$$

where N_{γ} is the number of γ-rays per collision, m is the DM particle mass and $\langle \sigma v \rangle$ is the velocity-weighted annihilation cross-section. The current cosmological DM density is set by their annihilation rate in the early Universe. This provides a natural value for the annihilation cross-section of $\langle \sigma v \rangle \sim 3 \times 10^{-26} cm^3 s^{-1}$.

Obviously, the higher ρ the better, so that the annihilation radiation is searched for in the most dense regions of the Milky Way (MW).

The improvement of the Galactic diffuse model as well as the potential contribution from other astrophysical sources could provide a better description of the data.

γ-ray observations, in particular, at very high energies where DM masses of a few 100 GeV and above are probed, are a promising way to detect DM by space-based experiments, such as Fermi and AMS, and by ground-based atmospheric Cherenkov telescopes.

H.E.S.S. is an array of four Imaging Atmospheric Cherenkov Telecopes (IACTs) that uses the atmosphere as a calorimeter and images electromagnetic showers induced by TeV γ-rays (Hofmannetal, 2004).

H.E.S.S. Collaboration performed searches for a very-high-energy (VHE; $\geq$ 100 GeV) γ-ray signal from annihilation of DM particles, towards the Galactic center, the Sagittarius Dwarf galaxy and hypothetical DM spikes that could have formed around Intermediate Mass Black Holes (IMBHs).

The γ-ray energy spectrum in the range between 300 GeV and 30 TeV after the background suppression shows no any convincing evidence for γ-ray emission from self-annihilating DM. Limits on annihilating cross section for χ and B_1 have been found constraining some models.

For instance, limits on $\langle \sigma v \rangle$ are derived as a function of the DM particle mass assuming Navarro-Frenk-White and Einasto density profiles. In particular, for the DM particle mass of $\sim$ 1 TeV, values for $\langle \sigma v \rangle$ above 3 $\times 10^{-25} cm^3 s^{-1}$ are excluded for the Einasto density profile (Abramowski et al., 2011).

The Galactic center (GC) region is the main target for DM search, because of its proximity and its expected large DM concentration. However, the search for DM induced γ rays in the GC is affected by a strong astrophysical background. In fact, in the GC there is the compact γ-ray source HESS J1745-290 (Abramowski et al., 2011), coincident with the position of the supermassive black hole Sgr A* and a nearby pulsar wind nebula, which represents other sources of VHE γ-rays in the observed region, thus complicating the detection of γ-ray emission from DM annihilation.

VERITAS is a ground based Cherenkov telescope for gamma ray astronomy and has among other things searched for an enhancement of gamma rays from the center of neighboring dwarf galaxies without a positive signal so far (Weinstein et al., 2008).

Data from the EGRET satellite in the energy range 20 MeV - 30 GeV, interpreted as evidence for DM (de Boer, 2005), are not confirmed by recent data from Fermi satellite (Abdo et al., 2010).

The new precise data from the FERMI satellite on the diffuse gamma ray data are well-described at intermediate latitudes by standard CR physics assuming isotropic propagation models , but towards the Galactic center a 20%-30% excess has been observed in the 2-4 GeV range or even more significantly, within $\approx 1^\circ$ from the Galactic Center (GC), that could be interpreted as a signal of DM annihilation (DMA) for a very light neutralino of about 8 GeV (de Boer, 2011).

Other products of DM annihilation are pairs of particles-antiparticles, like e^+e^-, $p\bar{p}$ and $d\bar{d}$ pairs.

Since anti-matter is a very rare product of conventional sources of cosmic radiation, is suitable as possible indicator for DM annihilation.

Actually, unlike neutrinos and γ, positrons and anti-protons do not point back to their sources making an unambiguous separation from backgrounds very difficult.

In contrast to direct detection, many anomalies have been reported in the indirect search with anti-matter, and some of these have been interpreted as possible evidence for DM. The most prominent recent example is the detection of positrons and electrons with energies between 10 GeV and 1 TeV by the PAMELA, ATIC, and Fermi LAT Collaborations (Adriani et al., 2009; ATIC website, n.d.; FERMILAT website, n.d.).

PAMELA (Picozza et al., 2007), a satellite based instrument specifically designed to search for anti-particles in cosmic radiation, has reported an enhancement of positrons around 100 GeV over the standard expectation from diffuse galactic cosmic ray secondary models, in which positrons result from inelastic collisions of primary protons on the intergalactic gas nuclei.

Since DM annihilation produces as many positrons as electrons, with particles of energies close to the DM particle mass (in the range 10 GeV - 100 GeV for best motivated particle models), the positron fraction is enhanced.

If such enhancement originates from DM annihilation, also a respective enhancement of anti-protons would be expected, but is not observed. Special DM models could avoid hadron production, but also astrophysical explanations exist as well, such as nearby pulsars.

Additionally, the ATIC balloon-borne experiment also reported an anomalous "bump" in the total flux of e^+e^-, at energies of $\sim$ 600 GeV (Panov et al., 2011) which also has been interpreted as possible evidence for DM annihilation. It would imply a large mass and a very large pair annihilation cross section for DM.

Recently the experimental information available on the Cosmic Ray Electron (CRE) spectrum has been dramatically expanded as the Fermi-LAT Collaboration has reported a high precision measurement of the electron spectrum from 7 GeV to 1 TeV performed with its Large Area Telescope (LAT) (Fermi LAT collaboration, 2010).

The Fermi experiment, like also ATIC, is unable to distinguish positrons from electrons, and so constrain the total $e^+ + e^-$ flux.

The ATIC "bump" is not confirmed by the Fermi LAT data. The Fermi-LAT spectrum shows no prominent spectral features and the spectral index of the cosmic ray background is significantly harder than that inferred from several previous experiments: depending on the diffusion model, best fit injection spectral indexes range between 2.3 and 2.4, as opposed to previous models with 2.54. These data together with the PAMELA data on the rise above 10 GeV of the positron fraction are quite difficult to explain with just secondary production . An additional primary positron source is required to match the high-energy positron fraction, but other astrophysical explanations could be considered.

Additional data from, for example, Fermi and the Alpha Magnetic Spectrometer (AMS) (Pohl et al., 2010), an anti-matter detector placed on the International Space Station, may be able to distinguish the various proposed explanations for the positron excesses, as well as be sensitive to canonical WIMP models, but it remains to be seen whether the astrophysical backgrounds may be sufficiently well understood for these experiments to realize their DM search potential.

Other promising indirect detection search strategies are based on anti-protons and anti-deuterons from WIMP annihilation in the galactic halo. This searches are sensitive to DM candidates that annihilate primarily to quarks, but, until now, no result is reported (Adriani et al., 2009; D.Casadei, 2006).

3.3 Particle colliders

Contributions to solve the mystery of the DM will come also from colliders. Collider experiments can access interactions not probed by direct detection searches. A suitable room to search for DM is SUSY search and presently LHC data and analyzes are advancing very quickly, and possible signal for direct detection of supersymmetry are strictly related to the DM discussion.

The LHC is a $p - p$ collider, designed to operate at a center of mass energy up to $\sqrt{s} = 14$ TeV (LHC website, n.d.). Two main general purpose detectors are installed at the LHC: ATLAS and CMS. They are designed to perform precision measurements of photons, electrons, muons and hadrons. At this writing the LHC is colliding p-p at $\sqrt{s} = 7$ TeV and analyzes of about 1 fb^{-1} has already excluded squarks with mass up to 1 TeV about, in many models, superseding results from Fermilab-Tevatron $p\bar{p}$ collider with $\sqrt{s} = 1.96$ TeV (CDF (CDF website, n.d.) and D0 (D0 website, n.d.) experiments). At hadron colliders heavy particles can be produced via quark-antiquark collisions, as at Tevatron, or via gluon-gluon and quark-gluon collision as at the LHC: the signals are an inclusive combination of all these processes. At Tevatron valence quark annihilation into virtual weak bosons give up to chargino and neutralino production with a large cross-section, unless squarks or gluino were lighter then 300 GeV, value ruled out by LHC. At LHC production of gluinos and squarks by gluon-gluon and gluon-quark fusion should be dominant. At hadron colliders, independently of whether it is a pp or $p\bar{p}$ kind, one can also have associated production of a chargino or neutralino together with a sqaurk or gluino, but most models predict diverse cross-section for several processes: slepton pair production was quite small at Tevatron, but might be observable eventually at the LHC.

Produced sparticles decay in final states with two neutralino LSPs, which escape the detectors, carrying away a part of the missing energy which is at least two times the mass of neutralinos. At hadron colliders only the component of the missing energy that is manifest as momenta transverse to the colliding beams is observable, so generally the observable signals for supersymmetry are: N leptons and M jets with a missing transverse energy, where N or M might be zero.

The main problem is that significant backgrounds to these signal come from Standard Model particle, in particular processes involving production of W and Z which decays into neutrinos, providing missing transverse energy.

If LHC discovers signatures which seems like R-parity conserving SUSY signatures, a SUSY mass scale has to be derived from the deviation from SM. Early LHC data had excluded a SUSY mass scale in the sub-TeV range (Martin, 2011). SUSY contribution to DM from excess above the SM, observed in observables, should be characterized by the decay chains isolated to disentangle this contribution. The amount of LSP relic density of the Universe today is determined by the efficiency of LSP annihilation before freeze-out, that is by the cross section for $\chi\chi \rightarrow f\bar{f}/W^+W^-/ZZ$ (Di Ciaccio, 2011). ATLAS/CMS data can determine the LSP relic density, once one have estimated the SUSY parameters which enter the dominating annihilation process/processes. In mSUGRA model (A. H. Chamseddine et al., 1982) the LSP is usually most bino and in this case the basic annihilation process $\chi\chi \rightarrow f\bar{f}$ proceed via exchange of a slepton or squark and has a too low cross-section. To determine the LSP composition, the mass of additional neutralinos are important, being SUSY models dependent from many parameters as mass of LSP, sleptons/quarks and ,whether they are superpartners of left-handed or right-handed SM particles, the neutral Higgs masses. Up to now no deviation from SM has been observed

4. Discussion of experimental results

As we have shortly reported, during the last years, several experiments have performed direct and indirect search investigating the possibility of DM presence.

Concerning the direct DM detection, recently, the CRESST-II and CoGeNT collaborations have reported the observation of low-energy events in excess over known backgrounds. This result has encouraged the hypothesis that such signals in addition to the long standing DAMA annual modulation signal might arise from the scattering of a light ($\lesssim$10 GeV) DM particle.

The rate of DM elastic scattering events as a result of Earth's motion around the Sun and relative to the rest frame of the DM halo, is predicted to vary with an annual cycle. The only experiment reporting the observations of an annual modulation is DAMA/LIBRA which detected with high significance ($\sim 8\sigma$) with a phase and period consistent with elastically scattering DM.

The spectrum of the signals reported by DAMA/LIBRA and CoGeNT seem to point toward a similar range of DM parameter space. Moreover, the range of DM mass implied by CoGeNT and DAMA/LIBRA is very similar to that required to explain the spectrum of gamma rays observed by the Fermi Gamma Ray Space Telescope from the the inner 0.5° around the Galactic Center (Hooper & Goodenough, 2011), and for the observed synchrotron emission known as the WMAP Haze (Hooper & Linder et al., 2011). Anyway much care must be taken for this interpretation since CoGeNT experiment does not discriminate between electron and nucleon recoils (Chang et al., 2010).

Against to the positive signals of DAMA and CoGeNT, several other experiments find no evidence for DM. In particular, the CDMS, Xenon10 and Xenon100 collaborations disfavor the parameter space indicated by DAMA and CoGeNT.

One remarkable difference between DAMA/CoGeNT and these experiments, is that the latter reject electronic interactions attempting to collect only nuclear recoil events. In fact, it would be possible that the anomalous signals arise from such electronic recoils, a possibility that would explain away the existing tension. A model of this type was considered in prior to the recent CoGeNT measurement and it remains to be seen whether this possibility is theoretically feasible. Concerning the inelastic DM interpretation the spin independent fit to the DAMA modulated rate allows two qualitatively different best-fit regions: one around $M_{DM} \approx 80GeV$ with $\sigma \approx 10^{-41}cm^2$ due to scattering on iodine (A = 127, Z = 53) and one around $M_{DM} \approx 10GeV$ with $\sigma \approx 10^{-40}cm^2$ due to scattering on sodium (A = 23, Z = 11). The first region is firmly excluded by many other experiments, such XENON100 and CDMS. The second region, while still disfavoured by other null searches, is not completely excluded due to many experimental uncertainties and due to the general difficulty of direct detection searches to deal with low recoil energy scatterings. A possibility to make the DAMA results consistent with other experiments is to include an effect called "channelling" which will be present only the NaI crystals with DAMA uses. Anyway though inclusion of this effect the situation does not improve significantly. For a WIMP mass close to $10GeV$ with the help of NaI channelling it is possible to explain the DAMA results in terms of spin independent inelastic DM nucleon scattering. In that case some relevant parameters as DM mass and splitting should be fine tuned, and also the WIMP velocity distribution in the Galaxy should be close to the escape velocity. Inelastic spin-dependent interpretation would be a possibility which do not receive significant constraints from other experiments. Anyway both interpretations make difficult to reconciling DAMA and CoGeNT results (Chang et al., 2010).

A possibility to avoid the conflict between DAMA results and other experiments is proposed in terms of different but not irrelevant DM candidates. Stable particles with charge -2, bound with primordial helium would represent a kind of nuclear interacting form of DM. Such particles cannot be detected using nuclear recoils by direct search experiments , but their low energy binding with Na nuclei can the annual variations of energy release from their radiative capture, in the energy range 2-4 KeV corresponding to the signal observed in DAMA and DAMA/LIBRA (Khlopov, 2010a; Khlopov et al., 2010b;c).

Regarding indirect detection four experiments on DM search have generated great excitement in the astroparticle community. Their results are briefly summarized below:

- the PAMELA experiment measured with high statistics the positron component in the cosmic ray flux in the energy range between 10 and 100 GeV, observing a positrons excess above 10 GeV claimed also by earlier experiments as HEAT and AMS whose experimental data extend up to 40 GeV but with larger uncertainties. Anyway the rise up to 20 GeV can be explained by solar modulation which depends on the charge sign of a particle affecting positrons and electrons in a different way (Adriani et al., 2009). No excess in the antiproton/proton fraction has been observed by PAMELA (Adriani et al., 2009). The positron excess cannot be explained by a purely secondary production (due to primary protons and He nuclei interacting with the interstellar medium), which is characterized by a not so hard spectrum suggesting the existence of other primary sources.

- The ATIC ballon born experiment has measured the total $e^+ + e^-$ spectrum in the energy range 3 GeV-2.5 TeV, finding an increase of the flux from 100 GeV up to 600 GeV where they observed a peak followed by a sharp fall to about 800 GeV. The rest of the data agree with the GALPROP theoretical predictions within the errors.
- The FERMI experiment is able to measure with high resolution and statistics γ-rays in the energy range 20 MeV-300 GeV and primary cosmic ray $e^+ + e^-$ spectrum between 20 GeV and 1 TeV. Recent results on a data sample of 6 months confirmed the excess in the total $e^+ + e^-$ flux seen by ATIC, even though with a flatter trend.
- The large array telescope HESS measured γ-rays up to 5 TeV. HESS's results do not confirm ATIC peak as well as the sharp fall around 800 GeV, but a suitable HESS data normalization leaves a room for an agreement with ATIC results within the uncertainties.

Several interpretations came out invoking different sources: from those purely astrophysical like nearby pulsars or SNR to more exotic such as DM annihilation or decay in the halo of our Galaxy, arising from annihilation or decay of DM particles. Presently we are not able to say which interpretation assumes more validity. Nearby pulsars have been proposed as accelerating mechanism of energetic particles to explain the observed positron excess. Primary electrons should be accelerated in the pulsar magnetosphere, emitting gamma rays by synchrotron radiation. In the high magnetic fields of the pulsar, such gammas induce and electromagnetic cascade. The electron-positron pairs are then accelerated and confined in the pulsar nebula before escaping into the interstellar medium, so enhancing the CR electron and positron components. The energy spectrum of such particles is expected to be harder than that of the positrons of secondary production , thus the positrons originated by pulsars may dominate the high energy end of the CR positron spectrum. Two nearby pulsars could enhance significantly the high energy electron and positron flux reaching the Earth: Monogem at a distance of 290 pc and Geminga at a distance of 160 pc. This because electrons and positrons, during their propagation loose energy mainly by inverse Compton scattering and synchrotron radiation so only sources at distance less than 1 kpc can give a significant contribute to the energy spectrum. The interpretation of the FERMI, PAMELA and HESS excesses in terms of DM signature is quite suggestive but encounters some difficulties arising from the PAMELA data which show an asymmetry between the hadronic (antiprotons) and leptonic (positrons) component.

A first source of disappointment for the neutralino devotees is, in this case, due to the fact that, assuming the usual local DM density $\rho = 0.3GeV/cm^3$, the annihilation cross section needed to explain the signal is about $\sigma v \simeq 1 \times 10^{-24} cm^3 s^{-1}$, while the value $\sigma v \simeq 3 \times 10^{-26} cm^3 s^{-1}$, is needed in the early universe for a thermal WIMP in order to provide the correct relic abundance.

In conclusion, it seems that we are on the track to find evidences for DM but no *experimentum crucis* has given final probes. A main challenge comes from cosmology since it seems that DM behaves in different ways depending on the astrophysical structures and cosmological scales. For example, DM dynamics is very different in the various self-gravitating systems as elliptical, spiral and dwarf galaxies. Besides, it seems that different DM components have to be considered to address local systems (galaxies) and large scale structures (clusters of galaxies and super clusters of galaxies). An alternative view, could be that the problem of missing matter does not require the introduction of new ingredient but a modification or an extension of GR. The point is that gravity could not work in the same way at any scale. This issue will be faced in the next section.

5. An alternative view

Both cosmic speed up (DE) and DM, instead of being related to the search of new ingredients, could be the signal of a breakdown in our understanding of the laws of gravitation at large (infra-red) scales. From this point of view, one should consider the possibility that the Hilbert - Einstein Lagrangian, on which GR relies and linear in the Ricci scalar R, should be generalized. Following this approach, the choice of the effective theory of gravity could be derived by means of the data and the "economic" requirement that no exotic ingredients have to be added. This is the underlying philosophy of what is referred to as Extended Theories of Gravity consisting in enlarging the geometric sector of GR and assuming the possibility that other curvature invariants could contribute to the dynamics (Capozziello & De Laurentis, 2011; Capozziello & Faraoni, 2010). From a theoretical standpoint, several issues from fundamental physics (quantum field theory on curved spacetimes, M-theory etc.) suggest that higher order terms must necessarily enter the gravity Lagrangian. On the other side, Solar System experiments show the validity of Einstein's theory at local scales so the problems could come out at galactic scales and beyond. The simplest extension of Einstein theory is the possibility to take into account actions of the form $f(R)$ where f is a function of the Ricci scalar R not necessarily linear in R as in the Hilbert case. This kind of theories have recently received much attention in cosmology, since they are naturally able to give rise to accelerating expansions (both in the late and the early Universe). However, it is possible to demonstrate that $f(R)$ theories can also play a major role at astrophysical scales (Capozziello & Faraoni, 2010). In fact, modifying the gravitational Lagrangian can affect the gravitational potential in the low energy limit, provided that the modified potential reduces to the Newtonian one at the Solar System scales. In fact, a corrected gravitational potential could offer the possibility to fit galaxy rotation curves without the need of DM. In addition, one could work out a formal analogy between the corrections to the Newtonian potential and the usually adopted DM models. The choice of an analytic function in term of Ricci scalar is physically corroborated by the Ostrogradski theorem, which states that this kind of Lagrangian is the only viable one which can be considered among the several that can be constructed by means of curvature tensor and possibly its covariant derivatives. The field equations of this approach can be recast in the Einstein form, that is:

$$G_{\alpha\beta} = R_{\alpha\beta} - \frac{1}{2}g_{\alpha\beta}R = T^{curv}_{\alpha\beta} + T^{M}_{\alpha\beta}/f'(R) \tag{5}$$

where the prime denotes derivative with respect to R, $T^{M}_{\alpha\beta}$ is the standard matter stress - energy tensor and

$$T^{curv}_{\alpha\beta} = \frac{1}{f'(R)}\left\{\frac{1}{2}g_{\alpha\beta}\left[f(R) - Rf'(R)\right] + f'(R)^{;\mu\nu}(g_{\alpha\mu}g_{\beta\nu} - g_{\alpha\beta}g_{\mu\nu})\right\}, \tag{6}$$

defines a *curvature stress-energy tensor*. The presence of the terms $f'(R)_{;\mu\nu}$ renders the equations of fourth order, while, for $f(R) = R$, Eqs.(5) reduce to the standard second-order Einstein field equations. As it is clear from Eq.(5), the curvature stress-energy tensor formally plays the role of a further source term in the field equations which effect is the same as that of an effective fluid of purely geometrical origin. Depending on the scales, it is such a curvature fluid which can play the role of DM and DE. From the cosmological viewpoint, in the standard framework of a spatially flat homogeneous and isotropic Universe, the cosmological dynamics is determined by its energy budget through the Friedmann equations. In particular, the cosmic acceleration is achieved when the r.h.s. of the acceleration equation remains positive. In physical units, we have $\ddot{a}/a = -(1/6)(\rho_{tot} + 3p_{tot})$, where a is the cosmic scale factor, $H = \dot{a}/a$ the Hubble parameter, the dot denotes derivative with respect to cosmic time, and the subscript *tot* denotes the sum of the curvature fluid and the matter contribution to the energy density and pressure. From the above relation, the acceleration condition, for a dust dominated model, leads to: $\rho_{curv} + \rho_M + 3p_{curv} < 0 \rightarrow w_{curv} < -\frac{\rho_{tot}}{3\rho_{curv}}$ so that a key role is played by the effective quantities:

$$\rho_{curv} = \frac{1}{f'(R)} \left\{ \frac{1}{2} \left[f(R) - R f'(R) \right] - 3H\dot{R} f''(R) \right\} , \tag{7}$$

$$w_{curv} = -1 + \frac{\ddot{R} f''(R) + \dot{R} \left[\dot{R} f'''(R) - H f''(R) \right]}{\left[f(R) - R f'(R) \right] / 2 - 3H\dot{R} f''(R)} . \tag{8}$$

As a direct simplest choice, one may assume a power-law form $f(R) = f_0 R^n$, with n a real number, which represents a straightforward generalization of the Einstein General Relativity in the limit $n = 1$. One can find power-law solutions for $a(t)$ providing a satisfactory fit to the SNeIa data and a good agreement with the estimated age of the Universe in the range $1.366 < n < 1.376$ (Capozziello & Faraoni, 2010). It is worth noticing, that even an inverse approach for the choice of $f(R)$ is in order. Cosmological equations derived from (5) can be reduced to a linear third order differential equation for the function $f(R(z))$, where z is the redshift. The Hubble parameter $H(z)$ inferred from the data and the relation between z and R can be used to finally work out $f(R)$. In addition, one may consider the expression for $H(z)$ in a given DE model as the input for the above reconstruction of $f(R)$ and thus work out a $f(R)$ theory giving rise to the same dynamics as the input model. This suggests the intriguing possibility to consider observationally viable DM models (such as ΛCDM and quintessence) only as effective parameterizations of the curvature fluid (Capozziello et al., 2008).

The successful results obtained at cosmological scales motivates the investigation of $f(R)$ theories even at astrophysical scales. In the low energy limit, higher order gravity implies a modified gravitational potential. Now, by considering the case of a pointlike mass m and solving the vacuum field equations for a Schwarzschild-like metric, one gets from a theory $f(R) = f_0 R^n$ the modified gravitational potential:

$$\Phi(r) = -\frac{Gm}{r} \left[1 + \left(\frac{r}{r_c} \right)^{\beta} \right] \tag{9}$$

where

$$\beta = \frac{12n^2 - 7n - 1 - \sqrt{36n^4 + 12n^3 - 83n^2 + 50n + 1}}{6n^2 + 4n - 2} \tag{10}$$

Id	β	$\log r_c$	f_g	$\Upsilon_\star$	χ^2/dof	σ_{rms}
UGC 1230	0.608	-0.24	0.26	7.78	3.24/8	0.54
UGC 1281	0.485	-2.46	0.57	0.88	3.98/21	0.41
UGC 3137	0.572	-1.97	0.77	5.54	49.4/26	1.31
UGC 3371	0.588	-1.74	0.49	2.44	0.97/15	0.23
UGC 4173	0.532	-0.17	0.49	5.01	0.07/10	0.07
UGC 4325	0.588	-3.04	0.75	0.37	0.20/13	0.11
NGC 2366	0.532	0.99	0.32	6.67	30.6/25	1.04
IC 2233	0.807	-1.68	0.62	1.38	16.29/22	0.81
NGC 3274	0.519	-2.65	0.72	1.12	19.62/20	0.92
NGC 4395	0.578	0.35	0.17	6.17	34.81/52	0.80
NGC 4455	0.775	-2.04	0.88	0.29	3.71/17	0.43
NGC 5023	0.714	-2.34	0.61	0.72	13.06/30	0.63
DDO 185	0.674	-2.37	0.90	0.21	6.04/5	0.87
DDO 189	0.526	-1.87	0.69	3.14	0.47/8	0.21
UGC 10310	0.608	-1.61	0.65	1.04	3.93/13	0.50

Table 2. Best fit values of the model parameters from maximizing the joint likelihood function $\mathcal{L}(\beta, \log r_c, f_g)$. We also report the value of $\Upsilon_\star$, the χ^2/dof for the best fit parameters (with $dof = N - 3$ and N the number of datapoints) and the root mean square σ_{rms} of the fit residuals. See also Fig.3 (Capozziello et al., 2007).

which corrects the ordinary Newtonian potential by a power-law term. In particular, this correction sets in on scales larger than r_c which value depends essentially on the mass of the system. The corrected potential (9) reduces to the standard $\Phi \propto 1/r$ for $n = 1$ as it can be seen from the relation (10). The generalization of Eq.(9) to extended systems is straightforward. We simply divide the system in infinitesimal mass elements and sum up the potentials generated by each single element. In the continuum limit, we replace the sum with an integral over the mass density of the system taking care of eventual symmetries of the mass distribution. Once the gravitational potential has been computed, one may evaluate the rotation curve $v_c^2(r)$ and compare it with the data. For the pointlike case we have:

$$v_c^2(r) = \frac{Gm}{r}\left[1 + (1-\beta)\left(\frac{r}{r_c}\right)^{\beta}\right] . \qquad (11)$$

Compared with the Newtonian result $v_c^2 = Gm/r$, the corrected rotation curve is modified by the addition of the second term in the r.h.s. of Eq.(11). For $0 < \beta < 1$, the corrected rotation curve is higher than the Newtonian one. Since measurements of spiral galaxies rotation curves signals a circular velocity higher than what is predicted on the basis of the observed luminous mass and the Newtonian potential, the above result suggests the possibility that our modified gravitational potential may fill the gap between theory and observations without the need of additional DM. It is worth noting that the corrected rotation curve is asymptotically vanishing as in the Newtonian case, while it is usually claimed that observed rotation curves are flat. Actually, observations do not probe v_c up to infinity, but only show that the rotation curve is flat within the measurement uncertainties up to the last measured point. This fact by no way excludes the possibility that v_c goes to zero at infinity. In order to observationally check the above result, we have considered a sample of LSB galaxies with well measured HI + Hα

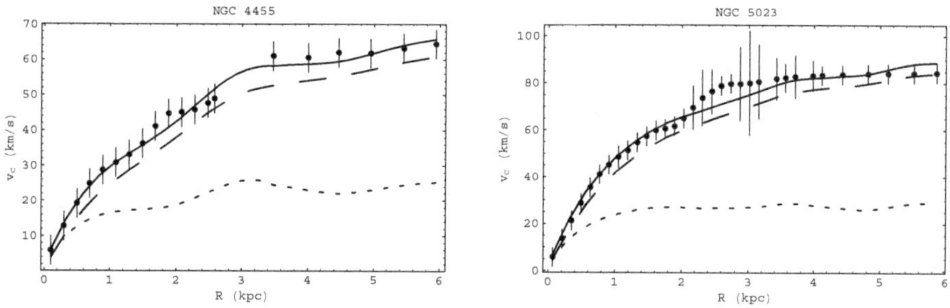

Fig. 3. Best fit theoretical rotation curve superimposed to the data for the LSB galaxy NGC 4455 (left) and NGC 5023 (right). To better show the effect of the correction to the Newtonian gravitational potential, we report the total rotation curve $v_c(R)$ (solid line), the Newtonian one (short dashed) and the corrected term (long dashed).

rotation curves extending far beyond the visible edge of the system. LSB galaxies are known to be ideal candidates to test DM models since, because of their high gas content, the rotation curves can be well measured and corrected for possible systematic errors by comparing 21 - cm HI line emission with optical Hα and [NII] data. Moreover, they are supposed to be DM dominated so that fitting their rotation curves without this elusive component is a strong evidence in favour of any successful alternative theory of gravity. Our sample contains 15 LSB galaxies with data on both the rotation curve, the surface mass density of the gas component and R - band disk photometry extracted from a larger sample selected by (de Blok & Bosma, 2002). We assume the stars are distributed in an infinitely thin and circularly symmetric disk with surface density $\Sigma(R) = \Upsilon_\star I_0 exp(-R/R_d)$ where the central surface luminosity I_0 and the disk scalelength R_d are obtained from fitting to the stellar photometry. The gas surface density has been obtained by interpolating the data over the range probed by HI measurements and extrapolated outside this range.

When fitting to the theoretical rotation curve, there are three quantities to be determined, namely the stellar mass - to - light (M/L) ratio, $\Upsilon_\star$ and the theory parameters (β, r_c). It is worth stressing that, while fit results for different galaxies should give the same β, r_c must be set on a galaxy - by - galaxy basis. However, it is expected that galaxies having similar properties in terms of mass distribution have similar values of r_c so that the scatter in r_c must reflect somewhat that on the terminal circular velocities. In order to match the model with the data, we perform a likelihood analyzis determining for each galaxy using as fitting parameters β, $\log r_c$ (with r_c in kpc) and the gas mass fraction[2] f_g. Considering the results summarized in Table 2, the experimental data are successfully fitted by the model. In particular, for the best fit range of β ($\beta = 0.58 \pm 0.15$), one can conclude that R^n gravity with $1.34 < n < 2.41$ (which well overlaps the above mentioned range of n interesting in cosmology) can be a good candidate to solve the missing matter problem in LSB galaxies without any DM (Capozziello et al., 2007). At this point, it is worth wondering whether a link may be found between R^n gravity and the standard approach based on dark matter haloes since both theories fit equally well the same data. As a matter of fact, it is possible to define an *effective DM halo* by imposing that its rotation curve equals the correction term to the

[2] This is related to the M/L ratio as $\Upsilon_\star = [(1 - f_g)M_g]/(f_g L_d)$ with $M_g = 1.4 M_{HI}$ the gas (HI + He) mass, $M_d = \Upsilon_\star L_d$ and $L_d = 2\pi I_0 R_d^2$ the disk total mass and luminosity.

Newtonian curve induced by R^n gravity. Mathematically, one can split the total rotation curve derived from R^n gravity as $v_c^2(r) = v_{c,N}^2(r) + v_{c,corr}^2(r)$ where the second term is the correction one. Considering, for simplicity a spherical halo embedding an infinitely thin exponential disk, we may also write the total rotation curve as $v_c^2(r) = v_{c,disk}^2(r) + v_{c,DM}^2(r)$ with $v_{c,disk}^2(r)$ the Newtonian disk rotation curve and $v_{c,DM}^2(r) = GM_{DM}(r)/r$ the DM one, $M_{DM}(r)$ being its mass distribution. Equating the two expressions, we get:

$$M_{DM}(\eta) = 2^{\beta-5}\eta_c^{-\beta}\pi(1-\beta)\Sigma_0 R_d^2 \eta^{\frac{\beta+1}{2}}\mathcal{I}_0(\eta,\beta)\,. \tag{12}$$

with $\eta = r/R_d$, $\Sigma_0 = \Upsilon_\star I_0$ and:

$$\mathcal{I}_0(\eta,\beta) = \int_0^\infty \mathcal{F}_0(\eta,\eta',\beta)k^{3-\beta}\eta'^{\frac{\beta-1}{2}}e^{-\eta'}d\eta' \tag{13}$$

with $\mathcal{F}_0$ only depending on the geometry of the system. Eq.(12) defines the mass profile of an effective spherically symmetric DM halo whose ordinary rotation curve provides the part of the corrected disk rotation curve due to the addition of the curvature corrective term to the gravitational potential. It is evident that, from an observational viewpoint, there is no way to discriminate between this dark halo model and R^n gravity. Having assumed spherical symmetry for the mass distribution, it is immediate to compute the mass density for the effective dark halo as $\rho_{DM}(r) = (1/4\pi r^2)dM_{DM}/dr$. The most interesting features of the density profile are its asymptotic behaviours that may be quantified by the logarithmic slope $\alpha_{DM} = d\ln\rho_{DM}/d\ln r$ which can be computed only numerically as function of η for fixed values of β (or n). The asymptotic values at the center and at infinity denoted as α_0 and α_∞ result particularly interesting. It turns out that α_0 almost vanishes so that in the innermost regions the density is approximately constant. Indeed, $\alpha_0 = 0$ is the value corresponding to models having an inner core such as the cored isothermal sphere and the Burkert model (Burkert, 1995). Moreover, it is well known that galactic rotation curves are typically best fitted by cored dark halo models (Gentile & Salucci, 2004). On the other hand, the outer asymptotic slope is between -3 and -2, that are values typical of most dark halo models in literature. In particular, for $\beta = 0.58$ one finds $(\alpha_0, \alpha_\infty) = (-0.002, -2.41)$, which are quite similar to the value for the Burkert model $(0, -3)$, that has been empirically proposed to provide a good fit to the LSB and dwarf galaxies rotation curves. The values of $(\alpha_0, \alpha_\infty)$ we find for our best fit effective dark halo therefore suggest a possible theoretical motivation for the Burkert-like models. Now, due to the construction, the properties of the effective DM halo are closely related to the disk one. As such, we do expect some correlation between the dark halo and the disk parameters. To this aim, exploiting the relation between the virial mass and the disk parameters , one can obtain a relation for the Newtonian virial velocity $V_{vir} = GM_{vir}/R_{vir}$:

$$M_d = \frac{(3/4\pi\delta_{th}\Omega_m\rho_{crit})^{\frac{1-\beta}{4}} R_d^{\frac{1+\beta}{2}} \eta_c^\beta}{2^{\beta-6}(1-\beta)G^{\frac{5-\beta}{4}}} \frac{V_{vir}^{\frac{5-\beta}{2}}}{\mathcal{I}_0(V_{vir},\beta)}\,. \tag{14}$$

We have numerically checked that Eq.(14) may be well approximated as $M_d \propto V_{vir}^a$ which has the same formal structure as the baryonic Tully-Fisher (BTF) relation $M_b \propto V_{flat}^a$ with M_b the total (gas + stars) baryonic mass and V_{flat} the circular velocity on the flat part of the observed rotation curve. In order to test whether the BTF can be explained thanks to the

effective DM halo we are proposing, we should look for a relation between V_{vir} and V_{flat}. This is not analytically possible since the estimate of V_{flat} depends on the peculiarities of the observed rotation curve such as how far it extends and the uncertainties on the outermost points. For given values of the disk parameters, we therefore simulate theoretical rotation curves for some values of r_c and measure V_{flat} finally choosing the fiducial value for r_c that gives a value of V_{flat} as similar as possible to the measured one. Inserting the relation thus found between V_{flat} and V_{vir} into Eq.(14) and averaging over different simulations, we finally get: $\log M_b = (2.88 \pm 0.04) \log V_{flat} + (4.14 \pm 0.09)$ while observational data give: $\log M_b = (2.98 \pm 0.29) \log V_{flat} + (3.37 \pm 0.13)$. The slope of the predicted and observed BTF are in good agreement thus leading further support to our approach. The zeropoint is markedly different with the predicted one being significantly larger than the observed one, but it is worth stressing, however, that both relations fit the data with similar scatter. A discrepancy in the zeropoint may be due to our approximate treatment of the effective halo which does not take into account the gas component. Neglecting this term, we should increase the effective halo mass and hence V_{vir} which affects the relation with V_{flat} leading to a higher than observed zeropoint. Indeed, the larger is M_g/M_d, the more the point deviate from our predicted BTF thus confirming our hypothesis. Given this caveat, we may therefore conclude with confidence that R^n gravity offers a theoretical foundation even for the empirically found BTF relation.
These results are referred to a simple choice of $f(R)$, while it is likely that a more complicated Lagrangian is needed to reproduce the whole dark sector phenomenology at all scales. Nevertheless, although not definitive, these achievements represent an intriguing matter for future more exhaustive investigations. In particular, exploiting such models can reveal a useful approach to motivate a more careful search for a single fundamental theory of gravity able to explain the full cosmic dynamics with the only two ingredients we can directly experience, namely the background gravity and the baryonic matter (S. Capozziello et al., 2006).

6. Conclusions

DM and DE can be considered among the biggest puzzles of modern physics. While their effects are evident from galactic to cosmological scales, their detection, at fundamental level, is a cumbersome task that, up to now, has no definitive answer. From a macroscopic viewpoint, DM is related to the clustering of astrophysical structures and DE is an unclustered form of energy (e.g. the cosmological constant Λ) which should be the source of cosmic speed up, detected by SNeIa and other cosmological indicators in the Hubble flow. Following the standard approach in physics, such *missing matter problem* and *cosmic acceleration* should be related to some fundamental ingredient (e.g. particles). Specifically, such particles should be not electromagnetically interacting (and then they are "dark") and, a part a few percent of further baryons, should be out of SM. In the case of DE, instead, they should be scalar fields that do not give rise to clustered structures.

Essentially, the DM (and DE) problem consists of three issues: *i*) the existence, *ii*) the detection, *iii*) the possible alternatives. In this review paper, we have tried to summarize the status of art with no claim to completeness.

After a discussion of the need of DM to address astrophysical and cosmological dynamics, we have reviewed the possible candidates. as SUSY, WIMP and extra dimension particles. The big issue is the detection of such particles that can be faced in three ways.

The direct detection (e.g. of WIMPs) consists, essentially, in identifying a certain mass m_X from scattering processes. Several underground experiments (e.g. DAMA) are running or have to start with this purpose. The approach is that standard interacting particles have to be shielded by some layer of rocks so then the weak interacting particles could be identified.

Indirect searches look for the excesses of annihilation products that should be DM remnants. Such experiments, like PAMELA and FERMI, are based on spacecrafts and are devoted to reconstruct the signature of DM.

Finally, DM evidences could come from particle colliders like LHC or Tevatron. In this case, DM should result from high-energy colliding processes where particles out of SM should be produced.

A radically alternative view is that DM (and DE) do not exist at all. Missing matter problem and cosmic speed up could be addressed by revising and extending the gravitational sector. Specifically, Einstein's GR should be a theory working only at local scale that should be revised at infrared (large scale structure and cosmology) and ultraviolet regimes (quantum gravity).

In conclusion, it seems that DM problem will be open until incontrovertible evidences (signatures) will be available at fundamental level. Very likely, the today working space- and ground-based experiments are going to reach the requested energy and precision levels in order to confirm or rule out the DM issue.

7. Acknowledgement

The authors thank G Barbarino for discussions and comments on the topics of this review paper.

8. References

C. E. Aalseth et al. [CoGeNT Collaboration], *Phys. Rev. Lett.* 107, 141301 (2011).
C. E. Aalseth et al. CoGeNT Collaboration, *Phys. Rev. Lett.* 106, 131301 (2011).
A. A. Abdo et al. (FermiLAT) ,*J. Cosmo. Astropart. Phys.* 04, 014 (2010).
A. Abramowski (Hamburg U.) et al. *Phys.Rev.Lett.* 106, 161301(2011).
O. Adriani et al., Phys.Rev.Lett. 102 (2009) 051101
O. Adriani et al., Nature 458 (2009) 607-609
Z. Ahmed et al., Phys. Rev. D 83, 112002 (2011).
B.C. Allanach, A. Dedes and H.K. Dreiner *Phys. Rev. D* 60 075014 (1999).
S. W. Allen, A. C. Fabian, R. W. Schmidt & H. Ebeling, *Mon. Not.Roy. Astron. Soc.* 342, 287 (2003).
J. Angle et al. (XENON10) *Phys.Rev.Lett.* 107, 051301 (2011).
Angle J. et al. (XENON10) *Phys. Rev. Lett.* 100, 021303 (2008).
E. Aprile et al., (XENON100 Coll.) arXiv:1104.3121v1 [astro-ph.CO] (2011).
C. Arina & N. Fornengo *JHEP* 11, 029 (2007).

E. Armengaud (EDELWEISS) *Proceeding conf. IDM2010* arXiv:1011.2319v1 [astro-ph.CO] (2010)
L. Baudis *SUSY07 proceedings* (2007).
R. Bernabei et al. *Int.J. Mod. Phys. D* 13, 2127 (2004).
R. Bernabei, et al., *Nucl. Phys. B* 480, 23 (2000).
R. Bernabei, et al., *Nucl. Phys. B (Proc. Suppl.)* 70, 79 (1999).
R. Bernabei, et al., *Phys. Lett. B* 424, 195 (1998).
R. Bernabei, et al., University of Rome preprint ROM2F/98/34 (27 August 1998).
A. Burkert,*Astroph. Journ.*, 447, L25, 1995
S. Capozziello & M. De Laurentis, *Extended Theories of Gravity*, to appear in *Phys. Rep* Phys. Rep DOI: 10.1016 j.physrep.2011.09.003 , arXiv:1108.6266v2 [gr-qc] (2011).
S. Capozziello, V. Faraoni, *Beyond Einstein Gravity: A Survey Of Gravitational Theories For Cosmology And Astrophysics*, Springer, New York, (2010).
S. Capozziello, V.F. Cardone & V. Salzano, *Phys. Rev. D 78*, 063504 (2008).
S. Capozziello, V.F. Cardone & A. Troisi *Mon. Not. R. Astron. Soc.* 375, 1423 (2007).
S. Capozziello, V.F. Cardone & A. Troisi *JCAP* 08, 001 (2006).
D.Casadei, arXiv:astro-ph/0609072v1
A. H. Chamseddine, R. Arnowitt and P. Nath, Phys. Rev. Lett. 49 (1982) 970.
S. Chang et al. *J. Cosmol. Astropart. Phys.* 1008, 018 (2010).
H.Y. Chiu, *Phys. Rev. Lett.* 17, 712 (1966).
D. Clowe et al., *Astrophys. J.* 648, L109 (2006)
R. R. de Austri et al. *JHEP* 05, 002 (2006).
W. de Boer et al., *Astron.Astrophys.* 444 (2005) 51
W. de Boer et al., *Prog.Part.Nucl.Phys.* 66 (2011) 197-201
W.J. G de Blok & A. Bosma, *Astron. Astroph.*, 385, 816 (2002).
A. Di Ciaccio, *Nucl. Instr. and Meth.*, A 630, 273-278, (2011).
A. K. Drukier, K. Freese, and D. N. Spergel, *Phys. Rev. D* 33, 3495 (1986).
J. Dunkley et al. [WMAP Collaboration], *Astrophys. J. Suppl.* 180, 306 (2009).
T. Falk, K. A. Olive & M. Srednicki, *Phys. Lett. B* 339, 248 (1994).
J. L. Feng et al. *JCAP* 0901, 032 (2009).
Fermi LAT collaboration, *Phys.Rev.D* 82, 092004 (2010).
B. D. Fields & S. Sarkar, *Phys. Lett. B* 667, 228 (2008).
A. L. Fitzpatrick, D. Hooper & K. M. Zurek, *Phys. Rev. D* 81, 115005 (2010).
G. Gentile & P. Salucci, *Mon. Not. Roy. Astron. Soc., 351*, 953, (2004).
H.E. Haber & G.L. Kane, *Phys. Rep.* 117, 75 (1985).
L.J. Hall & M. Suzuki *Nucl. Phys. B* 231, 419 (1984).
W.Hofmannetal.,in 28th *ICRC Proceedings*,1, 2811 (2004).
D. Hooper & C. Kelso, *FERMILAB-PUB* 11, 248 A (2011).
D. Hooper & T. Linden, *Phys. Rev. D* 83, 083517 (2011).
D. Hooper& L. Goodenough, *Phys. Lett. B* 697, 412 (2011).
D. Hooper et al. *Phys. Rev. D* 79, 015010 (2009).
http://amanda.uci.edu/
http://www-sk.icrr.u-tokyo.ac.jp/sk/pub/
http://icecube.wisc.edu/
http://atic.phys.lsu.edu/
http://www-glast.stanford.edu/

lhc.web.cern.ch
www-cdf.fnal.gov
www-d0.fnal.gov/
G. Jungman, M. Kamionkowski & K. Griest, *Phys. Reports* 267, 195 (1996).
M. Kamionkowski & A. Kinkhabwala, *Phys. Rev. D*57, 3256 (1998).
M.Yu.Khlopov, *AIP Conf.Proc.* 1241, 388 (2010).
M.Yu.Khlopov, A.G. Mayorov & E.Yu. Soldatov, *Prespacetime Journal* 1,1403 (2010).
M.Yu.Khlopov, A.G. Mayorov & E.Yu. Soldatov, Int.J.Mod.Phys.V. D 19, 1385, (2010).
E.W. Kolb & M.E. Turner, *The Early Universe*, Addison-Wesley (1990).
E. Komatsu et al., *Astrophys.J.Suppl.* 192, 18 (2011)
J. Kumar, J. G. Learned, and S. Smith, *Phys. Rev. D* 80, 113002 (2009).
Susmita Kundu, Pijushpani Bhattacharjee , arXiv:1106.5711v1 [astro-ph.GA] (2011).
J.D. Lewin & P.F. Smith it Astrop. Phys. 6 87 (1996).
A. D. Lewis, D. A. Buote, & J. T. Stocke, *Astrophys. J.* 586 135 (2003). arXiv:astro-ph/0209205.
M. Merrifield, *Astron. J.* 103, 1552 (1992).
Stephen P. Martin *http://arxiv.org/abs/hep-ph/9709356v6*
A.D. Panov et al., *Astrophys.Space Sci.Trans.* 7, 119 (2011).
S. Perlmutter et al., *Astrophys. J.* 517, 565 (1999).
P. Picozza et al., *Astropart.Phys.* 27 (2007) 296-315
M. Pohl et al., PoS ICHEP2010:443,2010.
A. Refregier, *Ann. Rev. Astron. Astrophys.* 41, 645, (2003).
A. G. Riess et al., *Astron. J.* 116,1009 (1998).
R. J. Scherrer & M. S. Turner, *Phys. Rev. D* 33,1585 (1986) .
G. Steigman, *Ann. Rev. Nucl. Part. Sci.* 29, 313 (1979).
J. A. Tyson, G. P. Kochanski, and I. P. Dell'Antonio *Astrophys. J.* 498, L107, (1998).
Y. B. Zeldovich Adv. Astron. Astrophys. 3 (1965) 241.
F. Zwicky *Helv. Phys. Acta* 6, 110 (1933). .
S. Weinberg *Phys. Rev. D* 26, 287 (1982).
A. Weinstein et al. VERITAS Collaboration , *Acta Phys.Polon.Supp* .1, 595, (2008).
Alex Wright, for the DarkSide Collaboration, *arXiv:1109.2979v1 [physics.ins-det].*

9

A Microscopic Equation of State for Neutron-Rich Matter and Its Effect on Neutron Star Properties

Francesca Sammarruca
University of Idaho
USA

1. Introduction

In this chapter, we will be concerned with hadronic interactions in the nuclear medium, particularly under conditions of extreme densities such as those encountered in some compact astrophysical objects. This issue goes to the very core of nuclear physics. In fact, our present knowledge of the nuclear force in free space is, in itself, the result of decades of struggle (Machleidt, 1989) which will not be reviewed in this chapter. The nature of the nuclear force in the medium is of course an even more complex problem, as it involves aspects of the force that cannot be constrained through free-space nucleon-nucleon (NN) scattering. Predictions of properties of nuclei are the ultimate test for many-body theories.

Nuclear matter is a convenient theoretical laboratory for many-body theories. By "nuclear matter" we mean an infinite system of nucleons acted on by their mutual strong forces and no electromagnetic interactions. Nuclear matter is characterized by its energy per particle as a function of density and other thermodynamic quantities, as appropriate (e.g. temperature). Such relation is known as the nuclear matter equation of state (EoS). The translational invariance of the system facilitates theoretical calculations. At the same time, adopting what is known as the "local density approximation", one can use the EoS to obtain information on finite systems. This procedure is applied, for instance, in Thomas-Fermi calculations within the liquid drop model, where an appropriate energy functional is written in terms of the EoS (Furnstahl, 2002; Oyamatsu et al., 1998; Sammarruca & Liu, 2009).

Isospin-asymmetric nuclear matter (IANM) simulates the interior of a nucleus with unequal densities of protons and neutrons. The equation of state of (cold) IANM is then a function of density as well as the relative concentrations of protons and neutrons.

The recent and fast-growing interest in IANM stems from its close connection to the physics of neutron-rich nuclei, or, more generally, isospin-asymmetric nuclei, including the very "exotic" ones known as "halo" nuclei. At this time, the boundaries of the nuclear chart are uncertain, with several thousands nuclei believed to exist but not yet observed in terrestrial laboratories. The Facility for Rare Isotope Beams (FRIB) has recently been approved for design and construction at Michigan State University. The facility will deliver intense beams of rare isotopes, the study of which can provide crucial information on short-lived elements normally not found on earth. Thus, this new experimental program will have widespread impact, ranging from the origin of elements to the evolution of the cosmos. It is estimated that the design and construction of FRIB will take ten years. In the meantime, systematic investigations to determine the properties of asymmetric nuclear matter are proliferating at existing facilities.

The equation of state of IANM is also the crucial input for the structure equations of compact stars, and thus establishes the connection between nuclear physics and compact astrophysical systems. It is the focal point of this chapter to present and discuss our approach to the devolopment of the EoS of nuclear and neutron-rich matter, with particular emphasis on the latter and its relation to the properties of neutron stars.

The chapter will articulate through the following sections: In Section **2**, we present a brief review of facts and phenomenology about IANM. We then proceed to describe our microscopic approach to calculate the energy per particle in IANM (Section **3**) and show the corresponding predictions. Section **4** will be dedicated to a review of neutron star structure and available empirical constraints. Microscopic predictions of neutron star properties will be presented and discussed there. Section **5** contains a brief discussion on the topic of polarized IANM. The chapter will end with our conclusions and an outlook into the future (Section **6**).

2. Facts about isospin-asymmetric nuclear matter

Asymmetric nuclear matter can be characterized by the neutron density, ρ_n, and the proton density, ρ_p, defined as the number of neutrons or protons per unit of volume. In infinite matter, they are obtained by summing the neutron or proton states per volume (up to their respective Fermi momenta, k_F^n or k_F^p) and applying the appropriate degeneracy factor. The result is

$$\rho_i = \frac{(k_F^i)^3}{3\pi^2}, \tag{1}$$

with $i = n$ or p.

It may be more convenient to refer to the total density $\rho = \rho_n + \rho_p$ and the asymmetry (or neutron excess) parameter $\alpha = \frac{\rho_n - \rho_p}{\rho}$. Clearly, α=0 corresponds to symmetric matter and α=1 to neutron matter. In terms of α and the average Fermi momentum, k_F, related to the total density in the usual way,

$$\rho = \frac{2k_F^3}{3\pi^2}, \tag{2}$$

the neutron and proton Fermi momenta can be expressed as

$$k_F^n = k_F(1+\alpha)^{1/3} \tag{3}$$

and

$$k_F^p = k_F(1-\alpha)^{1/3}, \tag{4}$$

respectively.

Expanding the energy per particle in IANM with respect to the asymmetry parameter yields

$$e(\rho,\alpha) = e_0(\rho) + \frac{1}{2}\left(\frac{\partial^2 e(\rho,\alpha)}{\partial\alpha^2}\right)_{\alpha=0}\alpha^2 + \mathcal{O}(\alpha^4)\,, \tag{5}$$

where the first term is the energy per particle in symmetric matter and the coefficient of the quadratic term is identified with the symmetry energy, e_{sym}. In the Bethe-Weizsäcker formula for the nuclear binding energy, it represents the amount of binding a nucleus has to lose when the numbers of protons and neutrons are unequal. A typical value for e_{sym} at nuclear matter density (ρ_0) is 30 MeV, with theoretical predictions spreading approximately between 26 and 35 MeV.

To a very good degree of approximation, the energy per particle in IANM can be written as

$$e(\rho, \alpha) \approx e_0(\rho) + e_{sym}(\rho)\alpha^2. \tag{6}$$

The effect of a term of fourth order in the asymmetry parameter ($\mathcal{O}(\alpha^4)$) on the bulk properties of neutron stars is very small, although it may impact the proton fraction at high density.

Equation (6) displays a convenient separation between the symmetric and aymmetric parts of the EoS, which facilitates the identification of observables that may be sensitive, for instance, mainly to the symmetry energy. At this time, groups from GSI (Sfienti et al., 2009; Trautmann et al., 2009), MSU (Tsang et al., 2009), Italy (Greco, 2010), France (Borderie & Rivet, 2008), China (Feng, 2010; Yong, 2010), Japan (Isobe, 2011), Texas A&M (Kohley et al., 2011), and more are investigating the density dependence of the symmetry energy through heavy-ion collisions. Typically, constraints are extracted from heavy-ion collision simulations based on transport models. Isospin diffusion and the ratio of neutron and proton spectra are among the observables used in these analyses.

These investigations appear to agree reasonably well on the following parametrization of the symmetry energy:

$$e_{sym}(\rho) = 12.5\, MeV\left(\frac{\rho}{\rho_0}\right)^{2/3} + 17.5\, MeV\left(\frac{\rho}{\rho_0}\right)^{\gamma_i}, \tag{7}$$

where ρ_0 is the saturation density. The first term is the kinetic contribution and γ_i (the exponent appearing in the potential energy part) is found to be between 0.4 and 1.0. Recent measurements of elliptic flows in ^{197}Au + ^{197}Au reactions at GSI at 400-800 MeV per nucleon favor a potential energy term with γ_i equal to 0.9 $\pm$ 0.4. Giant dipole resonance excitation in fusion reactions (Simenel et al., 2007) is also sensitive to the symmetry energy, since the latter is responsible for isospin equilibration in isospin-asymmetric collisions.

Isospin-sensitive observables can also be identified among the properties of normal nuclei. The neutron skin of neutron-rich nuclei is a powerful isovector observable, being sensitive to the slope of the symmetry energy, which determines to which extent neutrons will tend to spread outwards to form the skin.

Parity-violating electron scattering experiments are now a realistic option to determine neutron distributions with unprecedented accuracy. The neutron radius of ^{208}Pb is expected to be measured with a precision of 3% thanks to the electroweak program at the Jefferson Laboratory, the PREX experiment in particular, just recently completed at Jefferson Lab. This level of accuracy could not be achieved with hadronic scattering. Parity-violating electron scattering at low momentum transfer is especially suitable to probe neutron densities, as the Z^0 boson couples primarily to neutrons. With the success of this program, reliable empirical information on neutron skins will be able to provide, in turn, much needed *independent* constraint on the density dependence of the symmetry energy.

A measure of the density dependence of the symmetry energy is the symmetry pressure, closely related to the parameter

$$L = 3\rho_0 \Big(\frac{\partial e_{sym}(\rho)}{\partial \rho}\Big)_{\rho_0} \approx 3\rho_0 \Big(\frac{\partial e_{n.m.}(\rho)}{\partial \rho}\Big)_{\rho_0}, \tag{8}$$

where we have used Eq. (6) with α=1. Thus, L is sensitive to the gradient of the energy per particle in neutron matter ($e_{n.m.}$). As to be expected on physical grounds, the neutron skin, given by

$$S = \sqrt{< r_n^2 >} - \sqrt{< r_p^2 >}\,, \tag{9}$$

is highly sensitive to the same pressure gradient.

Values of L are reported to range from -50 to 100 MeV as seen, for instance, through the numerous parametrizations of Skyrme interactions, all chosen to fit the binding energies and the charge radii of a large number of nuclei, see (Li & Chen, 2005) and references therein. Heavy-ion data impose boundaries for L at 85 ± 25 MeV, with more stringent constraints being presently extracted. At this time constraints appear to favor lower values of L. . In fact, a range of L values given by 52.7 ± 22.5 MeV has emerged from recent analyses of global optical potentials (Xu et al., 2010).

Typically, parametrizations like the one given in Eq. (7) are valid at or below the saturation density, ρ_0. Efforts to constrain the behavior of the symmetry energy at higher densities are presently being pursued through observables such as π^-/π^+ ratio, K^+/K^0 ratio, neutron/proton differential transverse flow, or nucleon elliptic flow (Ko et al., 2010).

Another important quantity which emerges from studies of IANM is the symmetry potential. Its definition stems from the observation that the single-particle potentials experienced by the proton and the neutron in IANM, $U_{n/p}$, are different from each other and satisfy the approximate relation

$$U_{n/p}(k,\rho,\alpha) \approx U_{n/p}(k,\rho,\alpha=0) \pm U_{sym}(k,\rho)\,\alpha\,, \tag{10}$$

where the +(-) sign refers to neutrons (protons), and

$$U_{sym} = \frac{U_n - U_p}{2\alpha}\,. \tag{11}$$

(Later in the chapter we will verify the approximate linear behavior with respect to α displayed in Eq. (10).) Thus, one can expect isospin splitting of the single-particle potential to be effective in separating the collision dynamics of neutrons and protons. Furthermore, U_{sym}, being proportional to the gradient between the single-neutron and the single-proton potentials, should be comparable with the Lane potential (Lane, 1962), namely the isovector part of the nuclear optical potential. Optical potential analyses can then help constrain this quantity and, in turn, the symmetry energy.

Because of the fundamental importance of the symmetry energy in many systems and phenomena, it is of interest to identify the main contributions to its density dependence. In a recent work (Sammarruca, 2011) we discussed the contribution of the isovector mesons (π, ρ, and δ) to the symmetry energy and demonstrated the chief role of the pion. Note that the isovector mesons carry the isospin dependence by contributing differently in different partial waves, and that isospin dependence controls the physics of IANM. Hence, we stress the relevance of a microscopic model that contains all important couplings of mesons with nucleons.

3. Our microscopic approach to isospin-asymmetric nuclear matter

3.1 The two-body potential

Our approach is *ab initio* in that the starting point of the many-body calculation is a realistic NN interaction which is then applied in the nuclear medium without any additional free parameters. Thus the first question to be confronted concerns the choice of the "best" NN interaction. After the development of Quantum Chromodynamics (QCD) and the understanding of its symmetries, chiral effective theories (Weinberg, 1990) were developed as a way to respect the symmetries of QCD while keeping the degrees of freedom (nucleons and pions) typical of low-energy nuclear physics. However, chiral perturbation theory (ChPT) has definite limitations as far as the range of allowed momenta is concerned. For the purpose of applications in dense matter, where higher and higher momenta become involved with increasing Fermi momentum, NN potentials based on ChPT are unsuitable.

Relativistic meson theory is an appropriate framework to deal with the high momenta encountered in dense matter. In particular, the one-boson-exchange (OBE) model has proven very successful in describing NN data in free space and has a good theoretical foundation. Among the many available OBE potentials, some being part of the "high-precision generation" (Machleidt, 2001; Stocks et al., 1994; Wiringa et al., 1995), we seek a momentum-space potential developed within a relativistic scattering equation, such as the one obtained through the Thompson (Thompson, 1970) three-dimensional reduction of the Bethe-Salpeter equation (Salpeter & Bether, 1951). Furthermore, we require a potential that uses the pseudovector coupling for the interaction of nucleons with pseudoscalar mesons. With these constraints in mind, as well as the requirement of a good description of the NN data, Bonn B (Machleidt, 1989) is a reasonable choice. As is well known, the NN potential model dependence of nuclear matter predictions is not negligible. The saturation points obtained with different NN potentials move along the famous "Coester band" depending on the strength of the tensor force, with the weakest tensor force yielding the largest attraction. This can be understood in terms of medium effects (particularly Pauli blocking) reducing the (attractive) second-order term in the expansion of the reaction matrix. A large second-order term will undergo a large reduction in the medium. Therefore, noticing that the second-order term is dominated by the tensor component of the force, nuclear potentials with a strong tensor component will yield less attraction in the medium. For the same reason (that is, the role of the tensor force in nuclear matter), the potential model dependence is strongly reduced in pure (or nearly pure) neutron matter, due to the absence of isospin-zero partial waves.

Already when QCD (and its symmetries) were unknown, it was observed that the contribution from the nucleon-antinucleon pair diagram, Fig. 1, becomes unreasonably large if the

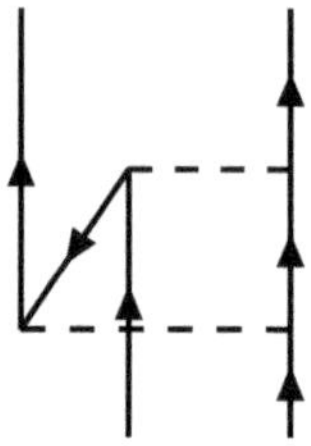

Fig. 1. Contribution to the NN interaction from virtual pair excitation. Upward- and downward-pointing arrows represent nucleons and antinucleons, respectively. Dashed lines denote mesons.

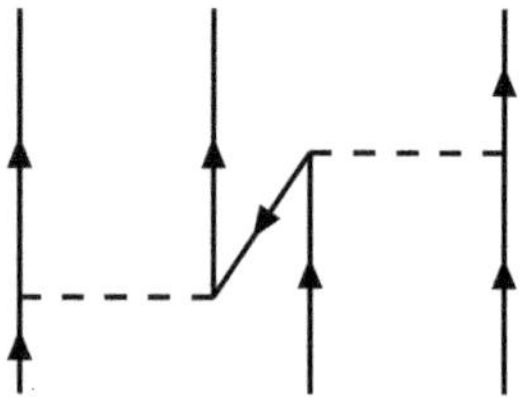

Fig. 2. Three-body force due to virtual pair excitation. Conventions as in the previous figure.

pseudoscalar (ps) coupling is used, leading to very large pion-nucleon scattering lengths (Brown, 1979). We recall that the Lagrangian density for pseudoscalar coupling of the nucleon field (ψ) with the pseudoscalar meson field (ϕ) is

$$\mathcal{L}_{ps} = -ig_{ps}\bar{\psi}\gamma_5\psi\phi. \tag{12}$$

On the other hand, the same contribution, shown in Fig. 1, is heavily reduced by the pseudovector (pv) coupling (a mechanism which became known as "pair suppression"). The reason for the suppression is the presence of the covariant derivative at the pseudovector vertex,

$$\mathcal{L}_{pv} = \frac{f_{ps}}{m_{ps}}\bar{\psi}\gamma_5\gamma^{\mu}\psi\partial_{\mu}\phi, \tag{13}$$

which reduces the contribution of the vertex for low momenta and, thus, explains the small value of the pion-nucleon scattering length at threshold (Brown, 1979). Considerations based on chiral symmetry (Brown, 1979) can further motivate the choice of the pseudovector coupling.

In closing this section, we wish to highlight the most important aspect of the *ab initio* approach: Namely, the only free parameters of the model (the parameters of the NN potential) are determined by fitting the free-space NN data and never readjusted in the medium. In other words, the model parameters are tightly constrained and the calculation in the medium is parameter free. The presence of free parameters in the medium would generate effects and sensitivities which are hard to control and interfere with the predictive power of the theory.

3.2 The Dirac-Brueckner-Hartree-Fock approach to symmetric and asymmetric nuclear matter

3.2.1 Formalism

The main strength of the DBHF approach is its inherent ability to account for important three-body forces through its density dependence. In Fig. 2 we show a three-body force (TBF) originating from virtual excitation of a nucleon-antinucleon pair, known as "Z-diagram". Notice that the observations from the previous section ensure that the corresponding diagram at the two-body level, Fig. 1, is moderate in size when the pv coupling, Eq. 13, is used. The main feature of the DBHF method turns out to be closely related to the TBF depicted in Fig. 2, as we will argue next. In the DBHF approach, one describes the positive energy solutions of the Dirac equation in the medium as

$$u^*(p,\lambda) = \left(\frac{E_p^* + m^*}{2m^*}\right)^{1/2} \begin{pmatrix} \mathbf{1} \\ \frac{\sigma\cdot\vec{p}}{E_p^*+m^*} \end{pmatrix} \chi_\lambda, \tag{14}$$

where the nucleon effective mass, m^*, is defined as $m^* = m + U_S$, with U_S an attractive scalar potential. (This will be derived below.) It can be shown that both the description of a single-nucleon via Eq. (14) and the evaluation of the Z-diagram, Fig. 2, generate a repulsive effect on the energy per particle in symmetric nuclear matter which depends on the density approximately as

$$\Delta E \propto \left(\frac{\rho}{\rho_0}\right)^{8/3}, \tag{15}$$

and provides the saturating mechanism missing from conventional Brueckner calculations. (Alternatively, explicit TBF are used along with the BHF method in order to achieve a similar result.) Brown showed that the bulk of the desired effect can be obtained as a lowest order (in p^2/m) relativistic correction to the single-particle propagation (Brown et al., 1987). With the in-medium spinor as in Eq. (14), the correction to the free-space spinor can be written approximately as

$$u^*(p,\lambda) - u(p,\lambda) \approx \begin{pmatrix} \mathbf{0} \\ -\frac{\sigma\cdot\vec{p}}{2m^2}U_S \end{pmatrix} \chi_\lambda, \tag{16}$$

where for simplicity the spinor normalization factor has been set equal to 1, in which case it is clearly seen that the entire effect originates from the modification of the spinor's lower component. By expanding the single-particle energy to order U_S^2, Brown showed that the correction to the energy consistent with Eq. (16) can be written as $\frac{p^2}{2m}(\frac{U_S}{m})^2$. He then proceeded to estimate the correction to the energy per particle and found it to be approximately as given in Eq. (15).

The approximate equivalence of the effective-mass description of Dirac states and the contribution from the Z-diagram has a simple intuitive explanation in the observation that Eq. (14), like any other solution of the Dirac equation, can be written as a superposition of positive and negative energy solutions. On the other hand, the "nucleon" in the middle of the Z-diagram, Fig. 2, is precisely a superposition of positive and negative energy states. In summary, the DBHF method effectively takes into account a particular class of TBF, which are crucial for nuclear matter saturation.

Having first summarized the main DBHF philosophy, we now proceed to describe the DBHF calculation of IANM (Alonso & Sammarruca, 2003; Sammarruca, 2010). In the end, this will take us back to the crucial point of the DBHF approximation, Eq. (14).

We start from the Thompson (Thompson, 1970) relativistic three-dimensional reduction of the Bethe-Salpeter equation (Salpeter & Bether, 1951). The Thompson equation is applied to nuclear matter in strict analogy to free-space scattering and reads, in the nuclear matter rest frame,

$$g_{ij}(\vec{q}',\vec{q},\vec{P},(\epsilon^*_{ij})_0) = v^*_{ij}(\vec{q}',\vec{q}) + \int \frac{d^3K}{(2\pi)^3} v^*_{ij}(\vec{q}',\vec{K}) \frac{m^*_i m^*_j}{E^*_i E^*_j} \frac{Q_{ij}(\vec{K},\vec{P})}{(\epsilon^*_{ij})_0 - \epsilon^*_{ij}(\vec{P},\vec{K})} g_{ij}(\vec{K},\vec{q},\vec{P},(\epsilon^*_{ij})_0) , \tag{17}$$

where g_{ij} is the in-medium reaction matrix (ij=nn, pp, or np), and the asterix signifies that medium effects are applied to those quantities. Thus the NN potential, v^*_{ij}, is constructed in terms of effective Dirac states (in-medium spinors) as explained above. In Eq. (17), $\vec{q}$, $\vec{q}'$, and $\vec{K}$ are the initial, final, and intermediate relative momenta, and $E^*_i = \sqrt{(m^*_i)^2 + K^2}$. The momenta of the two interacting particles in the nuclear matter rest frame have been expressed in terms of their relative momentum and the center-of-mass momentum, $\vec{P}$, through

$$\vec{P} = \vec{k}_1 + \vec{k}_2 \tag{18}$$

and

$$\vec{K} = \frac{\vec{k}_1 - \vec{k}_2}{2} . \tag{19}$$

The energy of the two-particle system is

$$\epsilon^*_{ij}(\vec{P},\vec{K}) = e^*_i(\vec{P},\vec{K}) + e^*_j(\vec{P},\vec{K}) \tag{20}$$

and $(\epsilon^*_{ij})_0$ is the starting energy. The single-particle energy e^*_i includes kinetic energy and potential energy contributions. The Pauli operator, Q_{ij}, prevents scattering to occupied nn, pp, or np states. To eliminate the angular dependence from the kernel of Eq. (17), it is customary to replace the exact Pauli operator with its angle-average. Detailed expressions for the Pauli operator and the average center-of-mass momentum in the case of two different Fermi seas can be found in (Alonso & Sammarruca, 2003).

With the definitions

$$G_{ij} = \frac{m^*_i}{E^*_i(\vec{q}')} g_{ij} \frac{m^*_j}{E^*_j(\vec{q})} \tag{21}$$

and

$$V^*_{ij} = \frac{m^*_i}{E^*_i(\vec{q}')} v^*_{ij} \frac{m^*_j}{E^*_j(\vec{q})} , \tag{22}$$

one can rewrite Eq. (17) as

$$G_{ij}(\vec{q}',\vec{q},\vec{P},(\epsilon^*_{ij})_0) = V^*_{ij}(\vec{q}',\vec{q}) + \int \frac{d^3K}{(2\pi)^3} V^*_{ij}(\vec{q}',\vec{K}) \frac{Q_{ij}(\vec{K},\vec{P})}{(\epsilon^*_{ij})_0 - \epsilon^*_{ij}(\vec{P},\vec{K})} G_{ij}(\vec{K},\vec{q},\vec{P},(\epsilon^*_{ij})_0) , \tag{23}$$

which is formally identical to its non-relativistic counterpart.

The goal is to determine self-consistently the nuclear matter single-particle potential which, in IANM, is different for neutrons and protons. To facilitate the description of the procedure, we will use a schematic notation for the neutron and proton potentials. We write, for neutrons,

$$U_n = U_{np} + U_{nn} , \tag{24}$$

and for protons

$$U_p = U_{pn} + U_{pp} , \tag{25}$$

where each of the four pieces on the right-hand-side of Eqs. (24-25) signifies an integral of the appropriate G-matrix elements (nn, pp, or np) obtained from Eq. (23). Clearly, the two equations above are coupled through the np component and so they must be solved simultaneously. Furthermore, the G-matrix equation and Eqs. (24-25) are coupled through the single-particle energy (which includes the single-particle potential, itself defined in terms of the G-matrix). So we have a coupled system to be solved self-consistently.

Before proceeding with the self-consistency, one needs an *ansatz* for the single-particle potential. The latter is suggested by the most general structure of the nucleon self-energy operator consistent with all symmetry requirements. That is:

$$\mathcal{U}_i(\vec{p}) = U_{S,i}(p) + \gamma_0 U^0_{V,i}(p) - \vec{\gamma}\cdot\vec{p}\, U_{V,i}(p) , \tag{26}$$

where $U_{S,i}$ and $U_{V,i}$ are an attractive scalar field and a repulsive vector field, respectively, with $U^0_{V,i}$ the timelike component of the vector field. These fields are in general density and momentum dependent. We take

$$\mathcal{U}_i(\vec{p}) \approx U_{S,i}(p) + \gamma_0 U^0_{V,i}(p) , \tag{27}$$

which amounts to assuming that the spacelike component of the vector field is much smaller than both $U_{S,i}$ and $U^0_{V,i}$. Furthermore, neglecting the momentum dependence of the scalar and vector fields and inserting Eq. (27) in the Dirac equation for neutrons or protons propagating in nuclear matter,

$$(\gamma_\mu p^\mu - m_i - \mathcal{U}_i(\vec{p}))u_i(\vec{p},\lambda) = 0 , \tag{28}$$

naturally leads to rewriting the Dirac equation in the form

$$(\gamma_\mu (p^\mu)^* - m^*_i)u_i(\vec{p},\lambda) = 0 , \tag{29}$$

with positive energy solutions as in Eq. (14), $m^*_i = m + U_{S,i}$, and

$$(p^0)^* = p^0 - U^0_{V,i}(p) . \tag{30}$$

The subscript "i" signifies that these parameters are different for protons and neutrons.

As in the symmetric matter case (Brockmann & Machleidt, 1990), evaluating the expectation value of Eq. (27) leads to a parametrization of the single particle potential for protons and neutrons [Eqs.(24-25)] in terms of the constants $U_{S,i}$ and $U^0_{V,i}$ which is given by

$$U_i(p) = \frac{m_i^*}{E_i^*} < \vec{p}|\mathcal{U}_i(\vec{p})|\vec{p} >= \frac{m_i^*}{E_i^*} U_{S,i} + U^0_{V,i} \, . \tag{31}$$

Also,

$$U_i(p) = \sum_{j=n,p} \sum_{p' \le k_F^j} G_{ij}(\vec{p}, \vec{p}') \, , \tag{32}$$

which, along with Eq. (31), allows the self-consistent determination of the single-particle potentials as explained below.

The kinetic contribution to the single-particle energy is

$$T_i(p) = \frac{m_i^*}{E_i^*} < \vec{p}|\vec{\gamma} \cdot \vec{p} + m|\vec{p} >= \frac{m_i m_i^* + \vec{p}^2}{E_i^*} \, , \tag{33}$$

and the single-particle energy is

$$e_i^*(p) = T_i(p) + U_i(p) = E_i^* + U^0_{V,i} \, . \tag{34}$$

The constants m_i^* and

$$U_{0,i} = U_{S,i} + U^0_{V,i} \tag{35}$$

are convenient to work with as they facilitate the connection with the usual non-relativistic framework (Haften & Tabakin, 1970).

Starting from some initial values of m_i^* and $U_{0,i}$, the G-matrix equation is solved and a first approximation for $U_i(p)$ is obtained by integrating the G-matrix over the appropriate Fermi sea, see Eq. (32). This solution is again parametrized in terms of a new set of constants, determined by fitting the parametrized U_i, Eq. (31), to its values calculated at two momenta, a procedure known as the "reference spectrum approximation". The iterative procedure is repeated until satisfactory convergence is reached.

Finally, the energy per neutron or proton in nuclear matter is calculated from the average values of the kinetic and potential energies as

$$\bar{e}_i = \frac{1}{A} < T_i > + \frac{1}{2A} < U_i > - m \, . \tag{36}$$

The EoS, or energy per nucleon as a function of density, is then written as

$$\bar{e}(\rho_n, \rho_p) = \frac{\rho_n \bar{e}_n + \rho_p \bar{e}_p}{\rho} \, , \tag{37}$$

or

$$\bar{e}(k_F, \alpha) = \frac{(1+\alpha)\bar{e}_n + (1-\alpha)\bar{e}_p}{2} \, . \tag{38}$$

Clearly, symmetric nuclear matter is obtained as a by-product of the calculation described above by setting α=0, whereas α=1 corresponds to pure neutron matter.

3.2.2 Microscopic predictions of the EoS and related quantities

In Fig. 3, we show EoS predictions for symmetric matter (solid red) and neutron matter (dashed black) as obtained from the Idaho calculation described in the previous section. The EoS from DBHF can be characterized as being moderately "soft" at low to medium density and fairly "stiff" at high densities. The predicted saturation density and energy for the symmetric matter EoS in Fig. 3 are equal to 0.185 fm^{-3} and -16.14 MeV, respectively, and the compression modulus is 252 MeV.

The increased stiffness featured by the DBHF EoS at the higher densities originates from the strongly density-dependent repulsion characteristic of the Dirac-Brueckner-Hartee-Fock method. In (Klähn et al., 2006), it is pointed out that constraints from neutron star phenomenology together with flow data from heavy-ion reactions suggest that such EoS behavior may be desirable.

The pressure as a function of density, as discussed in the next section, plays the crucial role in building the structure of a neutron star. In Fig. 4 we show the pressure in symmetric matter as predicted by the Idaho calculation compared with constraints obtained from flow data (Danielewicz et al., 2002). The predictions are seen to fall just on the high side of the constraints and grow rather steep at high density.

We show in Fig. 5 the pressure in neutron matter (red curve) and β-equilibrated matter (green) as predicted by DBHF calculations. The pressure contour is again from (Danielewicz et al., 2002).

Next we move on to the symmetry energy as defined from Eq. (6). In Fig. 6, we display the Idaho DBHF prediction for the symmetry energy by the solid red curve. The curve is seen to grow at a lesser rate with increasing density, an indication that, at large density, repulsion in the symmetric matter EoS increases more rapidly relative to the neutron matter EoS. This can be understood in terms of increased repulsion in isospin zero partial waves (absent from neutron matter) as a function of density. Our predicted value for the symmetry pressure L [see Eq. (8)], is close to 70 MeV.

The various black dashed curves in Fig. 6 are obtained with the simple parametrization

$$e_{sym} = C(\rho/\rho_0)^{\gamma}, \tag{39}$$

with γ increasing from 0.7 to 1.0 in steps of 0.1, and $C \approx 32$ MeV. Considering that all of the dashed curves are commonly used parametrizations suggested by heavy-ion data (Li & Chen, 2005), Fig. 6 clearly reflects our limited knowledge of the symmetry energy, particularly, but not exclusively, at the larger densities.

As already mentioned in Section **2**, from the experimental side intense effort is going on to obtain reliable empirical information on the less known aspects of the EoS. Heavy-ion reactions are a popular way to seek constraints on the symmetry energy, through analyses of observables that are sensitive to the pressure gradient between nuclear and neutron matter.

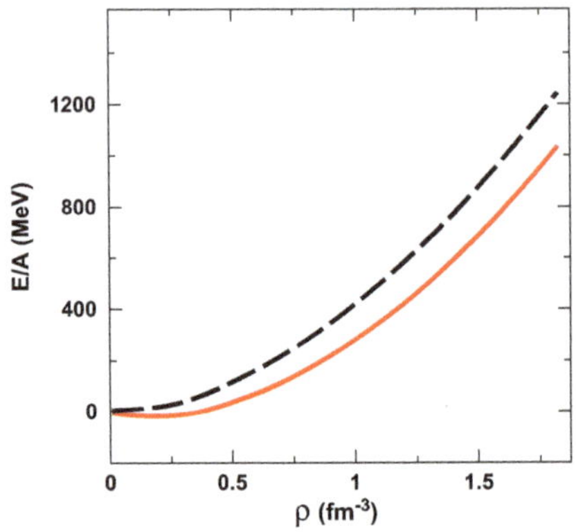

Fig. 3. DBHF predictions for the EoS of symmetric matter (solid red) and neutron matter (dashed black).

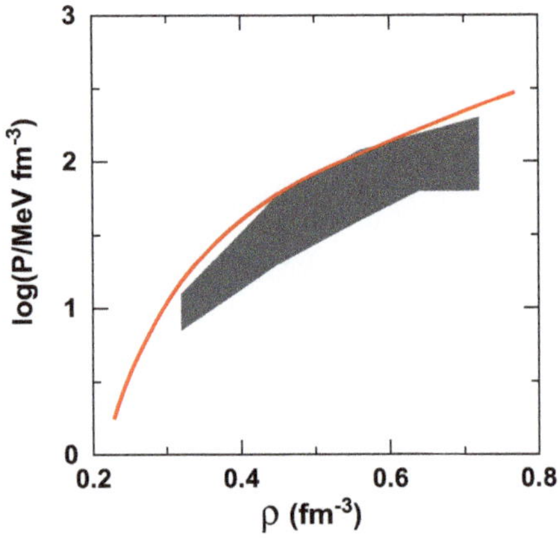

Fig. 4. Pressure in symmetric matter from the Idaho DBHF calculation. The shaded area corresponds to the region of pressure consistent with the flow data analysed in (Danielewicz et al., 2002).

We close this Section with demonstrating the approximately linear dependence on the asymmetry parameter of the single-nucleon potentials in IANM as anticipated in Eq. (10). We recall that this isospin splitting is the crucial mechanism that separates proton and neutron dynamics in IANM. In Fig. 7 we display predictions obtained with three different NN potentials based on the one-boson-exchange model, Bonn A, B, and C (Machleidt, 1989). These three models differ mainly in the strength of the tensor force, which is mostly carried by partial waves with isospin equal to 0 (absent from pure neutron matter) and thus should fade away in the single-neutron potential as the neutron fraction increases. In fact, the figure demonstrates reduced differences among the values of U_n predicted with the three potentials at large α.

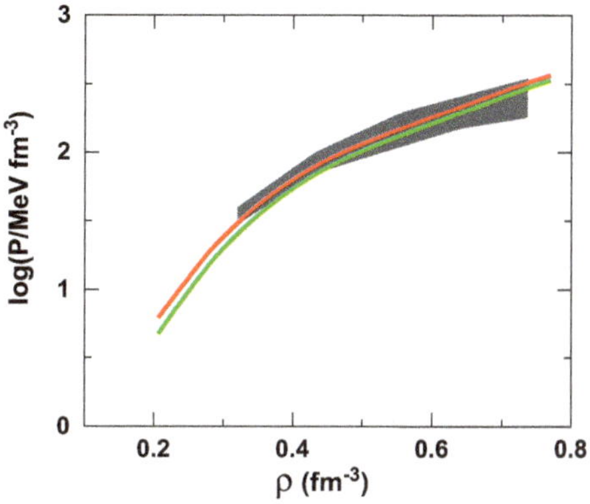

Fig. 5. Pressure in neutron (red curve) and baryon-lepton (green curve) matter from the Idaho DBHF calculation. The shaded area corresponds to the region of pressure consistent with flow data and the inclusion of strong density dependence in the asymmetry term (Danielewicz et al., 2002).

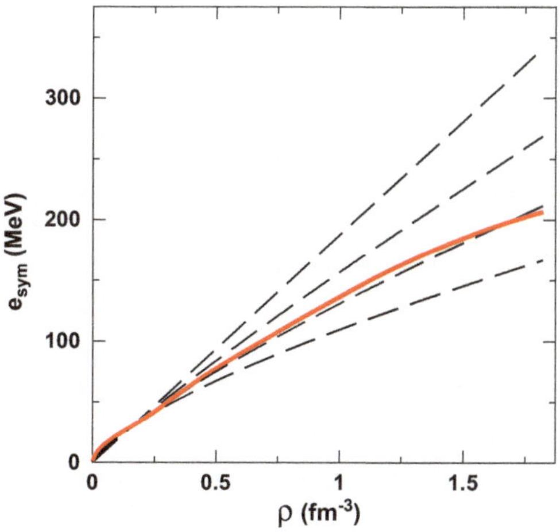

Fig. 6. DBHF prediction for the symmetry energy (solid red) compared with various phenomenological parametrizations (dashed black), as explained in the text.

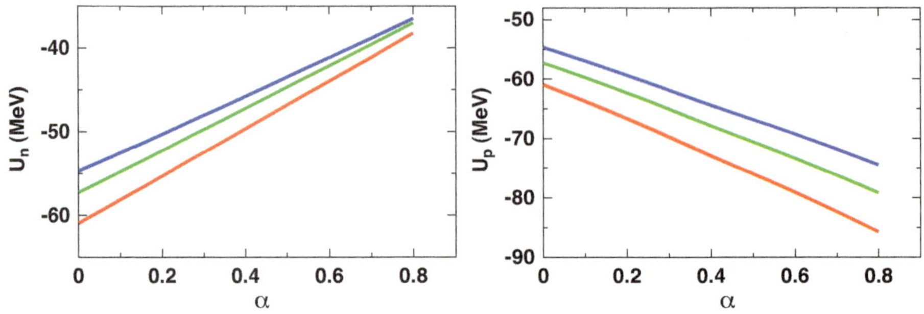

Fig. 7. The neutron and proton single-particle potentials as a function of the asymmetry parameter at fixed average density and momentum equal to the average Fermi momentum, which is chosen to be 1.4 fm^{-1}. The red, green, and blue lines represent the predictions from the Bonn A, B, and C potentials, respectively.

4. Neutron stars

4.1 A brief review of basic structure equations and available constraints

Fusion reactions in stars give rise to elements, but at the same time exhaust the nuclear fuel. After the fuel is exhausted, the star can die through four possible channels: it can become a black hole, a white dwarf, a neutron star, or it can disassemble completely. The ultimate outcome depends on the mass of the original star. If the mass is larger than about four solar masses, the star may become a supernova which, in turn, may result either in a neutron star or a black hole.

Neutron stars contain the most dense form of matter found in the universe and, therefore, are unique laboratories to study the properties of highly compressed (cold) matter. They are held together by gravity and neutron degeneracy pressure. (In contrast, white dwarfs are kept in hydrostatic equilibrium by gravity and electron degeneracy pressure.)

Although neutron stars were predicted as early as in the 1930's, hope for their observation remained slim for a long time. In 1967, strange new objects, outside the solar system, were observed at the University of Cambridge. They were named *pulsars*, as they emitted periodic radio signals. To date, about 1700 pulsars have been detected, many in binary systems.

Typically, detection of thermal radiation from the surface of a star is the way to access information about its properties. Furthermore, the possibility of exploring the structure of neutron stars via gravitational waves makes these exotic objects even more interesting.

The densities found in neutron stars range from the density of iron to several times normal nuclear density. Most of the mass consists of highly compressed matter at nuclear and supernuclear densities. The *surface* region is composed of normal nuclei and non-relativistic electrons, with typical mass densities in the range 10^4 g cm^{-3} $\leq \epsilon \leq 10^6$ g cm^{-3}. As density increases, charge neutrality requires matter to become more neutron rich. In this density range (about 10^6 g cm^{-3} $< \epsilon < 10^{11}$ g cm^{-3}), neutron-rich nuclei appear, mostly light metals, while electrons become relativistic. This is the *outer crust*. Above densities of approximately 10^{11} g cm^{-3}, free neutrons begin to form a continuum of states. The *inner crust* is a compressed solid with a fluid of neutrons which drip out and populate free states outside the nuclei, since those have become neutron-saturated. Densities in the inner crust range between 10^{11} g cm^{-3} and 10^{14} g cm^{-3}. At densities equal to approximately 1/2 of saturation density, clusters begin to merge into a continuum. In this phase, matter is a uniform fluid of neutrons, protons, and leptons. Above a few times nuclear matter density, the actual composition of stellar matter is not known. Strange baryons can appear when the nucleon chemical potential is of the order of their rest mass. Meson production can also take place. At even higher densities, transitions to other phases are speculated, such as a deconfined, rather than hadronic, phase. The critical density for such transition cannot be predicted reliably because it lies in a range where QCD is non perturbative (Sedrakian, 2007).

The possibility has been speculated that the most stable state at zero pressure may be u, d, s quark matter instead of iron. This would imply that strange quark matter is the most stable (in fact, the absolutely stable) state of strongly interacting matter, as originally proposed by Bodmer (Bodmer, 1971), Witten (Witten, 1984), and Terazawa (Terazawa, 1989). In such case, hyperonic and hybrid stars would have to be metastable with respect to stars composed of stable three-flavor strange quark matter (Weber, 1999), which is lower in energy than two-flavor quark matter due to the extra Fermi levels open to strange quarks. Whether or not strange quark stars can give rise to pulsar glitches (which are observed sudden small changes in the rotational frequency of a pulsar), may be a decisive test of the strange quark matter hypothesis (Weber, 1999).

The maximum gravitational mass of the star and the corresponding radius are the typical observables used to constraint the EoS. The gravitational mass is inferred mostly from observations of X-ray binaries or binary pulsars. Determination of the mass provide a unique test of both theories of nuclear matter and general relativity. The pulsar in the Hulse-Taylor binary system has a mass of 1.4408$\pm$ 0.0003 $M_\odot$, to date the best mass determination.

At this time, one of the heaviest neutron stars (with accurately known mass) has a mass of 1.671 $\pm$ 0.008 $M_\odot$ (Champion et al., 2008). The observation of an even heavier star has been confirmed recently, namely J1614-2230, with a mass of 1.97$\pm$0.04 $M_\odot$ (Demorest et al., 2010). This value is the highest measured with this certainty and represents a challenge for the softest EoS. We also recall that an initial observation of a neutron star in a binary system with a white dwarf had suggested a neutron star mass (PSR J0751+1807) of 2.1$\pm$0.2$M_\odot$ (Nice et al., 2005). Such observation, which would imply a considerable constraint on the high-density behavior of the EoS, was not confirmed.

The minumum mass of a neutron star is also a parameter of interest. For a cold, stable system, the minimum mass is estimated to be 0.09 $M_\odot$ (Lattimer & Prakash, 2007). The smallest reliably estimated neutron star mass is the companion of the binary pulsar J1756-2251, which has a mass of 1.18$\pm$ 0.02 $M_\odot$ (Faulkner et al., 2004).

Measurements of the radius are considerably less precise than mass measurements (Lattimer & Prakash, 2007). No direct measurements of the radius exist. Instead, the observed X-ray flux, together with theoretical assumptions (Weber, 1999), can provide information on the radiation or photospheric radius, R_∞, which is related to the actual stellar radius by $R_\infty = R(1 - 2GM/Rc^2)^{-1/2}$. Estimates are usually based on thermal emission of cooling stars, including redshifts, and the properties of sources with bursts or thermonuclear explosions at the surface. A major problem associated with the determination of radii is that the distance from the source is not well known, hence the need for additional assumptions. Much more stringent constraints could be imposed on the EoS if mass and radius were determined independently from each other.

Another bulk property of neutron stars is the moment of inertia, I. For softer EoS, both mass and radius are smaller and so is I. From observations of the Crab nebula luminosity, a lower bound on the moment of inertia was inferred to be $I \geq$4-8 $\times$ 10^{44} g cm^2, see (Weber, 1999) and references therein. A measurement of the moment of inertia within 10%, together with the information on the mass, would be able to discriminate among various EoS (Lattimer & Prakash, 2007). To date, the best determination of the moment of inertia is the one for the Crab pulsar (Beiger & Haensel, 2003) which would rule out only very soft EoS (Lattimer & Prakash, 2007).

A proton-neutron star is the result of a supernova explosion resulting from the gravitational collapse of a massive star core. Nearly all of the remaining binding energy is carried away by neutrinos during the first few tens of seconds of the evolution. Thus neutrino emission is very efficient as a cooling mechanism, with the internal temperature dropping to about 10^{10} K within a few days. Cooling through neutrino emission continues for a long time (in the order of 1,000 years), until the temperature drops to about 10^8 K, at which point photon emission becomes the dominant cooling mechanism. Neutrino luminosity and emission timescale are controlled by several factors including the total mass of the (proton-neutron) star and the opacity of neutrinos at high densities, which is sensitive to the EoS of dense hadronic matter.

Gravitational waves are a less conventional way to probe neutron star properties. Compact stars in binary systems are epected to produce gravitational radiation. In turn, emission of gravitational waves causes decay of the mutual orbits and eventually merger of the binary system. Because of the merger timescale (250 million years for PSR B1913+16, for instance, and 85 million years for PSR J0737-3039), it can be expected that many such decaying binary systems exist in the galaxy and emit large amounts of gravitational radiation. The observation of gravitational waves has the potential to set strong constraints on masses and radii, see see (Lattimer & Prakash, 2007) and references therein.

A theoretical estimate of the maximum possible mass of a neutron star was performed by Rhoades and Ruffini (Rhoades & Ruffini, 1974) on the following assumptions: 1) General relativity is the correct theory of gravitation; 2) the EoS satisfy the Le Chatelier's principle ($\partial P/\partial \epsilon \geq 0$) and the causality condition, $\partial P/\partial \epsilon \leq c^2$; and 3) the EoS below some matching density is known. On this basis, they determined that the maximum mass of a neutron star cannot exceed 3.2 solar masses. Abandoning the causality condition, which would hold exactly only if stellar matter is neither dispersive nor absorptive, this limit can be as high as 5 solar masses due to the increased stiffness of the EoS at supernuclear densities.

The maximum mass and the radius of a neutron star are sensitive to different aspects of the EoS. The maximum mass is mostly determined by the stiffness of the EoS at densities greater than a few times saturation density. The star radius is mainly sensitive to the slope of the symmetry energy. In particular, it is closely connected to the internal pressure (that is, the energy gradient) of matter at densities between about 1.5ρ_0 and 2-3ρ_0 (Lattimer & Prakash, 2007). Non-nucleonic degrees of freedom, which typically make their appearance at those densities, are known to have a considerable impact on the maximum mass of the star. The latter is predicted by the equation of hydrostatic equilibrium for a perfect fluid.

In general relativity, the invariant interval between two infinitesimally close space-time events is given by

$$ds^2 = g_{\alpha\beta} dx^\alpha dx^\beta \,, \tag{40}$$

where $g_{\alpha\beta}$ is the space-time metric. For a spherically symmetric space-time, the most general static line element consistent with all required symmetries has the form

$$ds^2 = -f(r)dt^2 + g(r)dr^2 + h^2(r)(sin^2\theta \, d\phi^2 + d\theta^2) \,. \tag{41}$$

Choosing a radial coordinate r such that $h^2(r) = r^2$ yields the so-called "standard" form of the metric.

The equation of hydrostatic equilibrium (the TOV equation) determines the form of the metric functions along with the pressure and the total mass-energy density as a function of the radial coordinate in the interior of the star. It reads

$$\frac{dP(r)}{dr} = -\frac{G}{c^2} \frac{(P(r) + \epsilon(r))(M(r) + 4\pi r^3 P(r)/c^2)}{r(r - 2GM(r)/c^2)} \,, \tag{42}$$

with

$$\frac{dM(r)}{dr} = 4\pi r^2 \epsilon(r) \,, \tag{43}$$

where ϵ is the total mass-energy density. The star *gravitational* mass is

$$M(R) = \int_0^R 4\pi r^2 \epsilon(r) dr \,, \tag{44}$$

where R is the value of r where the pressure vanishes. It's worth recalling that no mass limit exists in Newtonian gravitation.

The pressure is related to the energy per particle through

$$P(\rho) = \rho^2 \frac{\partial e(\rho)}{\partial \rho} \,. \tag{45}$$

The structure equations of rotationally deformed compact stars are much more complex than those of spherically symmetric stars (Weber, 1999) presented here. The most rapidly rotating pulsar, PSR J1748-2446 (Hessels et al., 2006), is believed to rotate at a rate of 716 Hz, although an X-ray burst oscillation at a frequency of 1122 Hz was reported (Kaaret et al., 2006), which may be due to the spin rate of a neutron star. Naturally, the maximum mass and the (equatorial) radius become larger with increasing rotational frequency.

4.2 Composition of β-stable matter

Assuming that only neutrons, protons, and leptons are present, the proton fraction in stellar matter under conditions of β-equilibrium is calculated by imposing energy conservation and charge neutrality. The resulting algebraic equations can be found in standard literature (Glendenning, 1997). The contribution to the energy density from the electrons is written as

$$e_e = \frac{\hbar c}{4\pi^2} (3\pi^2 \rho_e)^{4/3} \,, \tag{46}$$

whereas for muons we write

$$e_\mu = \rho_\mu m_\mu c^2 + (\hbar c)^2 \frac{(3\pi^2\rho_\mu)^{5/3}}{10\pi^2 m_\mu c^2} \,. \tag{47}$$

These contributions are added to the baryonic part to give the total energy density. The derivative of the total energy per particle with respect to the fraction of a particular species is the chemical potential of that species. The conditions

$$\mu_p + \mu_e = \mu_n \;;\; \mu_\mu = \mu_e \;;\; \rho_p = \rho_\mu + \rho_e \,, \tag{48}$$

allow to solve for the densities (or fractions) of protons, electrons, and muons. Near the saturation density, when the muon fraction is close to zero, one can estimate the equilibrium proton fraction, x_p, to be (Lattimer & Prakash, 2007)

$$x_p \approx \left(\frac{4e_{sym}(\rho_0)}{\hbar c}\right)^3 / (3\pi^2\rho_0). \tag{49}$$

The fractions of protons, electrons, and muons as predicted with the DBHF equation of state are shown in Fig. 8. The critical density for the proton fraction to exceed approximately 1/9 and, thus, allow cooling through the direct Urca processes,

$$n \rightarrow p + e + \bar{\nu}_e \quad and \quad p + e \rightarrow n + \nu_e \,, \tag{50}$$

is about 0.36 – 0.39 fm^{-3}. Notice that, due to the relation between symmetry energy and proton fraction, large values of the symmetry energy would make the star cool rapidly. In fact, already in earlier studies (Boguta, 1981) tha rapid cooling of neutron stars and the corresponding high neutrino luminosity was understood in terms of neutron β decay and large proton fractions.

At densities close to normal nuclear density, protons and neutrons are the only baryonic degrees of freedom. As density increases, other baryons begin to appear, such as strange baryons or isospin 3/2 nucleon resonances. Hyperonic states can be classified according to the irreducible representation of the $SU(3)$ group. The octet of baryons that can appear in neutron matter includes nucleons, Λ, $\Sigma^{0,\pm}$, and $\Xi^{0,-}$.

Neglecting the nucleon-hyperon interaction, the threshold for stable hyperons to exist in matter is determined by comparing the hyperon mass with the neutron Fermi energy, which is the largest available energy scale in neutron-rich matter. We consider cold neutron stars, after neutrinos have escaped. Strange baryons appear at about 2-3 times normal density (Baldo et al., 1998), an estimate which is essentially model independent, through the processes $n + n \rightarrow p + \Sigma^-$ and $n + n \rightarrow n + \Lambda$. The equilibrium conditions for these reactions are

$$2\mu_n = \mu_p + \mu_{\Sigma^-} \;;\; \mu_n = \mu_\Lambda \,. \tag{51}$$

Also, we have

$$\mu_e = \mu_\mu \;;\; \mu_n = \mu_p + \mu_e \,, \tag{52}$$

the equations above being special cases of

$$\mu = b\mu_n - q\mu_e, \tag{53}$$

where b and q are the baryon number and the charge (in units of the electron charge) of the particular species with chemical potential μ. Together with the charge neutrality condition and baryon number conservation,

$$\rho_p = \rho_e + \rho_\mu + \rho_{\Sigma^-} \; ; \; \rho = \rho_n + \rho_p + \rho_{\Sigma^-} + \rho_\Lambda, \tag{54}$$

the above system allows to determine the various particle fractions.

Naturally, the composition of matter at supra-nuclear densities determines the behavior of stellar matter. It is also speculated that a transition to a quark phase may take place at very high densities, the occurrence of which depends sensitively on the properties of the EoS in the hadronic (confined) phase. The presence of hyperons in the interior of neutron stars is reported to soften the equation of state, with the consequence that the predicted neutron star maximum masses become considerably smaller (Schulze et al., 2006). Strange baryons are not included in the predictions shown below.

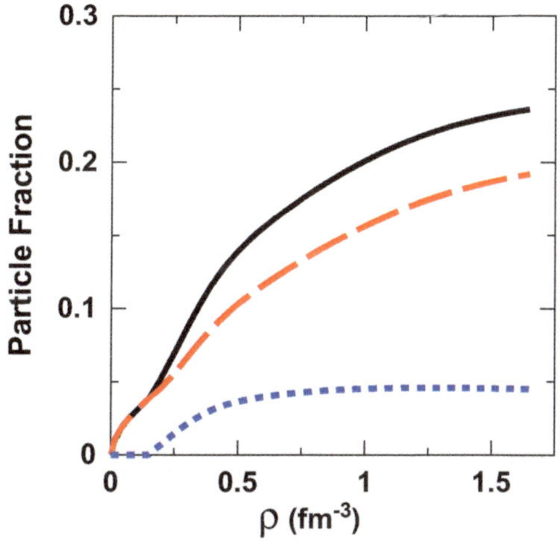

Fig. 8. Proton (solid black), electron (dashed red), and muon (dotted blue) fractions in β-stable matter as a function of total baryon density as predicted by the DBHF model.

4.3 Microscopic predictions of neutron star properties

We are now ready to move to applications of our EoS to compact stars.

As explained in Section **3**, the DBHF model does not include three-body forces explicitely, but effectively incorporates the class of TBF originating from the presence of nucleons and antinucleons (the "Z-diagrams" in Fig. 2), see discussion in Section **3.1**. In order to broaden our scopes, we will compare our predictions with those of other microscopic models. As the other element of our comparison, we will take the EoS from the microscopic approach of (Li

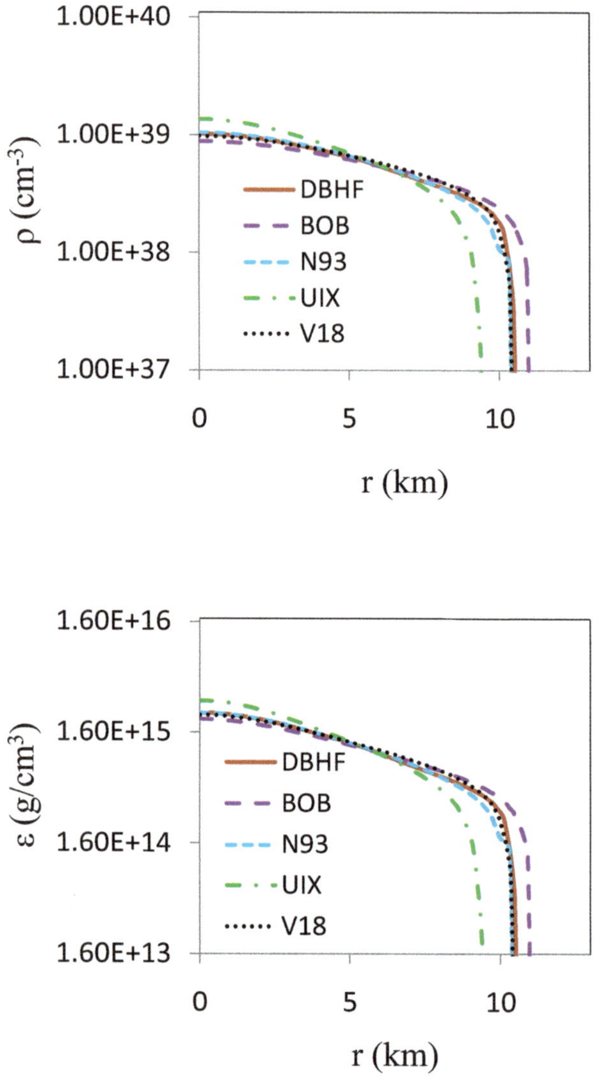

Fig. 9. The baryon density and the mass-energy density profile for a neutron star with the maximum mass allowed by each EoS model.

et al., 2008). There (and in previous work by the same authors), the Brueckner-Hartree-Fock (BHF) formalism is employed along with microscopic three-body forces. In particular, in (Li & Schulze, 2008) the meson-exchange TBF are constructed applying the same parameters as used in the corresponding nucleon-nucleon potentials, which are: Argonne V18 (Wiringa et al., 1995) (V18), Bonn B (Machleidt, 1989) (BOB), Nijmegen 93 (Stocks et al., 1994) (N93). The popular (but phenomenological) Urbana TBF (Pieper et al., 2001) (UIX) is also utilized in (Li &

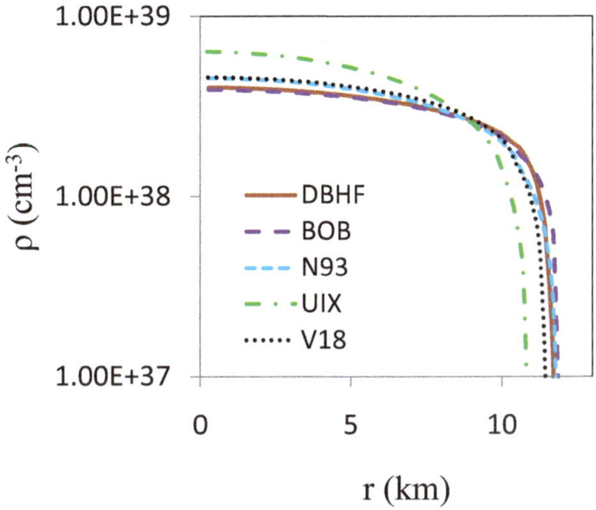

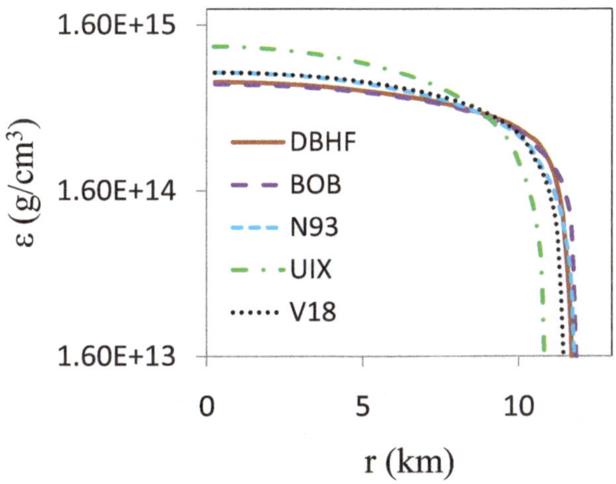

Fig. 10. The baryon density and the mass-energy density profile for a neutron star with a mass of 1.4 solar masses.

Schulze, 2008). Convenient parametrizations in terms of simple analytic functions are given in all cases and we will use those to generate the various EoS. We will refer to this approach, generally, as "BHF + TBF".

At subnuclear densities all the EoS considered here are joined with the crustal equations of state from Harrison and Wheeler (Harrison et al., 1965) and Negele and Vautherin (Negele &

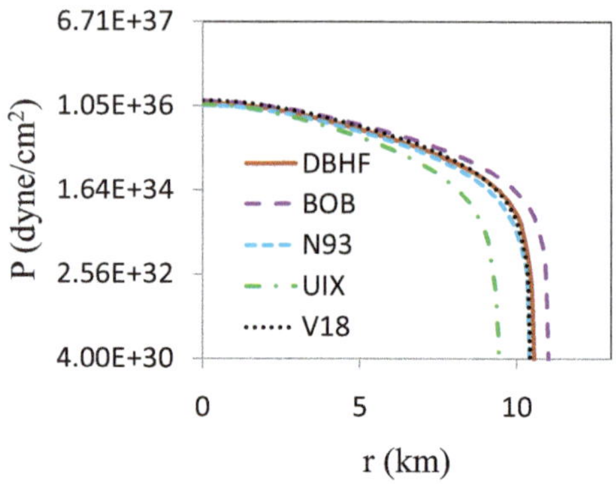

Fig. 11. Pressure profile for the maximum-mass star allowed by each EoS model.

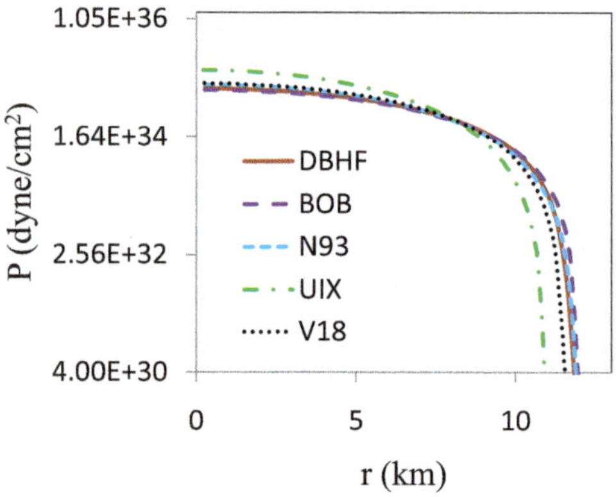

Fig. 12. Pressure profile predicted by the various models for a 1.4 solar mass star.

Vautherin, 1973). The composition of the crust is crystalline, with light (Harrison et al., 1965) or heavy (Negele & Vautherin, 1973) metals and electron gas.

We begin with showing the baryon number density and the mass-energy density profile of the star, see Fig. 9. For each EoS model, the maximum mass configuration is considered. Thus the models differ in their central density, which, in turn, impact the radius. The relations shown in Fig. 9 are insightful, as they reveal the detailed structure of the star at each radial position. Furthermore, the compactness of the star, whose density profile is reminescent of the one in a nucleus, a system 55 orders of magnitude lighter, is apparent. In Fig. 10, the same quantities are shown for a star with a mass of 1.4 solar masses, the most probable mass of a neutron star.

The models labeled as UIX and BOB have the smallest and largest radius, respectively, as can be seen from the figure. We also see that the star's outer regions, that is, for energy densities less than about 10^{14} g cm^{-3}, are influenced quite strongly by differences in the various EoS models. Note that the UIX model, with the smallest radius, can tolerate larger central densities.

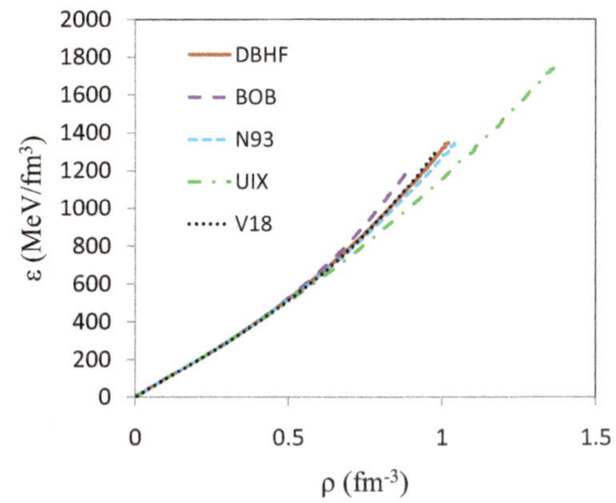

Fig. 13. Energy density *vs.* baryon number density for the various EoS being addressed in the text. The maximum-mass model is considered in each case.

Of interest is also the pressure profile for the maximum-mass star in each model, which is shown in Fig. 11 for the maximum mass and in Fig. 12 for a 1.4 solar mass star.

The $\epsilon(r)$ and $\rho(r)$ relations from Fig. 9 are combined to provide the $\epsilon(\rho)$ relation within the star as shown in Fig. 13 for the maximum mass. Again, we see that the stiffest (BOB) and softest EoS (UIX) support the smallest and largest central densities, respectively. At the same time, these two EoS predict the largest (BOB) and smallest (UIX) maximum mass, see below.

In Fig. 14, we show the mass-radius relation for a sequence of static neutron stars as predicted by the various models. All models besides DBHF share the same many-body approach (BHF+TBF) but differ in the two-body potential and TBF employed. The differences resulting from the use of different NN potentials can be larger than those originating from emplying different many-body approaches. This can be seen by comparing the DBHF and BOB curves, both employing the Bonn B interaction (although in the latter case the non-relativistic, r-space version of the potential is adopted). Overall, the maximum masses range from $1.8M_\odot$ (UIX) to $2.5M_\odot$ (BOB). Radii are less sensitive to the EoS and range between 10 and 12 km for all models under consideration, DBHF or BHF+TBF. Concerning consistency with present constraints, the observations reported in Section **4.1** would appear to invalidate only the model with the smallest maximum mass, UIX. Notice, further, that phenomena such as condensation of mesons may soften the EoS considerably at supernuclear density as condensation would bring loss of pressure.

Also of interest is the star baryon number, A, which is obtained by integrating the baryon density over the proper volume (Weber, 1999). Namely,

$$A = 4\pi \int_0^R dr\, r^2 \frac{\rho(r)}{\sqrt{1 - 2GM(r)/(rc^2)}} \,. \tag{55}$$

Defining the star's baryon mass as

$$M_A = m_n\, A \,, \tag{56}$$

where m_n is the mass of the baryon, one can calculates the star's binding energy, defined as

$$E_B = M - M_A \,. \tag{57}$$

The baryon number and the star binding energy as a function of the central density (in units of nuclear matter density) are shown in Fig. 15 and Fig. 16 for the various models. We see that the baryon number for stable stars is approximately equal to $10^{56} - 10^{57}$. A much higher value would make the star unstable with respect to gravitational collapse. The binding energy displayed in Fig. 16 is defined in units of the solar mass. Typically, the binding energy changes the sign for masses less than 0.1 solar masses. The binding energy is a potentially observable quantity, since neutrinos from a supernova carry information about the gravitational binding energy of the neutron star that has resulted from the explosion.

Next, we calculate the gravitational redshift predicted by each model. The redshift is defined as

$$z = \frac{\nu_E}{\nu_\infty} - 1 \,, \tag{58}$$

where ν_E and ν_∞ are the photon frequencies at the emitter and at the infinitely far receiver. The photon frequency at the emitter is the inverse of the proper time between two wave crests in the frame of the emitter,

$$\frac{1}{d\tau_E} = (-g_{\mu\nu} dx^\mu dx^\nu)_E^{-1/2} \,, \tag{59}$$

with a similar expression for the frequency at the receiver. Then

$$\frac{\nu_\infty}{\nu_E} = \frac{((-g_{00})^{1/2} dx^0)_E}{((-g_{00})^{1/2} dx^0)_\infty} \,. \tag{60}$$

Assuming a static gravitational field, in which case the time dx^0 between two crests is the same at the star's surface and at the receiver, and writing g_{00} as the metric tensor component at the surface of a nonrotating star yield the simple equation

$$z = \left(1 - \frac{2MG}{Rc^2}\right)^{-1/2} - 1 \,. \tag{61}$$

Notice that simultaneous measurements of R_∞ and z determines both R and M, since

$$R = R_\infty (1+z)^{-1} \,, \tag{62}$$

and

$$M = \frac{c^2}{2G} R_\infty (1+z)^{-1} [1 - (1+z)^{-2}] \,. \tag{63}$$

In Fig. 17 we show the gravitational redshift as a function the mass for each model. Naturally the rotation of the star modifies the metric, and in that case different considerations need to be applied which result in a frequency dependence of the redshift. We will not consider the general case here.

We conclude this section with showing a few predictions for the case of rapidly rotating stars. The model dependence of the mass-radius relation is shown in Fig. 18. The 716 Hz frequency corresponds to the most rapidly rotating pulsar, PSR J1748-2446, (Hessels et al., 2006) although

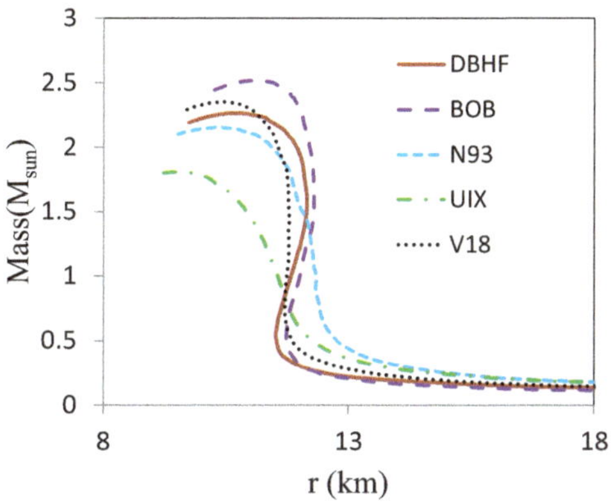

Fig. 14. Static neutron star mass-radius relation for the models considered in the text.

recently an X-ray burst oscillation at a frequency of 1122 Hz has been reported (Kaaret et al., 2006) which may be due to the spin rate of a neutron star. As expected, the maximum mass and the (equatorial) radius become larger with increasing rotational frequency.

In Fig. 19, we show the moment of inertia at different rotational speeds (again, for all models). These values are not in contradiction with observations of the Crab nebula luminosity, from which a lower bound on the moment of inertia was inferred to be $I \geq$4-8 $\times$ 10^{44} g cm^2, see (Weber, 1999) and references therein.

Clearly, at the densities probed by neutron stars the model dependence is large, but presently available constraints are still insufficient to discriminate among these EoS. The model dependence we observe comes from two sources, the two-body potential and the many-body approach, specifically the presence of explicit TBF or Dirac effects. The dependence on the two-body potential is very large. Typically, the main source of model dependence among NN potentials is found in the strength of the tensor force. Of course, differences at the two-body level impact the TBF as well, whether they are microscopic or phenomenological.

5. Polarized isospin-asymmetric matter

Before concluding this chapter, we like to touch upon the issue of polarization in IANM.

When both isospin and spin asymmetries are present, constraints are much more difficult to obtain and predictions regarding magnetic properties of nuclear matter are sometimes found to be in qualitative disagreement with one another. This is especially the case with regard to the possibility of spontaneous phase transitions into spin ordered states, ferromagnetic (FM, with neutron and proton spins alligned), or antiferromagnetic state (AFM, with opposite spins for neutrons and protons). Notice that the presence of polarization would impact neutrino cross section and luminosity, resulting into a very different scenario for neutron star cooling, which is why we find it appropriate to briefly discuss this issue here.

Recently, we have extended the framework described in Section **3** (Sammarruca, 2011) to include both spin and isospin asymmetries of nuclear matter and calculated the energy per particle under extreme conditions of polarization. The existence (or not) of a possible phase

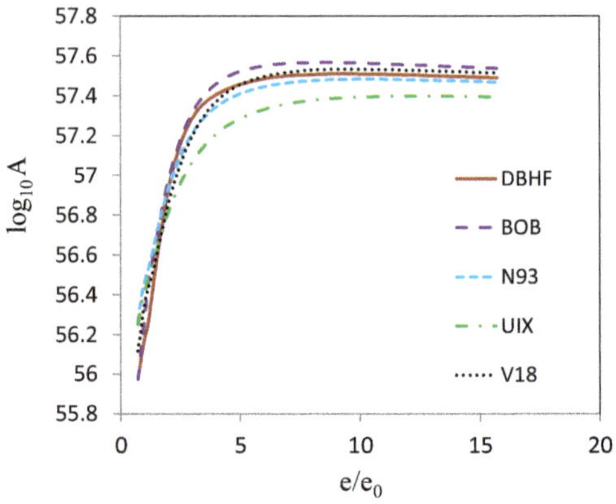

Fig. 15. Baryon number as a function of the central density (in units of $e_0 = 2.5\ 10^{14}$ g cm^{-3}) for the models considered in the text.

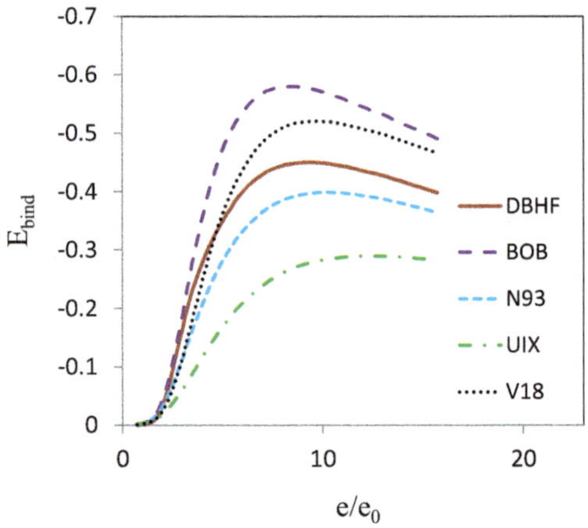

Fig. 16. Binding energy (in units of the solar mass) as a function of the central density for the EoS models considered in the text.

transition can be argued by comparing the energies of the fully polarized and the unpolarized phases.

In a spin-polarized and isospin asymmetric system with fixed total density, ρ, the partial densities of each species are

$$\rho_n = \rho_{nu} + \rho_{nd}\,, \quad \rho_p = \rho_{pu} + \rho_{pd}\,, \quad \rho = \rho_n + \rho_p\,, \tag{64}$$

where u and d refer to up and down spin-polarizations, respectively, of protons (p) or neutrons (n). The isospin and spin asymmetries, α, β_n, and β_p, are defined in a natural way:

$$\alpha = \frac{\rho_n - \rho_p}{\rho}\,, \quad \beta_n = \frac{\rho_{nu} - \rho_{nd}}{\rho_n}\,, \quad \beta_p = \frac{\rho_{pu} - \rho_{pd}}{\rho_p}\,. \tag{65}$$

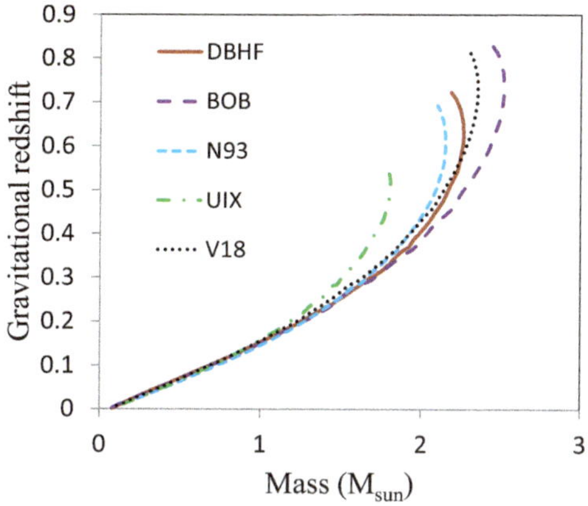

Fig. 17. Gravitational redshift for all models. For each model, the corresponding sequence of static stars is considered.

The single-particle potential of a nucleon in a particular $\tau\sigma$ state, $U_{\tau\sigma}$, is now the solution of a set of four coupled equations, which are the appropriate extension of Eqs. (24-25). They read

$$U_{nu} = U_{nu,nu} + U_{nu,nd} + U_{nu,pu} + U_{nu,pd} \quad (66)$$

$$U_{nd} = U_{nd,nu} + U_{nd,nd} + U_{nd,pu} + U_{nd,pd} \quad (67)$$

$$U_{pu} = U_{pu,nu} + U_{pu,nd} + U_{pu,pu} + U_{pu,pd} \quad (68)$$

$$U_{pd} = U_{pd,nu} + U_{pd,nd} + U_{pd,pu} + U_{pd,pd}\ , \quad (69)$$

to be solved self-consistently along with the two-nucleon G-matrix. In the above equations, each $U_{\tau\sigma,\tau'\sigma'}$ term contains the appropriate (spin and isospin dependent) part of the interaction, $G_{\tau\sigma,'\tau'\sigma'}$. More specifically,

$$U_{\tau\sigma}(\vec{k}) = \sum_{\sigma'=u,d}\sum_{\tau'=n,p}\sum_{q\leq k_F^{\tau'\sigma'}} < \tau\sigma,\tau'\sigma'|G(\vec{k},\vec{q})|\tau\sigma,\tau'\sigma' >, \quad (70)$$

where the third summation indicates integration over the Fermi seas of protons and neutrons with spin-up and spin-down. Notice that this equation is the extension of Eq. (32) in the presence of spin polarization.

In the left panel of Fig. 20, we show, in comparison with unpolarized symmetric matter (solid line): the EoS for the case of fully polarized neutrons and completely unpolarized protons (dashed line); the EoS for the case of protons and neutrons totally polarized in the same direction, that is, matter in the ferromagnetic (FM) state (dashed-dotted line); the EoS for the case of protons and neutrons totally polarized in opposite directions, namely matter in the antiferromagnetic (AFM) state (dotted line). A similar comparison is shown in the right panel of Fig. 20, but for isospin asymmetric matter. (Notice that all predictions are invariant under a global spin flip.)

We conclude that, for both symmetric and asymmetric matter, the energies of the FM and AFM states are higher than those of the corresponding unpolarized cases, with the AFM state

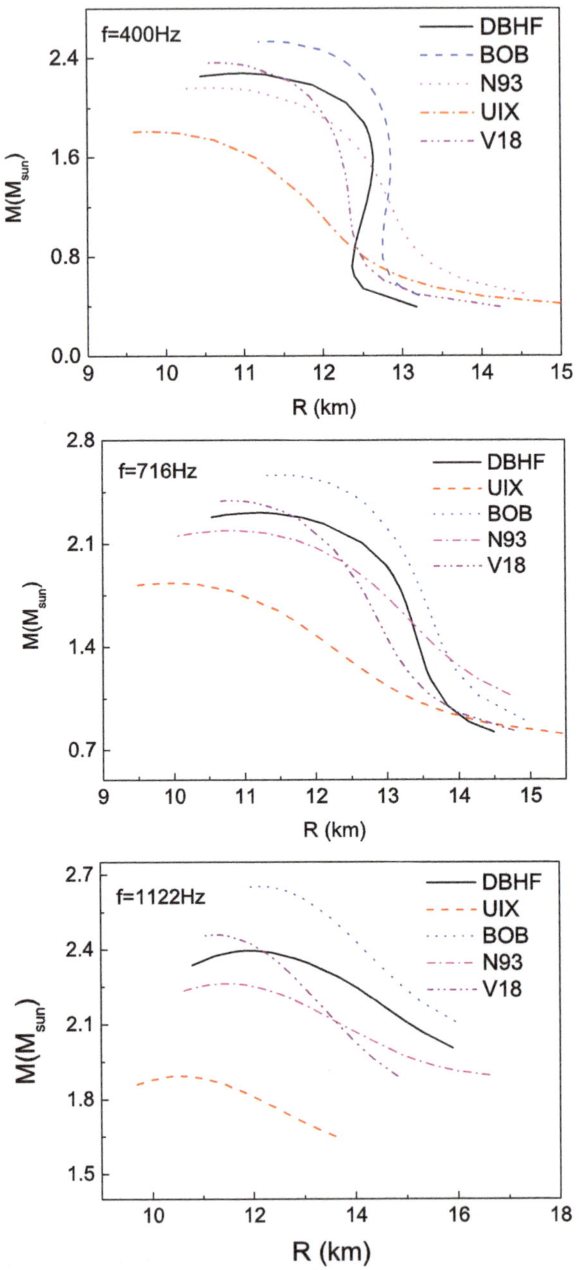

Fig. 18. Mass-radius relation for the models considered in the text and for different rotational frequencies.

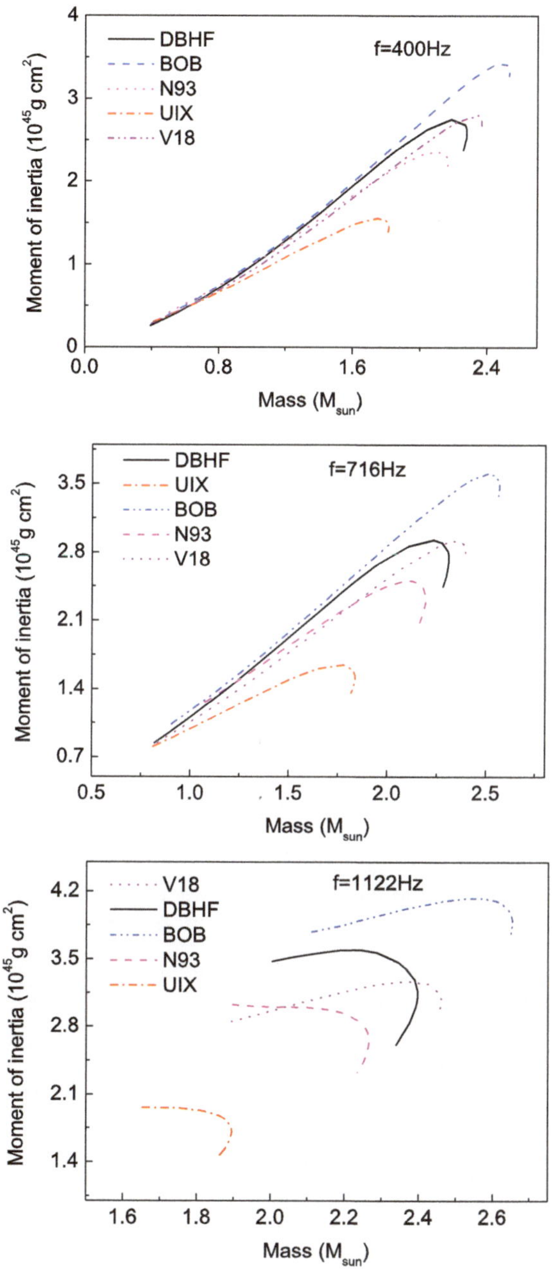

Fig. 19. Moment of inertia for the models considered in the text and for different rotational frequencies.

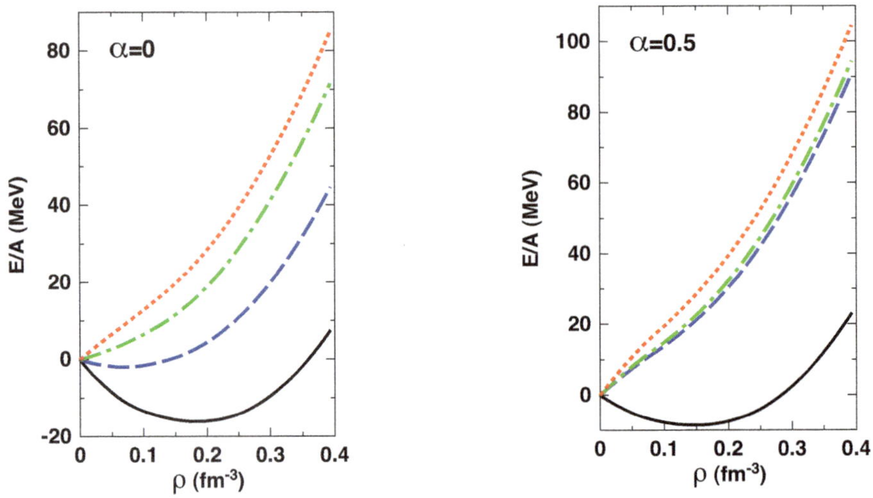

Fig. 20. The energy per particle as a function of density and variuos degrees of proton and neutron polarizations in symmetric matter (left) and asymmetric matter (right). In both frames, the (blue) dashed line corresponds to totally polarized neutrons and unpolarized protons (β_n=1, β_p=0); the (green) dash-dotted line is the prediction for the FM state (β_n=1, β_p=1); the (red) dotted line shows the energy of the AFM state (β_n=1, β_p=-1). The (black) solid line shows the predictions for unpolarized matter.

being the most energetic. Thus, a phase transition to a spin-ordered state is not anticipated in our model. This conclusion seems to be shared by predictions of microscopic models, such as those based on conventional Brueckner-Hartree-Fock theory (Vidaña & Bombaci, 2002). On the other hand, calculations based on various parametrizations of Skyrme forces result in different conclusions. For instance, with the *SLy4* and *SLy5* forces and the Fermi liquid formalism a phase transition to the AFM state is predicted in asymmetric matter at a critical density equal to about 2-3 times normal density (Isayev & Yang, 2004).

In closing this brief section, it is interesting to remark that models based on realistic nucleon-nucleon potentials, whether relativistic or non-relativistic, are at least in qualitative agreement with one another in predicting more energy for totally polarized states (FM or AFM) up to densities well above normal density.

6. Summary and conclusions

In this chapter, we have been concerned with the nuclear equation of state of isospin asymmetric nuclear matter, the main input for calculations of the properties of compact stars as well as a variety of other systems, such as the neutron skin of neutron-rich nuclei.

After describing our microscopic approach to the development of the equation of state for nuclear matter and neutron-rich matter, we presented a brief review of the structure equations leading to the prediction of neutron star properties. Microscopic predictions from different models employing three-body forces along with the non-relativistic Brueckner-Hartree-Fock method have also been shown for comparison. Large model dependence is seen among predictions, especially those involving the highest densities.

Rich and diverse effort is presently going on to improve the available constraints on the EoS or find new ones. These constraints are usually extracted through the analysis of selected heavy-ion collision observables. At the same time, partnership between nuclear physics and astrophysics is becoming increasingly important towards advancing our understanding of exotic matter. The recently approved Facility for Rare Isotope Beams (FRIB), thanks to new powerful technical capabilities, will forge tighter links between the two disciplines, as it will allow access to rare isotopes which play a critical role in astrophysical processes but have not yet been observed in terrestrial laboratories.

7. Acknowledgements

Support from the U.S. Department of Energy is gratefully acknowledged. I am deeply indebted to Prof. F. Weber for the use of his TOV code and many helpful advises and suggestions. I like to thank Mr. Boyu Chen for help with the preparation of this manuscript. Some of the neutron star properties have been calculated using public software downloaded from the website *http://www.gravity.phys.uwm.edu/rns*.

8. References

Alonso, D. & Sammarruca, F. (2003). Microscopic Calculations in Asymmetric Nuclear Matter. *Phys. Rev. C* Vol. 67, 054301, pp. 1-16.

Baldo, M., Burgio, G.F. & Schulze, H.-J. (1998). Onset of Hyperon Formation in Neutron Star Matter from Brueckner Theory. *Phys. Rev. C* Vol. 58, pp. 3688-3695.

Bejger, M. & Haensel, P. (2003). Accelerated expansion of the Crab nebula and evolution of its neutron-star parameters. *Astron. Astrophys.* Vol. 405, pp. 747-752.

Bodmer, A.R. (1971). Collapsed Nuclei. *Phys. Rev. D* Vol. 4, pp. 1601-1606.

Boguta, J. (1981). Remarks on the Beta Stability in Neutron Stars. *Phys. Lett. B* Vol. 106, pp. 255-258.

Borderie, B. & Rivet, M.F. (2008). Nuclear Multifragmentation and phase transition for hot nuclei. *Prog. Part. Nucl. Phys.* Vol. 61, pp. 551-601.

Brockmann, R. & Machleidt, R. (1990). Relativistic Nuclear Structure I. Nuclear Matter. *Phys. Rev. C* Vol. 42, pp. 1965-1980.

Brown, G.E. (1979). Chiral Symmetry and the Nucleon-Nucleon Interaction, *in* Rho, M. & Wilkinson, D.H. (Ed.), *Mesons in Nuclei*, Vol. I, North-Holland, Amsterdam, pp. 330-356.

Brown, G.E., Weise, W., & Baym, G. (1987). Relativistic effects in Nuclear Physics. *Comments Nucl. Part. Phys.* Vol. 17, pp. 39-62.

Champion, D.J. et al. (2008). An eccentric Binary Millisecond Pulsar in the Galactic Plane. *Science* Vol. 320, pp. 1309-1312.

Danielewicz, P., Lacey, R. and Lynch, W.G. (2002). Determination of the equation of state of dense matter. *Science* Vol. 298, pp. 1592-1596.

Demorest, P. et al. (2010). Shapiro delay measurement of a two solar mass neutron star. *Nature* Vol. 468, pp. 1081-1083.

Faulkner, A.J. et al. (2004). A new relativistic double neutron star system. *Astrophys. J.* Vol. 618, pp. L119-L122.

Feng, Z.-Q. & Jin, G.-M. (2010). Probing the Symmetry Energy at Supra-Saturation Densities from Pion Emission in Heavy-Ion Collisions, *Proceedings of the International Workshop on Nuclear Dynamics in Heavy-Ion Reactions and the Symmetry Energy*, pp. 1686-1693, Shanghai, China, August 23-25, 2009, Special Issue of Int. J. Mod. Phys. E Vol. 19, World Scientific, Singapore.

Furnstahl, R.J. (2002). Neutron radii in mean-field models. *Nucl. Phys. A* Vol. 706, pp. 85-110.

Glendenning, N.K. (1997). *Compact Stars, Nuclear Physics, Particle Physics, and General Relativity*, Springer-Verlag, New York.

Greco, V. et al. (2010). Probing the Symmetry Energy at High Baryon Density with Heavy-Ion Collisions, *Proceedings of the International Workshop on Nuclear Dynamics in Heavy-Ion Reactions and the Symmetry Energy*, pp. 1664-1674, Shanghai, China, August 23-25, 2009, Special Issue of Int. J. Mod. Phys. E Vol. 19, World Scientific, Singapore.

Haftel, M.I. & Tabakin, F. (1970). Nuclear Saturation and the Smoothness of Nucleon-Nucleon Potentials. *Nucl. Phys. A* Vol. 158, pp. 1-42.

Harrison, B.K., Thorne, K.S., Wakano, M., and Wheeler, J.A. (1965). *Gravitation Theory and Gravitational Collapse*, University of Chicago Press, Chicago.

Hessels, J.W.T. et al. (2006). A radio pulsar spinning at 716-Hz. *Science* Vol. 311, pp. 1901-1904.

Isayev, A.A. & Yang, J. (2004). Antiferromagnetic spin phase transition in nuclear matter with effective Gogny interaction. *Phys. Rev. C* Vol. 70, 064310, pp. 1-6.

Isobe, T. (2011). Overview of SAMURAI/TPC Project. Talk presented at the International Symposium on Nuclear Symmetry Energy, Smith College, June 17-20, 2011.

Kaaret, P. et al. (2006). Discovery of 1122-Hz X-Ray Burst Oscillations from the Neutron Star X-Ray Transient XTE J1739-285. arXiv:0611716 [astro-ph].

Klähn, T. et al. (2006). Constraints on the high-density nuclear equation of state from the phenomenology of compact stars and heavy-ion collisions. *Phys. Rev. C* Vol. 74, 035802, pp. 1-15.

Ko, C.M. et al. (2010). Medium Effects on Charged Pion Ratio in Heavy Ion Collisions. *Proceedings of the International Workshop on Nuclear Dynamics in Heavy-Ion Reactions and the Symmetry Energy*, pp. 1763-1772, Shanghai, China, August 23-25, 2009, Special Issue of Int. J. Mod. Phys. E Vol. 19, World Scientific, Singapore.

Kohley, Z. et al. (2011). Transverse collective flow and midrapidity emission of isotopically identified light charged particles. *Phys. Rev. C* Vol. 83, 044601, pp. 1-10.

Lane, A.M. (1962). Isobaric spin dependence of the optical potential and quasi-elastic (p,n) reactions. *Nucl. Phys.* Vol. 35, p. 676-685.

Lattimer, J.M. & Prakash, M. (2007). Neutron star observations: Prognosis for equation of state constraints. *Phys. Rep.* Vol. 442, pp. 109-165.

Li, B.-A. & Chen, L.-W. (2005). Nucleon-nucleon cross sections in neutron-rich matter and isospin transport in heavy-ion reactions at intermediate energies. *Phys. Rev. C* Vol. 72, 064611, pp. 1-12.

Li, Z.H. et al. (2008). Consistent nucleon-nucleon potentials and three-body forces. *Phys. Rev. C* Vol. 77, p. 034316 (1-10).

Li, Z.H. & Schulze, H.-J. (2008). Neutron star structure with modern nucleonic three-body forces. *Phys. Rev. C* Vol. 78, 028801, pp. 1-4.

Machleidt, R. (1989). The meson theory of nuclear forces and nuclear structure, *in* Negele, J.W. & Vogt, E. (ed.), *Adv. Nucl. Phys.*, Plenum Press, New York, Vol. 19, pp. 189-376.

Machleidt, R. (2001). High-precision, charge-dependent Bonn nucleon-nucleon potential. *Phys. Rev. C* Vol. 63, 024001, pp. 1-32.

Negele, J.W. & Vautherin, D. (1973). Neutron star matter at subnuclear densities. *Nucl. Phys. A* Vol. 207, pp. 298-320.

Nice, D.J. et al. (2005). A 2.1 solar mass pulsar measured by relativistic orbital decay. *Astrophys. J.* Vol. 634, pp. 1242-1249.

Oyamatsu, K., Tanihata, I., Sugahara, Y., Sumiyoshi, K., & Toki, H. (1998). Can the equation of state of asymmetric nuclear matter be studied using unstable nuclei? *Nucl. Phys. A* Vol. 634, p. 3-14.

Pieper, S.C., Pandharipande, V.R., Wiringa, R.B., & Carlson, J. (2001). Realistic models of pion-exchange three-nucleon interactions. *Phys. Rev. C* Vol. 64, 014001, pp. 1-21.

Rhoades, C.D. & Ruffini, R. (1974). Maximum mass of a neutron star. *Phys. Rev. Lett.* Vol. 32, pp. 324-327.

Salpeter, E.E. & Bethe, H.A. (1951). A Relativistic Equation for Bound-State Problems. *Phys. Rev.* Vol. 84, pp. 1232-1242.

Sammarruca, F. & Liu, P. (2009). Neutron skin of ^{208}Pb and density dependence of the symmetry energy. *Phys. Rev. C* Vol. 79, 057301, pp. 1-4.

Sammarruca, F. (2010). The microscopic approach to nuclear matter and neutron star matter. *Int. J. Mod. Phys. E* Vol. 19, pp. 1259-1313.

Sammarruca, F. (2011). Spin- and isospin-polarized states of nuclear matter in the Dirac-Brueckner-Hartree-Fock model. *Phys. Rev. C* Vol. 83, p. 064304 (1-6).

Sammarruca, F. (2011). Contribution of isovector mesons to the symmetry energy in a microscopic model. arXiv:1107.3339 [nucl-th].

Schulze, H.-J., Polls, A., Ramos, A., & Vidaña, I. (2006). Maximum mass of neutron stars. *Phys. Rev. C* Vol. 73 058801, pp. 1-4.

Sedrakian, A. (2007). The physics of dense hadronic matter and compact stars. *Prog. Part. Nucl. Phys.* Vol. 58, pp. 168-246.

Sfienti, C. *et al.* (2009). Isotopic Dependence of the Nuclear Caloric Curve. *Phys. Rev. Lett.* Vol. 102, 152701, pp. 1-4.

Simenel, C., Chomaz, Ph., & de France, G. (2007). Fusion process with a preequilibrium giant dipole resonance in time-dependent Hartree-Fock theory. *Phys. Rev. C* Vol. 76, 024609, pp. 1-14.

Stocks, V.G.J., Klomp, R.A.M., Terheggen, C.P.F., & de Swart, J.J. (1994). Construction of high-quality NN potential models. *Phys. Rev. C* Vol. 49, pp. 2950-2962.

Terazawa, H. (1989). Superhypernuclei in the Quark Shell Model. *J. Phys. Soc. Japan* Vol. 58, pp. 3555-3563.

Thompson, R.H. (1970). Three-Dimensional Bethe-Salpeter Equation Applied to the Nucleon-Nucleon Interaction. *Phys. Rev. D* Vol. 1, pp. 110-117.

Trautmann, W. et al. (2009). Neutron-proton elliptic flow in Au+Au. arXiv:0907.2822 [nucl-ex].

Tsang, M.B. et al. (2009). Constraints on the density dependence of the symmetry energy. *Phys. Rev. Lett.* Vol. 102, 122701, pp. 1-12.

Vidaña, I. & Bombaci, I. (2002). Equation of state and magnetic susceptibility of spin polarized isospin asymmetric nuclear matter. *Phys. Rev. C* Vol. 66, 045801, pp. 1-11.

Weber, F. (1999). *Pulsars as Astrophysical Laboratories for Nuclear and Particle Physics*, Institute of Physics Publishing, Bristol and Philadelphia.

Weinberg, S. (1990). Nuclear forces from chiral Lagrangians. *Phys. Lett. B* Vol. 251, pp. 288-292.

Wiringa, R.B., Stocks, V.G.J., & Schiavilla, R. (1995). Accurate nucleon-nucleon potential with charge-independence breaking. *Phys. Rev. C* Vol. 51, pp. 38-51.

Witten, E. (1984). Cosmic separation of phases. *Phys. Rev. D* Vol. 30, pp. 272-284.

Xu, C., Li, B.-A., & Chen, L.-W. (2010). Symmetry energy, its density slope, and neutron-proton effective mass splitting at normal density extracted from global nucleon optical potentials. *Phys. Rev. C* Vol. 82, 054607, pp. 1-5.

Yong, G.-C., Li, B.-A., & Chen, L.-W. (2010). Triton-^{3}HE Relative and Differential Flows and the High Density Behavior of Nuclear Symmetry Energy. *Proceedings of the International Workshop on Nuclear Dynamics in Heavy-Ion Reactions and the Symmetry Energy*, pp. 1647-1652, Shanghai, China, August 23-25, 2009, Special Issue of Int. J. Mod. Phys. E Vol. 19, World Scientific, Singapore.

10

Methods for Image Recognition of Charged Particle Tracks in Track Detector Data Automated Processing

A.B. Aleksandrov, N.G. Polukhina and N.I. Starkov
Lebedev Physical Institute, Russian Academy of Sciences, Moscow, Russia

1. Introduction

The completely automated measuring facility PAVICOM has been created at the Lebedev Physical Institute and is successfully used for processing the data of track detectors (nuclear emulsions, X-ray films, olivine crystals from meteorites, mylar, CR-309) (Aleksandrov et al., 2005a, 2005b, 2006, 2007a, 2007b, 2007c, 2010; Alexandrov et al., 2008; Belovitsky et al., 2006; Vladymyrov et al., 2008).

Track detectors have been widely used in particle physics experiments for many decades now. Such a long-standing life of the method is, certainly, due to the unique spatial resolution and the possibility of separating particle tracks.

The first track detectors were extremely simple: they were common photographic plates used by photographers at that time (early 20th century), as well as electroscopes. The Wilson cloud chamber makes use of the condensation of liquid from supersaturated vapour (under suitable conditions, ionization performed by a charged particle in a substance may cause a phase transition in it). The device was invented in 1912 by Charles T.R. Wilson (Wilson, 1897), who for many years had studied the physics of cloud formation in the atmosphere. The bubble chamber was invented and refined in early 1950s by D.A. Glaser (Glaser, 1952) (use is made of a superheated liquid, which boils near the nucleation centres - local energy-release sites $\geq$ 0.1 keV in the trajectory of a particle in the superheated liquid). Cloud chambers and bubble chambers make it possible to directly observe tracks of particles. This means that the position of a particle can be determined accurate to the size of a drop or a bubble, i.e., approximately up to 1 mm. However, the drawbacks of these thermodynamic chambers are their low-rate response, low spatial resolution and, above all, impossibility of automating data collection/processing on a real-time basis. These drawbacks were eliminated in detectors of another (electronic) type - gas-discharge counters with gas amplification, proportional and drift chambers, scintillation and Cherenkov detectors, solid-state detectors.

The simplicity of track detectors secures them a significant advantage over many other detection systems. The advantage of track detectors as integral instruments accumulating information under conditions of small fluxes of particles has been used not only in aerostat experiments but also in satellite experiments with cosmic rays (Chan & Price, 1975). Track detectors (nuclear emulsions including) played an outstanding role in the development of nuclear physics due to their visual character and the possibility to obtain a comprehensive spatial pattern of processes studied. Nuclear decays and reactions, as well as novel particles (positron, muon, charged pions, strange and charmed particles) have been discovered owing to these detectors.

The track detection technique is also widely used in quite a number of applied works. Thus, data on the energy spectrum of reactor neutrons are obtained by means of neutron dosimeters containing dividing layers and track detectors (Burger et al., 1970). Using track detectors, information is obtained on the distribution of radiologically essential α-emitting natural nuclei of inert gases ^{222}Rn and ^{230}Rn (from, respectively, decays of ^{238}U and ^{232}Th), which, diffusing from a solid (a rock or a structural material) get into the atmosphere and may create a hazardous level of radiation (Fremlin & Abu-Jarad, 1980). Radon irradiation time in uranium mines is controlled by strips of cellulose nitrate fixed on miners' helmets (Frank & Benton, 1975). Registration of α-particles from radon gas has been used in attempts to predict earthquakes because an increase of seismic hazard before earthquakes is often noted to be accompanied by the emergence of cracks and stresses; herewith, a large amount of radon is released from uranium and thorium occurring in the Earth's crust (Savvides et al., 1985). The track detection method is also used in studies of the exchange processes in the troposphere, where radon is used as an indicator (Birot et al., 1970). Track detectors are used in beams of negative pions in radio therapy to study events with high linear losses of energy (Benton et al., 1970; Fowler & Perkins, 1961). Thus, the method of track detectors is constantly developing, its protocol is being refined, and at present there is hardly a field of science and technology where it is not used. These are high energy physics, cosmic ray physics, reactor physics, metallurgy, geology, archeology, medicine, biology, studies of meteorites and lunar samples (Benton et al., 1970; Birot et al., 1970; Burger et al., 1970; Chan & Price, 1975; Frank & Benton, 1975; Fowler & Perkins, 1961; Fowler et al., 1970; Fremlin & Abu-Jarad, 1980; O'Sullivan et al., 1980; Price et al., 1967, 1968; Savvides et al., 1985).

However, none of the elementary particle detectors currently used can provide for the spatial resolution given by nuclear emulsion: at a grain size of 0.3–1 μm, the deviation of grains from the recovered trajectory of a particle does not on average exceed 0.8 μm, and under certain conditions can be reduced to 0.2 μm. The use of a double-sided emulsion makes it possible to determine the particle-movement direction accurate to less than one milliradian. Nuclear emulsion is used in quite a number of experiments by physicists in Europe, America and Asia. The largest of them can contain tons of nuclear emulsion, which corresponds to thousands of square metres of emulsion surface (Acquafredda et al., 2009; Eskut et al., 2008; Kodama et al., 2008).

In this context, of paramount importance are methods, which should provide for a fast and high-quality retrieval of information from data obtained by means of track detectors. Browsing of large areas of detectors (as a rule, at high magnification) poses a rather complex technical problem. Manual examination of track detector data in optical microscopes by an operator required enormous input of labour and time. The measurement rate proved not high, which determined the low statistics of processed events. Besides, the probability of hard-to-find errors is sufficiently high in such measurements, so the obtained results poorly yielded to checks for possible failures occurring in the processing of material (e.g., losses of particle tracks by measurers and other errors).

In recent years, this drawback has been largely overcome owing to progress in production of precision instruments and the development of optical tables with high accuracy of computer-aided displacement, extensive use of modern CCD and CMOS video cameras for recording and digitizing optical images, and computing power of modern computers. Complete automation of microscopists' labour became a reality owing to the use of these achievements in precision mechanics, potential of computer means and development of required software. In such automatic measurements, digitized images of tracks of charged particles and nuclei in track detectors obtained by means of video cameras are entered into computers, whose software makes possible the search for, and recognition and examination of tracks, and enables the reconstruction of their spatial position. Such an automated method of measurements excludes almost totally the use of microscopist's wearisome visual work and accelerates processing by about three orders of magnitude as compared with the use of the so called semiautomatic devices. The new method makes it possible to process large arrays of experimental data and to significantly increase the statistics of events, which was earlier unrealistic in practice. The development of similar automated facilities is of current interest, as it enables transition to a higher level of experimentation making use of the track detector technique and significantly expands the sphere of problems, where it can be efficiently used.

PAVICOM-1

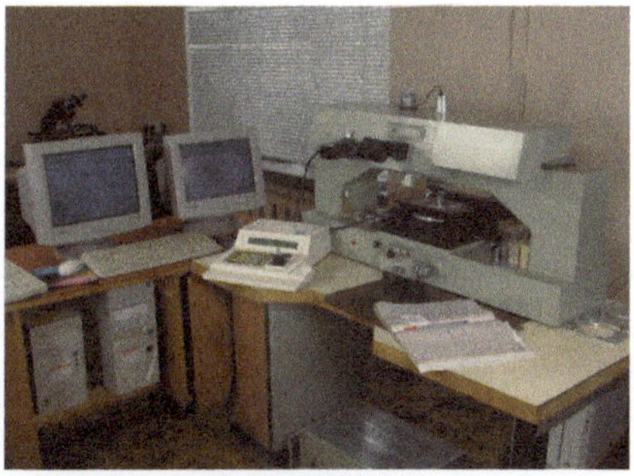

PAVICOM-2

PAVICOM-3

Fig. 1. Appearance of the three microscopes of the PAVICOM facility.

The development of automated facilities was pioneered by Japan, where such a facility has been created well back in mid 1980s. But the real impetus to the development of automated facilities throughout the world were CHORUS and DONUT experiments (Eskut et al., 2008;

Kodama et al., 2008), which made use of significant amounts of nuclear emulsion. At present, only the OPERA experiment employs hundreds of tons of nuclear emulsion, which corresponds to hundreds of thousands square metres of emulsion surface. About 40 automated facilities in total are operated at present throughout the world, 20 of which are in Europe, where in the recent years the number of such facilities rapidly increases.

2. Equipment of the PAVICOM facility

In Russia, there is only one facility of such a level, which satisfies modern international standards. This is the high-tech completely automated measuring facility PAVICOM. It is intended for processing the data of emulsion and solid-state track detectors used in various physical studies and satisfies the most modern world standards (Feinberg et al., 2004). Initially, the facility was developed to process materials of LPI's EMU-15 experiment (Astafyeva et al., 1997; Boos et al., 1996; Chernavskaya et al., 1996; Dobrotin et al., 1999; Dremin et al., 2001a), in which the emulsion chamber including a lead target 0.4 mm thick with the succeeding stack of 38 layers of nuclear emulsion was irradiated by a beam of lead nuclei with energy of 158 GeV/nucleon. However, the versatility and potentially high hardware potential of the PAVICOM facility provided for in the course of its development enabled its use for a significantly broader sphere of problems (Polukhina, 2006). Staff members of about 10 Russian and several foreign institutes together with LPI's PAVICOM group are involved in processing and analysis of experimental data.

PAVICOM consists of three independent completely automated installations, which differ, first and foremost, in the optical tables' displacements and, respectively, admissible sizes of the processing detectors (Fig. 1).

The main units of the PAVICOM-1 installation are the automatic precision table (MICOS, Germany); the optical system, which is set up in accordance with the problem solved; Pentium computer. The MICOS precision mechanical system consists of the massive metal platform and movable object table, whose displacement limits in the horizontal plane are 400 mm × 800 mm. The CCD camera fixed over the table can be moved vertically within the limits of 0 to 200 mm. The coordinates in all three axes are measured accurate to 0.5 μm. The optical table and vertical ruler are displaced by means of the step motors driven by the controller operated by the computer or (manually) by the joystick. The optical system of the microscope was created at the LPI using the components from LOMO (Russia). The microscope's optics project the image of an object on the CCD matrix of the VS-CTT system (Videoscan, Russia). The CCD pixel resolution is 1360 × 1024; colour depth, 10 bit. The installation is arranged in a clean room with glass walls.

The automated microscope PAVICOM-2 was created on the basis of the microscope MPE-11 (LOMO, Russia). The main units of PAVICOM-2 are the precision table (Carl Zeiss) with the control module; two digital CCD cameras; computer. The automatic displacement limits of the table along the X axis are 0–120 mm; along the Y axis, 0–100 mm; the vertical displacement limits are about 1 cm. The optical table is displaced along all the axes by step motors driven by a controller operated by computer. The coordinate measurement accuracy is 0.25 μm. The characteristics of the video equipment: (1) The pixel resolution of the first CCD camera is 768 × 576; colour depth, 8 bit. An image in the camera is produced by the objective lens of the microscope, with a magnification of up to 60×. An analog signal formed by the video camera is fed to the input of the analog-to-digital converter - the image capture and digitization board - which transfers the digitized signal to computer memory. (2) The second video camera: CMOS MC1310 with 10-bit colour depth and pixel resolution of 1280 × 1024; shooting speed, up to 500 frames per second.

The third installation, PAVICOM-3, has the following main characteristrics: the motorized movable table (MICOS); the microcontroller to operate the step motors (National Instruments); the illumination system (Nikon); the computer-controlled power supply unit for the illumination system, the computer, a dual-processor workstation based on Intel Pentium 4 Xeon 3.6 GHz; the video camera, CMOS MC1310 with 1024 grey levels, shooting speed up to 500 frames per second.

3. Principle of PAVICOM software construction

The PAVICOM facility makes use of two types of commercial image-digitization plates: VS-CTT and Matrox Odyssey XPro, using which an image is digitized and its primary processing is done. The entire subsequent image analysis is performed using specially developed software. The search for and digitization of the coordinates of charged particles' tracks in the material of the detector, computer-aided track recognition and tracing, data systematization and primary processing are done here in the completely automated mode. Mathematical processing of digitized images is performed by computer at a rate of up to 500 fps using a library of image processing programs written in C++. An image in the CMOS video camera is formed by the objective lens of the microscope. An analog signal formed by the video camera is fed to the input of the Matrox Odyssey XPro image, digitization and processing board.

The use of multiprocessor systems in automated scanning facilities enables a significant increase of their performance due to the simultaneous execution of various tasks. PAVICOM makes use of a dual-processor workstation based on Intel Xeon processors with Hyperthreading Technology support. The Matrox Odyssey image digitization and processing board is also a separate multiprocessor system. The complete use of PAVICOM's computing potential is possible only at the realization of an efficient multithreaded model in the software. Another way to increase the computing power of the system for solving labour-intensive computing tasks is to combine two or more computers into a common network. The computing load in this case is distributed between all computers of the network.

The modular approach to the construction of the software was used to implement all the requirements to its versatility and fast response. This approach makes it possible to incapsulate the implementation of the modules, i.e., none of the modules depends on the operation and design of the other modules. Correspondingly, a change in one module in no way affects the other modules. Thus, the required flexibility in the adjustment of the program is achieved, which enables it to perform scans on microscopes of various types. For this, it is only necessary to replace the software-dependent parts of the modules, the adaptors. The program can also be used for processing the data of various experiments. For this, the respective handler should only be implemented. Herewith, all the other modules of the program, not dependent on the hardware and the character of the experiment, would remain invariable at various variants of the adjustment.

4. Methods for image recognition of elementary particle tracks for EMU-15 experiment

In LPI's EMU-15 experiment performed at CERN, which made use of nuclear emulsion as a detector, the main aim was to search for possible signals of quark–gluon plasma formation at ultrahigh temperatures and superdense states of matter. The experiment made use of 16 emulsion chambers shaped as cylinders 260 mm in length and 95 mm in diameter. Each chamber contained a thin (400 μm) lead target and 38 layers of nuclear emulsion (each 50 μm thick) poured on to a 25-μm thick mylar support. One layer of emulsion was positioned immediately before the target; the others were behind the target. The total thickness of each chamber was only 0.07 cascade lengths, which is extremely important at the registration of the central collisions of very high energy lead nuclei, in which thousands of secondary particles are generated. The chamber was placed into a transverse magnetic field of intensity 2 T and positioned such that the planes of the target and nuclear emulsions be perpendicular to the beam. The total number of Pb nuclei in irradiation of each of the 16 emulsion chambers was 10^4. About 10 central Pb–Pb interactions with multiplicity of secondary particles of more than 10^3 was found in each chamber (the criterion for selecting such events was the large multiplicity of secondary charged particles and the absence of fragments with charge $Z \geq 2$).

The image processing and microtrack search methods developed for this experiment enabled the first Russian automated search for tracks of secondary charged particles in central nucleus–nucleus interactions with multiplicity of >1500 in nuclear emulsion.

The automated processing of tracks in nuclear emulsion required the development of software to control the displacement of the microscope table, video shooting of images, their analysis and reconstruction of the geometric pattern of the projection of secondary charged particles.

The image processing task was divided into several stages:

- Preliminary search for and elimination of extraneous spots ("black" pixels of the video camera)
- Filtration of images
- Binarization of images
- Singling-out of connected regions (clusterization)
- Selection of clusters by size.

The result of image processing are sets of clusters assigned to one event.

The task of reconstructing the geometry of the projection of particles born as the result of nucleus–nucleus interaction was also split into several subtasks:

- Reconstruction of a microtrack (i.e., a segment of a microtrack in one of the emulsion layers) from a chain of clusters (tracking)
    - Calculation of the parameters of the microtrack by the given set of clusters (fitting)
    - Selection of clusters for the fitting without consideration of information on the interaction vertex (free tracking)
    - Search for the interaction vertex
    - Selection of clusters for the fitting with consideration of information on the interaction vertex (vertex tracking)
- Extension of the microtrack to the adjacent emulsion layer
    - Calculation of the track parameters by the given set of microtracks (linking)
    - Selection of microtracks for linking
- Determination of the direction of movement of the initial nucleus (search for the axis of the event).

The result of the reconstruction of particles' projection geometry after nucleus-nucleus interaction is a pseudorapid distribution of secondary particles, which is then analyzed in detail by various mathematical methods.

Images of particle tracks at the video camera output look like groups of blobs against the grey background (Fig. 2). Each pixel carries some information on the extent of image blackening (called colour). For the initial images the colour of a pixel is within the range of 0 to 255 (8 bit). For images converted by the filter the colour is described by a real number, and for binarized images by 0 or 1. At given depths of emulsion, 25 fields of vision were successively shot by the video camera by the spiral around the assumed centre of the event.

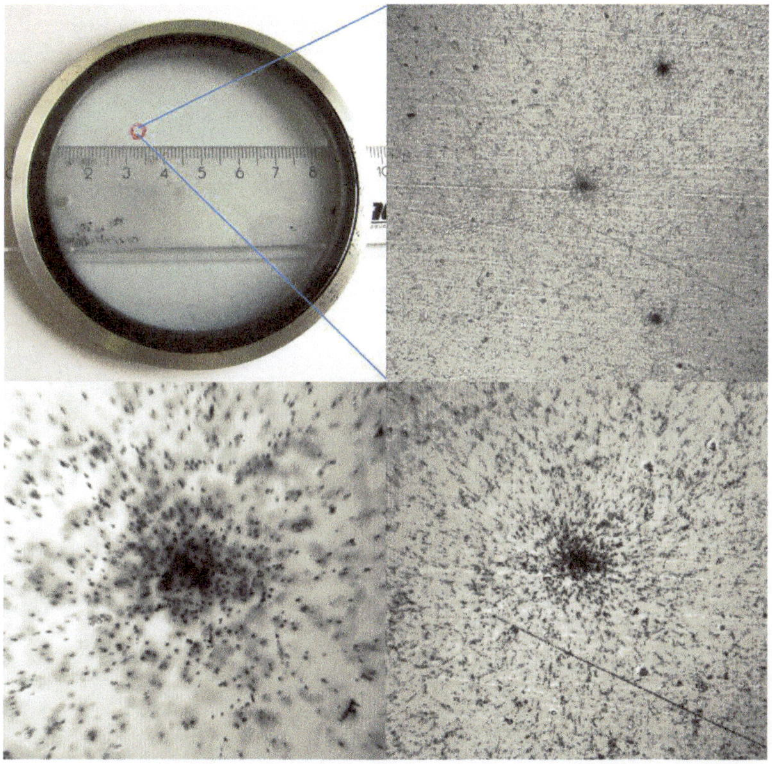

Fig. 2. *Top left*: an emulsion plate of EMU-15 experiment. *Top right*: the central field of vision at an objective-lens magnification of 8× (820 × 820 μm²). *Centre*, the tracks had not yet drifted apart and concentrate very densely, which makes the central region darker. Two tracks of noninteracting lead nuclei are also well seen on the image. *Bottom*: the same field of vision at a magnification of 20× (*right*, 330 × 330 μm²) and 60× (*left*, 115 × 115 μm²).

As nuclear emulsion was processed at a magnification, in practice it was impossible to completely rule out accidental specks (e.g., micron-size villuses on the video camera screen) in the field of vision, which led to the occurrence of dark spots on images (sets of "black" pixels) corresponding to background effects and undistinguishable from images of real blobs. They may hinder the correct reconstruction of microtracks or lead to the emergence of false microtracks. (For instance, in Fig. 3 a Z-form speck is well seen at the right edge but two other specks are undistinguishable from the blobs.)

Preliminary search for such spots (sets of "black" pixels) is based on the fact that on all images they are at the same place. If on any image a "black" pixel has the coordinates (*i, j*), then on all the other images a pixel with the coordinates (*i, j*) shall be "black". Therefore, if the colours of pixels with the same coordinates are averaged by 25 images at the same depth, the defect regions should become more intensive: the colour of "black" pixels proves considerably more black than the colour of the other pixels (Fig. 3). Now the pixels assigned to defect regions can be removed by setting the same colour threshold for all fields of vision. Further on, the colours of the singled-out pixels are averaged by the neighbourhood of 20 × 20 pixels in size, after which they become practically undistinguishable from the background and, thereby, eliminated from the further processing (Fig. 4). Here and further, the background is meant to be the image background - the pixels not belonging to the images of blobs and, therefore, carrying no useful information.

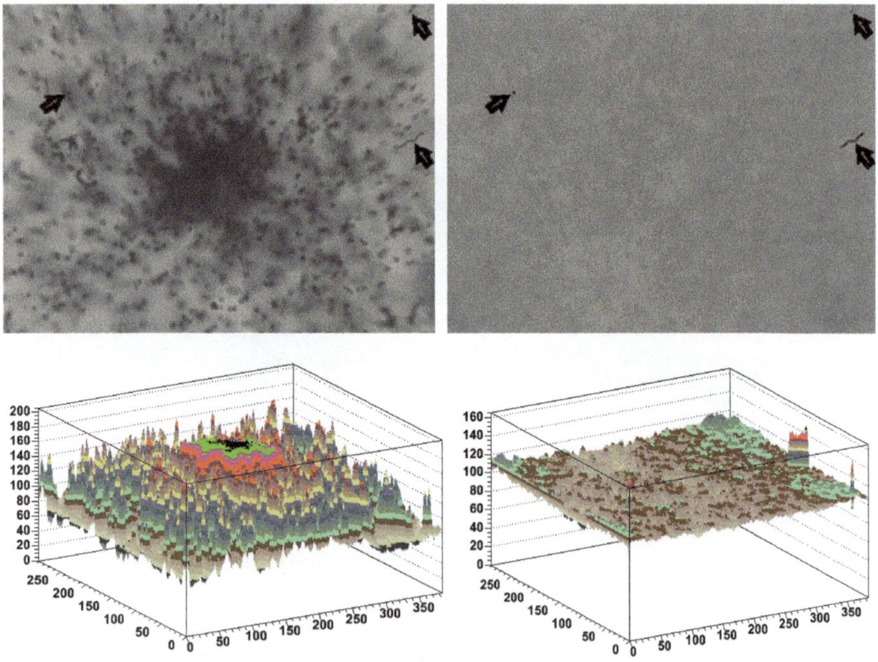

Fig. 3. *Top left*: initial image. Black arrows show spots formed from sets of "black" pixels. *Bottom left*: coordinate colour histogram of the initial image. *Top right*: image obtained as the result of averaging 25 images. *Bottom right*: its coordinate colour histogram. It is seen that in the case of the averaging the colours of "black" pixels do not change, which enables their easy finding by setting the pixel-colour threshold.

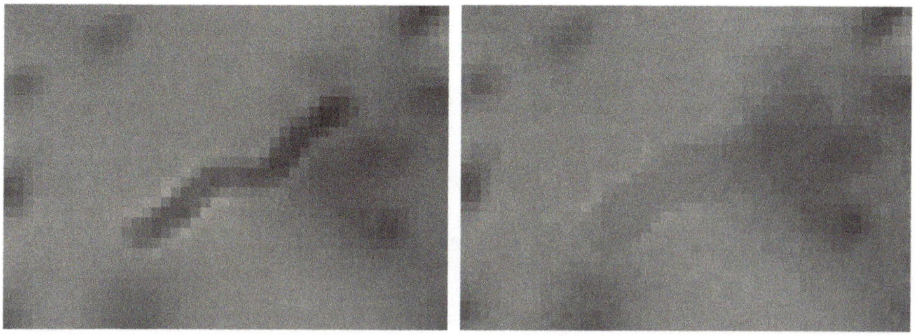

Fig. 4. Image of a Z-form speck before (left) and after (right) the removal of "black" pixels.

Filtration is the preliminary image-processing stage. Its aim is to single out spots which correspond to images of individual blobs. The expected intervals of sizes and extent of blackening of such spots are set beforehand.

To reduce the nonuniformity of the image background and to increase the quality of filtration in the darker central region of an image, the image is preliminarily processed. First, it is processed by a low-frequency averaging filter (Pratt, 1982). The filter blurs the images of separate blobs, owing to which the summary image becomes similar to the background, which would have been obtained in the absence of all blobs (Fig. 5).

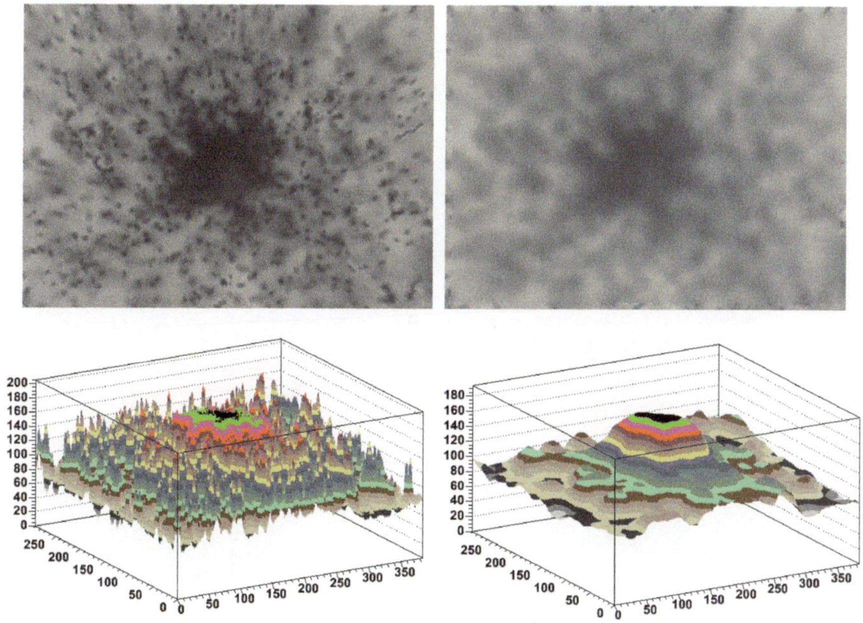

Fig. 5. *Top left*: initial image. *Bottom left*: coordinate colour histogram of the initial image. *Top right*: image processed by the averaging filter. *Bottom right*: its coordinate colour histogram. The filter blurs the images of all blobs to leave only the distribution of the darkening close to the background.

After this the pixel-colour conversion is done according to the formula:

$$c'_{i,j} = 255 \cdot \left(1 - \frac{c_{i,j}}{b_{i,j}}\right),$$

where $c_{i,j}$ is the pixel colour of the initial image and $b_{i,j}$ is the colour of the corresponding pixel of the averaged image. This procedure converts the pixel colours of the initial image into relative colours, thereby intensifying the images of the blobs in darker regions of the image (with lower $b_{i,j}$).

Subsequent filtration consists in the conversion of the image by the scanning high-frequency filter (Pratt, 1982), which reacts to the change of the extent of blackening on the edge of the spot. The size of the filter is chosen such that its response to spots similar to images of blobs forming the track of a particle were maximal. The result of this property of the filter is that the edges of blobs become more sharp, and blobs themselves more distinct, and their size and number do not change. Herewith, large spots formed by merged blobs (e.g., the large spot in the centre of Fig. 6) almost vanish after filtration, blending into the background of the image.

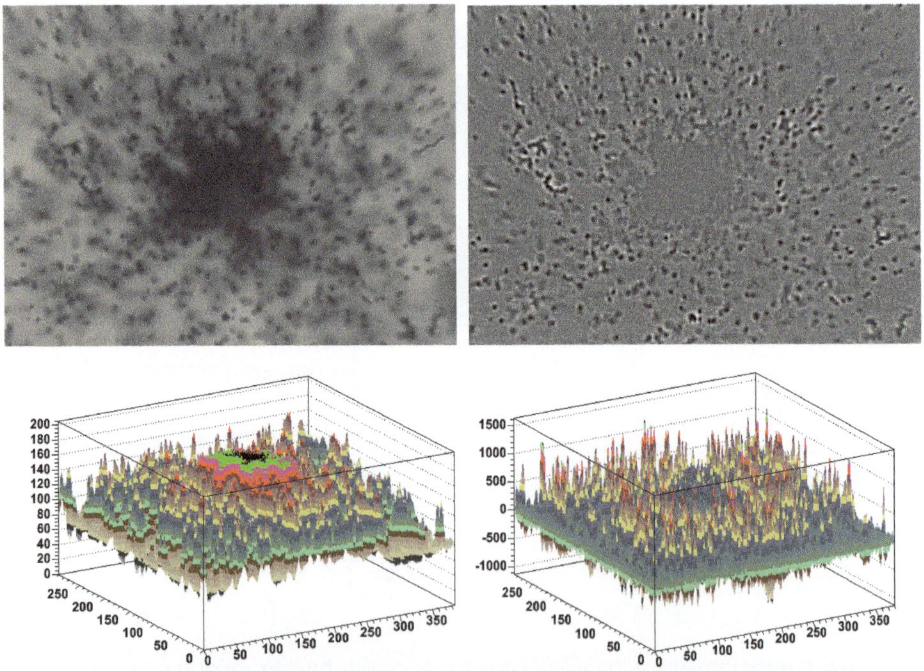

Fig. 6. *Top left*: initial image. *Bottom left*: coordinate colour histogram of the initial image. *Top right*: image processed by a high-frequency filter. *Bottom right*: its coordinate colour histogram. It is seen that the filter intensifies only spots of certain size and shape.

Mathematically, filtration is a matrix operation over pixel colours:

$$g_{i,j} = TrF^T C_{i,j},$$

where $g_{i,j}$ is the new pixel colour *(i,j)*, *F* is the filter matrix and $C_{i,j}$ is the matrix composed of the pixel colour *(i,j)* and the colours of the enclosing pixels such that the colour $g_{i,j}$ of the pixel *(i,j)* is the central element of the matrix $C_{i,j}$.

After filtration, the coordinate colour histogram of an image represents a multitude of narrow high peaks in those places where on the image there were characteristic spots – images of blobs. Comparison of the right- and left-hand parts of Fig. 6 shows the extent of blob intensification in this operation.

Then the procedures of binarization and clusterization were performed. Binarization is the process of singling-out the pixels, which correspond to the images of blobs (assigning colour 1 to these pixels). For this, a threshold restriction is imposed on an image: the pixels whose colour exceeds the threshold are assigned the value of 1, the others are assigned colour 0. Clusterization is the singling-out of connected regions – integration into clusters, which correspond to spots of blackening (blobs) in nuclear emulsion, of pixels with colour 1 singled out in the process of binarization. Knowing the coordinates (x_i, y_i) of each pixel in a cluster, we can find the centre of masses, the mean radius and the area of the cluster.

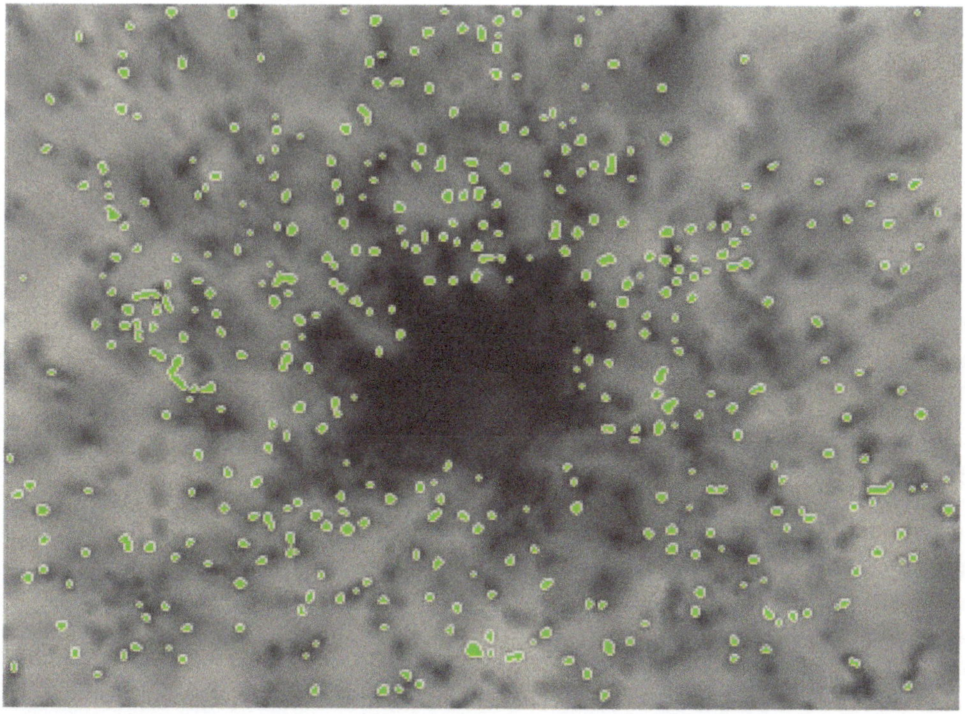

Fig. 7. The result of clusterization of the central field of vision. The clusters found are highlighted.

Thus, physical objects (blobs) are put in correspondence with mathematical objects (clusters). The results of clusterization of the central field of vision are given in Fig. 7.

Clusterization yields sets of points - centres of masses of clusters on the plane corresponding to a certain depth. By adding the results of the measurements along the Z axis (i.e., along the depth of the fields of vision), we obtain sets of clusters distributed in the volume studied. Further, it is necessary to reconstruct the microtrack of a particle from the sets of clusters. (The microtrack is part of the track of a particle, which part lies inside a particular layer of emulsion.) For this purpose, the microtrack is depicted by a straight line given by a point with the coordinates (x, y, z) and slopes (tx, ty) of the projections of this line on the plane XZ and YZ to the axis. The Z axis is perpendicular to the plane of emulsion and is codirected to the beam of particles. The coordinate origin along the Z axis is chosen on the front plane (relative to the beam) of the target. Reconstruction of the microtrack by the given set of clusters (drawing the straight line through the known set of coordinates of the centres of mass of the clusters - the fitting) was done by the method of least squares (Bock et al., 2000). When selecting the clusters for microtrack recognition (the so called "tracking"), several requirements were to be satisfied: at each depth inside emulsion (i.e., on the plane of the clusters), no more than one cluster is assigned to a track; the distance between two successive clusters should not exceed a given value; the number of clusters in a microtrack should be no less than the preset value.

Due to a large number of clusters, it does not appear to be possible to consider all combinations of their chains in full measure over reasonable time. Therefore, the search for tracks is done in several stages: free tracking, search of the interaction vertex and vertex tracking.

The algorithm of the free tracking is illustrated in Fig. 8. An arbitrary cluster $(x^0{}_i, y^0{}_i)$ on the emulsion plane z^0 nearest to the target is considered. The perpendicular (tx_0, ty_0) is taken as the initial direction, and a cone with the vertex at $(x^0{}_i, y^0{}_i, z^0{}_i)$ and preset angles $(s^0{}_x, s^0{}_y)$ in the XZ and YZ planes between the generatrices and the axis (tx_0, ty_0) is constructed around it. Around the selected cluster a sphere is constructed, which *a fortiori* captures several depths, and clusters getting into this sphere are picked out. By these clusters, all possible combinations of microtracks in the cone are made up (set C). Successively adding clusters one by one from C, we construct a set of chains of clusters, which is sorted out by the criterion χ^2.

This procedure is performed for each chain of clusters. Herewith, the direction of the microtrack, which corresponds to a chain, is taken to be the axis of the cone, and the vertex of the cone is constructed at the point of intersection of this microtrack with the plane corresponding to the Z coordinate of the last cluster in the chain.

As the result, we obtain a set of microtracks (chains of clusters), which originate from one common cluster $(x^0{}_i, y^0{}_i)$. Then we choose a track with the lowest χ^2 and with the number of clusters not smaller than that given for the chain corresponding to it. The constituent clusters of the microtrack are marked as used and are not involved in further tracking. The same procedure applies to all clusters on the next planes, so that the origin of the microtrack can be at any depth along Z. The tracking is stopped if all layers are passed or if no candidates are available in several layers.

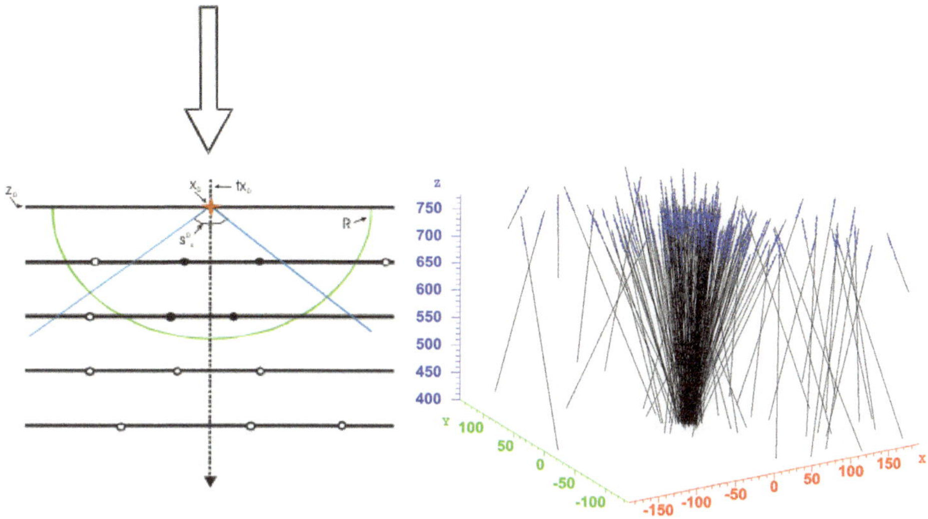

Fig. 8. *Left*: an illustration of free tracking (see text). Solid horizontal lines correspond to various depths of emulsion. The Z axis is directed downwards and is codirected with the beam of particles (arrow). The vertex of the cone is in the current cluster. The direction around which the cone is constructed is shown by a dashed line. Clusters, which got into the cone and the distance to which from the vertex does not exceed R (marked with black circles) will be used in building-up the track, the other clusters (white circles) are ignored. *Right*: the result of free tracking. It is seen that most particles escape from the common centre with $z \sim 400$ μm.

As the result of this procedure, we have a set of microtracks as shown in Fig. 8.

In the search for the interaction vertex, all found microtracks are initially the components of microtrack set A. To find the initial values of the vertex coordinates, for each pair of microtracks from set A we searched for pairs of points of closest approach (set B). Then an average value μ of the coordinates of all points from B is found.

For each track from set A, a point of closest approach with μ is found (set C), and the coordinate distributions of such points are constructed. From these distributions, the new values of μ and their root-mean-square errors σ_μ are determined. From set A, microtracks are excluded for which the points of closest approach with μ lie at distances exceeding $3\sigma_\mu$. The procedure is repeated until all microtracks from A are within $3\sigma_\mu$, i.e., when none of the microtracks could not be excluded from A. An elliptic cylinder with the sizes of the axes of the ellipsis along X and Y, depending on the measuring errors, and the height Z equal to the thickness of the target, is considered to be the region of the vertex.

The final stage of microtrack recognition – the vertex tracking – takes into consideration that the EMU-15 experiment investigates central collisions. In this case, the projection of all tracks of one event to the plane perpendicular to the beam has a pronounced centre with the coordinates (x^c, y^c).

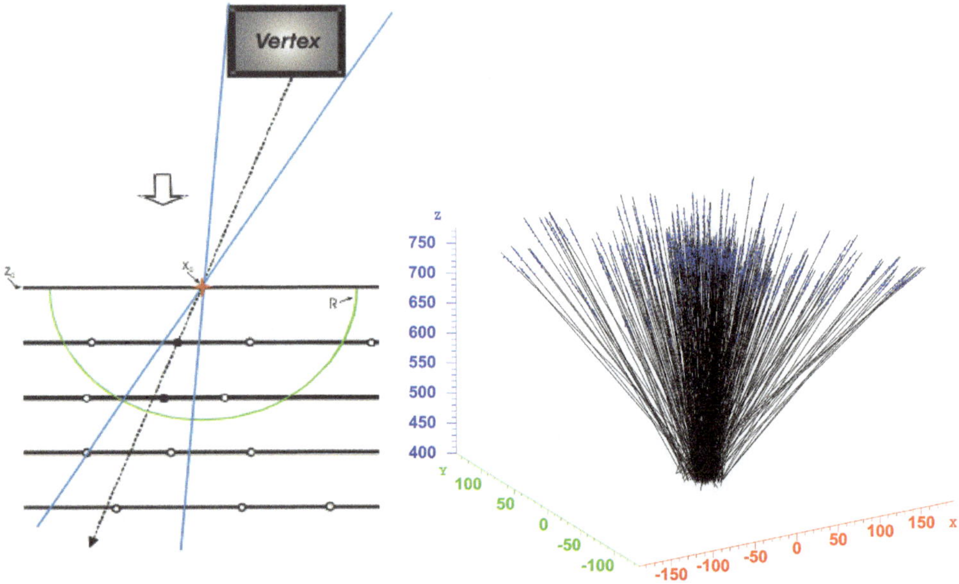

Fig. 9. *Left*: an illustration of vertex tracking. Solid horizontal lines correspond to various depths of emulsion. The *Z* axis is directed downwards and is codirected with the beam (arrow). The vertex of the cone is in the current cluster. The direction around which the cone is constructed is shown by a dashed line. The clusters which got into the cone and the distance to which does not exceed *R* (shown by black circles) are used in building up the track. The other clusters (white circles) are ignored. *Right*: the result of vertex tracking.

The algorithm sets up a cylinder whose axis is parallel to the *Z* axis and passes through point $(x^c,y^c,0)$. From above and from below the cylinder is limited by the planes perpendicular to the *Z* axis ($Z = z^c_1$ and $Z = z^c_2$) (Fig. 9). Further, an arbitrary cluster (x^0_i,y^0_i) on the upper plane of emulsion, $Z = z^0$, is selected. From the lower layers, clusters (set *C*) are picked out, for which the straight line connecting any cluster from *C* with the cluster (x^0_i,y^0_i) passes through the preset cylinder. By successively adding clusters one by one from *C*, a set of chains of clusters is constructed. The fitting of each chain is performed, and the obtained set of microtracks is sorted out with respect to χ^2. The described procedure is repeated for each chain of clusters. As the result, a set of microtracks (chains of clusters) is obtained, which originate from one common cluster (x^0_i,y^0_i). Then, microtracks with the lowest value of χ^2 and the limited number of clusters in the respective chain are successively selected. The clusters that form a microtrack are not involved in the further tracking. Then the procedure is repeated for the next layer.

The efficiency of tracking was checked in two ways: manually and by modelling the events of passing a charged particle through emulsion. The efficiency of microtrack recognition was 90% (for the central field of vision with high particle density) and 99% (for the other regions).

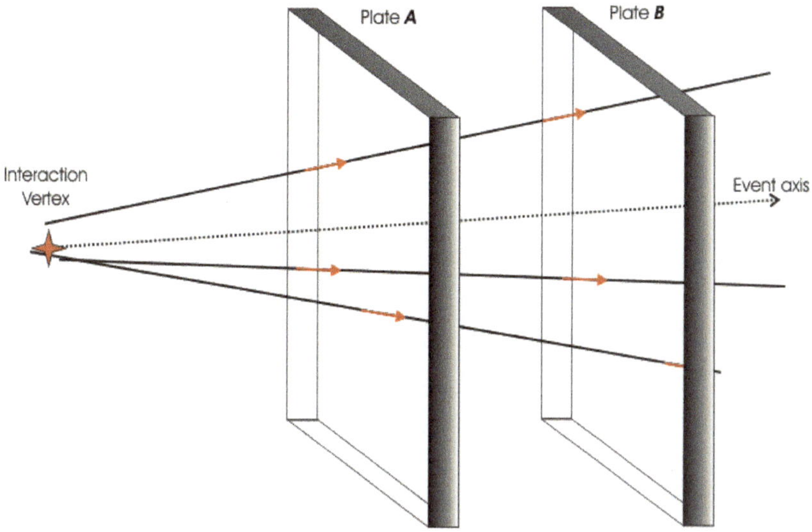

Fig. 10. Extension of microtracks to an adjacent layer of emulsion.

For microtrack linking, it is necessary to calculate the parameters of a track, which fits a given set of microtracks the best. This task is solved by the least squares method and is largely similar to the fitting task, see (Aleksandrov, 2009).

Finally, to determine the angular characteristics (pseudorapidity) of secondary particles formed in nuclear interaction, microtracks from adjacent emulsion layers are linked (Fig. 10). At the initial stage of this procedure, each microtrack from emulsion layer A is linked with each microtrack from adjacent layer B. Of all the diversity of variants, a track with minimal χ^2, passing through the interaction vertex, is chosen. As the result, a set of tracks (the so called target diagram), which consist of two parts and have a lower angular uncertainty than each particular microtrack, is obtained (Fig. 11).

For the search of the axis of an event, it should be taken into account that the direction of travel of the initial accelerated Pb nucleus can differ from the normal to the plane of the event owing to the divergence of the beam. At the same time, the axis of the event is given by the direction of travel of the initial nucleus. The algorithm of the search for the axis of an event is largely similar to the algorithm of the search for the interaction vertex. Based on the data on its coordinates x_c, y_c, the set of points of the maximal approach of tracks from set A with vertex c is constructed. From the initial set of tracks A, we exclude all tracks whose points of the maximal approach with vertex c are beyond $3\sigma_c$, where σ_c is the standard error of determining the position of point c. The angles of the remaining tracks with the axis of the event are averaged, and the initial direction of the axis is found. Then histograms of the angular distribution of tracks relative to this axis are plotted (Fig. 12), from which the new values of the direction of the axis are found. From A, we exclude tracks the angle between the direction of which and the direction of the axis exceeds $3\sigma_t$, where σ_t is the angle determination error. Then again histograms of angular distributions are plotted, and the procedure is repeated until the angles between the directions of all tracks prove to be inside $3\sigma_t$, i.e., when none of the tracks could be excluded from A.

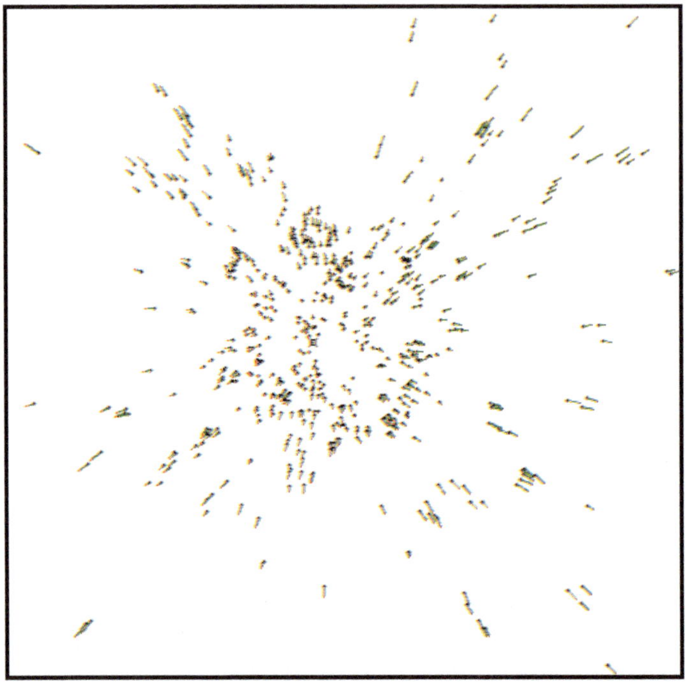

Fig. 11. The result of tracking and linking of two emulsion layers: a target diagram for 647 particles (event 5c15e).

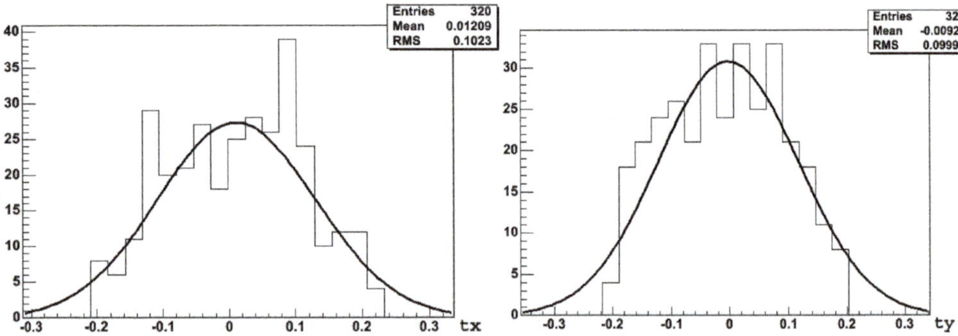

Fig. 12. An example of angular distributions of tracks relative to the axis of an event (*left, tx; right, ty*).

The procedure described was implemented as a library of programs (Aleksandrov, 2009). The first program, which includes algorithms of filtration, binarization, clustering, search for the interaction vertex, free and vertex tracking was implemented as the image handler for the automation program. This enabled the processing of images, search for clusters, interaction vertex and microtracks in real time mode over a time of less than 1 min. The algorithms of linking, search for the axis of an event and physical analysis were implemented in a separate program. Both programs were implemented using C++.

Thus, in accordance with the algorithm of the image processing program for EMU-15 experiment, we successively carried out discrimination of "black" (background) pixels, filtration of images, binarization of images, isolation of connected regions (clusterization), filtration of clusters. All stages of the program were thoroughly checked, for which the real interaction was methodically filmed with a step of ¼ micron by depth. A practically total correspondence of the program-operation results to the characteristics of real emulsion confirms the correctness of the developed and executed algorithm of clusterization.

The performance of the tracking software package was checked by visual measurements and on model interactions. The efficiency of charged-particle track reconstruction was greater than 90%. The experimental data obtained in this way make it possible to study the peculiarities of the angular distributions of particles, to search for multi-particle correlations, the emergence of which can be expected in hadronization of excited nuclear matter, and to analyse these distributions with the view of searching for signals specific for quark–gluon plasma. The processing results were demonstrated using event 5c15e as an example (Fig. 13); a databank was created and a preliminary processing of 100 events was done.

The successfully solved image recognition problem made it possible to find microtracks in nuclear emulsion with high efficiency. Herewith, original image-processing algorithms were developed: filtration, binarization and clusterization (search for images of metal silver grains in nuclear emulsion). A detailed study of individual interactions can help find new rare phenomena, and conclusions obtained therewith will be statistically significant owing to a large number of particles born in an individual event: a large number of secondary particles are born in collisions of high-energy nuclei in EMU-15 experiment. Analysis of their distribution in phase space is the main aim of the study, as it enables conclusions on the dynamics of the process. This is not an easy task even if only particles' escape angles (polar and azimuthal) are measured, because the number of secondary particles sometimes exceeds 1000. The task is reduced to the recognition of images generated on the plane of the target diagram by this number of points. Wavelet analysis made it possible to solve this problem. Daubechies wavelet was first used in the processing of EMU-15 data for two-dimensional wavelet analysis of particle spectra. The result of this analysis was demonstrated by the example of studying the internal structure of the escape of secondary generated particles locally and on various scales. It was shown that on the distribution of secondary charged particles by pseudorapidity the correlated groups tended to be arranged as a ring around the centre of the diagram. This corresponds to groupings of particles at a constant polar angle, i.e., fixed pseudorapidity. The mechanism of the emergence of these structures in strong interactions can be explained by either an analog of Cherenkov radiation (gluons are discussed as an analog of photons), or else by the emergence of Mach waves (Dremin, 2006; Dremin et al., 2000, 2001b). Both discussed mechanisms have a similar nature - the emergence of radiation in the motion of a body in a medium at a velocity exceeding the phase velocity of the propagation of a disturbance in this medium.

The most significant indication of the existence of ring structures is the occurrence of two peaks in the pseudorapidity distribution (Fig. 13). Herewith, ring structures are seen on the plane of the target diagram (Fig. 14).

Thus, the processing of EMU-15 data taken as an example confirmed the nonsymmetric character of the distribution of secondary charged particles by the azimuthal angle in combination with pseudorapidity peaks in an individual event. This is indicative of the

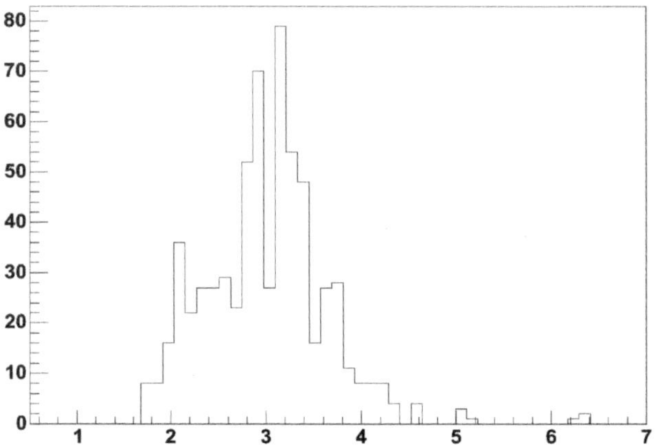

Fig. 13. Distribution of pseudorapidities of secondary charged particles in event 5c15e after automated processing.

peculiar features of the escape, caused by multiparticle correlations, the emergence of which, in particular, was expected at a comparatively small number of emitted Cherenkov gluons in each ring.

5. Study of the characteristics of galactic cosmic ray nuclei in automated recognition of tracks in olivine crystals from meteorites

Processing of EMU-15 data is only one of the projects implemented at PAVICOM. The multipurpose unique measuring facility PAVICOM is successfully used for high-technology processing of experimental data in nuclear physics, cosmic ray physics, high energy physics obtained in studies with emulsion and solid state detectors. The facility is used to solve problems associated with high-speed information support of the analysis of intravolume and surface defects of the crystal lattice and the rupture of materials on the nanostructural level. Its main distinction from other similar systems in the world and an advantage is versatility - the automated installations of the facility process data obtained using both nuclear emulsions and plastic detectors, and also crystals of olivines from meteorites. The use of this automated facility makes it possible to process large arrays of experimental data and significantly increase the statistics of events, which had been practically unrealistic earlier. A broad range of topical problems of modern nuclear physics is being covered.

In particular, in 2006 the project OLIMPIYA (Russian acronym for **OLI**viny iz **M**eteoritov: **P**oisk Tyazhelykh **I** Sverkhtyazhelykh **YA**der = Olivines from Meteorites: Search for Heavy and Superheavy Nuclei) was started to be implemented at PAVICOM. Its task are studies of heavy and superheavy nuclei in cosmic rays and search for trans-Fermi nuclei with charges $Z \geq 110$ in crystals of olivines from pallasites. The issue of the existence of superheavy nuclei is of greatest significance for understanding the properties of nuclear matter. First and foremost, of interest is to verify the prediction (Strutinsky, 1967) of a significant increase in the stability of nuclei near the magic numbers $Z = 114$ and $N = 184$ (N, the number of neutrons), which could lead to the existence of "stability islands" of superheavy nuclei in this region. Confirmations of this prediction have been obtained in experiments under the direction of Yu.Ts. Oganesyan at a JINR accelerator, where recently the nuclei of elements

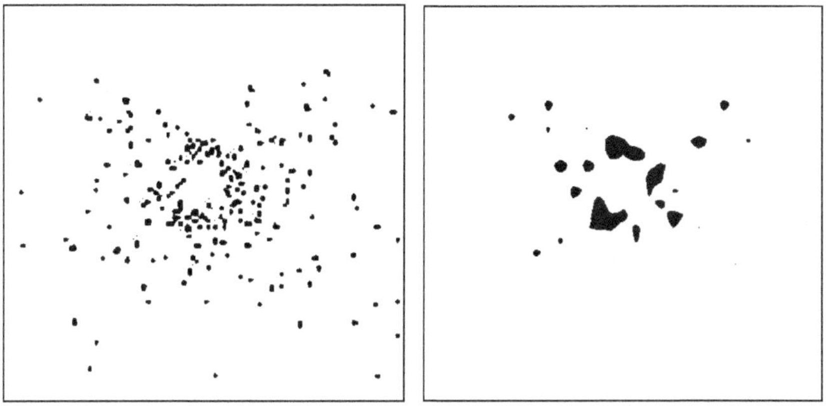

Fig. 14. Target diagrams of event 5c15e after, respectively, two and four iterations of inverse wavelet transformation.

114 and 116 have been discovered (Oganesyan, 2001) and the discovery of element 117 has been declared. The lifetimes of some of these nuclei are several seconds and even minutes, which exceeds tens of thousand times the lifetimes of nuclei with smaller charge.

According to the existing views, elements starting from carbon and heavier are formed in stellar interiors and at supernova explosions (Ginzburg, 1999). Heavy and superheavy elements being in Mendeleev's Periodic Table after bismuth are formed as the result of r (rapid) processes, which occur at a high concentration of neutrons and can lead to the formation of superheavy nuclei with the number of neutrons up to $N = 184$. Besides these traditional mechanisms, the possibility of the formation of very heavy nuclei (with the mass number of up to 500) at a high density of neutrons (of the order of 10^{30} cm^{-3}) and moderate temperature $T < 10^8$ K is discussed (Zeldovich, 1960). This situation can be realized in unequilibrium shells of neutron stars, ejections from which will lead to the emergence of super heavy elements in interstellar medium, stars and planets (Bisnovatyi-Kogan & Chechetkin, 1979; Kramarovsky & Chechev, 1987).

Measurements of the fluxes and spectra of heavy and superheavy nuclei in cosmic rays are an efficient way to study the composition of cosmic ray sources, processes occurring both in the sources themselves and in interstellar medium, in which cosmic rays propagate, and models of cosmic ray retention in the galaxy. The currently available experimental data on the abundance of heavy nuclei (with $Z > 50$) in the Universe, as well as on the spectra and fluxes of these nuclei in cosmic rays, are rather scarce, and for trans-Fermi nuclei no sufficiently reliable data are available at all. In the same way, there are no data on the possible existence of exotic superheavy nuclei, either.

The use of the factor of long-time exposure of meteorites in space leads to a great advantage of the method for the search of superheavy elements in crystals of olivine from meteorites as compared with methods based on the use of various satellite and aerostat detectors. The search for relict tracks left by particles of cosmic rays in meteorite minerals makes use of the ability of silicate crystals in meteorites (olivines, pyroxenes) to register and preserve for a long time (>10^8 years) tracks of nuclei with $Z > 20$. A typical age of meteorites and, therefore, their exposure time in the flux of cosmic rays is assessed to be 10^7–10^9 years. For this reason, they may contain a large number of tracks of cosmic nuclei. As estimates show, 10^2–10^3 tracks of nuclei with $Z > 90$ can be formed over 10^8 years in 1 cm^3 of such crystals located at a depth of <5 cm from the preatmospheric meteorite surface, and up to 10^4 tracks in crystals from meteorite's surface layers (depth, < 1 cm). By measuring the track parameters, it is possible not only to identify particles but also to determine their energy spectra. Pallasite meteorites consist of iron–nickel "matrix", with inclusions of crystals of olivine - a yellow-colour semitransparent mineral up to 1–2 cm in size. As the previous works with olivines, the project OLIMPIYA is based on the used of the method of solid-state track detectors, in which particles are registered by the radiation damage they cause in the bulk of the detector material. A specific feature of the authors' method of the treatment of olivines from meteorites is the necessity of grinding, i.e., the irreversible disintegration of part of the crystal. Within the framework of the project OLIMPIYA, based on the PAVICOM facility, the technique for the recognition of tracks of nuclei in olivines was developed, the technique of scanning the entire area of the crystal, as well as the database of images to preserve information on tracks in the crystal.

Above we described in detail, using the processing of EMU-15 data as an example, the main stages of discrimination of tracks of particles as independent objects on individual images. The same methods were also used in studies of the characteristics of tracks of nuclei in olivines. First of all, coordinates of all pixels of each track should be obtained in particle track recognition. Then, based on this information, the characteristics of tracks are determined in accordance with the space of features chosen for this task. This procedure is performed in several stages.

First, depending on the quality of an image, a preliminary processing is carried out. Its aim is to maximally differentiate by the gradation of blackening the pixels of noise and the pixels belonging to tracks of particles, and to increase the efficiency of the binarization procedure.

The purpose of binarization is to separate analyzed objects from the background and noise. The discrimination of a subset on an image is equivalent to the setting of its "characteristic function", i.e., the function equal to unity in points of the subset and to zero in the other points. The generally recognized method of discrimination of some subset on the initial image is reduced to finding its characteristic function by quantification of the initial image to two levels. This procedure consists in determining the threshold of blackenings. All pixels with the blackening above the threshold are considered to belong to tracks of particles and are marked by black colour on the image. The other pixels are assigned to the background and noise and are marked with white colour.

The last stage of processing consists in discrimination of selected clusters (tracks) as independent objects (acquisition of the coordinates of pixels), determination of their characteristics and elimination of some of the remaining noise.

The set of transformations required at the stage of preliminary processing depends on the aims of this task, the quality of respective images and feature space. The preliminary processing results are present in several classes (in terms of C++), the volume of information in which can be chosen depending on the task and feature space. For this purpose, in the software developed for the project OLIMPIYA the classes for the description of clusters are constructed by the hierarchical principle. The main class Cluster0 containing minimal information about a cluster includes the number of pixels (cluster area) and their coordinates (*X*, *Y*, *Z*) with indication of the cluster boundary. The derivative classes Cluster1, Cluster2 etc. contain additional information. Selection of a class to be used in a particular task depends on the requirements of the subsequent processing. Information about all clusters is gathered up in class Clusters, which also contains the functions of images' and particular clusters' processing. Due to distinctions in tracks of particles, the feature space can strongly differ for different tasks. Correspondingly, track images are processed differently in different experiments (Starkov, 2010).

The characteristics of clusters play an important role in the analysis of images. We adduce here techniques for determining the major of them.

Cluster area is simply equal to the number of pixels of which it consists, and is determined at the stage of cluster discrimination. A set of special algorithms was developed in the software package to obtain the other characteristics.

Each cluster is processed separately. First, the maximal and minimal coordinates *x* and *y* limiting the cluster are determined (see Fig. 15; it should be taken into account that in computer graphics it is accepted to direct the Y axis from top downwards). This cluster-limiting rectangle is required to assess the extent of its oblongness. Further, the *coordinates of the centre of masses* of the cluster pixels are calculated; they are its most important characteristic, which determines the position in space.

To find the *direction of axis of a cluster*, the coordinates of the midpoints of its lines are first determined. Further, using the least squares method, the obtained set of points is approximated by a segment of the straight line passing through the centre of masses (Fig. 15). This straight line is considered to be the axis of the cluster. Sometimes this procedure has to be repeated for the midpoints of the columns. This is due to the fact that at the first stage we can obtain not a longitudinal axis but a transverse axis. Then at the second stage the axis will for certain be longitudinal. An important characteristic of the axis is its slope. (For instance, in the problem of the scattering of neutron-excess nuclei (Starkov, 2010) it determines the direction of the search for the extension of the tracks of particles scattered at a large angle.)

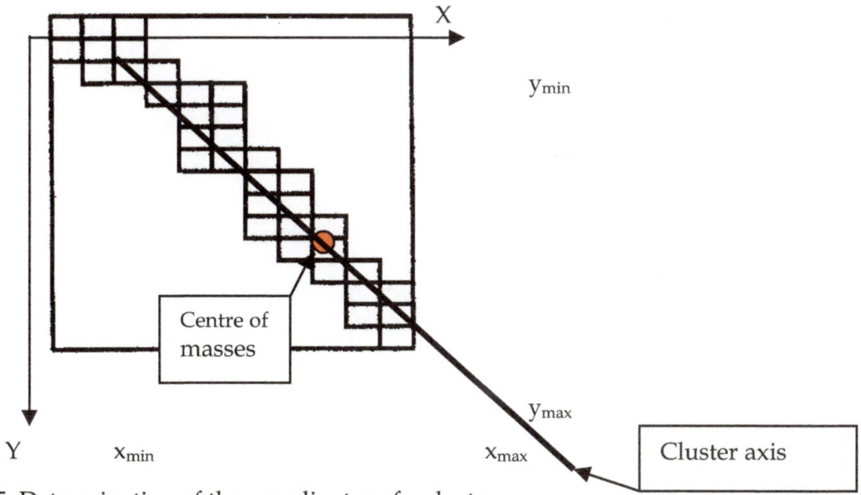

Fig. 15. Determination of the coordinates of a cluster.

There is a possibility to find the position of the axis and its directions by minimizing the momenta of inertia of the pixels relative to the straight line passing through the centre of masses of the cluster:

$$S_x = \sum x_i \qquad M_x = S_{xx} - \frac{S_x^2}{Area}$$

$$S_y = \sum y_i \qquad M_y = S_{yy} - \frac{S_y^2}{Area}$$

$$S_{xx} = \sum x_i^2 \qquad M_{xy} = S_{xy} - \frac{S_x \cdot S_y}{Area}$$

$$S_{yy} = \sum y_i^2$$

$$S_{xy} = \sum x_i y_i \qquad \Theta = \tan^{-1}\left\{\frac{M_{xx} - M_{yy} + \sqrt{(M_{xx} - M_{yy})^2 + 4 \cdot M_{xy}^2}}{2 \cdot M_{xy}}\right\}$$

The *width of the cluster* is determined as a doubled mean distance of the pixels of the boundary to its axis. The *length* is determined as the length of part of the axis lying inside the cluster.

The above given definitions for the axis, width and length are sufficiently exact for clusters of convex or close to convex shape. If clusters are bent or have a more complex shape, these concepts cease to be exact and are not used. The shape of a cluster in this case is reflected more accurately, e.g., by the ratio of its area to a circle of a diameter equal to the length of the cluster.

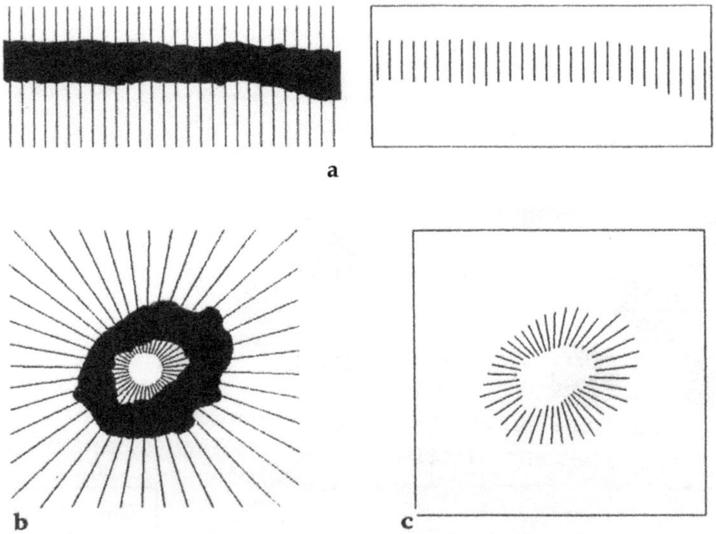

Fig. 16. Assessment of cluster parameters using logic operations.

Sometimes logic operations help to assess the characteristics of tracks (Russ, 2006). For instance, Fig. 16a shows how the alignment of two images - the initial cluster and a set of parallel lines - and the application of AND operation to them makes it possible to assess the width of the cluster and its profile. Another example (Fig. 16b) shows the application of AND operation to a cluster and a family of semistraight lines emanating from the centre of masses, for determining the profile of a convex cluster with a hole. PAVICOM's software package is provided with a block enabling similar operations.

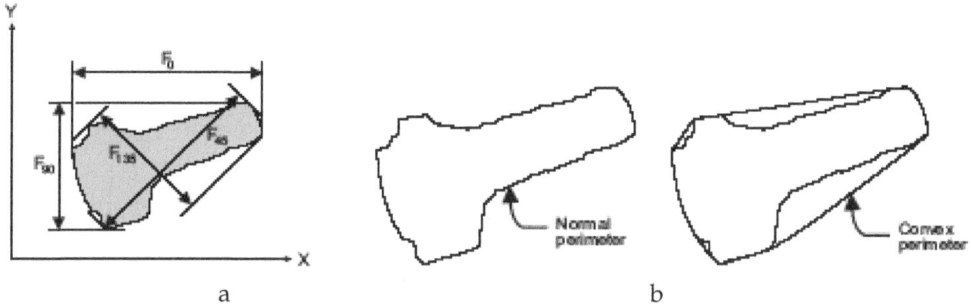

Fig. 17.

The shape of a cluster can also be assessed using Feret diameters - the distances between the outer tangents to the cluster drawn at a certain angle to the *X* axis. Figure 17a presents examples of Feret diameters F_0, F_{45}, F_{90} and F_{135}. Another important concept is the convex perimeter, which is a perimeter of such a convex figure, which is circumscribed around the cluster and has the least perimeter of all similar figures (Fig. 17b).

Tracks of nuclei in olivine have complex shapes. Etching of olivine in places of heavy nuclei passage forms channels shaped as "syringes" (see Fig. 18a). It should be noted that details of the geometric shape of a track can be important for the determination of the charge of a particle. For this reason, information on the pixels of the cluster boundary should be preserved during the clusterization. The result of clusterization in the case of a syringe-shaped cluster is given in Fig. 18b.

Proceeding from the physico-chemical pattern of interaction, the scheme of the formation of such an etchable channel can be described as follows. There is a threshold value of energy losses, *D*, to which E_{max} corresponds. For olivine, D = 18 MeV/(mg·cm^{-2}). If the energy of a particle is greater than E_{max}, no visible channel emerges after etching.

As it decreases, starting from E_{max}, when losses become greater than D - 2 MeV/(mg·cm^{-2}), the rate of etching begins to exceed the rate of etching of non-damaged sections and gradually rises until the value of losses D + 2 MeV /(mg·cm^{-2}) (energy E_1) is reached. A narrow etched channel (Fig. 20) emerges at this segment of the track. In the further deceleration of the particle ($E < E_1$) the etching rate sharply increases and, respectively, the etchable region of this segment acquires a larger diameter. At the end of the track before the particle is stopped the energy losses again drop down below the threshold *D*, and the channel ends with a narrow spike (Fig. 20).

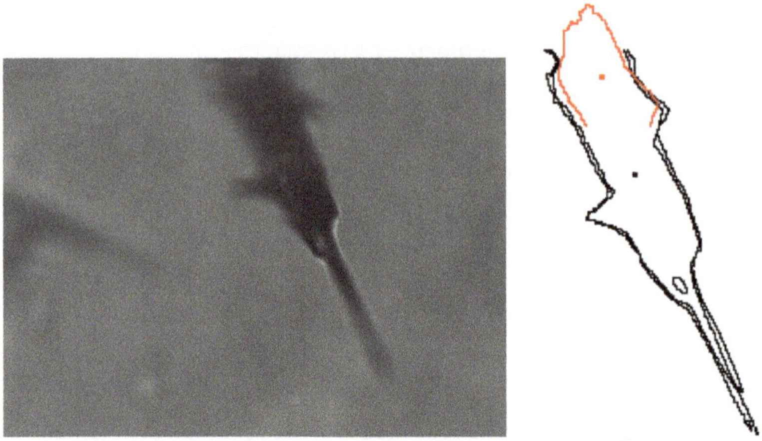

a b

Fig. 18. An example of a track of a galactic cosmic-ray heavy nucleus in olivine (a); the result of its recognition by the program (b). The track shown in Fig. 18a has an extension to the higher adjacent field of vision. In Fig. 18b, this extension is found by means of a special algorithm of searching for track extensions on adjacent fields of vision and is highlighted in red. It can be seen that the fields of vision are partially overlapped, which facilitates the alignment of adjacent fields of vision in the search for track extensions.

Thus, the etchable channels for heavy nuclei are shaped as "syringes". Figure 21 presents examples of images of real channels in olivine, which confirm the existence of the above described mechanism.

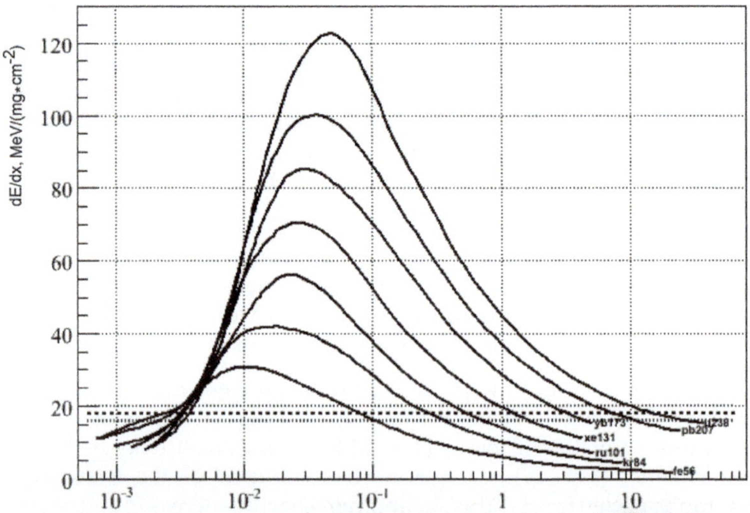

Fig. 19. Ionization losses of $_{26}^{56}Fe$, $_{36}^{84}Kr$, $_{44}^{101}Ru$, $_{54}^{131}Xe$, $_{70}^{173}Yb$, $_{82}^{207}Pb$, $_{92}^{238}U$ nuclei in olivine depending on the residual run.

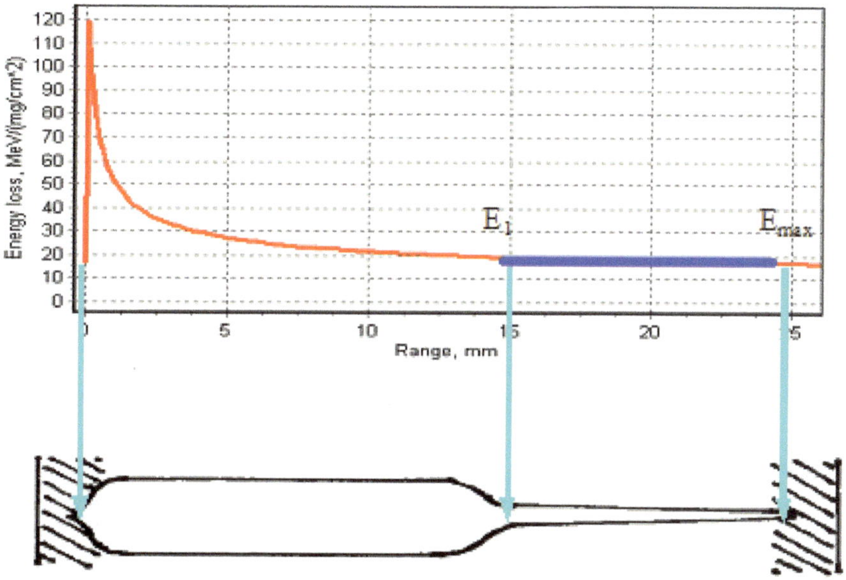

Fig. 20. A scheme of the formation of an etched section of a track in olivine. *Top*, the value of ionization losses. *Bottom*, shapes of various sections of the track.

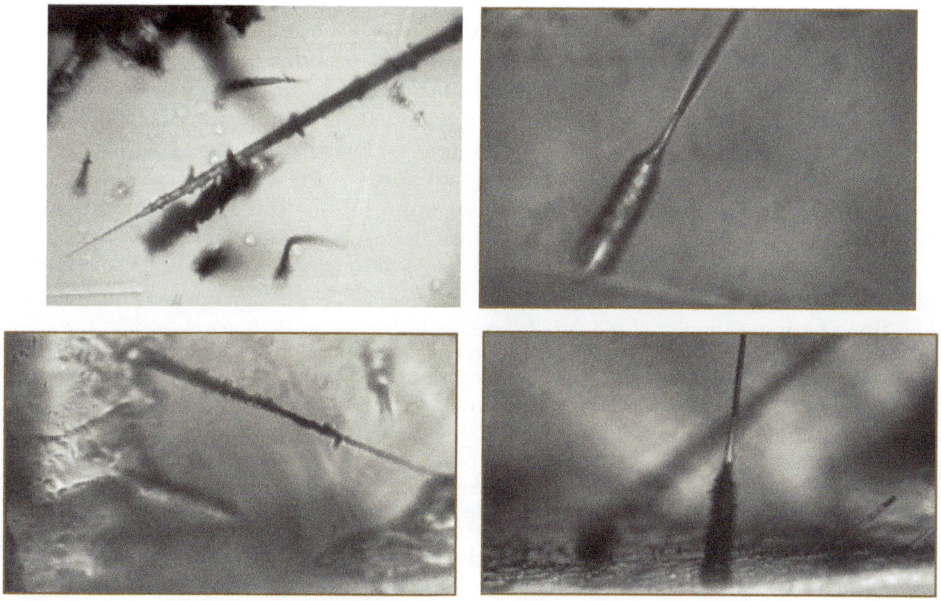

Fig. 21. Photographs of the tracks of nuclei in olivine.

As seen in Fig. 19, the length of the etchable channel can reach several millimetres or even more. At the same time, the size of most olivine crystals available for analysis is 2–3 mm.

This indicates that in the case of very heavy nuclei part of the etchable segment of a track proves to be outside the crystal.

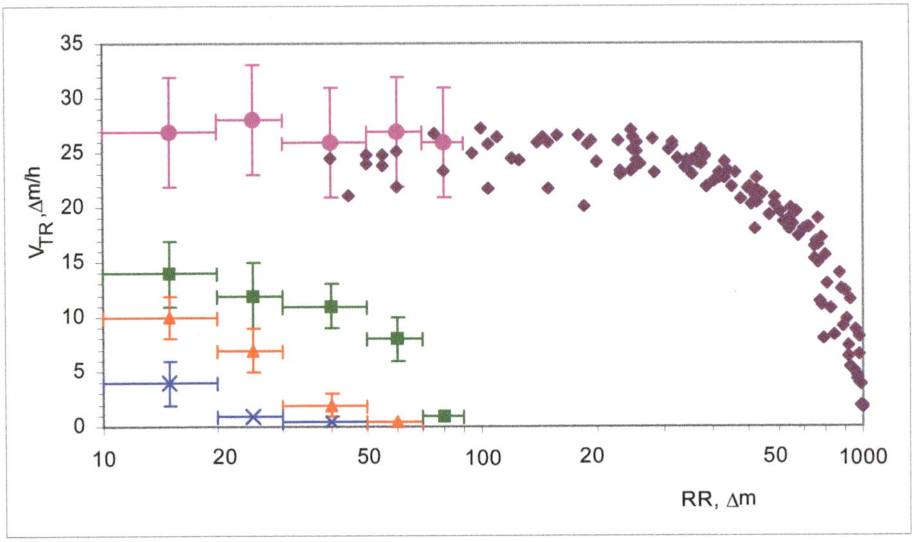

Fig. 22.

As it was already pointed out, the main aim of the project OLIMPIYA is to determine the charge composition of cosmic rays in the region of heavy and superheavy nuclei. The value of the charge is related to the characteristics of the etched track. The major of them is the etchable length L_{etch}, however, as we already mentioned, for very heavy nuclei it exceeds the sizes of olivine specimens and could not be found by available means. As a way out of the situation, it was proposed to use an additional value - the etching velocity. This value changes during the etching and increases as we approach the stoppage point of the particle (Perron & Bourot-Denise, 1986; Perron & Maury, 1986). Figure 22 shows the dependence of the etching velocity on charge and residual run *RR* (Perron & Bourot-Denise, 1986), i.e., practically before the stoppage point. (The PAVICOM facility processed the data of the calibration of olivines by the U, Xe, Kr, Au nuclei.)

In the general form, the dependence of the etching velocity *V* on charge *Z* and residual length *L* can be presented as a surface shown in Fig. 23, whose values for the missing charges are obtained by interpolation.

The method of assessing the value of charge consists in the two-dimensional interpolation of the surface (Fig. 23) by the measured values of *V* and *L* for a track etched in olivine.

Figure 24 shows the result of such a procedure for the heaviest nuclei. The family of the curves shown by coloured squares, represents the dependence of *Z*(*V*, *L*). Blue circles with errors are the results of our measurements. The size of errors along the vertical and horizontal axes reflects the accuracy of the measurements. It is seen that the accuracy of charge determination depends on the position of points on the plane (*V*, *L*) and is from ±1 up to ±2.

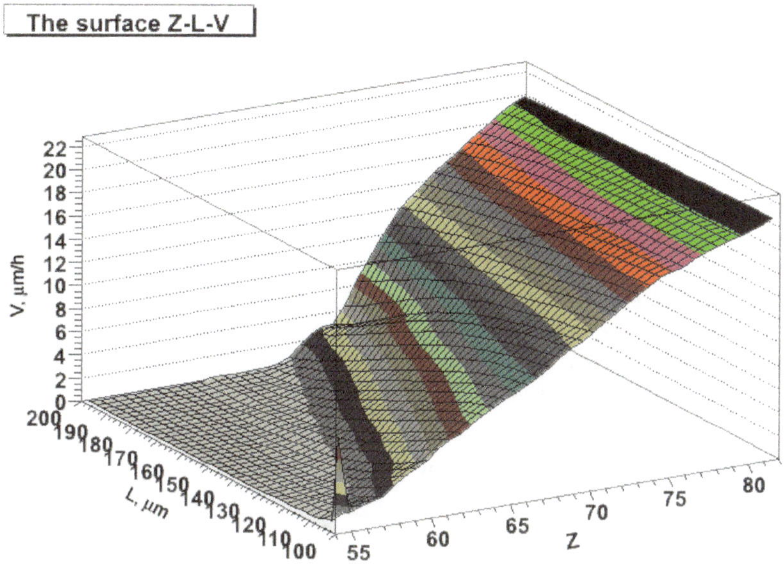

Fig. 23.

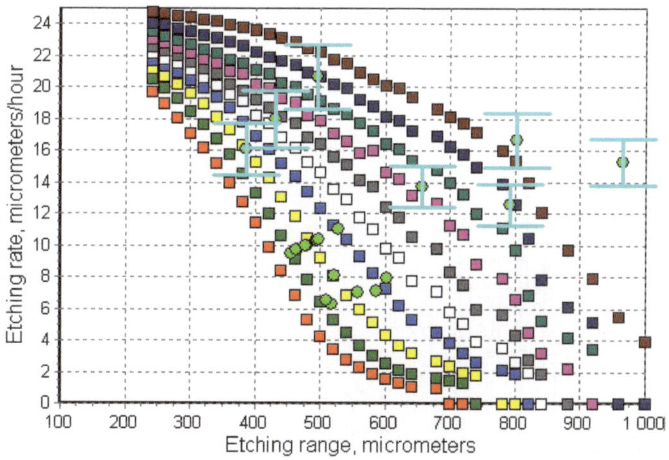

Fig. 24.

To date, 133 crystals from the Marjalahti meteorite and 37 from the Eagle Station meteorite have been processed; 4900 and 1839 tracks, respectively, have been found and identified. Of them, about 40 tracks have been assigned to nuclei with $Z \geq 88$. Among them, 5 events with $Z \geq 92$ have been found. The ratio of the abundance of nuclei with $Z \geq 88$ to that of nuclei with $74 \leq Z \leq 87$ is 0.045 ± 0.015 (Marjalahti) and 0.025 ± 0.02 (Eagle Station). These values are slightly larger than in the UHCRE experiment (0.0147 ± 0.0032) (Donnelly et al., 1999), but are well consistent with the data of the TREK, HEAO and Ariel experiments (Binns et al., 1989; Fowler et al., 1987; Westphal et al., 1998). Figure 25 shows the charge distribution of tracks of nuclei we obtained in the course of the project OLIMPIYA, as compared with the abundance of galactic nuclei obtained in the HEAO and Ariel experiments (Binns et al., 1989; Fowler et al., 1987; Westphal et al., 1998).

Three superlong tracks (L_{tr} > 700 μm), whose etching speed was V_{etch} > 35 μm/h were found in the course of the project OLIMPIYA. If it is taken into account that the experimentally measured maximal etching rate of tracks in olivine for uranium nuclei before their stop is $V_{etch,\ U}$ = 26 ± 1 μm/h, then it becomes clear that the charges of these nuclei significantly exceed Z = 92. As in this region of charges the function $Z(RR, V_{etch})$ is unknown, the function $Z(RR \approx 50, V_{etch})$ at a residual range $RR \approx 50$ nuclei, for which the experimental data of the calibration measurements are available, was extrapolated to assess the charge of the transuranium nuclei in the first approximation. In this case, the lower boundary of the charge was assessed to be Z = 105 (Aleksandrov et al., 2011).

Thus, the data on the charge composition of approximately 600 nuclei with charge more than 55, whose distribution is in agreement with the data of other experiments, were obtained in the project OLIMPIYA. Besides, tracks of three superheavy nuclei of galactic cosmic rays were found and identified in the studied crystals of olivines from meteorites. In the first approximation, their charge is within the interval of 105 < Z <130. There is no doubt that the results obtained in the project OLIMPIYA confirm the hypothesis on the existence of "stability islands" for natural trans-Fermi nuclei.

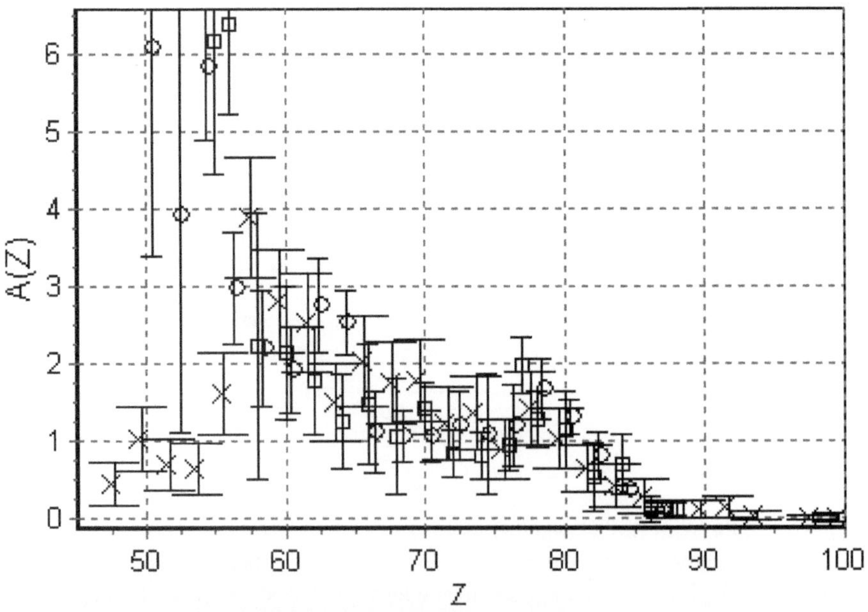

Fig. 25. Abundance of superheavy nuclei $A(Z)$ ($A_{Fe} = 10^6$). Squares - HEAO (Binns et al., 1989; Fowler et al., 1987; Westphal et al., 1998), circles - Ariel (Binns et al., 1989; Fowler et al., 1987; Westphal et al., 1998), crosses - own results.

6. Conclusions

Thus, the worked out image recognition methods were successfully used in processing the data of various track detectors. An important advantage of the developed software package for the PAVICOM facility is its universal character, which makes it possible to combine separate components into a user program without any significant modifications and corrections. At present, work is underway to elaborate the prospects of using PAVICOM and accumulated experience of automated recognition of images for studies of neutrinoless double beta decay, research into the inner structure of industrial objects by means of muon radiography, expansion of the PAVICOM potential for recognition of objects on images in various innovation works, e.g., in automation of blood parameters in medicine.

7. References

Acquafredda, R. et al. (2009). The OPERA Experiment in the CERN to Gran Sasso Neutrino, *JINST* P04018, Vol. 4, April.

Aleksandrov, A.B.; Apacheva, I.Yu.; Goncharova, L.A.; Konovalova, N.S.; Orlova, G.I.; Polukhina, N.G.; Starkov, N.I.; Feinberg, E.L. & Chernyavsky, M.M. (2005a). The Method of Measuring Charges of Relativistic Nuclei in Nuclear Emulsion on the Automated Facility PAVICOM, LPI Preprint No. 29 (in Russian).

Aleksandrov, A.B.; Apacheva, I.Yu.; Goncharova, L.A.; Merzon, G.I.; Polukhina, N.G.; Starkov, N.I.; Feinberg, E.L. (2005b). The Method of Automated Processing the Data of Emulsion Trackers for Studies of Pb–Pb Interactions at the Energy of 158 GeV/nucleon, LPI Preprint No. 23 (in Russian).

Aleksandrov, A.; Polukhina, N. & Starkov, N. (2006). The Pattern Recognition Software for Automatic Treatment of Track Detector Data at the PAVICOM Completely Automated Measuring Facility, *23rd International Conference on Nuclear Tracks in Solids*; Program&Abstracts; p. 84; Beijing, China, September 11-15, 2006.

Aleksandrov, A.B.; Goncharova, L.A.; Davydov, D.A.; Feinberg, E.L.; Polukhina, N.G.; Publichenko, P.A. & Roganova, T.M. (2007a). Automated Methods of Processing Track Detectors Based on the PAVICOM Facility, Pis'ma v *Fiz. Elementarn. Chastits i Atomn. Yadra,* Vol. 4, No. 1 (137), pp. 170–175 (in Russian).

Aleksandrov, A.B.; Goncharova, L.A.; Konovalova, N.S.; Orlova, G.I.; Peresad'ko, N.G.; Polukhina, N.G.; Starkov, N.I.; Tarasova, I.Yu.; Chernyavsky, M.M. & Shchelkanov, A.O. (2007b) The Method of Measuring Charges of Relativistic Nuclei in Nuclear Emulsion at the Automated Facility PAVICOM, *Pribory I Tekhnika Eksperimenta,* Vol. 50, No. 4, p. 469 (in Russian).

Aleksandrov, A.B.; Vladymyrov, M.S.; Goncharova, L.A.; Konovalova, N.S.; Orlova, G.I.; Polukhina, N.G.; Starkov, N.I.; Chernyavsky, M.M. & Shchelkanov, A.O. (2007c) Automation of Measurements in Thick-Layer Nuclear Emulsions in Longitudinal Irradiation by Nuclei with Energy of 1 GeV per Nucleon with the View to Obtain Survey Information on Charge States of Secondary Particles, LPI Preprint No. 4 (in Russian).

Aleksandrov, A.B. (2009). *PhD Thesis*, Lebedev Physical Institute, Moscow (in Russian).

Aleksandrov, A.B.; Bagulya, A.V.; Vladymyrov, M.S.; ... Polukhina, N.G. et al. (2010). Charge Spectrum of the Nuclei of Galactic Cosmic Rays in Olivines from Meteorites, *Uspekhi Fiz. Nauk,* Vol. 180, No. 8, pp. 839–842 (in Russian).

Aleksandrov, A.B.; Bagulya, A.V.; Vladimirov, M.S. et al. (2011). Results of Investigations Associated with the Search for Tracks of Relict Galaxy Nuclei in Olivine Crystals from Meteorites. *Proceedings of 32nd International Cosmic Ray Conference*, Beijing, 2011, in print.

Alexandrov, A.; Kashkarov, L.; Polukhina, N. & Starkov, N. (2008). The Pattern Recognition Software for Automatic Treatment of Track Detector Data at the PAVICOM Completely Automated Measuring Facility, *Radiation Measurements*, Vol. 43, Suppl. 1, August 2008, Pages S120-S124 Proceedings of the 23rd Int. Conference on Nuclear Tracks in Solids.

Astafyeva, N.M.; … Feinberg, E.L.; Polukhina, N.G. et al. (1997). Peculiarities of Secondary Particle Generation Process in Pb–Pb Interactions at 158 A GeV, *6th Conference on the Intersections of Particle and Nuclear Physics*, pp. 269-273.

Belovitsky, G.E.; Konobeevsky, E.S.; Zavrazina, V.P.; Zuev, S.V.; Polukhina, N.G.; Starkov, N.I.; Aleksandrov, A.B.; Lukyanov, S.M. & Sobolev, Yu.G. (2006). Discrimination of the trajectories of charged particles in nuclear emulsions. *Izvestiya RAN*, Vol. 70, No. 5, pp. 646–649 (in Russian).

Benton, E.V.; Curtin, S.B.; Raju, M.R. & Tobias, C.A. (1970) Studies of Negative Pion Beams by Means of Plastic Nuclear Track Detectors. In: *Proc. 7th Int. Colloq. Corpuscular Photography and Visual Solid Detectors*, Barselona, pp. 423–428.

Binns, E.V. et al. (1989). *ApJ*, Vol. 346, p. 997.

Birot, A.; Adroguer, B. & Fontan, J. (1970) Vertical Distribution of Radon 222 in the Atmosphere and its Use for Study of Exchange in the Lower Troposphere, *J. Geophysics Res.*, Vol. 75, pp. 2373–2383.

Bisnovatyi-Kogan, G.S. & Chechetkin, V.M. (1979). *Uspekhi Fiz. Nauk*, Vol. 127, p. 263 (in Russian).

Bock, R.K. et al. (2000). *Data Analysis Techniques for High-energy Physics*, 2nd ed., Cambridge University Press.

Boos, E.G.; Feinberg, E.L.; … Polukhina, N.G. et al. (1996). Investigation of Central Pb–Pb Interactions at Energies of 160 GeV/Nucleon with the Help of the Emulsion Magnetic Chamber, *Experiments at CERN in 1996*, Geneva, ISSN 0259-093X, pp. 122–123.

Burger, G.; Frunauer, F. & Paretzke, H. (1970) The Applicability of Track Detectors in Neutron Dosimetry. In: *Proc. Symp. Adv. Rad. Detectors*. International Atomic Energy Agency, Vienna, paper Sm-143.17.

Chan, J.H. & Price, P.B. (1975) Composition and Energy Spectra of Heavy Nuclei of Unknown Origin Detected on Skylab, *Phys. Rev. Lett.*, Vol. 35, pp. 539–542.

Chernavskaya, O.D. et al. (1996). The Central Pb–Pb Interactions with Energies 158 GeV/nucleon, *Proc. of the 28th International Conference on High Energy Physics*, Warsaw, Vol. 1, pp. 941–942.

Dobrotin, N.A. et al. (1999). Studies of Central Pb–Pb Interactions at High Energies, *Izvestiya AN, Ser. Phys.*, Vol. 63, No. 3, pp. 485–488 (in Russian).

Donnelly, J. et al. (1999). Proc. of the 26th Int. Cosmic Ray Conf., OG1.1.30.

Dremin, I.M. (2006). Ring-like Events: Cherenkov Gluons or Mach Waves? *Nucl. Phys. A*; hep-ph/0507167.

Dremin, I.M.; Ivanov, O.V.; Kalinin, S.A; .... Polukhina, N.G. et al. (2000). Wavelet Patterns in Nucleus-Nucleus Collisions at 158 A GeV, *Proc. of the 4th Rencontres du Vietnam*, pp. 531-533.

Dremin, I.M.; Ivanov, O.V.; Kalinin, S.A.; Kotelnikov, K.A.; Nechitailo, V.A. & Polukhina, N.G. (2001a). Wavelet-Patterns in Nucleus-Nucleus Collisions at 158 A GeV, *Phys. Lett. B*, Vol. 499, No. 1-2, pp. 97-103, hep-ph/0007060.

Dremin, I.M.; Ivanov, O.V.; Kalinin, S.A; .... Polukhina, N.G. et al. (2001b). Wavelet Patterns in Nucleus-Nucleus Collisions at 158 A GeV, *Phys. Lett. B*, Vol. 499, No. 1-2, pp. 97-103.

Eskut, E. et al. (2008). Final Results from a Search for Nu-Mu to Nu-Tau Oscillations with the CHORUS Experiment, Publ. Ref.: CERN-PH-EP/2007-034, arXiv:0710. 3361, *Nucl. Phys. B*, Vol. 793, pp. 326-343.

Feinberg, E.L.; Polukhina, N.G. & Kotelnikov, K.A. (2004). The Completely Automated Measuring Facility PAVICOM for Processing Experimental Material of Track Detectors, *Fizika Elementarnykh Chastits i Atomnogo Yadra*, Vol. 35, No. 3, pp. 763-787 (in Russian).

Fowler, F.H. & Perkins, D.H. (1961) The Possibility of Therapeutic Applications of Beams of Negative π Mesons, *Nature*, Vol. 189, pp. 524-528.

Fowler, P.H.; Adams, R.V.; Cowen, V.G. & Kidd, J.M. (1970) The Charge Spectrum of Very Heavy Cosmic Ray Nuclei, *Proc. Roy. Soc. Lond.*, A318, 1-43.

Fowler, P.H. et al. (1987). *ApJ*, Vol. 314, p. 739.

Frank, A.L. & Benton, E.V. (1975) Active and Passive Radon-Daughter Dosimeters Using Track-Etch Detectors. Dept. of Physics, Univ. of San Francisco, Tech. Report 39.

Fremlin, J.H. & Abu-Jarad, F. (1980) Alpha Emitters in the Environment. I: Natural Sources, *Nucl. Instr. Meth.*, Vol. 173, pp. 197-200.

Ginzburg, V.L. (1999). Which Problems of Physics and Astrophysics Appear Today the Most Important and Interesting? *Uspekhi Fiz. Nauk*, Vol. 169, p. 419 (in Russian).

Glaser, D.A. (1952) Some Effects of Ionizing Radiation on the Formation of Bubbles in Liquids, *Phys. Rev.*, Vol. 87, No. 4.

Kodama, K. et al. (2008). Detection and Analysis of Tau-Neutrino Interactions in DONUT Emulsion Target, *Nuclear Inst. And Meth. A*, Vol. 493 (202), pp. 45-66.

Kramarovsky, Ya.M. & Chechev, V.P. (1987). *Synthesis of Elements in the Universe*, Moscow, Nauka (in Russian).

Oganesyan, Yu.Ts. (2001). *Vestnik RAN*, Vol. 71, p. 590 (in Russian).

O'Sullivan, D.; Thompson, A.; Daly, J.; O'Ceallaigh, C. et al. (1980). A Solid State Track Detector Array for the Study of Ultraheavy Cosmic Ray Nuclei in Earth Orbit. In: *Proc. 10th Int. Conf. Solid State Nucl. Track Detectors*, Lyon, and Suppl. 2, Nucl. Tracks, Pergamon, Oxford, pp. 1011-1019.

Perron, C. & Bourot-Denise, M. (1986). Heavy Ion Track Etch Rate Measurements and Track Structure in a Mineral, *Int. J. Radiat. Appl. Instrum., D Nuclear Track*, Vol. 12, No. 1-6, p. 29.

Perron, C. & Maury, M. (1986). Very Heavy Ion Track Etching in Olivin, *Int. J. Radiat. Appl. Instrum., D Nuclear Track*, Vol. 11, No 1/2, p. 73.

Polukhina, N.G. (2006). *DSc Thesis*, Lebedev Physical Institute, Moscow (in Russian).

Pratt, U. (1982). *Digital Processing of Images, Vol. 2*, Mir, Moscow (in Russian).

Price, P.B.; Fleischer, R.L.; Peterson, D.D. et al. (1967). Identification of Isotopes of Energetic Particles with Dielectric Track Detectors, *Phys. Rev.*, Vol. 164, pp. 1618–1620.

Price, P.B.; Fleischer, R.L.; Peterson, D.D. et al. (1968). High Resolution Study of Low Energy Cosmic Rays with Lexan Track Detectors, *Phys. Rev. Lett.*, Vol. 21, pp. 630–633.

Russ, John C. (2006). The Image Processing Handbook, Fourth edition, CRC Press, London, New York.

Savvides, E.; Manolopoulou, M.; Papastefanou, C. & Charalambous, S. (1985). A Simple Device for Measuring Radon Exhalation from the Ground, *Int. J. Appl. Radiat. Isotop.* Vol. 36, pp. 79–81.

Starkov, N.I. (2010). DSc Thesis, Lebedev Physical Institute, Moscow (in Russian).

Strutinsky, V.M. (1967). *Nucl. Phys.*, Vol. A95, p. 420.

Vladymyrov, M.S.; Aleksandrov, A.B.; Chernyavsky, M.M.; Goncharova, L.A.; Orlova, G.I.; Polukhina, N.G.; Starkov, N.I.; Tsarev, V.A.; Galkin, V.I.; Publichenko, P.A.; Roganova, T.M. & Sazhina, G.P. (2008). PAVICOM Facility for Treatment of OPERA Experimental Data, *Proceedings of Science*, Nufact08, p. 143, Valencia, June 2008 http://pos.sissa.it/archive/conferences/074/143/Nufact08_143.pdf.

Westphal, A.J. et al. (1998). *Nature*, Vol. 396, p. 50.

Wilson, C.T.R. (1897). *Phil. Trans.*, Vol. 189, p. 265.

Zeldovich, Ya.B. (1960). *Zhurnal Eksp. Teor. Fiziki*, Vol. 38, p. 1123 (in Russian).

11

Implementation of Dynamic Logic Algorithm for Detection of EM Fields Scattered by Langmuir Soliton

V.I. Sotnikov, R.W. Deming and L. Perlovsky
Air Force Research Laboratory, Sensors Directorate, Hanscom AFB, MA, USA

1. Introduction

In plasma diagnostics, scattering of electromagnetic waves is widely used for identification of density and wave field perturbations. Very often difficulties with implementations of this type of diagnostics are connected with the fact that scattered signals have small amplitudes comparable with the noise level. Under such conditions it is very important to implement the algorithm which allows us to identify and separate the scattered wave spectrum from the noise pattern. In the present work we use a powerful mathematical approach, dynamic logic (DL) [1], to identify the spectra of electromagnetic (EM) waves scattered by a Langmuir soliton - a one-dimensional localized wave structure. Such wave structures are understood to be basic elements of strong plasma turbulence.

We are interested in the implementation of the Dynamic Logic algorithm for diagnostics of electromagnetic waves scattered by Langmuir soliton in the presence of random noise with amplitudes comparable with the amplitudes of the scattered waves. Scattering of EM waves on Langmuir soliton was previously analyzed by many authors (see for example Sinitzin, 1978, 1979; Mendoca, 1983).

For our purposes we will use the results for electromagnetic wave (EM) scattering on Langmuir soliton obtained in Mendonca, 1983, where scattering on low frequency part of density perturbations (density well) as well as scattering on high frequency part of density perturbations was examined and corresponding expressions for the amplitudes of the scattered waves were obtained.

2. EM scattering by Langmuir soliton

Langmuir soliton is a solution to the well known Zakharov equations (Zakharov, 1972):

$$2i\frac{\partial E}{\partial t}+3\omega_{pe}r_{De}^{2}\frac{\partial^{2}E}{\partial x^{2}}-\omega_{pe}\frac{\delta n}{n_{0}}E \tag{1}$$

$$\frac{\partial^{2}\delta n}{\partial t^{2}}-c_{s}^{2}\frac{\partial^{2}\delta n}{\partial x^{2}}=\frac{1}{32\pi M}\frac{\partial^{2}|E|^{2}}{\partial x^{2}} \tag{2}$$

Here n_0 is the plasma density, δn the density of perturbation, $c_s = \sqrt{\frac{T_e}{M}}$ the ion-sound velocity (T_e is electron temperature and M is ion mass), E the complex amplitude of the Langmuir waves.

Fundamental object of the theory of strong Langmuir turbulence, Langmuir soliton (Rudakov, 1972), can be obtained from the equations (1)-(2) when we are looking for traveling wave solutions in the form:

$$\tilde{\mathbf{E}}_{LS}(\mathbf{r},t) = E_{LS}(x-ut)\mathbf{e}_x \exp[i(k_{LS}x - \delta\omega_{LS}t)] + c.c \tag{3}$$

with imposed boundary conditions $E(x \to \pm\infty) \to 0$ and with $\delta\omega_{LS} = \omega - \omega_{pe}$. In (3) $\mathbf{e}_x$ is the unit vector along the x-axis.

Below we will use a well known solution to (1)-(2) which is given by:

$$\tilde{\mathbf{E}}_{LS}(\mathbf{r},t) = E_{LS}\mathbf{e}_x \operatorname{sech}(\frac{x-ut}{\Delta})\exp(ik_{LS}x - i\delta\omega_{LS}t) + c.c. \tag{4}$$

Where Δ is the soliton half-width, which can be expressed through the electrostatic energy density normalized to the plasma thermal energy density $W_{LS} = {|E_{LS}|^2}/{4\pi n_0 T_e}$:

$$\Delta = \frac{r_{De}}{\sqrt{3W_{LS}}} \tag{5}$$

Frequency $\delta\omega_{LS}$ and wave number k_{LS} are defined as:

$$\delta\omega_{LS} = \omega_{pe}[1 + \frac{1}{2}(W_{LS} - \frac{u^2}{3\mathrm{v}_{Te}^2})] \tag{6}$$

$$k_{LS} = \frac{u}{3\mathrm{v}_{Te}}\frac{1}{r_{De}} \tag{7}$$

where v_{Te} and r_{De} are electron thermal speed and Debye radius.

Electric field of an EM wave interacting with a Langmuir soliton can be written in the following form:

$$\tilde{\mathbf{E}}_0(\mathbf{r},\mathbf{t}) = E_0\mathbf{e}_p e^{i(\mathbf{k}_0\mathbf{r} - \omega_0 t)} + c.c. \tag{8}$$

where $\mathbf{e}_p$ is the polarization unit vector, ω_0 and $\mathbf{k}_0$ are the frequency and the wave vector of the EM wave. These are connected through the dispersion relation:

$$\omega_0^2 = k_0^2c^2 + \omega_{pe}^2 \tag{9}$$

In the process of scattering the EM wave will interact with electron density perturbations associated with a Langmuir soliton. These density oscillations consist of high and low frequency parts:

$$\delta n = \delta n_1 + \delta n_2 \tag{10}$$

The high frequency oscillations can be presented as:

$$\delta n_1 = i(\frac{2}{3}W_{LS})^{1/2}\frac{u}{\mathrm{v}_{te}}n_0 \operatorname{sec}h(\frac{x-ut}{\Delta})e^{i(k_{LS}x-\omega_{LS}t)} + c.c. \tag{11}$$

For the low frequency part (density well) we can use:

$$\delta n_2 = -W_{LS}n_0 \operatorname{sec}h^2(\frac{x-ut}{\Delta}) \tag{12}$$

It is well known that the ratio $\delta n_1 / \delta n_2$ is of the order of unity and we need to take into account contributions to the scattered EM fields from both high and low frequency parts of density perturbations. As a result far away from the soliton it is possible to obtain the scattered wave field in the following asymptotic form (Mendonca, 1983):

$$E_f(x) = 2\pi A(\omega)E_{LS}E_0e^{iqx}\delta(\omega-\omega_{s1}) + 2\pi B(\omega)W_{LS}E_0e^{iqx}\delta(\omega-\omega_{s2}) \tag{13}$$

In (13) $A(\omega)$ and $B(\omega)$ are given by:

$$A(\omega) = \pm\pi\Delta\frac{\omega}{\omega_0}\frac{\omega_{pe}^2}{c^2}\frac{u}{\mathrm{v}_{Te}}(\frac{\varepsilon_0}{3n_0T_e})^{1/2}(\mathbf{e}_0\cdot\mathbf{e}_{s1})\operatorname{sec}h(\frac{\pi}{2}\frac{\Omega_1}{u}\Delta) \tag{14}$$

$$B(\omega) = -\pi\Delta\frac{\omega}{\omega_0}\frac{\omega_{pe}^2}{c^2}(\mathbf{e}_0\cdot\mathbf{e}_{s1})co\operatorname{sec}h(\frac{\pi}{2}\frac{\Omega_2}{u}\Delta) \tag{15}$$

where Ω_1 and Ω_2 are defined as:

$$\Omega_1 = \omega - \omega_0 \pm \omega_{LS} \tag{16}$$

$$\Omega_2 = \omega - \omega_0$$

As it follows from (13)-(16) at large distances from the soliton it is possible to detect four scattered waves. Below for illustrative purposes we are presenting these four scattered waves which appear due to interaction of the incident EM wave, propagating at the 45^0 angle in x-y plane with the moving along the x - axis soliton.

The pattern of the total normalized electric field amplitudes of scattered EM waves given in (13) is presented in Fig. 2.

The Fourier spectrum of scattered EM waves is presented in Figure 3. Below we will use this spectrum in implementation of the Dynamic Logic algorithm.

We are interested in implementation of Dynamic Logic algorithm for diagnostics of electromagnetic waves scattered by Langmuir soliton in the presence of random noise with amplitudes comparable with the amplitudes of the scattered waves. Imposing the random

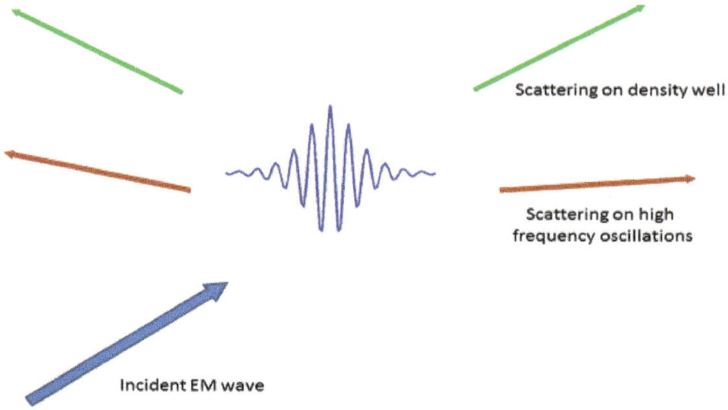

Fig. 1. Scattered EM waves produced due to interaction of the incident EM wave with Langmuir soliton.

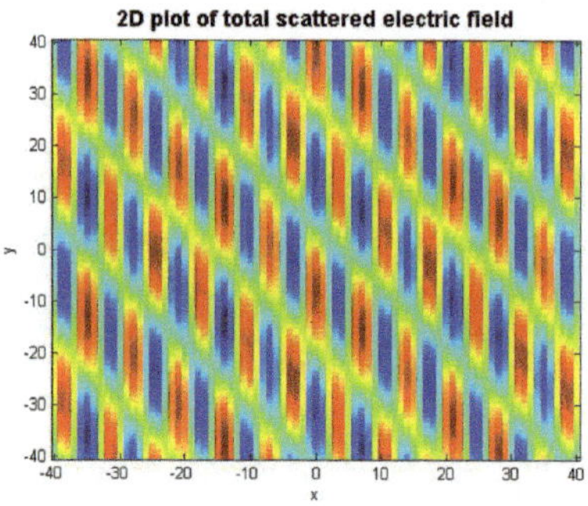

Fig. 2. The pattern of the total normalized electric field amplitudes of scattered EM waves.

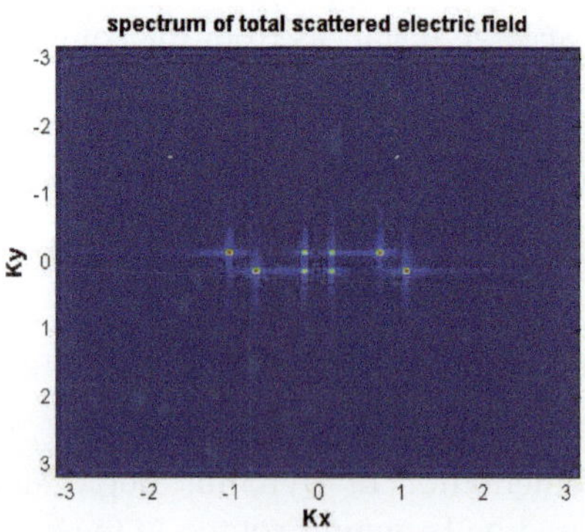

Fig. 3. The Fourier spectrum of scattered EM waves from (13).

noise pattern on top of the Fourier spectrum in Figure 3 with amplitudes comparable with that of the scattered EM waves we will obtain the following 2D pattern in the Fourier space:

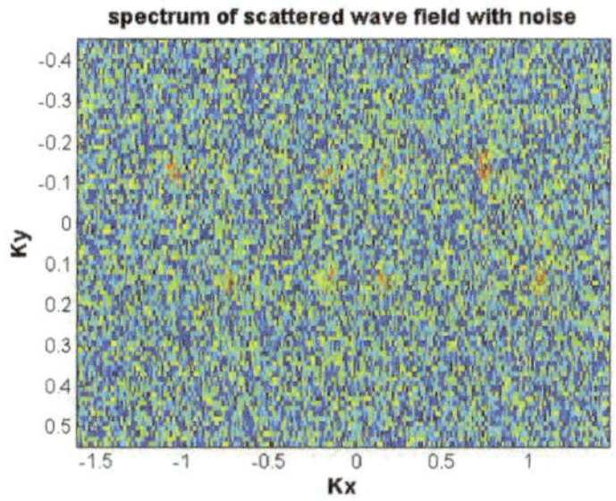

Fig. 4. Combined Fourier spectrum, consisting from the spectrum of scattered EM waves and random noise pattern with comparable amplitudes.

As it clearly seen in Figure 4, scattered EM wave spectrum is hidden in the noise pattern. Our aim is to apply Dynamic Logic algorithm in order to extract information on scattered EM waves from the pattern in Figure 4.

3. Implementation of the dynamic logic algorithm

In order to demonstrate the basics of the DL algorithm we will present below one illustrative example. Suppose we have two classes ω_1 and ω_2. We also have samples $\mathbf{X}_k$ where samples with k=1 belong to class ω_1 and with k=2 to class ω_2. Every sample k has two features $\mathbf{X}_k = (x_{1k}, x_{2k})$. In Figure 5 two groups of samples are presented. Red dots define samples from class k=1 and blue dots - from class k=2.

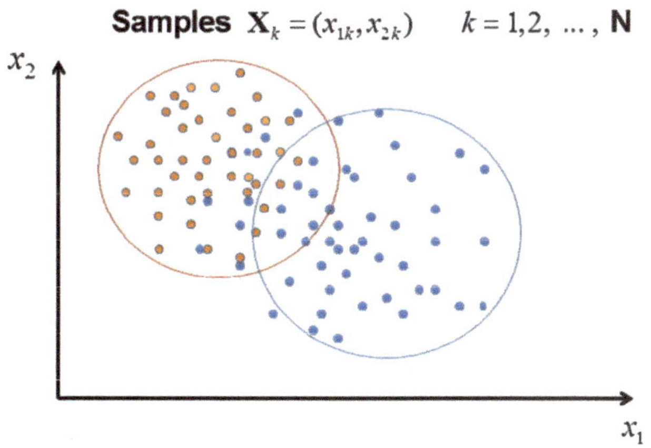

Fig. 5. Two groups of samples. Samples defines as red dots belong to the class k=1 and samples defined as blue dots to class k=1.

Typically, due to random variations there is good reason to expect the features in every class to be distributed normally (that is, according to the well-known Gaussian function, or "bell curve"). A normal distribution is completely specified by its set of parameters $\Theta = \{\mathbf{M}, \Sigma\}$, where $\mathbf{M}$ is the mean vector and Σ is the covariance matrix (related to the width, or "spread", of the distribution). The equation for a normally-distributed probability density function (pdf) for the class k=1 is:

$$p(\mathbf{X}_n, \Theta_1) = \frac{1}{\sqrt{(2\pi)^d |\Sigma_1|}} \exp[-\frac{1}{2}(\mathbf{X}_n - \mathbf{M}_1)^T \Sigma_1^{-1} (\mathbf{X}_n - \mathbf{M}_1)] \quad (17)$$

In (17) $p(\mathbf{X}_n, \Theta_1)$ is the probability that sample n with the feature vector $\mathbf{X}_n = (X_{n,1}, X_{n,2})$ belongs to process ω_1 with $\Theta_1 \equiv (\mathbf{M}_1, \Sigma_1)$. In (17) $\mathbf{M}_1$ is the mean vector for the class ω_1 and components of the mean vector are defined as:

$$M_{1,k} = \int_{-\infty}^{\infty} X_k p(\mathbf{X} | \Theta_1) d\mathbf{X} \quad (18)$$

Components of the covariance matrix Σ_1 for the class ω_1 are defined as:

$$\sigma_{1,ij} = \int_{-\infty}^{\infty} (X_i - M_{1,i})(X_j - M_{1,j}) p(\omega_1 | \mathbf{X}, \Theta_1) d\mathbf{X} \quad (19)$$

When we have N training samples $\mathbf{X}_{1,n}$ which belong only to class ω_1 the procedure to find the parameters of the Gaussian distribution Θ_1={$\mathbf{M}_1, \Sigma_1$} is as follows [6]. We need to introduce the likelihood function $L(\Theta_1)$:

$$L(\Theta_1) = \prod_{n=1}^{N} p(\mathbf{X}_{1,n} | \Theta_1)$$

Logarithm of the likelihood function :

$$LL(\Theta_1) = \ln L(\Theta_1) = \sum_{n=1}^{N} \ln p(\mathbf{X}_{1,n} | \Theta_1) \quad (20)$$

can be maximized by setting to zero its partial derivatives with respect to the parameters $\Theta_1 = \{\mathbf{M}_1, \Sigma_1\}$. Resulting maximum likelihood values for the mean and covariance in the class ω_1 are:

$$\widehat{\mathbf{M}}_1 = \frac{1}{N} \sum_{n=1}^{N} \mathbf{X}_{1,n} \quad (21)$$

Components of the covariance matrix Σ for the class ω_1 are defined as:

$$\widehat{\Sigma}_1 = \frac{1}{N} \sum_{n=1}^{N} (\mathbf{X}_{1,n} - \widehat{\mathbf{M}}_1)(\mathbf{X}_{1,n} - \widehat{\mathbf{M}}_1)^T \quad (22)$$

Now we will return to the problem formulated at the beginning of this section. Suppose we have two processes (classes) running ω_1 and ω_2. We also have identified N samples $\{\mathbf{X}_{n,1}\}$ which belong to class ω_1 and another N samples $\{\mathbf{X}_{n,2}\}$ which belong to class ω_2. Our aim is to design a classifier for sorting out the remaining unlabelled samples which belong either to the class k=1 or to the class k=2, assuming that we know the features $\mathbf{X}_n = (x_{1n}, x_{2n})$ of remaining samples, but do not know to which class do they belong. The procedure we should follow is as follows. First we gather a set of training samples N for each of the two classes k=1 and k=2. Next we assign feature values $\mathbf{X}_n = (x_{1n}, x_{2n})$ to each sample. Then for each class, separately, we estimate the mean and covariance using (22), to give us a set of parameters $\Theta_k = \{\mathbf{M}_k, \Sigma_k\}$ with k = 1,2. Knowing these parameters we can compute the pdf $p(\mathbf{X} \mid \Theta_k)$ of the data according to equation (17). Then we can compute, for each unlabelled sample, the probability

$$P(k \mid n) = \frac{p(\mathbf{X}_n \mid \Theta_k) P(k)}{p(\mathbf{X}_n)} \tag{23}$$

where $p(\mathbf{X}_n) = \sum_{k=1}^{k=2} p(\mathbf{X}_n \mid \Theta_k) P(k)$

Equation (23) gives the probability $P(k \mid n)$ that sample n belongs to class k. To classify the unlabelled sample n, we simply choose the class k', for which probability is the highest.

Equations (23) is just an expression of Baye's rule. Note that in equation (23) it is assumed we know the a priori probabilities P(k) for each class. Obviously, for each sample the probabilities must add up to one, i.e.

$$\sum_{k=1}^{k=2} P(k \mid n) = 1 \tag{24}$$

After we described how to estimate the mean and covariance of a single normal (Gaussian) distribution from a collection of data samples, below we will describe how to estimate parameters when the pdf is a weighted mixture, or summation, of Gaussian functions. Gaussian mixtures can be used to model complicated data sets. In such a case, each source corresponds to a class k which is parameterized by its own mean and covariance which we need to estimate. In addition to the mean and covariance, we also need to estimate the a priori probabilities for each class, i.e. what fraction of the samples belong to each class. The estimation of parameters for mixture models is complicated by the fact that the data samples are unlabelled, i.e. we don't know which data samples correspond to which class. Because of this inconvenience, it will be impossible to compute direct analytical expressions for the maximum-likelihood parameter estimates. However, we will be able to derive a convergent, iterative process to estimate the parameters. To do so we will perform the following basic procedure. First, a generic model for the data is defined in which the parameters for each class k are assumed to be the only unknowns. Next, a metric is selected which will be used to evaluate the "goodness of fit" between the model and the data for certain set of parameter values. Finally, an algorithm is developed to maximize the "goodness of fit" with respect to the parameters. Below we will describe maximum likelihood (ML) parameter estimation, in which we seek to maximize the log-likelihood of the parameters with respect to a collection of data samples. This approach is appropriate for standard data in which all samples are weighted equally in terms determining model parameters.

Next, we introduce an alternative known as maximum entropy (ME) parameter estimation. Here the parameters are chosen to maximize the Einsteinian log-likelyhood. This metric is appropriate for processing digitized image data, in which samples are weighted according to their pixel values.

4. Maximum likelihood (ML) estimation

We now describe maximum-likelihood (ML) parameter estimation for a Gaussian Mixture [1, 6]. Mathematically, the mixture model for two classes k=1 and k=2 is represented by the equation:

$$p(\mathbf{X} \mid \Theta) = \sum_{k=1}^{k=2} r_k p(\mathbf{X} \mid, \Theta_k) \tag{25}$$

where r_k is the mixture weight for class k and $\Theta = \{\Theta_1, r_1, \Theta_2, r_2\}$ is the total set of parameters for the mixture. Also $\Theta_k = (\mathbf{M}_k, \Sigma_k)$ is the subset of mean and covariance parameters for class k. Each class k has the normal (Gaussian) distribution:

$$p(\mathbf{X}, \Theta_k) = \frac{1}{\sqrt{(2\pi)^d \mid \Sigma_k \mid}} \exp[-\frac{1}{2}(\mathbf{X} - \mathbf{M}_k)^T \Sigma_k^{-1} (\mathbf{X} - \mathbf{M}_k)]$$

Since r_k denotes the fraction of the total samples that belong to class k, it is, by definition, equivalent to the *a priori* probability $P(k)$ for class k. It is a well-known fact that these probabilities should sum to one, i.e.:

$$\sum_{k=1}^{K} P(k) \equiv \sum_{k=1}^{K} r_k = 1 \tag{26}$$

Our goal is to compute the maximum likelihood (ML) parameter estimates $\hat{\Theta} = \{\hat{\Theta}_1, \hat{r}_1, \hat{\Theta}_2, \hat{r}_2\}$ given the additional constraint in equation (25). As in the previous Section, this is accomplished by a process that incorporates the derivatives of the log-likelihood with respect to each parameter:

$$LL(\Theta) = \sum_{n=1}^{N} \ln p(\mathbf{X}_n \mid \Theta) = \sum_{n=1}^{N} \ln \sum_{k=1}^{K} r_k p(\mathbf{X}_n \mid \Theta_k) \tag{27}$$

The maximum-likelihood parameter estimates exactly satisfy the equations:

$$\hat{\mathbf{M}}_k = \frac{\sum_{n=1}^{N} P(k \mid n) \mathbf{X}_n}{\sum_{n=1}^{N} P(k \mid n)} \tag{28}$$

$$\widehat{\Sigma}_k = \frac{\sum_{n=1}^{N} P(k \mid n)(\mathbf{X}_n - \widehat{\mathbf{M}}_k)(\mathbf{X}_n - \widehat{\mathbf{M}}_k)^T}{\sum_{n=1}^{N} P(k \mid n)} \tag{29}$$

$$\widehat{r}_k = \frac{1}{N}\sum_{n=1}^{N} P(k \mid n) \tag{30}$$

$$P(k \mid n) = \frac{r_k p(\mathbf{X}_n \mid \Theta_k)}{p(\mathbf{X} \mid \Theta)} = \frac{r_k p(\mathbf{X}_n \mid \Theta_k)}{\sum_{j=1}^{K} r_j p(\mathbf{X} \mid \Theta_j)} \tag{31}$$

Equation (31) is actually a form of Baye's and can be interpreted as the association probability, i.e. the probability that sample n belongs to class k. Equations (28) and (29) are intuitively pleasing in the sense that they are identical to their counterparts Eqs. (21) and (22) for the normal distribution, except that here each sample is weighted by the association probability $P(k/n)$.

Although Eqs. (28)-(31) look simple enough, they are actually a set of coupled nonlinear equations, because the quantity $P(k/n)$ on the right-hand sides is a function of all the parameters $\widehat{\mathbf{\Theta}}$, as can be seen from equations (31) and (25). If a person could solve these equations directly, they would yield the exact values for the maximum-likelihood parameters - unfortunately a direct solution is intractable. However, we can compute the parameters indirectly, by iterating back and forth between the parameter estimation equations (28)-(30) and the probability estimation equation (31). In other words, the parameter estimates $[\mathbf{M}_k^{(I)}, \Sigma_k^{(I)}, r_k^{(I)}]$ in iteration I are computed using the association probabilities $P^{(I-1)}(k / n)$ computed from the previous iteration (I -1). Next, the association probabilities $P^{(I)}(k / n)$ in iteration I are computed using the parameter estimates $[\mathbf{M}_k^{(I)}, \Sigma_k^{(I)}, r_k^{(I)}]$ from iteration I. Mathematically, this procedure corresponds to an approximate set of equations, analogous to Eqs. (28)-(31), which are iteration procedure to compute parameter estimates:

$$\widehat{\mathbf{M}}_k^{(I)} = \frac{\sum_{n=1}^{N} P^{(I-1)}(k \mid n)\mathbf{X}_n}{\sum_{n=1}^{N} P^{(I-1)}(k \mid n)} \tag{32}$$

$$\widehat{\mathbf{\Sigma}}_k^{(I)} = \frac{\sum_{n=1}^{N} P^{(I-1)}(k \mid n)(\mathbf{X}_n - \widehat{\mathbf{M}}_k^{(I)})(\mathbf{X}_n - \widehat{\mathbf{M}}_k^{(I)})^T}{\sum_{n=1}^{N} P^{(I-1)}(k \mid n)} \tag{33}$$

$$\widehat{r}_k^{(I)} = \frac{1}{N}\sum_{n=1}^{N} P^{(I-1)}(k \mid n) \tag{34}$$

and iteration procedure to compute probabilities :

$$P^{(I)}(k \mid n) = \frac{r_k^{(I)} p(\mathbf{X}_n \mid \Theta_k^{(I)})}{p(\mathbf{X} \mid \Theta^{(I)})} = \frac{r_k^{(I)} p(\mathbf{X}_n \mid \Theta_k^{(I)})}{\sum_{j=1}^{K} r_j^{(I)} p(\mathbf{X} \mid \Theta_j^{(I)})} \tag{35}$$

To initialize the procedure, parameter values are assigned at random, or by other means. Then the above set of equations is repeated until convergence, typically on the order of 100 iterations.

Below in Figures 6 - 7 results of implementation of the DL algorithm for the problem of identification of samples which belong to two different classes ω_1 and ω_2 described above and illustrated in Figure 5, are presented.

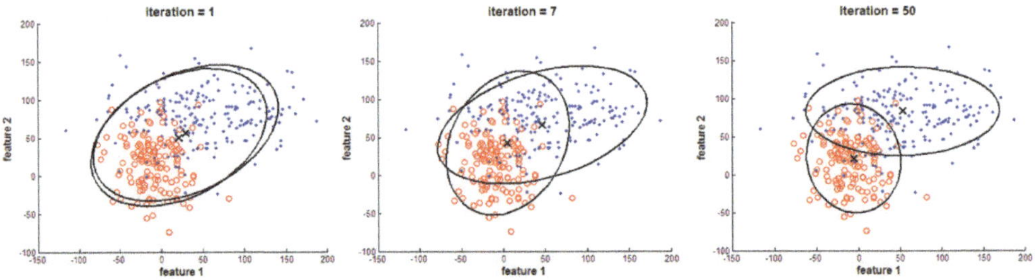

(a) (b) (c)

Fig. 6. With increasing iterations, the Dynamic Logic (DL) model adapts to segment data samples drawn from multiple random processes. In this example, the data samples are drawn from two Gaussian random processes, and the DL model consists of two components. The dots and circles represent the data samples, while the black ellipses represent the 2-sigma extent of the DL covariance, centered at the mean for each DL component. (a) Initially, the mean parameters for the components are assigned random initial values, and the covariance parameters are initialized in order to roughly span the sample data. (b) After several iterations the DL components begin to partition the data in a fuzzy manner as their evolving mean and covariance parameters cause them to "shrink" to fit each sample cluster. (c) Upon convergence, the DL model correctly segments the samples, and the mean and covariance parameters of the DL components provide good estimates for the mean and covariance of the underlying random processes.

In practice, one can decide whether convergence has been obtained by plotting the log-likelihood vs. iteration number and looking for a leveling of in the curve. The procedure is guaranteed to converge, in the sense that *LL* will never increase between successive iterations. However, convergence to an extraneous local minimum is sometimes possible. If this error occurs, and can be detected, the user must repeat the procedure after choosing a new set of initial conditions. The iterative procedure described above was pioneered by Perlovsky [1] and, independently, by others. In fact, the procedure has been shown to be an application of the well-known *expectation-maximization* (or EM) algorithm which is described, for example, in [20] and the references therein.

5. Maximum entropy (ME) estimation

In ML estimation, discussed above, we utilized the *log-likelihood* to determine the "goodness of fit" between the model and the data. This well-known metric is appropriate for standard data in which all samples are weighted equally in terms determining model parameters. However for some data sets, for example those involving digitized images, we require a different metric that will incorporate the natural weighting of samples corresponding to their pixel values [1,6]. For these problems we now introduce the metric known as the *Einsteinian log-likelihood:*

$$LL_E(\Theta) = \sum_{n=1}^{N} p_0(\mathbf{X}_n) \ln p(\mathbf{X}_n \mid \Theta) \tag{36}$$

Here, the function $p_0(\mathbf{X}_n)$ denotes the distribution of pixel values as a function of pixel position within the image. Note that $\mathbf{X}_n$ is no longer a vector of classification features, as has been the case up until now, but rather a vector giving the position $\mathbf{X}_n = [x_{1n}, x_{2n}]^T$ of the nth pixel in the image.

The Einsteinian log-likelihood LL_E, which was introduced by Perlovsky[1], has some interesting physical interpretations relating to photon ensembles, etc., as discussed in [1]. It is also related to some standard measures of similarity used in the information theory arena. In fact, maximizing LL_E with respect to the parameters is equivalent to minimizing the *relative entropy*, or *Kullback-Leibler distance* [5].

$$D(p_0 \mid p) = \sum_{x} p_0(x) \ln[\frac{p_0(x)}{p(x)}] \tag{37}$$

In any case, it is easy to see that if $p_0(\mathbf{X}_n) = 1$, then the Einsteinian log-likelihood LL_E in equation (36) reduces to the standard log-likelihood LL given by equation (27).

The equation describing our mixture distribution is the same as equation (25):

$$p(\mathbf{X} \mid \Theta) = \sum_{k=1}^{K} r_k p(\mathbf{X} \mid, \Theta_k) \tag{38}$$

Where

$$p(\mathbf{X}, \Theta_k) = \frac{1}{\sqrt{(2\pi)^d \mid \Sigma_k \mid}} \exp[-\frac{1}{2}(\mathbf{X} - \mathbf{M}_k)^T \Sigma_k^{-1} (\mathbf{X} - \mathbf{M}_k)]$$

Also, as before, $\Theta = (\Theta_k, r_k)$, k = 1; 2; ::: ;K, is the total set of parameters for the mixture, while $\Theta_k = \{\mathbf{M}_k, \Sigma_k\}$ is the subset of mean and covariance parameters for class k, and r_k is the mixture weight for class k. Substituting equation (38) into equation (37), we obtain the following expression for the Einsteinian log-likelihood:

$$LL_E(\Theta) = \sum_{n=1}^{N} p_0(\mathbf{X}_n) \ln \sum_{k=1}^{K} r_k p(\mathbf{X}_n \mid \Theta_k) \tag{39}$$

For convenience we will assume that the data $p_0(\mathbf{X}_n)$ is normalized, i.e.:

$$\sum_{n=1}^{N} p_0(\mathbf{X}_n) = 1 \tag{40}$$

Therefore, to insure that also:

$$\sum_{n=1}^{N} p(\mathbf{X}_n \mid \Theta) = 1 \tag{41}$$

so that the data and model have the same total energy, the r_k must sum to one, i.e.:

$$\sum_{k=1}^{K} r_k = 1 \tag{42}$$

Our goal is to compute the maximum entropy (ME) parameters $\hat{\Theta}$ which maximize LL_E given by equation (39) subject to the constraint in equation (42). The mathematical derivation for the ME parameter estimates is almost identical to the mathematics governing the ML estimates. The result is that the ME parameters exactly satisfy the equations:

$$\hat{\mathbf{M}}_k = \frac{\sum_{n=1}^{N} P(k \mid n) p_0(\mathbf{X}_n) \mathbf{X}_n}{\sum_{n=1}^{N} P(k \mid n) p_0(\mathbf{X}_n)} \tag{43}$$

$$\hat{\Sigma}_k = \frac{\sum_{n=1}^{N} P(k \mid n) p_0(\mathbf{X}_n)(\mathbf{X}_n - \hat{\mathbf{M}}_k)(\mathbf{X}_n - \hat{\mathbf{M}}_k)^T}{\sum_{n=1}^{N} P(k \mid n) p_0(\mathbf{X}_n)} \tag{44}$$

$$\hat{r}_k = \frac{1}{N} \sum_{n=1}^{N} P(k \mid n) p_0(\mathbf{X}_n) \tag{45}$$

$$P(k \mid n) = \frac{r_k p(\mathbf{X}_n \mid \Theta_k)}{\sum_{j=1}^{K} r_j p(\mathbf{X} \mid \Theta_j)} \tag{46}$$

As described in Section IV, the parameter estimation equations, above, are actually a set of coupled nonlinear equations, due to the implicit dependence within the bracket notation on the unknown parameters. However, we can compute the ME parameter estimates using a convergent, iterative process, identical to the process described in Section IV for the ML parameter estimates. In this process we alternate, in each iteration, between updating the probabilities $P(k \mid n)$ and the parameters $\{\mathbf{M}_k, \ \Sigma_k, \ r_k\}$.

Applying maximum entropy estimation method to the problem of detection of scattered by a Langmuir soliton EM waves hidden in a noise pattern (see Fig. 5) we will obtain the following result, presented in Fig. 7.

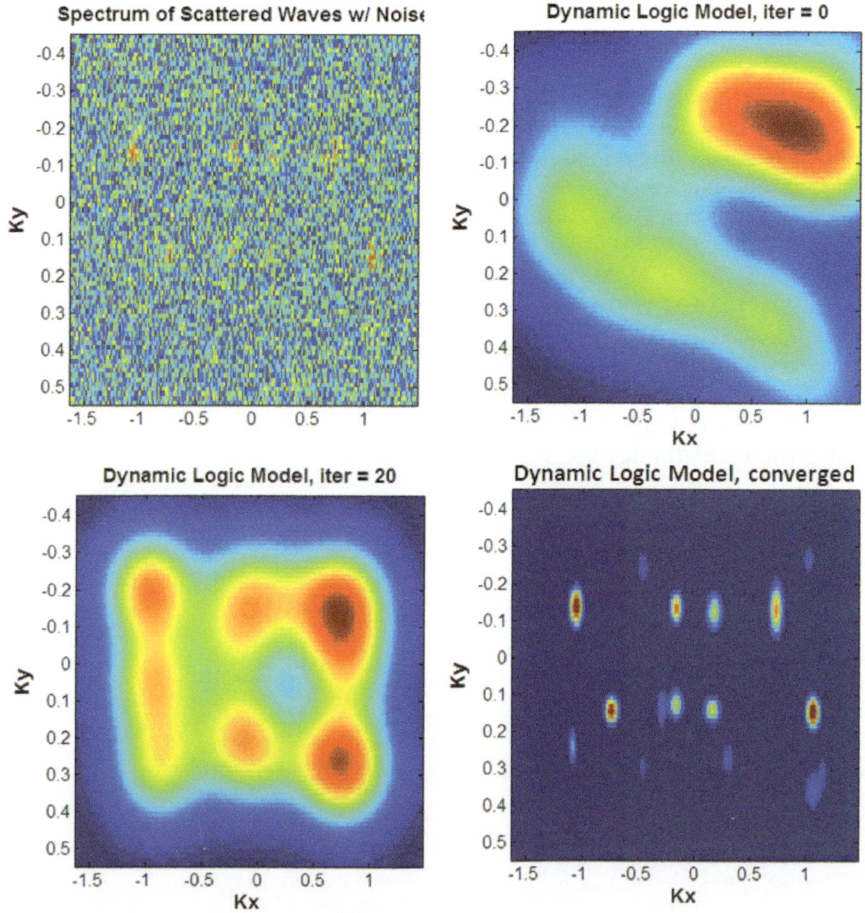

Fig. 7. (upper left) The combined measured spectra of several soliton waves are completely obscured by measurement noise. (upper right) The DL model is initialized with random model parameters. (lower left) After a few iterations the DL model begins to learn the structure of the data, and model components begin to adapt to fit the soliton spectra. (lower right) Upon convergence, the DL model is able to adaptively segment the data into background noise vs. multiple soliton spectra, despite the low signal-to-clutter ratio.

6. References

[1] Perlovsky, L.I. (2001). Neural Networks and Intellect: using model-based concepts. Oxford University Press, New York, NY (3rd printing). [2] Rudakov, L.I., DAN Sov. Acad. Sci, vol. 207, p. 821-823, 1972.

[2] Rudakov, L.I., Langmuir soliton, DAN Sov. Acad. Sci, vol. 207, p. 821-823, 1972.
[3] Mendonca J.T., Scattering of waves by Langmuir soliton, J. Plasma Physics, Vol.30, part 1, pp. 65-73, 1983.
[4] Sotnikov V. J.N. Leboeuf, S. Mudaliar, Scattering of electromagnetic waves in the presence of wave turbulence excited by a flow with velocity shear, Trans. Plasma Sci. IEEE, Vol. 38, No. 9, September 2010.
[5] MacKay, D.J.C., *Information Theory, Inference and Learning Algorithms,* Cambridge University Press, Cambridge, U.K., 2003.
[6] R.W. Deming, Parameter estimation for mixtures, August, 2005.
[7] Moon, T.K., The Expectation-Maximization Algorithm, IEEE Signal Processing Magazine, November, 1996, pp. 47-60.
[8] Duda, R. O. and Hart, P. E., Pattern Classification and Scene Analysis, section 6.4, New York: Wiley, 1973.
[9] Dempster, A. P., Laird, N. M., and Rubin, D. B., "Maximum Likelihood from incomplete data via the EM algorithm", Journal of Royal Statistical Society, Series B, vol. 39, (1977), 1-38.
[10] McLachlan, G. J., and Krishnan, T., The EM Algorithm and Extensions, New York: Wiley, 1997.
[11] Redner, R. A., and Walker, H. F., "Mixture densities, maximum likelihood and the EM algorithm", SIAM Review, vol. 26, (1984), 195-239.

12

Asteroseismology of Vibration Powered Neutron Stars

Sergey Bastrukov[1,2], Renxin Xu[1], Junwei Yu[1],
Irina Molodtsova[2] and Hsiang-Kuang Chang[3]
[1]State Key Laboratory of Nuclear Physics and Technology, School of Physics, Peking University, Beijing
[2]Joint Institute for Nuclear Research, Dubna
[3]Institute of Astronomy and Department of Physics, National Tsing Hua University, Hsinchu
[1]China
[2]Russia
[3]Taiwan

1. Introduction

There is a general recognition today that basic features of asteroseismology of non-convective final stage (FS) stars, such as white dwarfs, neutron stars and strange quark star, can be properly understood working from the model of vibrating solid star, rather than the liquid star model lying at the base of asteroseismology of convective main-sequence (MS) stars. In accord with this, most of current investigations of an FS-star vibrations rests on principles of solid-mechanical theory of continuous media, contrary to the study of the MS-star vibrations vibrations which are treated in terms of fluid-mechanical theory. This means that super dense matter of FS-stars (whose gravitational pressure is counterbalanced by degeneracy pressure of constituting Fermi-matter), possesses elasticity and viscosity generic to solid state of condensed matter, whereas a fairly dilute matter of the MS-stars (whose internal pressure of self-gravity is opposed by radiative pressure) possesses property of fluidity which is generic to the liquid state of a highly conducting condensed matter. This feature of the MS-star matter plays crucial role in generation of their magnetic fields in the dynamo processes involving macroscopic flows which are supported by energy of nuclear reactions in the central, reactive zone, of these stars. In the meantime, in the finale-stage (FS) stars, like white dwarfs and neutron stars, there are no nuclear energy sources to support convection. The prevailing today view, therefore, is that a highly stable to spontaneous decay dipolar magnetic fields of neutron stars are fossil. The extremely large intensity of magnetic fields of degenerate solid stars is attributed to amplification of fossil magnetic field in the magnetic-flux-conserving process of core-collapse supernova.

Even still before discovery of neutron stars, it has been realized that, for absence of nuclear sources of energy, the radiative activity of these of compact objects should be powered by energy of either rotation or vibrations and that the key role in maintaining the neutron star radiation should play an ultra strong magnetic field. As is commonly know today, the neutron star capability of accommodating such a field is central to understanding pulsating character of magneto-dipole radiation of radio pulsars whose radiative power is provided, as is commonly believed, by the energy of rigid-body rotation. The discovery of soft gamma-ray repeaters and their identification with magnetars (1) – quaking neutron stars endowed with ultra strong magnetic fields experiencing decay – has stimulated enhanced interest in the study of models of quake-induced magneto-mechanical seismic vibrations of neutron star and resultant electromagnetic radiation. Of particular interest in this study are torsional magneto-mechanical vibrations about axis of magnetic dipole moment of the star driven by forces of magnetic-field-dependent stresses in a perfectly conducting matter and in a permanently magnetized non-conducting matter (2). Most, if not all, reported up to now computations of frequency spectra of poloidal and toroidal Alfvén vibration modes in pulsars and magnetars rest on tacitly adopted assumption about constant-in-time magnetic field in which a perfectly conducting neutron star matter undergoes Lorentz-force-driven oscillations (3–12) (see, also, references therein). A special place in the study of the above Alfvén modes of pure shear magneto-mechanical seismic vibrations (a-modes) occupies a homogeneous model of a solid star with the uniform density ρ and frozen-in poloidal static magnetic field of both homogeneous and inhomogeneous internal and dipolar external configuration. The chapter is organized as follows. In section 2 with a brief outline of the solid-star model undergoing node-free torsional Alfvén vibrations in uniform internal (and dipolar external) magnetic field of constant intensity. In section 3, a model is extended to the case of vibrations in magnetic field experiencing decay. Here emphasis is placed on the loss of vibration energy caused by depletion of internal magnetic field pressure and resulting vibration-energy powered magneto-dipole radiation of vibrating neutron star. The decreasing of magnetic field pressure in the star is presumed to be caused by coupling between vibrating star and outgoing material which is expelled by quake, but mechanisms of star-envelope interaction resulting in the decay of magnetic field, during the time of vibrational relaxation, are not considered. It is shown that physically meaningful inferences regarding radiative activity of quaking neutron star can be made even when detailed mechanisms of depletion of magnetic field pressure in the process of vibrations triggered by quakes are not exactly known. This statement is demonstrated by a set of representative examples of magnetic field decay. The basic results of the model of vibration powered neutron star are summarized in section 4 with emphasis of its relevance to astrophysics of magnetars.

2. Lorentz-force-driven torsion vibrations of neutron star

To gain better understanding of the basic physics behind the interconnection between seismic and radiative activity of quaking neutron star, we start with a brief outline a fiducial model of a solid star with frozen-in homogeneous internal magnetic field

$$\mathbf{B}(\mathbf{r}) = B\,\mathbf{b}(\mathbf{r}), \quad B = \text{constant}, \tag{1}$$
$$\mathbf{b}(\mathbf{r}) = [b_r = \cos\theta, b_\theta = -\sin\theta, b_\phi = 0]$$

and dipolar external magnetic field, as is shown in Fig.1. In the last equation, B is the field intensity [in Gauss] and $\mathbf{b}(\mathbf{r})$ stands for the dimensionless vector-function of spatial distribution of the field.

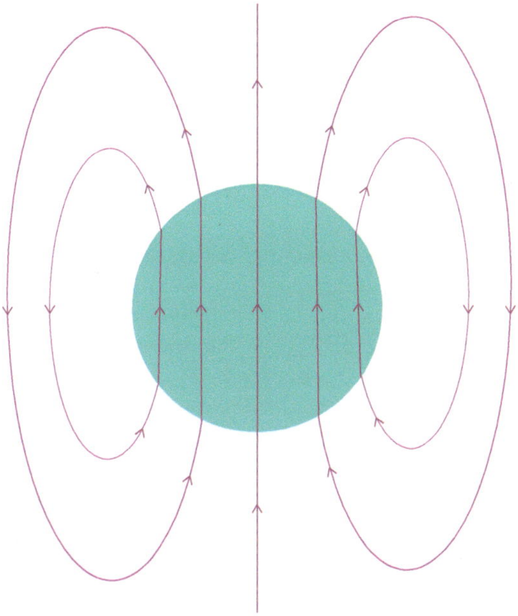

Fig. 1. The lines of magnetic field in fiducial model of neutron star with dipolar external and homogeneous internal magnetic field.

The above form of $\mathbf{B}(\mathbf{r})$ has been utilized in recent works(10–12) in which the discrete spectra of frequencies of node-free torsional Alfvén oscillations has been computed in analytic form on the basis of equations of linear magneto-solid mechanics

$$\rho\ddot{\mathbf{u}} = \frac{1}{c}[\delta\mathbf{j} \times \mathbf{B}], \quad \nabla \cdot \mathbf{u} = 0, \tag{2}$$

$$\delta\mathbf{j} = \frac{c}{4\pi}[\nabla \times \delta\mathbf{B}], \; \delta\mathbf{B} = \nabla \times [\mathbf{u} \times \mathbf{B}], \; \nabla \cdot \mathbf{B} = 0. \tag{3}$$

Here $\mathbf{u} = \mathbf{u}(\mathbf{r},t)$ is the field of material displacement, the fundamental dynamical variable of solid-mechanical theory of elastically deformable (non-flowing) material continua. These equations describe Lorentz-force-driven non-compressional vibrations of a perfectly conducting elastic matter of a non- convective solid star with Ampére form of fluctuating current density $\delta\mathbf{j}$. Equation for $\delta\mathbf{B}$ describing coupling between fluctuating field of material displacements $\mathbf{u}$ and background magnetic field $\mathbf{B}$ pervading stellar material is the mathematical form of Alfvén theorem about frozen-in lines of magnetic field in perfectly conducting matter. The adopted for Alfvén vibrations of non-convective solid stars terminology [which is not new of course, see for instance(13), namely, equations of magneto-solid mechanics and/or solid-magnetics is used as a solid-mechanical counterpart of well-known terms like equations of magneto-fluid mechanics, magnetohydrodynamics (MHD) and hydromagnetics $\rho\delta\dot{\mathbf{v}} = (1/c)[\delta\mathbf{j} \times \mathbf{B}]$, $\delta\mathbf{j} = (c/4\pi)[\nabla \times \delta\mathbf{B}]$, $\delta\dot{\mathbf{B}} = \nabla \times [\delta\mathbf{v} \times \mathbf{B}]$, describing flowing magento-active plasma in terms of the velocity of fluctuating flow $\delta\mathbf{v} = \dot{\mathbf{u}}$ and fluctuating magnetic field $\delta\mathbf{B}$. The MHD approach is normally utilized in astrophysics of convective main-sequence (MS) liquid stars such, for instance, as rapidly oscillating Ap (roAp) stars, chemically peculiar magnetic stars exhibiting high-frequency oscillations that have been and still remain the subject of extensive investigation(14; 15).

Inserting (2) in (3) the former equation of solid-magnetics takes the form

$$\rho\,\ddot{\mathbf{u}}(\mathbf{r},t) = \frac{1}{4\pi}[\nabla \times [\nabla \times [\mathbf{u}(\mathbf{r},t) \times \mathbf{B}(\mathbf{r})]]] \times \mathbf{B}(\mathbf{r}). \tag{4}$$

The analogy between perfectly conducting medium pervaded by magnetic field (magneto-active plasma) and elastic solid, regarded as a material continuum, is strengthened by the following tensor representation of the last equation $\rho\,\ddot{u}_i = \nabla_k \delta M_{ik}$, where $\delta M_{ik} = (1/4\pi)[B_i\delta B_k + B_k\delta B_i - B_j\delta B_j\delta_{ik}]$ is the Maxwellian tensor of magnetic field stresses with $\delta B_i = \nabla_k[u_iB_k - u_kB_i]$. This form is identical in appearance to canonical equation of solid-mechanics $\rho\,\ddot{u}_i = \nabla_k\sigma_{ik}$, where $\sigma_{ik} = 2\mu\,u_{ik} + [\kappa - (2/3)\mu]\,u_{jj}\delta_{ik}$ is the Hookean tensor of mechanical stresses and $u_{ik} = (1/2)[\nabla_i\,u_k + \nabla_k\,u_i]$ is the tensor of shear deformations in an isotropic elastic continuous matter with shear modulus μ and bulk modulus κ (having physical dimension of pressure). The most prominent manifest of physical similarity between these two material continua is their capability of transmitting non-compressional perturbation by transverse waves. Unlike incompressible liquid, an elastic solid can respond to impulsive non-compressional load by transverse waves of shear material displacements traveling with the speed $c_t = \sqrt{\mu/\rho}$. The unique feature of an incompressible perfectly conducting and magnetized continuous matter (in liquid or solid aggregated state) is the capability of transmitting perturbation by transverse magneto-mechanical, Alfvén, wave in which mechanical displacements of material and fluctuations of magnetic field undergo coupled oscillations traveling with the speed $v_A = B/\sqrt{4\pi\rho} = \sqrt{2P_B/\rho}$ (where $P_B = B^2/8\pi$ is magnetic field pressure) along magnetic axis. It is stated, therefore, that magnetic field pervading perfectly conducting medium imparts to it a supplementary portion of solid-mechanical elasticity (16; 17). This suggests that hydromagnetic Alfvén vibrations of a spherical mass of a perfectly conducting matter with frozen-in magnetic field can be specified in a manner of eigenstates of elastic vibrations of a solid sphere. As for the general asteroseismology of compact objects is concerned, the above equations seems to be appropriate not only for neutron stars but also white dwarfs(18) and quark stars. The superdense material of these latter yet hypothetical compact stars is too expected to be in solid state(19; 20).

Equation (4) serves as a basis of our further analysis. The studied in above works regime of node-free torsion vibrations under the action of Lorentz restoring force is of some interest in that the rate of differentially rotational material displacements

$$\dot{\mathbf{u}}(\mathbf{r},t) = [\boldsymbol{\omega}(\mathbf{r},t) \times \mathbf{r}], \quad \boldsymbol{\omega}(\mathbf{r},t) = [\nabla\chi(\mathbf{r})]\,\dot{\alpha}(t), \tag{5}$$

$$\nabla^2\chi(\mathbf{r}) = 0, \quad \chi(\mathbf{r}) = f_\ell\,P_\ell(\cos\theta), \quad f_\ell(r) = A_\ell\,r^\ell \tag{6}$$

has one and the same form as in torsion elastic mode of node-free vibrations under the action of Hooke's force of mechanical shear stresses. Hereafter $P_\ell(\cos\theta)$ stands for Legendre polynomial of degree ℓ specifying the overtone of toroidal a-mode. Fig.2 shows quadrupole and octupole overtones of such vibrations. The time-dependent amplitude $\alpha(t)$ describes temporal evolution of above vibrations; the governing equation for $\alpha(t)$ is obtained form

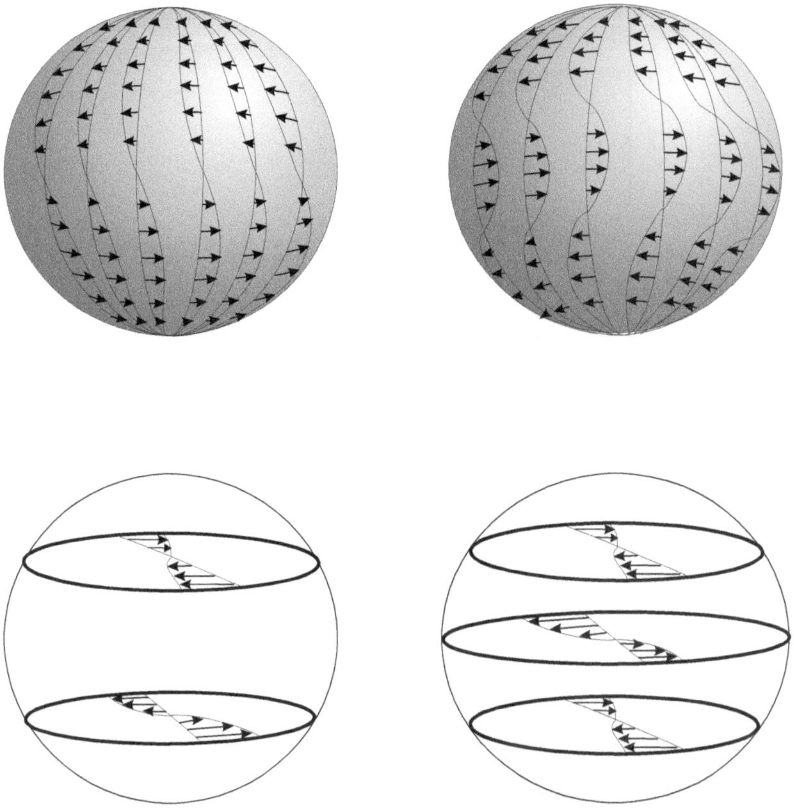

Fig. 2. The fields of material displacements in a neutron star undergoing torsional quadrupole (left) and octupole (right) node-free vibrations about magnetic axis.

equation (4). The prime purpose of above works was to get some insight into difference between spectra of discrete frequencies of toroidal *a*-modes in neutron star models having one and the same mass M and radius R, but different shapes of constant-in-time poloidal magnetic fields. By use of the energy method, it was found that each specific form of spatial configuration of static magnetic field about axis of which the neutron star matter undergoes nodeless torsional oscillations is uniquely reflected in the discrete frequency spectra by form of dependence of frequency upon overtone ℓ of nodeless vibration. It worth noting that first computation of discrete spectra of frequencies of toroidal Alfvén stellar vibrations in the standing wave-regime, has been reported by Chandrasekhar(23). The extensive review of other earlier computations of discrete frequency spectra of *a*-modes, $\omega_\ell = \omega_A s_\ell = B\kappa_\ell$, is given in well-known review of Ledoux and Walraven(14). The assumption about constant in time undisturbed magnetic field means that the internal magnetic field pressure, P_B, the velocity v_A of Alfvén wave in the star bulk

$$P_B = \frac{B^2}{8\pi}, \quad v_A = \sqrt{\frac{2P_B}{\rho}} \tag{7}$$

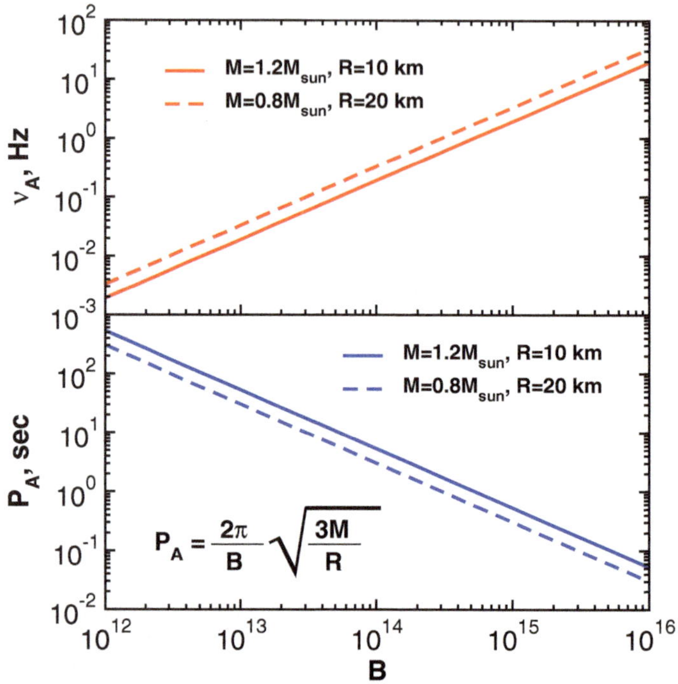

Fig. 3. The basic frequency and period of global Alfvén oscillations, equations (8), as functions of magnetic field intensity in the neutron star models with indicated mass and radius.

and, hence, the frequency $\nu_A = \omega_A/2\pi$ (where $\omega_A = v_A/R$) and the period $P_A = \nu_A^{-1}$ of global Alfvén oscillations

$$\nu_A = \frac{B}{2\pi}\sqrt{\frac{R}{3M}}, \quad P_A = \frac{2\pi}{B}\sqrt{\frac{3M}{R}} \tag{8}$$

remain constant in the process of vibrations whose amplitude $\alpha(t)$ subjects to standard equation of undamped harmonic oscillator(10; 11). The allow for viscosity of stellar material leads to exponential damping of amplitude, but the frequency ν_A and, hence, the period P_A preserve one and the same values as in the case of non-viscous vibrations(12). In Fig.3 these latter quantities are plotted as functions of intensity B of undisturbed poloidal magnetic field in the neutron star models with indicated mass M and radius R. The practical usefulness of chosen logarithmic scale in this figure is that it shows absolute vales of ν_A and P_A for global vibrations of typical in mass and radius neutron stars.

In what follows, we relax the assumption about constant-in-time magnetic field and examine the impact of its decay on the vibration energy and period. A brief analysis of such a case has been given in the context of magnetic white dwarfs(21). In this paper we present a highly extensive consideration of this problem in the context of neutron stars with emphasis on its relevance to the post-quake radiation of magnetars.

3. Vibration powered neutron star

In the following we consider a model of a neutron star with time-dependent intensity of homogeneous poloidal magnetic field which can be conveniently represented in the form

$$\mathbf{B}(\mathbf{r},t) = B(t)\,\mathbf{b}(\mathbf{r}), \quad [b_r = \cos\theta, b_\theta = -\sin\theta, b_\phi = 0]. \tag{9}$$

On account of this the equation of solid-magnetics, (4), takes the form

$$\rho\,\ddot{\mathbf{u}}(\mathbf{r},t) = \frac{B^2(t)}{4\pi}\,[\nabla\times[\nabla\times[\mathbf{u}(\mathbf{r},t)\times\mathbf{b}(\mathbf{r},t)]]]\times\mathbf{b}(\mathbf{r}). \tag{10}$$

Inserting here the following separable form of fluctuating material displacements

$$\mathbf{u}(\mathbf{r},t) = \mathbf{a}(\mathbf{r})\,\alpha(t) \tag{11}$$

we obtain

$$\{\rho\,\mathbf{a}(\mathbf{r})\}\,\ddot{\alpha}(t) = 2P_B(t)\,\times\{[\nabla\times[\nabla\times[\mathbf{a}(\mathbf{r})\times\mathbf{b}(\mathbf{r})]]]\times\mathbf{b}(\mathbf{r})\}\,\alpha(t). \tag{12}$$

Scalar product of (12) with the time-independent field of instantaneous displacements $\mathbf{a}(\mathbf{r})$ followed by integration over the star volume leads to equation for amplitude $\alpha(t)$ having the form of equation of oscillator with depending on time spring constant

$$\mathcal{M}\ddot{\alpha}(t) + \mathcal{K}(t)\alpha(t) = 0, \tag{13}$$

$$\mathcal{M} = \rho\, m_\ell, \quad \mathcal{K}(t) = 2P_B(t)\,k_\ell, \tag{14}$$

$$m_\ell = \int \mathbf{a}(\mathbf{r})\cdot\mathbf{a}(\mathbf{r})\,d\mathcal{V},\ \mathbf{a} = A_t\nabla\times[\mathbf{r}\,r^\ell\,P_\ell(\cos\theta)], \tag{15}$$

$$k_\ell = \int \mathbf{a}(\mathbf{r})\cdot[\mathbf{b}(\mathbf{r})\times[\nabla\times[\nabla\times[\mathbf{a}(\mathbf{r})\times\mathbf{b}(\mathbf{r})]]]]d\mathcal{V}. \tag{16}$$

The solution of equation of non-isochronal (non-uniform in duration) and non-stationary vibrations with time-dependent frequency $[\ddot{\alpha}(t) + \omega^2(t)\alpha(t) = 0$ where $\omega^2(t) = \mathcal{K}(t)/\mathcal{M}]$ is non-trivial and fairly formidable task(22). But solution of such an equation, however, is not a prime purpose of this work. The main subject is the impact of depletion of magnetic-field-pressure on the total energy of Alfvén vibrations $E_A = (1/2)[\mathcal{M}\dot{\alpha}^2 + \mathcal{K}\alpha^2]$ and the discrete spectrum of frequency of the toroidal a-mode

$$\omega_\ell^2(t) = \omega_A^2(t)\,s_\ell^2,\ \omega_A^2(t) = \frac{v_A^2(t)}{R^2},\ s_\ell^2 = \frac{k_\ell}{m_\ell}\,R^2, \tag{17}$$

$$\omega_\ell^2(t) = B^2(t)\kappa_\ell^2, \quad \kappa_\ell^2 = \frac{s_\ell^2}{4\pi\rho R^2}, \quad s_\ell^2 = \left[(\ell^2-1)\frac{2\ell+3}{2\ell-1}\right], \quad \ell\geq 2. \tag{18}$$

It is to be stated clearly from the onset that it is not our goal here to speculate about possible mechanisms of neutron star demagnetization and advocate conceivable laws of magnetic field decay. The main purpose is to gain some insight into the effect of arbitrary law of magnetic field decay in quaking neutron star on period of Lorentz-force-driven torsional seismic vibrations (whose quadrupole and octupole overtones are pictured in Fig.2) and radiative activity of the star brought about by such vibrations. In the reminder of the paper we focus on the case of torsional Alfvén vibrations in quadrupole ($\ell = 2$) overtone. In so doing we omit index ℓ putting $\omega(t) = \omega_{\ell=2}(t)$.

3.1 Magnetic-field-decay induced loss of vibration energy

The total energy stored in quake-induced Alfvén seismic vibrations of the star is given by

$$E_A(t) = \frac{\mathcal{M}\dot{\alpha}^2(t)}{2} + \frac{\mathcal{K}(B(t))\alpha^2(t)}{2}, \tag{19}$$

$$\mathcal{K}(B(t)) = \omega^2(B(t))\mathcal{M}. \tag{20}$$

Perhaps most striking consequence of the magnetic-field-pressure depletion during the post-quake vibrational relaxation of neutron star is that it leads to the loss of vibration energy at a rate proportional to the rate of magnetic field decay

$$\begin{aligned} \frac{dE_A(t)}{dt} &= \dot{\alpha}(t)[\mathcal{M}\ddot{\alpha}(t) + \mathcal{K}(B(t))\alpha(t)] + \frac{\alpha^2(t)}{2}\frac{d\mathcal{K}(B(t))}{dt} \\ &= \frac{\mathcal{M}\alpha^2(t)}{2}\frac{d\omega^2(B)}{dB}\frac{dB(t)}{dt} = \mathcal{M}\kappa^2\alpha^2(t)B(t)\frac{dB(t)}{dt}. \end{aligned} \tag{21}$$

In the model under consideration, a fairly rapid decay of magnetic field during the time of post-quake vibrational relaxation of the star is though of as caused, to a large extent, by coupling of the vibrating star with material expelled by quake. In other words, escaping material removing a part of magnetic flux density from the star is considered to be a most plausible reason of depletion of internal magnetic field pressure in the star. It seems quite likely that, contrary to viscous dissipation, the loss of vibration energy due to decay of magnetic field must be accompanied by coherent (non-thermal) electromagnetic radiation. Adhering to this supposition in the next section special consideration is given to the conversion of the energy of Lorentz-force-driven seismic vibrations into the energy of magneto-dipole emission whose flux oscillates with frequency of torsional Alfvén magneto-mechanical vibrations of the final stage solid stars, like magnetic white dwarf and neutron stars. Fig.4 replicates seismic torsional Alfvén vibrations of the star which are accompanied by oscillations of lines of dipolar magnetic field defining the beam direction of outburst X-ray emission.

3.2 Conversion of vibration energy into power of magneto-dipole radiation

The point of departure in the study of vibration-energy powered magneto-dipole emission of the star (whose radiation power, $\mathcal{P}$, is given by Larmor's formula) is the equation

$$\frac{dE_A(t)}{dt} = -\mathcal{P}(t), \quad \mathcal{P}(t) = \frac{2}{3c^3}\delta\ddot{\mu}^2(t). \tag{22}$$

Consider a model of quaking neutron star whose torsional magneto-mechanical oscillations are accompanied by fluctuations of total magnetic moment preserving its initial (in seismically

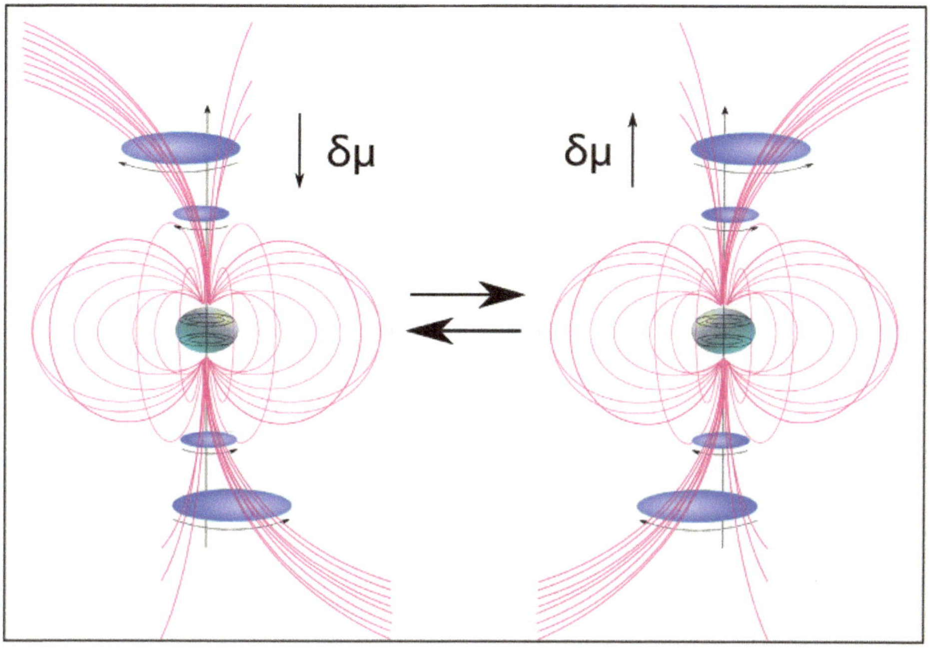

Fig. 4. Schematic view of quadrupole overtone of seismic torsional Alfvén vibrations of neutron star with homogeneous internal and dipolar external field whose lines, defining direction of outburst beam, oscillate with the frequency of this seismic a-mode.

quiescent state) direction: $\boldsymbol{\mu} = \mu\, \mathbf{n} =$ constant. The total magnetic dipole moment should execute oscillations with frequency $\omega(t)$ equal to that for magneto-mechanical vibrations of stellar matter which are described by equation for $\alpha(t)$. This means that $\delta\boldsymbol{\mu}(t)$ and $\alpha(t)$ must obey equations of similar form, namely

$$\delta\ddot{\boldsymbol{\mu}}(t) + \omega^2(t)\delta\boldsymbol{\mu}(t) = 0, \tag{23}$$

$$\ddot{\alpha}(t) + \omega^2(t)\alpha(t) = 0, \quad \omega^2(t) = B^2(t)\kappa^2. \tag{24}$$

It is easy to see that equations (23) and (24) can be reconciled if

$$\delta\boldsymbol{\mu}(t) = \boldsymbol{\mu}\,\alpha(t). \tag{25}$$

Then, from (23), it follows $\delta\ddot{\boldsymbol{\mu}} = -\omega^2\boldsymbol{\mu}\alpha$. Given this and equating

$$\frac{dE_A(t)}{dt} = \mathcal{M}\kappa^2\alpha^2(t)B(t)\frac{dB(t)}{dt} \tag{26}$$

with

$$-\mathcal{P} = -\frac{2}{3c^3}\mu^2\kappa^4B^4(t)\alpha^2(t) \tag{27}$$

we arrive at the equation of time evolution of magnetic field

$$\frac{dB(t)}{dt} = -\gamma B^3(t), \quad \gamma = \frac{2\mu^2\kappa^2}{3\mathcal{M}c^3} = \text{constant} \tag{28}$$

which yields the following law of its decay

$$B(t) = \frac{B(0)}{\sqrt{1+t/\tau}}, \quad \tau^{-1} = 2\gamma B^2(0). \tag{29}$$

The lifetime of magnetic field τ is regarded as a parameter whose value is established from below given relations between the period P and its time derivative $\dot{P}$ which are taken from observations. Knowing from observations $P(t)$ and $\dot{P}(t)$ and estimating τ one can get information about the magnitude of total magnetic moment and the strength of undisturbed magnetic field.

It is worth noting that in the model of vibration-energy powered magneto-dipole emission under consideration, the equation of magnetic field evolution is obtained in similar fashion as equation for the angular velocity Ω does in the standard model of rotation-energy powered emission of neutron star. As is shown in the next section, the substantial physical difference between models of rotation-energy and vibration-energy powered pulsating emission of neutron stars is that in the model of quaking neutron star vibrating in toroidal a-mode, the elongation of period of pulses is attributed to magnetic field decay, whereas in canonical Pacini-Gold model of radio-pulsar the lengthening of period of pulses is ascribed to the slow down of the neutron star rotation(39; 40; 49).

3.3 Lengthening of vibration period

The immediate consequence of above line of argument is the magnetic-field-decay induced lengthening of vibration period

$$B(t) = \frac{B(0)}{\sqrt{1+t/\tau}} \rightarrow P(t) = \frac{C}{B(t)}, \; C = \frac{2\pi}{\kappa}, \tag{30}$$

$$P(t) = P(0)\,[1+(t/\tau)]^{1/2}, \; P(0) = \frac{C}{B(0)}, \tag{31}$$

$$\dot{P}(t) = \frac{1}{2\tau}\frac{P(0)}{[1+(t/\tau)]^{1/2}}. \tag{32}$$

It follows that lifetime τ is determined by

$$P(t)\dot{P}(t) = \frac{P^2(0)}{2\tau} = \text{constant}. \tag{33}$$

This inference of the model under consideration is demonstrated in Fig.5. The difference between periods evaluated at successive moments of time $t_1 = 0$ and $t_2 = t$ is given by

$$\begin{aligned} \Delta P(t) &= P(t) - P(0) \\ &= -P(0)\left[1 - \frac{B(0)}{B(t)}\right] > 0, \quad B(t) < B(0). \end{aligned} \tag{34}$$

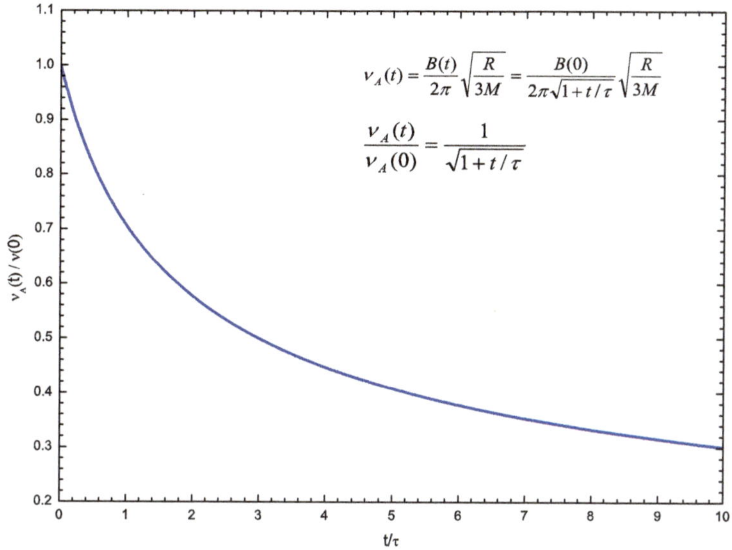

Fig. 5. Time evolution of Alfvén frequency of Lorentz-force-driven vibrations converting vibration energy into power of magneto-dipole radiation.

The practical usefulness of these general relations is that they can be used as a guide in search for fingerprints of Alfvén seismic vibrations in data on oscillating emission from quaking neutron star.

3.4 Time evolution of vibration amplitude

The considered model permits exact analytic solution of equation for vibration amplitude $\alpha(t)$ which is convenient to represent as

$$\ddot{\alpha}(t) + \omega^2(t)\alpha(t) = 0, \tag{35}$$

$$\omega^2(t) = \frac{\omega^2(0)}{1+t/\tau}, \quad \omega(0) = \omega_A\kappa. \tag{36}$$

The procedure is as follows. Let us introduce new variable $s = 1 + t/\tau$. In terms of $\alpha(s)$, equation (35) takes the form

$$s\alpha''(s) + \beta^2\alpha(s) = 0, \quad \beta^2 = \omega^2(0)\tau^2 = \text{const} \tag{37}$$

This equation permits exact analytic solution(24)1.5mm

$$\alpha(s) = s^{1/2}\{C_1 J_1(2\beta s^{1/2}) + C_2 Y_1(2\beta s^{1/2})\} \tag{38}$$

where $J_1(2\beta s^{1/2})$ and $Y_1(2\beta s^{1/2})$ are Bessel functions(25)

$$J_1(z) = \frac{1}{\pi}\int_0^{\pi} \cos(z\sin\theta - \theta)d\theta, \tag{39}$$

$$Y_1(z) = \frac{1}{\pi}\int_0^{\pi} \sin(z\sin\theta - \theta)d\theta, \quad z = 2\beta s^{1/2}. \tag{40}$$

The arbitrary constants C_1 and C_2 can be eliminated from two conditions

$$\alpha(t=0) = \alpha_0, \quad \alpha(t=\tau) = 0. \tag{41}$$

where zero-point amplitude

$$\alpha_0^2 = \frac{2\bar{E}_A(0)}{M\omega^2(0)} = \frac{2\bar{E}_A(0)}{K(0)}, \quad \omega^2(0) = \frac{K(0)}{M} \tag{42}$$

is related to the average energy $\bar{E}_A(0)$ stored in torsional Alfvén vibrations at initial (before magnetic field decay) moment of time $t = 0$. Before magnetic field decay, the star oscillates in the harmonic in time regime, that is, with amplitude $\alpha = \alpha_0\cos(\omega(0)t)$, so that $<\alpha^2>= (1/2)\alpha_0^2$ and $<\dot{\alpha}^2>= (1/2)\omega^2(0)\alpha_0^2$. The average energy of such oscillations is given by equation $\bar{E}_A(0) = (1/2)M<\dot{\alpha}^2> +(1/2)K(0)<\alpha^2>= (1/2)M\omega^2(0)\alpha_0^2 = (1/2)K(0)\alpha_0^2$ which relates the energy stored in vibrations with vibration amplitude α_0. As a result, the general solution of (35) can be represented in the form

$$\alpha(t) = C\,[1+(t/\tau)]^{1/2}$$

$$\times\{J_1(2\beta\,[1+(t/\tau)]^{1/2}) - \eta\,Y_1(2\beta\,[1+(t/\tau)]^{1/2})\}, \tag{43}$$

$$\eta = \frac{J_1(z(\tau))}{Y_1(z(\tau))}, \quad C = \alpha_0[J_1(z(0)) - \eta\,Y_1(z(0)]^{-1}. \tag{44}$$

The vibration period lengthening in the process of vibrations is illustrated in Fig.6 and Fig.7, where we plot $\alpha(t)$, equation (43), at different values of parameters β and η pointed out in the figures. Fig.7 shows that η is the parameter regulating magnitude of vibration amplitude, the larger η, the higher amplitude. However this parameter does not affect the rate of period lengthening. As it is clearly seen from Fig.6 both the elongation rate of vibration period and magnitude of vibration amplitude are highly sensitive to parameter β. All the above shows that the magnetic-field-decay induced loss of vibration energy is substantially different from the vibration energy dissipation caused by shear viscosity of matter resulting in heating of stellar material(26; 27). As was noted, the characteristic feature of this latter mechanism of vibration energy conversion into the heat (i.e., into the energy of non-coherent electromagnetic emission responsible for the formation of photosphere of the star) is that the frequency and, hence, period of vibrations are the same as in the case of viscous-free vibrations(12). However, it is no longer so in the case under consideration. It follows from above that depletion of magnetic field pressure resulting in the loss of total energy of Alfvén vibrations of the star causes its vibration period to lengthen at a rate proportional to the rate of magnetic field decay.

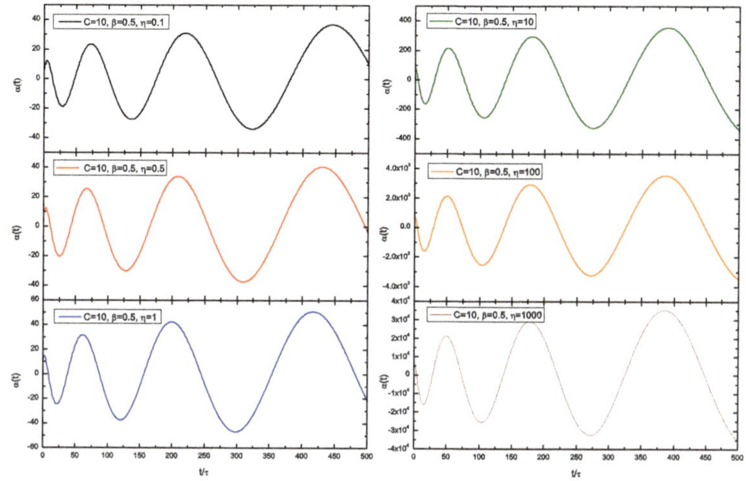

Fig. 6. Vibration amplitude $\alpha(t)$ computed at fixed C and β and different values of η. This shows that variation of this parameter is manifested in change of magnitude of $|\alpha|$, but period elongation is not changed.

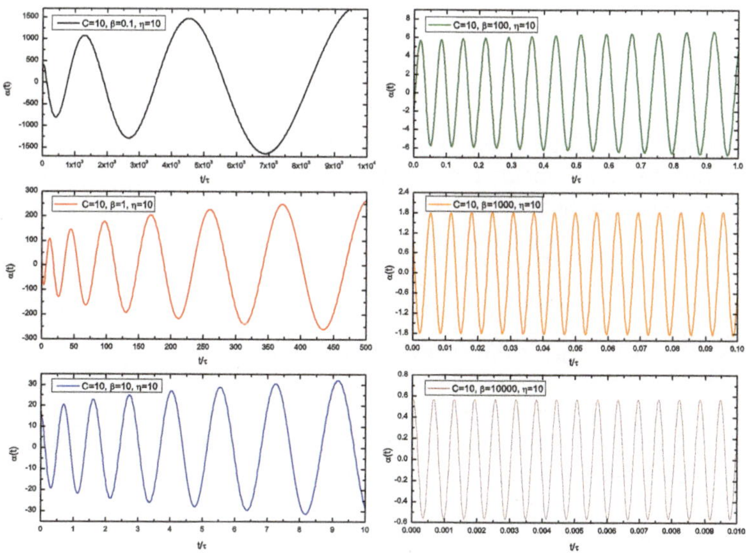

Fig. 7. Vibration amplitude $\alpha(t)$ computed at fixed C and η and different values of β.

	$M(M_\odot)$	R(km)	B(G)	ν_A(Hz)	τ(yr)
Pulsars	0.8	20	10^{12}	3.25×10^{-3}	4.53×10^{10}
	1.0	15	10^{13}	2.52×10^{-2}	2.98×10^{7}
Magnetars	1.1	13	10^{14}	0.22	7.4×10^{3}
	1.2	12	10^{15}	2.06	1.31
	1.3	11	10^{15}	1.89	2.38
	1.4	10	10^{16}	17.4	4.44×10^{-4}

Table 1. The Alfvén frequency of Lorentz-force-driven torsion vibrations, ν_A, and their lifetime equal to decay time of magnetic field, τ, in neutron stars with magnetic fields typical for pulsars and magnetars.

4. Oscillating luminosity of vibration powered neutron star

To get an idea of the magnitude of characteristic parameter of vibrations providing energy supply of magneto-dipole radiation of a neutron star, in Table 1 we present results of numerical computations of the fundamental frequency of neutron star oscillations in quadrupole toroidal a-mode and time of decay of magnetic field τ as functions of increasing magnetic field. As was emphasized, the most striking feature of considered model of vibration powered radiation is the lengthening of periods of pulsating emission caused by decay of internal magnetic field. This suggests that this model is relevant to electromagnetic activity of magnetars - neutron stars endowed with magnetic field of extremely high intensity the radiative activity of which is ultimately related to the magnetic field decay. Such a view is substantiated by estimates of Alfvén frequency presented in the table. For magnetic fields of typical rotation powered radio pulsars, $B \sim 10^{12}$ G, the computed frequency ν_A is much smaller than the detected frequency of pulses whose origin is attributed to lighthouse effect. In the meantime, for neutron stars with magnetic fields $B \sim 10^{14}$ G the estimates of ν_A are in the realm of observed frequencies of high-energy pulsating emission of soft gamma repeaters (SGRs), anomalous X-ray pulsars (AXPs) and sources exhibiting similar features. According to common belief, these are magnetars - highly magnetized neutron stars whose radiative activity is related with magnetic field decay. The amplitude of vibrations is estimated as (28)

$$\alpha_0 = \left[\frac{2\bar{E}_{\rm A}(0)}{\mathcal{M}\omega^2(0)}\right]^{1/2} = 3.423 \times 10^{-3}\bar{E}_{\rm A,40}^{1/2}B_{14}^{-1}R_6^{-3/2}. \quad (45)$$

where $\bar{E}_{\rm A,40} = \bar{E}_{\rm A}/(10^{40}\ {\rm erg})$ is the energy stored in the vibrations. $R_6 = R/(10^6\ {\rm cm})$ and $B_{14} = B/(10^{14}\ {\rm G})$. The presented computations show that the decay time of magnetic field (equal to duration time of vibration powered radiation in question) strongly depends on the intensity of initial magnetic field of the star: the larger magnetic field B the shorter time of radiation τ at the expense of energy of vibration in decay during this time magnetic field. The effect of equation of state of neutron star matter (which is most strongly manifested in different values mass and radius of the star) on frequency ν_A is demonstrated by numerical vales of this quantity for magnetars with one and the same value of magnetic field $B = 10^{15}$ G but different values of mass and radius. The luminosity powered by neutron star vibrations

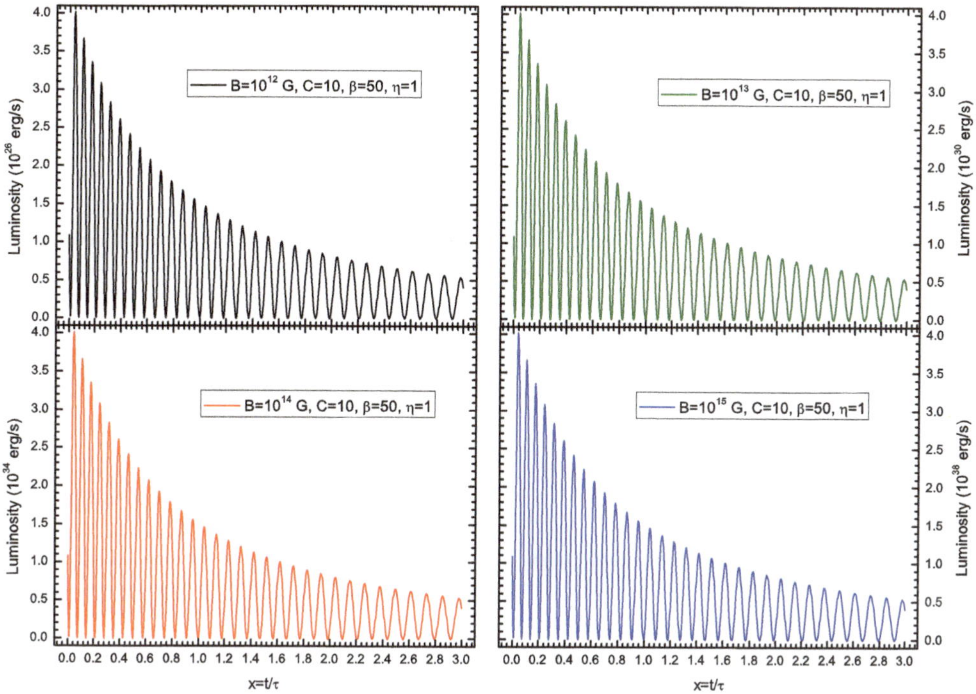

Fig. 8. Time evolution of luminosity of magneto-dipole radiation powered by energy of torsional Alfvén seismic vibrations of neutron star with mass $M = 1.2M_{\odot}$ and radius $R = 15$ km with intensities of magnetic field.

in quadrupole toroidal a-mode is given by

$$\mathcal{P} = \frac{\mu^2}{c^3}B^4(t)\alpha^2(t), \quad \mu = (1/2)B(0)R^3, \quad B(t) = B(0)[1 + t/\tau]^{-1/2}, \tag{46}$$

$$\alpha(t) = C\, s^{1/2}\{J_1(z(t)) - \eta\, Y_1(z(t))\}, \quad z = 2\omega(0)\tau\,(1 + t/\tau) \tag{47}$$

The presented in Fig.8 computations of power of magneto-dipole radiation of a neutron stars (of one and the same mass and radius but different values of magnetic fields) exhibit oscillating character of luminosity. The frequency of these oscillations equal to that of torsional Alfvén seismic vibrations of neutron star. All above suggests that developed theory of vibration-energy powered emission of neutron star is relevant to electromagnetic activity of magnetars - neutron stars endowed with magnetic field of extremely high intensity the radiative activity of which is ultimately related to the magnetic field decay. This subclass of highly magnetized compact objects is commonly associated with soft gamma repeaters and anomalous X-ray pulsars(29–32) – young isolated and seismically active neutron stars(33; 34). The magnetar quakes are exhibited by short-duration thermonuclear gamma-ray flash followed by rapidly oscillating X-ray flare of several-hundred-seconds duration. During this latter stage of quake-induced radiation of magnetar, a long-periodic (2-12 sec) modulation of brightness was observed lasting for about 3-4 minutes. Such long period of pulsed emission might be expected from old rotation-energy powered pulsars (due to slow down of rotation), but not from magnetars which are young neutron stars, as follows from their association with pretty young supernovae. Taking this into account (and also the fact that energy release during X-ray flare is much larger than the energy of rigid-body rotation with so long periods) it has been suggested(2) that detected long-periodic pulsating emission of magnetars is powered by the energy of torsional magneto-elastic vibrations triggered by quake. In view of key role of ultra-strong magnetic field it is quite likely that quasi-periodic oscillations (QPOs) of outburst flux from SGR 1806-20 and SGR 1900+14 detected in(35–37) are produced by torsional seismic vibrations predominately sustained by Lorentz force(10–12) which are accompanied, as was argued above, by monotonic decay of background magnetic field. If so, the predicted elongation of QPOs period of oscillating outburst emission from quaking magnetars should be traced in existing and future observations.

4.1 Rotation powered neutron star

For a sake of comparison, in the considered model of vibration powered radiation, the equation of magnetic field evolution is obtained in similar fashion as that for the angular velocity $\Omega(t)$ does in the standard model of rotation powered neutron star which rests on equations (38)

$$\frac{dE_R}{dt} = -\frac{2}{3c^3}\delta\ddot{\mu}^2(t), \tag{48}$$

$$E_R(t) = \frac{1}{2}I\,\Omega^2(t), \quad I = \frac{2}{5}MR^2 \tag{49}$$

One of the basic postulates of the model of rotation-energy powered emission is that the time evolution of total magnetic moment of the star is governed by the equation

$$\delta\ddot{\boldsymbol{\mu}}(t) = [\boldsymbol{\Omega}(t) \times [\boldsymbol{\Omega}(t) \times \boldsymbol{\mu}]], \quad \boldsymbol{\mu} = \text{constant}. \tag{50}$$

It follows

$$\delta\ddot{\mu}^2(t) = \mu_\perp^2\Omega^4(t), \quad \mu_\perp = \mu\,\sin\theta \tag{51}$$

where θ is angle of inclination of $\boldsymbol{\mu}$ to $\boldsymbol{\Omega}(t)$. The total magnetic moment of non-rotating neutron star is parametrized by equation of uniformly magnetized (along the polar axis) sphere (38)

$$\boldsymbol{\mu} = \mu\mathbf{n},\ \mu = \frac{1}{2}BR^3 = \text{constant}. \tag{52}$$

This parametrization presumes that in the rotation powered neutron star, the frozen-in the star magnetic field operates like a passive promoter of magneto- dipole radiation, that is, intensity of the internal magnetic field remain constant in the process of radiation. As a result, the equation of energy conversion from rotation to magnetic dipole radiation is reduced to equation of slow down of rotation

$$\dot{\Omega}(t) = -K\Omega^3(t), \quad K = \frac{2\mu_\perp^2}{3Ic^3}, \tag{53}$$

$$\Omega(t) = \frac{\Omega(0)}{\sqrt{1+t/\tau}}, \quad \tau^{-1} = 2K\Omega^2(0). \tag{54}$$

where θ is angle of inclination of $\boldsymbol{\mu}$ to $\boldsymbol{\Omega}(t)$. From the last equation it follows

$$P(t) = P(0)\,[1+(t/\tau)]^{1/2}, \ P(0) = \frac{2\pi}{\Omega(0)}, \tag{55}$$

$$\dot{P}(t) = \frac{1}{2\tau}\frac{P(0)}{[1+(t/\tau)]^{1/2}} \tag{56}$$

and, hence, the lifetime τ is related with $P(t)$ and $\dot{P}(t)$ as

$$P(t)\dot{P}(t) = \frac{P^2(0)}{2\tau} = \text{constant}. \tag{57}$$

Equating two independent estimates for τ given by equations (54) and (57) we arrive at widely utilized analytic estimate of magnetic field on the neutron star pole:

$$B = [3Ic^3/(2\pi^2 R^6)]^{1/2}\sqrt{P(t)\,\dot{P}(t)}.$$

For a neutron star of mass $M = M_\odot$, and radius $R = 13$ km, one has

$$B = 3.2\,10^{19}\sqrt{P(t)\,\dot{P}(t)}, \text{ G}.$$

The outlined treatment of rotation-powered magneto-dipole radiation of a neutron star emphasizes kinematic nature of variation of magnetic moment of pulsar whose magnetic field is regarded as independent of time. Thus, the substantial physical difference between vibration-powered and rotation-powered neutron star models is that in the former the elongation of pulse period is attributed to magnetic field decay, whereas in the latter the period lengthening is ascribed to slow down of rotation (39–41).

4.2 Comment on magnetic field decay in quaking neutron star

The considered law of magnetic field decay cannot be, of course, regarded as universal because it reflects a quite concrete line of argument regarding the fluctuations of magnetic moment of the star. The interrelation between quake-induced oscillations of total magnetic moment of the star, $\delta\boldsymbol{\mu}(t)$, and the amplitude, $\alpha(t)$, of its seismic magneto-mechanical oscillations can be consistently interpreted with the aid of the function of dipole demagnetization $\mathbf{f}(B(t))$ which is defined by the following condition of self-consistency in α of right and left hand sides of equation (22), namely

$$\delta\ddot{\boldsymbol{\mu}}(t) = \mathbf{f}(B(t))\alpha(t). \tag{58}$$

The vector-function of dipole demagnetization $\mathbf{f}(B(t))$ depending on decaying magnetic field reflects temporal changes of electromagnetic properties of neutron star matter as well as evolution of magnetic-field-promoted coupling between neutron star and its environment. This means that specific form of this phenomenological function should be motivated by heuristic arguments taking into account these factors. With this form of $\delta\ddot{\boldsymbol{\mu}}(t)$, equation (22) is transformed to magnetic field decay of the form

$$\frac{dB(t)}{dt} = -\eta\,\frac{\mathbf{f}^2(B(t))}{B(t)}, \quad \eta = \frac{2}{3\mathcal{M}\kappa^2c^3} = \text{const.} \tag{59}$$

In the above considered case, this function is given by

$$\mathbf{f}(B(t)) = \beta\,B(t)\,\mathbf{B}(t), \quad \beta = \kappa^2\mu = \text{constant.} \tag{60}$$

The practical usefulness of the dipole demagnetization function, $\mathbf{f}(B(t))$, consists in that it provides economic way of studying a vast variety of heuristically motivated laws of magnetic field decay, $B = B(t)$, whose inferences can ultimately be tested by observations. In this subsection, with no discussing any specific physical mechanism which could be responsible for magnetic field decay, we consider a set of representative examples of demagnetization function $\mathbf{f}(B(t))$ some of which have been regarded before, though in a somewhat different context [42-49]. Here we stress again that the model under consideration deals with magnetic field decay in the course of vibrations triggered by starquake, not with long-term secular decay which has been the subject of these latter investigations (see also references therein). 1. As a first representative example, a model of quaking neutron star whose function of dipole

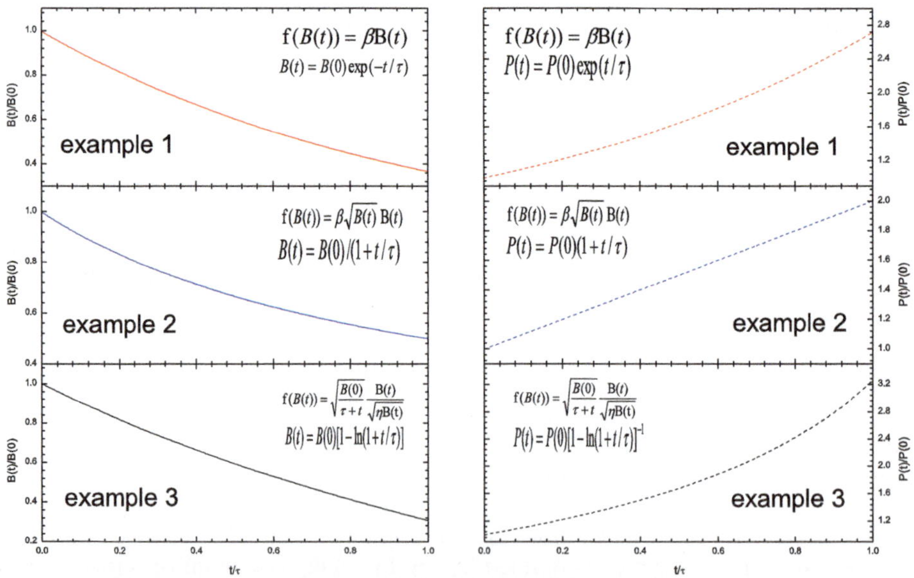

Fig. 9. Three representative examples with different laws of the magnetic field decay resulting in the elongation of the vibration period for each case.

demagnetization has the following form

$$\mathbf{f}(B(t)) = \beta\mathbf{B}(t) \tag{61}$$

is been considered. In such demagnetization model, the temporal evolution of the magnetic field and the lengthening of the vibration period obey the following laws

$$B(t) = B(0)\,e^{-t/\tau},\ P(t) = P(0)\,e^{t/\tau},\ \tau = \frac{P(t)}{\dot{P}(t)}. \tag{62}$$

where $B(0)$ is the intensity of magnetic field before quake. 2. For a quaking neutron star model whose function of demagnetization is given by

$$\mathbf{f}(B(t)) = \beta\sqrt{B(t)}\,\mathbf{B}(t) \tag{63}$$

the resultant equation of magnetic field evolution and the vibration period elongation read

$$B(t) = B(0)\left(1 + \frac{t}{\tau}\right)^{-1},$$

$$P(t) = P(0)\left(1 + \frac{t}{\tau}\right),\ \tau = \frac{P(0)}{\dot{P}(0)}. \tag{64}$$

Similar analysis can be performed for the demagnetization function of the form

$$\mathbf{f}(B(t)) = \beta\sqrt{B^{m-1}(t)}\,\mathbf{B}(t), \quad m = 3, 4, 5...$$

Namely,

$$\frac{dB(t)}{dt} = -\gamma\,B^m(t), \quad \gamma = \eta\beta^2,$$

$$B(t) = \frac{B(0)}{[1 + t/\tau_m]^{1/(m-1)}}, \quad \tau_m^{-1} = \gamma(m-1)B^{m-1}(0).$$

3. Finally, let's consider a model with quite sophisticated function of dipole demagnetization

$$\mathbf{f}(B(t)) = \sqrt{\frac{B(0)}{\tau + t}}\frac{\mathbf{B}(t)}{\sqrt{\eta B(t)}}. \tag{65}$$

which lead to a fairly non-trivial logarithmic law of magnetic field decay and vibration period lengthening

$$B(t) = B(0)\left[1 - \ln\left(1 + \frac{t}{\tau}\right)\right], \tag{66}$$

$$P(t) = P(0)\left[1 - \ln\left(1 + \frac{t}{\tau}\right)\right]^{-1}, \quad \tau = \frac{P(0)}{\dot{P}(0)}. \tag{67}$$

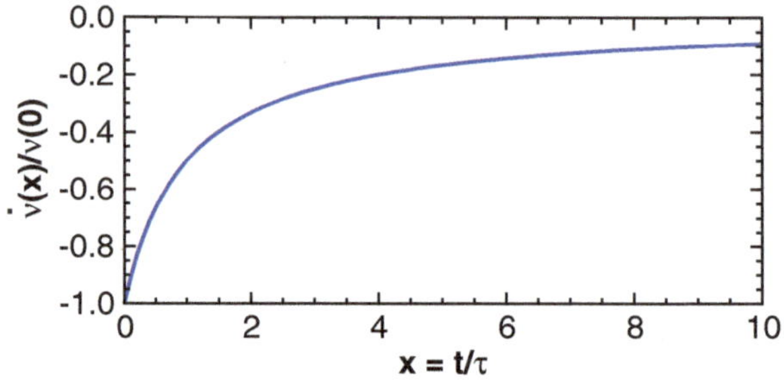

Fig. 10. The rate of frequency $\dot{\nu}(x)$ normalized to $\nu(0)$ as a function of $x = t/\tau$ for logarithmic law of magnetic filed decay.

For each of above examples, the magnetic field and the resultant lengthening of vibration period, computed as functions of fractional time t/τ, are shown in Fig. 9. This third example is interesting in that computed in this model ratio $\dot{\nu}(t)/\nu(0)$, pictured in Fig.10, as a function of t, is similar to that which exhibit data on post-glitch emission of PSR J1846-0258 [50]. These examples show that the period elongation is the common effect of magnetic field decay. The interrelations between periods and its derivatives substantially depend on specific form of the magnetic field decay law although physical processes responsible for magnetic field decay remain uncertain. Practical significance of above heuristic line of argument is that it leads to meaningful conclusion (regarding elongation of periods of oscillating magneto-dipole emission) even when detailed mechanisms of magnetic field decay in the course of vibrations are not exactly known.

5. Summary

It is generally realized today that the standard model of inclined rotator, lying at the base of our understanding of radio pulsars, faces serious difficulties in explaining the long-periodic ($2 < P < 12$ s) pulsed radiation of soft gamma repeaters (SGRs) and anamalous X-ray pulsars (AXPs). Observations show that persistent X-ray luminosity of these sources ($10^{34} < L_X < 10^{36}$ erg s^{-1}) is appreciably (10-100 times) larger than expected from neutron star deriving radiation power from energy of rotation with frequency of detected pulses. It is believed that this discrepancy can be resolved assuming that AXP/SGR-like sources are magnetars – young, isolated and seismically active neutron stars whose energy supply of pulsating high-energy radiation comes not from rotation (as is the case of radio pulsars) but from different process involving decay of ultra strong magnetic field, $10^{14} < B < 10^{16}$ G. Adhering to this attitude we have presented the model of quaking neutron star deriving radiation power from the energy of torsional Lorentz-force-driven oscillations. It is appropriate to remind early works of the infancy of neutron star era (51; 52) in which it has been pointed out for the first time that vibrating neutron star should operate like Hertzian magnetic dipole deriving radiative power of magneto-dipole emission from the energy of magneto-mechanical vibrations(53). What is newly disclosed here is that the main prerequisite of the energy conversion from vibrations into radiation is the decay of magnetic field in the star. Since the magnetic field decay is one of the most conspicuous features distinguishing magnetars from rotation powered pulsars, it seems meaningful to expect that at least some of AXP/SGR - like sources are magnetars deriving power of pulsating magnetic dipole radiation from the energy of quake-induced Alfvén torsion vibrations about axis of dipole magnetic field experiencing decay.

6. Acknowledgment

This work is supported by the National Natural Science Foundation of China (Grant Nos. 10935001, 10973002), the National Basic Research Program of China (Grant No. 2009CB824800), the John Templeton Foundation and also National Science Council of Taiwan, grant number NSC 99-2112-M-007-017-MY3.

7. References

[1] R. C. Duncan and C. Thompson, Astrophys. J. 392, L9 (1992).
[2] S. Bastrukov, J. Yang, M. Kim, and D. Podgainy, in *Current high-energy emission around black holes* eds. C.-H. Lee, H.-Y. Chang (World Scientific, Singapore, 2002) p.334.
[3] B. W. Carroll, E. G. Zweibel, C. J. Hansen, P. N. McDermott P. N., M. P. Savedoff, J. H. Thomas, and H. M. van Horn, Astrophys. J. 305, 767 (1986).
[4] S. I. Bastrukov, I. V. Molodtsova, V. V. Papoyan, and D. V. Podgainy, Astrophysics 40, 46 (1997).
[5] S. I. Bastrukov, I. V. Molodtsova, D. V. Podgainy, F. Weber, and V. V. Papoyan, Phys. Part. Nucl. 30, 436 (1999).
[6] U. Lee, Mon. Not. R. Astron. Soc. 374, 1015 (2007).
[7] U. Lee, Mon. Not. R. Astron. Soc. 385, 2069 (2008).
[8] S. I. Bastrukov, H.-K. Chang, Ş. Mişicu, I. V. Molodtsova, and D. V. Podgainy, Int. J. Mod. Phys. A 22, 3261 (2007).
[9] S. I. Bastrukov, H.-K. Chang, J. Takata, G.-T. Chen, and I. V. Molodtsova, Mon. Not. R. Astron. Soc. 382, 849 (2007).
[10] S. I. Bastrukov , K.-T. Chen, H.-K. Chang, I. V. Molodtsova, and D. V. Podgainy. Astrophys. J. 690, 998 (2009).
[11] S. I. Bastrukov, H.-K. Chang, I. V. Molodtsova, E.-H. Wu, K.-T. Chen, and S.-H. Lan, Astrophys. Space Sci. 323, 235 (2009).
[12] S. I. Bastrukov, I. V. Molodtsova, J. Takata, H.-K. Chang, R. X. Xu, Physics of Plasmas, 17, 112114 (2010).
[13] F. C. Moon, *Magneto-Solid Mechanics* (Wiley, New York, 2004).
[14] P. Ledoux and T. Walraven, *Handbuch Der Physik* 51, 353 (1958).
[15] F. Rincon and M. Rieutord, 2003, Astron. Astrophys. 398, 663 (2003).
[16] S. Chandrasekhar, *Hydromagnetic and Hydrodynamic Stability*, Clarendon, Oxford, (1961).
[17] H. Alfvén, and C.-G., Fälthammar, *Cosmical Electrodynamics. Fundamental Principles*, Clarendon, Oxford (1963).
[18] I. V. Molodtsova, S. I. Bastrukov, K.-T. Chen, and H.-K. Chang, Astrophys. Space Sci. 327, 1 (2010).
[19] R. X. Xu, Astrophys. J. 596, L59 (2003).
[20] R. X. Xu, J. Phys. G 36, 064010 (2009).
[21] S. I. Bastrukov, J. W. Yu, R. X. Xu, and I. V. Molodtsova, Mod. Phys. Lett. A 26, 359 (2011).
[22] D. E. Vakman and L. A. Vainshtein,, Sov. Phys. Uspekhi, 20, 1002 (1977).
[23] S. Chandrasekhar, Astrophys. J. 124, 571 (1956).

[24] A. D. Polyanin and V. F. Zaitsev, *Handbook of Nonlinear Partial Differential Equations* (Chapman and Hall, Boca Raton, 2004).
[25] M. Abramowitz and I. Stegun, *Handbook of mathematical functions* (Dover, New York, 1972).
[26] Frank H. Shu, *The Physics of Astrophysics, Vol. II: Gas Dynamics* (University Science Books, 1992).
[27] L. Mestel, *Stellar Magnetism* (Clarendon, Oxford, 1999).
[28] S. I. Bastrukov, J. W. Yu, R. X. Xu and I. V. Molodtsova, Research in Astron. and Astrophys. 11, 1085 (2011).
[29] A. K. Harding, Astronomische Nachrichten, 320, 260 (1999).
[30] C. Kouveliotou, Proc. Nat. Acad. Sci., 96, 5351 (1999).
[31] P. M. Woods and C. Thompson, C. in *Compact Stellar X-ray Sources* eds. Lewin W., van der Klis M. (Cambridge University Press, 2006).
[32] S. Mereghetti S., 2008, Astron. Astrophys. Rev. 15, 225 (2008).
[33] B. Cheng R. I., Epstein, R. A. Guyer, and A. C. Young, Nature 382 518 (1996).
[34] L. M. Franco, B. Link, and Epstein R. I., 2000, Astyrophys. J. 543, 987 (2000).
[35] G. L. Israel, T. Belloni, L. Stella, Y. Rephaeli, D. E. Gruber, P. Casella, S. Dall'Osso, N. Rea, M. Persic, and R. E. Rothschild, Astrophys. J. 628, L53 (2005).
[36] A. Watts and T. E. Strohmayer, Astrophys. J. 637, L117 (2006).
[37] T. Terasawa, Y. T. Tanaka, I. Yoshikawa, and N. Kawai, J. Phys.: Conf. Ser 31, 76 (2006).
[38] S. I. Bastrukov, R. X. Xu, J. W. Yu, and I. V. Molodtsova, Astrophys. Space Sci. 334, 150 (2011).
[39] R. N. Manchester and J. H. Taylor, *Pulsars* (Freeman, San Francisco, 1977).
[40] D. R. Lorimer and M. Kramer M., *Handbook of Pulsar Astronomy* (Cambridge University Press, Cambridge, 2004).
[41] G. S. Bisnovatyi-Kogan *Stellar Physics: 2 Stellar Evolution and Stability* (Springer-Verlag, Berlin, 2010).
[42] Y. Sang and G. Chanmugam, Astrophys. J. 323, L6 (1987).
[43] G. Srinivasan, D. Bhachataria, A.G. Muslimov and A. L. Tsygan, Current Sci. 59, 31 (1990).
[44] P. Goldreich and A. Reisenegger, Astrophys. J. 395, 250 (1992).
[45] V. A. Urpin, G. Chanmugam, Y. Sang, Astrophys. J. 433, 780 (1994).
[46] J. C. L. Wang, Astrophys. J. 486, L119 (1997).
[47] M. Livio, C. Xu, and J. Frank, Astrophys. J. 492, 298 (1998).
[48] M. Colpi, U. Geppert, and D. Page, Astrophys. J. 529, L29 (2000).
[49] G. S. Bisnovatyi-Kogan, Mem. Soc. Astron. Ital. 73, 318 (2002).
[50] M. A. Livingstone., V. M. Kaspi, and F. P. Gavriil, Astrophys. J., 710, 1710 (2010).
[51] F. Hoyle, J. V. Narlikar, and J. A. Wheeler, Nature 203, 914 (1964).
[52] A.G.W. Cameron, Nature 206, 1342 (1965).
[53] F. Pacini, in *Proc. MEASRIM No1* eds. A. Hady and M.I. Wanas., (2008) p. 16; Online at http://www.mearim.cu.edu.eg/new/Proceeding.htm, p.16.

13

Visualization Methods for Numerical Astrophysics

Werner Benger[1,2], Markus Haider[1], Harald Höller[1], Dominik Steinhauser[1], Josef Stöckl[1], Biagio Cosenza[1] and Marcel Ritter[1]
[1]*Department of Astro- and Particle Physics, Distributed and Parallel Systems Group, University of Innsbruck, Innsbruck*
[2]*Center for Computation & Technology at Louisiana State University, Baton Rouge*
[1]*Austria*
[2]*USA*

1. Introduction

Numerical simulations of astrophysical processes and subsequent predictions are an important complement to observational astronomy. With increasing computational performance available the demands on accuracy and functionality of the simulations grow steadily, leading to increasing amounts of complex data sets requiring three-dimensional visualization for proper analysis. Modern graphics hardware provides suitable rendering technology, however, many visually impressive approaches used in computer games, for instance, do not immediately transport over to application upon scientific data sets since priorities are different. Specific software for scientific visualization of astrophysical data does not necessarily need to provide photo-realistic rendering, but should rather be able to comprehensively display all features of a data set. False-color imaging has long been an established approach in observational astronomy. For three-dimensional time-dependent data sets the optional parameter space is even larger. As similar requirements occur in other scientific domains beyond astrophysics, data visualization benefits extraordinarily well from interdisciplinary approaches, for instance describing curved space in general relativity is similar to diffusion tensor fields occurring in magneto-resonance imaging in medical visualization (Benger et al., 2006).

Visualization of numerical astrophysics faces certain challenges because of the different data types involved together. For instance, a typical visualization data set may involve both particle-based and volumetric data, thus requiring complex techniques to handle transparency, see e.g. Kaehler et al. (2007).

Interdisciplinary approaches face various difficulties, starting with incompatibilities of the data format and data model (Benger, 2009). Common use is to provide specific solutions to specific problems. Unifying approaches to cover various needs under one denominator are rare, but essential in the longer term (Dougherty et al., 2009). In this article we will review a systematic approach for modeling and visualization of data sets across scientific disciplines, demonstrating this approach upon a set of state-of-the art astrophysical simulations.

2. Open issues in numerical astrophysics

The demand for sophisticated numerical simulations and hence for high performance visualization is somewhat exceptional in the field of astrophysics. Other than related disciplines, astrophysical research is very limited in the feasibility to conduct controlled experiments with their objects of interest. Experimental data and deducted knowledge is mainly gained from extensive observational studies. However they generally lack two important pieces of information: Due to the vast time scales of astrophysical processes (typically $10^6 - 10^9$ years), imaging comes practically without any data on temporal evolution, since even a decade of astronomical observation is nothing but a snatch on these scales. Moreover, we always find entirely two dimensional spatial information in astrophysical images, since - especially for extragalactic objects - the viewing angle between the observatory and the astrophysical object is fixed. Hence, scientific insight to the evolution and development of astrophysical targets comes from comprehensive statistical analysis of a large number of single observations that are interlinked to a big picture.

From a purely theoretical point of view, the physics that governs astrophysical processes and objects are very well known; a multitude of systems and configurations are described with concepts from hydrodynamics coupled with gravitation, radiation and magnetism. Mathematically, radiation hydrodynamics (RHD) and magnetohydrodynamics (MHD) are systems of coupled nonlinear partial differential equations. The necessity for enormous computational efforts lies in the nonlinearity of the equations on the one hand and the diversity of time- and spatial scales that are involved and to be resolved on the other hand. Supercomputing has thus become a major driver for research in astrophysics and even has opened entirely new fields that defy astronomical observations at the present state of technology. In the following two sections we want to present two case studies of astrophysical numerical simulations where data examinations benefit immensely from visualization techniques.

2.1 Galaxy clusters

Groups and clusters of galaxies have formed cosmologically *recently* in the hierarchical structure formation process of the universe. They are the largest gravitationally bound objects known and were mainly built during the last four billion (10^9) years by successive merging of galaxies, a picture that is supported both by observations (White & Frenk, 1991) as well as simulations (Bertschinger, 1998). The common understanding of cosmological structure formation is based on the so called *concordance model* (Ostriker & Steinhardt, 1995), which assumes hypothetical massive particles in the form of cold dark matter (CDM), a collisionless species of material that exclusively interacts gravitationally. In this CDM model, primordial fluctuations briefly after the big bang acted as gravitational potential seeds for slightly inhomogeneous mass accumulations that were not felt by normal baryonic matter until the universe had cooled down substantially. The length scales and amplitudes of these anisotropies can be observed in the cosmic microwave background (de Oliveira-Costa et al., 2004) and are used for the generation of initial conditions for CDM simulations on cosmological scales.

2.1.1 Simulation setup

For initial conditions we are using COSMICS (Bertschinger, 1995), a package that generates appropriate Gaussian random initial conditions with the feasibility to enforce statistical overdensities of dark and baryonic matter at designated places in the domain. This way, groups and clusters of dark matter halos evolve naturally within a N-body simulation of collisionless CDM. In our setup we model dark matter particles with the N-body implementation of GADGET-2 (Springel, 2005). This sophisticated parallelized tree-code is applied at resolutions of $128^3 - 512^3$ particles in comoving volumes of 40Mpc/h which produces up to $>$ 2TB of binary data for $>$ 130 million particles traced. We are primarily interested in secondary processes (rather than structure formation itself) that affect the baryonic matter fraction of such as the formation of actual galaxies within dark matter halos and their interaction with the surrounding gas, the so called intra cluster medium (ICM). Our premier research topic deals with the enrichment of this ICM with chemical elements that are produced in stars and transported to the intra cluster gas via a number of distinct mechanisms. One main challenge in such simulations is a proper modeling of the astrophysical processes on the wide range of spatial and temporal scales that are at play. While we are interested in gas dynamics up to several Mpc radius from the cluster center, those mechanisms that cause the enrichment of the ICM with heavier elements occur on practically stellar dimensions ($<$pc), such as stellar winds, the loss of gaseous envelopes at the AGB phase of stellar evolution and supernovae explosions. Even with the most advanced technical facilities imaginable today we would still be far away from being able to resolve such a wide range of scales consistently.

In order to overcome this dilemma, semi-analytical galaxy formation models (SAM) are introduced. These SAMs deduce characteristic galactic quantities by a recipe based on the merger history of the dark matter halos that were simulated beforehand. In other words, the SAM assigns collective quantities such as total gas mass, angular momentum, starformation rate in the galaxies etc. to the halos regarding their merger history by observationally motivated, heuristic formulae and prescriptions. We use our own semi-analytical model (van Kampen et al., 1999) while there is also a relevant publicly available open source project called Galacticus (Benson, 2010). On top of all specified preliminary simulations there is a hydrodynamical grid-code with PPM Riemann solver that treats the intra cluster medium on the gravitational background of the dark matter simulation. The output from the SAM provides for two main contributions to ICM enrichment with chemical elements, namely ram-pressure stripping and galactic winds (see section 2.1.2), which are used as input for the hydrodynamical simulation. The number of independent points in computation is $11/8 \cdot 128^3$ up to $11/8 \cdot 512^3$ due to the fact that we apply four nested grids each with doubling resolution. The galaxy cluster is forming in the domain center of the hydrodynamical simulation which captures a smaller volume than the CDM simulation in order to have adequate boundary conditions for the gas flows (mainly inflow). All of the simulations are conducted on a non-steady scale, the aforementioned comoving volume of the cosmic expansion. The physical variables are scaled accordingly to this expansion, in cosmology sometimes also called the *Hubble flow*.

2.1.2 Metal enrichment by ram-pressure stripping and galactic winds

The chemical evolution of the universe is quite well understood in principle. The theory of primordial nucleosynthesis provides abundances that fill the early universe merely with hydrogen and helium homogeneously. However we know that today's universe is high in heavier elements, such as carbon, silicon, oxygen etc. Furthermore we know that these heavy elements (in astrophysics also called metals) have to be synthesized in stars but the question remains: How are they transported into the ICM of a galaxy group or a galaxy cluster where we can observe them (Blanton et al., 2003; Lovisari et al., 2009)?

There are two main astrophysical mechanisms that enrich the intra cluster medium, namely ram-pressure stripping and galactic winds while the latter is to be understood as collective of numerous distinct astronomical processes. Ram-pressure is the force a galactic plane of gas experiences while falling towards the center of a cluster due to the surrounding ICM as described in section 2.2. Ram-pressure stripping occurs whenever a certain pressure threshold is reached, and leads to the separation of the galaxy's planar gas from the stellar component of stars in the disk. Its contribution to metal enrichment in galaxy clusters is mainly influenced by the density of the ICM which naturally shows a gradient towards the cluster center. Also the in-falling velocities of the galaxies increase towards the main potential well in the center. Hence, ram-pressure stripping is primarily governed by circumstances of kinematics and the spatial distribution of the galaxies (Domainko et al., 2008). On the other hand, galactic winds are dominantly affected by the factors of galactic evolution such as the history of merger events and subsequent star-burst phases which lead to enormous galactic winds due to temporarily concentrated collective supernovae explosions (Finoguenov et al., 2000). Generally, galactic winds are driven by the star formation rate which is in principle higher in early times of cluster formation since merging events occur much more frequently. Such disturbances trigger density fluctuations in the gas which then leads to increased star formation.

We are interested in the efficiency of different transport processes in enriching the ICM and in the spatial distribution of metal abundances in the cluster at a certain epoch of evolution. Due to the complexity of the involved physical processes and the richness of our data, we are keen to visualize the manifold parameter space both preferably exhaustive and as clearly as possible, which will be discussed exemplary in section 4.1.

2.2 Ram-pressure stripping of disk galaxies

As described in section 2.1, galaxy clusters consist of dark matter, galaxies and a hot, thin gas, the ICM. These clusters of galaxies have been formed through the hierarchical structure formation process from dark matter and primordial gas. Part of the primordial gas, the ICM, can be observed by X-ray observations. This plasma, even though it is very hot ($\sim 10^8 K$) and thin ($10^{-29} - 10^{-26}$ g cm^{-3}), exerts a pressure on the inter-stellar medium (ISM). According to (Gunn & Gott, 1972), the ISM is stripped if this ram pressure exceeds the gravitational restoring force. Moreover, ram pressure compresses the gas in the spiral arms of the galaxy and enhances the star formation within the galaxy. External pressure of the ICM also causes the formation of dense gas knots where lots of new stars are formed (e.g. (Hester et al., 2010)). Yet it is unclear and matter of research how efficient these processes are with uncertainties being mainly mixing of different gas phases and instabilities.

2.2.1 Simulation setup

To simulate ram-pressure stripping of a galaxy, we use the N-body/hydrodynamic simulation code GADGET-2 (Springel, 2005). The N-body part treats the collisionless dynamics of dark-matter and stars. To this end, GADGET-2 uses a sophisticated tree code (as first introduced by Barnes & Hut, 1986) and treePM code respectively to calculate the gravitational force. The gas of the galaxies and the ICM are treated hydrodynamically via smoothed particle hydrodynamics (SPH, see Gingold & Monaghan, 1977; Lucy, 1977)). Furthermore, our special version of GADGET-2 includes radiative cooling according to (Katz et al., 1996) and a recipe for star formation, stellar feedback and galactic winds from (Springel & Hernquist, 2003) because we are mainly interested in the influence of ram-pressure stripping on the star formation. As the resolution in our simulations is far too low to resolve gas clouds forming single stars, the star formation in the ISM is described in a statistical manner. The continuous distribution of the ISM is represented by SPH particles, holding spatially averaged hydrodynamic quantities, the star formation is calculated for each such particle. The gas mass is divided into two gas phases, hot ambient medium used for hydrodynamic computations and cold gas clouds which form new stars on a characteristic time scale when a density threshold is exceeded. During star formation, at the same time, a fraction of 10 % according to a Salpeter IMF (Salpeter initial mass function, see Salpeter, 1955) is immediately released to the hot ambient medium due to supernovae from massive stars exceeding 8 $M_{\odot}$. The inclusion of the cooling of hot ambient medium leads to the growth of the cold gas clouds making star formation self regulating.

To simulate ram-pressure stripping of disk galaxies, we need a model galaxy as well as a way to simulate the ICM, interacting with the model galaxy. Model galaxies include the ISM, the stellar disk, a bulge and dark matter. For the stellar and gaseous component a disk with exponentially decreasing density along the radius is used whereas for the bulge and the dark-matter halo a NFW profile (Navarro et al., 1996) is applied. Consequently, the distribution and velocity of the particles are set according to (Mo et al., 1998). To avoid numerical artifacts and to get a realistic model, we evolve the galaxies in isolation for 1 billion years. Thereafter, the star formation rate has settled and spiral arms are forming in the disk.

Finally, the model galaxies interact with the ICM. For that purpose, the setup described in Kapferer et al. (2009) and Kronberger et al. (2008) is used. The idea is to simulate the effect of a wind tunnel. The galaxy is moving instead of the wind in our setup, leading to the same result. We fill a cube with gas particles to represent the ICM. To keep the gas density stable over time, periodic boundary conditions are imposed. To avoid numerical artifacts due to the SPH scheme, the same gas mass-resolution as for the ISM has to be used. Since only cubic simulation domains are supported and a side-length of at least 800 kpc for the cube is needed, a huge amount of particles would be necessary. To avoid lowering the resolution in the ISM, just a cuboid of gas particles with the required resolution is built up where the model galaxy flies through. The rest of the cube is filled with gas particles having a 100 times lower mass resolution. Anyway, the same gas density is used. With this configuration, the density of the wind-tunnel changes only by 4 % over a simulation time of 800 million years. We evaluate the results, finally, by removing the ICM particles from the data to investigate the morphology of the gas and stellar disk as well as the influence of ram-pressure stripping on the star formation rate.

3. Visualization methodology

In many visualization systems solutions to a particular problem are based on specific data structures. Such diverse approaches are sensitive to the growing demands of new datasets, as such may likely require design changes leading to incompatibilities within the same application as well as among older and younger data sets. Even worse, an extensive set of specific solutions effects the learning curve of both the application developer and end-user of the visualization system since each solution has to be learned anew on a case-by-case basis. Our approach, based on the Vish Visualization Shell (Benger et al., 2007), contrasts such ad-hoc approaches on the design level by routing all data structures and visualization methods based on them through a common denominator, providing a systematic approach for modeling data and visualizing them. While this concrete data structure is limited in applicability to a certain type of data, this class of data types is broad enough to address most cases occurring in scientific visualization. This data model will be reviewed in this section and discussed for its applicability specifically to data sets from astrophysical simulations.

3.1 Using a common data model

Expressing all types of data occurring in astrophysical simulations within the same data organization framework is the first step towards a systematic approach contrasting ad-hoc visualization on case-by-case basis. No commonly agreed solution exist yet, but an approach based on the mathematics of fiber bundles has been proposed early by (Butler & Pendley, 1989). Inspired by its successful implementation in the OpenDX data explorer (Treinish, 1997), the data model has been advanced to cast data into a hierarchy of six levels (Benger, 2004). This skeleton forms the basis for visualizing data in a very generic and reusable way (Benger, 2008).

3.1.1 Mathematical background

The central idea is to organize data by their mathematical commonalities, following the theories of topology, differential geometry, and geometric algebra (Hestenes, 1999). The model is based on the assumption that data for scientific visualization can be organized as a fiber bundle, which considers a total space E as being constructed from a base space B and a fiber space F such that locally $E = B \times F$. The base space in this context is described by a discretized manifold, which in topology is modeled via a $CW - complex$ as a hierarchical set of spaces, the so-called k-skeletons, each of them describing vertices, edges, faces, cells of some spatial domain (see Benger et al., 2011, for a more detailed review). Differential geometry introduces charts as mappings from manifolds to real numbers, allowing to represent physical quantities given at each point of the underlying manifold numerically in one or more coordinate systems. Having the choice of alternative numerical representations of the same physical quantity is essential for computational accuracy since not all coordinate systems are equivalently suited for a particular problem. However, the numerical representation of some quantity is frequently confused with the object itself, though differential geometry demonstrates that objects with same numerical representation (for instance, three numbers) have different mathematical properties (e.g. covariant and contra-variant vectors). Not many data models (and file formats) support expressing such distinctions as required by accurate treatment of differential geometry. One step further is even the consideration of Geometric Algebra (Hestenes, 1999), which intents to unify various branches of mathematics itself through a common treatment. It is thus a natural complement to be considered in a data model (Benger et al., 2010).

3.1.2 Conceptual data organization

In the model used for our work data are cast into a hierarchy of seven levels. These levels are:

1. **Slice** - bundles all data by their parameter space such as physical time
2. **Grid** - bundles all data related to a geometric entity
3. **Skeleton** - bundles all data related to a topological property
4. **Representation** - bundles all data related to a specific coordinate system
5. **Field** - contains data representing a physical quantity
6. **Fragment** - contains a subset of a data field
7. **Component** - contains a component of a multi-valued field

From these seven levels only two are exposed to the end-user: the **Grid** and the **Field**; the remaining levels serve to internally describe the mathematical properties of the data. The two last levels, **Fragment** and **Component**, are optional, and used to express practical extensions beneficial to domain decomposition in HPC computations and different memory layouts (Benger, 2008). Actual data in the form of numerical arrays reside only in the last level of this hierarchy. Each of these arrays is accessed via a path through this hierarchy, similar to a file in a filesystem. This path defines the semantics of the given data values and the existence of entries in the hierarchy defines the properties of the entire data set. For instance, diagram (1) demonstrates how the coordinates of a particle data set data set describing dark matter at a time 3.4Gy are modeled using a Cartesian chart ("/" denotes the "root" node which bundles all data):

$$/ \xrightarrow{\text{Slice}} 3.4Gy \xrightarrow{\text{Grid}} \text{Dark Matter} \xrightarrow{\text{Skeleton}} \text{Vertices} \xrightarrow{\text{Representation}} \text{Cartesian Chart} \xrightarrow{\text{Field}} \text{Positions} \tag{1}$$

Similarly arbitrary fields such as velocity of the particles can be added to the data:

$$/ \xrightarrow{\text{Slice}} 3.4Gy \xrightarrow{\text{Grid}} \text{Dark Matter} \xrightarrow{\text{Skeleton}} \text{Vertices} \xrightarrow{\text{Representation}} \text{Cartesian Chart} \xrightarrow{\text{Field}} \text{Velocity} \tag{2}$$

The *Skeleton* level carries a set of topological invariants described by integer values: *dimensionality*, *index depth* and *refinement*. Hereby *index depth* denotes the "dereference" information of a Skeleton, similar to a pointer indirection in C, allowing to build agglomerations of points: Vertices are assigned index depth zero, k-cells constructed from vertices are index depth one, complexes of cells are index depth two and so forth. The *dimensionality* is an intrinsic property of a topological space, which is uniquely defined by its neighborhood information. This neighborhood information could be explicitly specified by providing a map from each element to its neighbors (such as required for a particle set), or implicitly in case of a structured data set given by an n-dimensional array. Such a gridded data set (a regular grid) is described by two entries, one defining the coordinate information (physical coordinates), the second one the actual data, for instance a density field:

$$/ \to 3.4Gy \to \text{Dark Matter} \to \text{Vertices:}32 \times 32 \times 32 \to \text{Cartesian Chart} \to \text{Positions} \tag{3}$$

$$/ \to 3.4Gy \to \text{Dark Matter} \to \text{Vertices:}32 \times 32 \times 32 \to \text{Cartesian Chart} \to \text{Density} \tag{4}$$

The positions may be given explicitly at each point (curvilinear grid), via a tensor product of one-dimensional arrays (rectilinear grid) or by specifying just the bounding box of a uniform grid. This infrastructure allows to naturally exploit commonalities among data sets describing particle sets and uniform grids as the mere difference is the dimensionality of the skeleton describing the vertices. A specific visualization algorithm may or may not consider this property, but in any case will be based on the same data management functions. In a similar manner as distinguishing particle sets and uniform (or curvilinear, rectilinear) grids within the same framework more complex and other data types can be built, for instance by adding refinement levels as additional Skeletons (adaptive mesh refinement), or utilizing the Fragment level to specify a distribution of a data set by many blocks (multiblock data) instead of one contiguous data volume.

3.2 Visualization on the fiber bundle

3.3 Base space

The base space of a data set refers to the topological properties of how data are given in space, in particular if (and what kind of) neighborhood information among data points is known. For instance, for a particle set we might not know such neighborhood information and we may thus consider it to be "zero-dimensional" even though these particles may be embedded in some two- or three-dimensional space. This embedding would allow to construct neighborhood information (e.g. via triangulation), but such is secondary information as it does not come with the data set per se.

3.3.1 Zero-dimensional base space

Display of a point sets' coordinates is the most common denominator across all data types and can be done for any data set providing a global coordinate system (which is not necessarily the case for general relativistic data where a curved spacetime does not necessarily exhibit a global coordinate system, but must rather be described by coordinate patches - "fragments" in the data model).

Splatting (Westover, 1990) is a technique for displaying smooth volumetric information on a point set, appropriate both for particle sets as well as for unstructured and structured grids. In its basic implementation it draws billboards with a point spread function at each coordinate location, a technique which is well supported by graphics hardware. Using a Gaussian intensity function is a frequent choice for the point spread function, but other functions have their merits as well. We use the same spline kernel (Monaghan & Lattanzio, 1985) as point spread function for the visualization as is used in the SPH simulations performed by GADGET-2 (Springel, 2005), which has the advantage of vanishing at the boundaries of the finite billboard geometry (in contrast to a Gaussian function which extends to infinity):

$$I(r) \propto \begin{cases} 1 - 6r^2 + 6r^3 & 0 \le r \le \frac{1}{2} \\ 2(1-r)^3 & \frac{1}{2} < r < 1 \\ 0 & r \ge 1 \end{cases} \tag{5}$$

Here, r is the normalized radius per billboard (point splat) ranging from zero to one and $I(r)$ the intensity at each pixel. The size of the splat may be constant or scaled with a numerical value given at each point (for instance the mass of a galaxy's stellar disc). This intensity function forms the basics for displaying scalar-, vector- and tensor fields given at each point. Point splatting is most effective when rendering the billboards transparently; this however turns out to be quite challenging, as will be discussed in section 3.5.

3.3.2 One-dimensional base space

In a one-dimensional base space each vertex has exactly two neighbors, described via a Skeleton of dimension one and index depth one for the "edges" of this data set. A sequence of edges is modeled by a Skeleton of index depth two, which defines a set of lines. It is described in the data model by three nodes:

$$/ \to 3.4Gy \to \text{Dark Matter} \to \text{Vertices} \to \text{Cartesian Chart} \to \text{Positions} \quad (6)$$

$$/ \to 3.4Gy \to \text{Dark Matter} \to \text{Edges} \to \text{Vertices} \to \text{Positions} \quad (7)$$

$$/ \to 3.4Gy \to \text{Dark Matter} \to \text{LineSet} \to \text{Edges} \to \text{Positions} \quad (8)$$

Hereby the edge and lineset information is just a collection of arrays of indices. In the context of astrophysical SPH simulations lines occur as trajectories of the involved particles, showing their evolution over time. Given a line we may explore intrinsic scalar (fig. 1) or vectorial (fig. 2) quantities such as the velocity, curvature (which is related to the acceleration) and torsion (which is related to the change of the acceleration), see e.g. (Benger & Ritter, 2010) for a discussion of line properties for visualization. Such may help verifying the dynamic model of the simulation and provide further insight to the dynamics.

Difficulties arise if the number of particles varies over time, for instance when considering mergers of galaxies. In such case a particle given at one time does not have a unique predecessor which leads to trajectories becoming non-manifolds where derivative operations are no longer defined. In the data model it is possible to model such topologies by providing explicit "future" and "past" skeletons that map the index of one particle at a time to the set of successors or predecessors. In order to provide this information, this topology has to be defined already in the previous simulation step, namely the link between the N-body simulation and the semi-analytical galaxy model as described in section 2.1.1. This intermediate step is realized via so called *halo finders* (Knebe et al., 2011) which identify gravitationally bound structures but often fail to represent a holistic merger tree with unique mappings over several time steps.

3.3.3 Two-dimensional base space

Surfaces introduce a Skeleton of dimension two and index depth one to the data model. Each element of this Skeleton describes a *face* element, which are triangles for a triangular surface. Such a Skeleton may co-exist with a line set Skeleton, thereby allowing to describe paths within a surface. Evolution surfaces tracing the dynamic behavior of some initial surface may be of interest, as well as iso-surfaces of a scalar quantity in a volume. For mere visual depiction of three-dimensional isolevels of a scalar field the explicit computation of isosurfaces (for instance using the marching cubes (Lorensen & Cline, n.d.) algorithm) is not required, modern graphics hardware allows to yield fast and even visually superior results via volume rendering with transparency set to "peaks" in the alpha channel, as will be discussed in the next section.

3.3.4 Three-dimensional base space

In the data model a three-dimensional base space is formally described by a one-dimensional Skeleton for the edges, a two-dimensional Skeleton for the faces and a three-dimensional Skeleton for the cells, extending the description of a two-dimensional base space by a third dimension. In practice this full description is only required for irregular meshes, e.g. tetrahedral, hexahedral or mixed grids. For regular grids, where each vertex has eight neighbors these Skeletons can be omitted as specifying the number of vertices in each dimension is sufficient, allowing for extensions such as Adaptive Mesh Refinement (AMR) schemes (Berger & Oliger, 1984), multiblock and domain decomposition as well as curvilinear and rectilinear coordinates. Each of these properties can be independently introduced in the data layout, thus allowing all possible combinations (e.g. rectilinear multiblock, curvilinear AMR, etc.).

3.4 Fiber Space

3.4.1 One-dimensional fiber space

Scalar fields are an example of a one-dimensional fiber space as they provide one quantity per data point. Depending on the dimensionality of the base space, point splatting, colorized lines or surface, or volume rendering are established techniques. Fig. 1 shows lines colorized by intrinsic line quantities, norm of acceleration and torsion (change of acceleraton). For

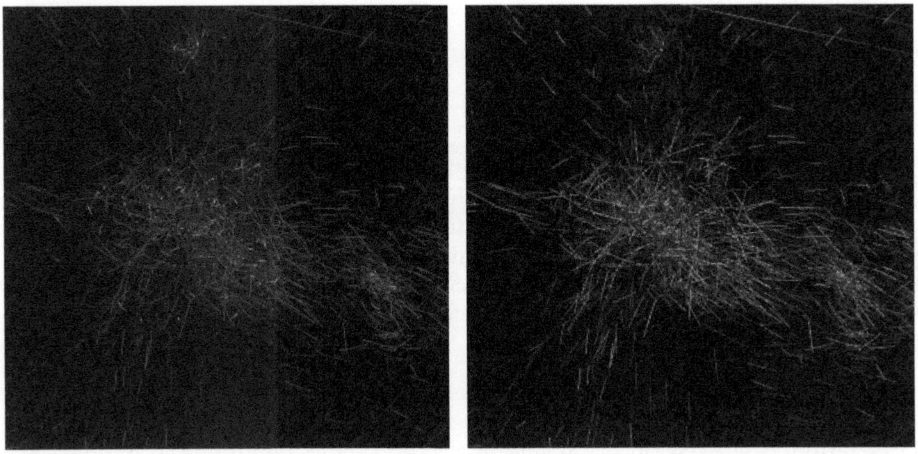

(a) Curvature (b) Torsion

Fig. 1. Scalar properties of trajectories in the galaxy cluster.

three-dimensional base spaces one of the most popular hardware-accelerated technique used to visualize one-dimensional Fiber Space (e.g. scalar field) is texture-based volume rendering. First introduced by Cullip & Neumann (1994), the basic idea is to approximate the volume rendering equation, eq. 13, by sampling the volume data using a stack of 2D textures, and to exploit graphics hardware capabilities to extract sets of slices from the volume. The volume rendering evaluation is then approximated by blending the textured slices in back-to-front order into the frame buffer. The high parallelism exposed by the pixel-processing unit during rasterization and the exploitation of bi- or trilinear interpolation capabilities of modern GPUs are the rationale behind of the success of this method. The technique has been further extended in several ways, e.g to support multiresolution representations of the volume data, to increase depth perception using shading (Stöckl et al., 2010) or to use an advanced optical model (see e.g. Kniss et al., 2002; Weiler et al., 2000).

3.4.2 Two-dimensional fiber space

It is often desirable to visualize the correlation of scalar fields, such as mass versus potential or other combinations. For particle sets (Benger, 2008) demonstrated the mapping of such scalar quantities to colorization, size and coordinate displacement of a splat-based representation. In situations where texture-based volume rendering can be used, the usual color-mapping algorithm can be extended to use the transparency values from another scalar field, thereby allowing to display two scalar fields at once in a volume ("dual volume rendering", see section 4.2).

3.4.3 Three-dimensional fiber space

Vector fields constitute a three-dimensional fiber space. The intuitive way to visualize a vector field via arrows fails for massive amounts of data, in particular when given within a three-dimensional volume. The splatting technique for scalar fields can well be extended to incorporate directional information by using anisotropic splats (Crawfis & Max, 1993). In our work we extended this idea by adding a colorization mimicking doppler-shift such that the velocity component in view direction is mapped to red or blue. This technique of "'Doppler Speckles" (Benger et al., 2009) is particularly effective for astrophysical data. It is of general usability independent from the dimensionality of the base space (as any splat-based technique) and allows to display e.g. the tangential vectors (fig. 2(a)) or the acceleration (fig. 2(b)) of galaxy trajectories.

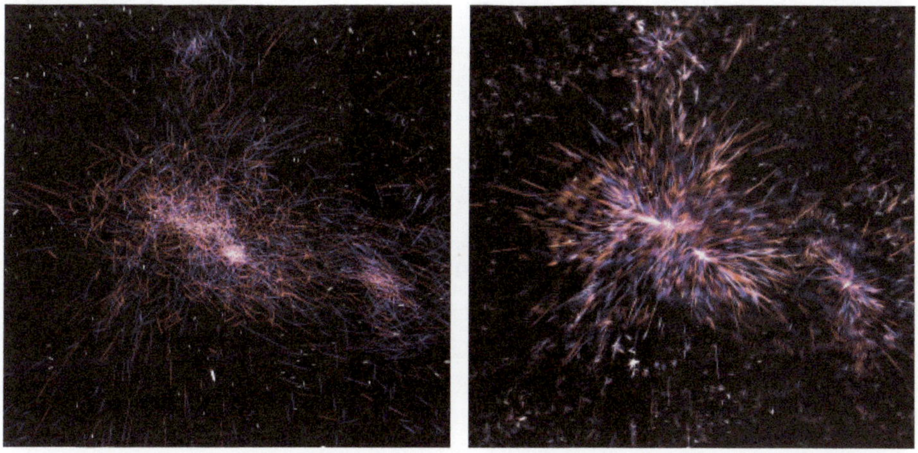

(a) Galaxy Trajectory Velocity (b) Galaxy Trajectory Acceleration

Fig. 2. Vectorial properties of galaxy cluster trajectories.

3.4.4 Higher-dimensional fiber space

Tensor fields extent the presented dimensionalities. Second order tensor fields are commonly used in computational physics, for example, to describe tension in continuum mechanics, viscosity in fluid dynamics, or space time curvature in numerical relativity. Direct and integration methods exist for visualization. An overview of tensor visualization can be found in Benger & Hege (2006). An integration method by computing geodesics is, for example, described in Ritter (2011). In figure 3 a direct splat based technique is applied, utilizing orientation, scaling, color and texture (presented in Benger & Hege, 2003). Three shape factors linearity, planarity and spherity, see Westin et al. (1997), are the basis for the appeareance of a splat capturing the relationship of the eigenvalues. One dominanant eigenvector results in high linearity, illustrated as green lines. Two dominant eigenvectors result in high planarity, illustrated as red plane. Three dominant eigenvectors result in high sphericity, illustrated through increased transparency, see fig. 3.

Fig. 3 shows tensor splats of the point distribution tensor illustrating geometrical properties of a point cloud. The point distribution tensor at a point i with position P_i is given by

$$T_{dist}(i) = \sum_{k=0}^{N-1} (P_k - P_i) \otimes (P_k - P_i)^t \tag{9}$$

where N is the number of points to consider.

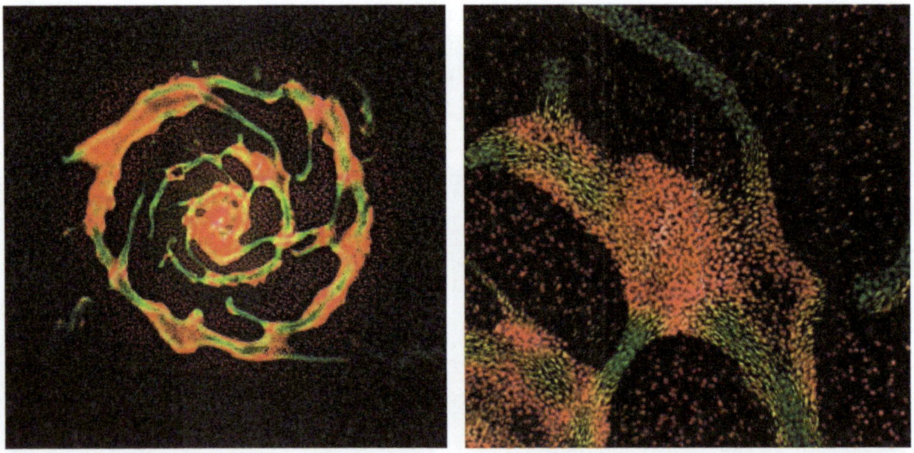

(a) Distribution Tensor Splats (b) Distribution Tensor Splats Detail

Fig. 3. Distribution tensor field of the RAM-pressure data set at $T = 0$ reveals geometric regions of linear, planar and volumetric structure. The neighborhood was chosen four times the SPH particle radius.

3.5 Fusing of methods via global transparency

Transparency has been enlisted as one of the five major challenges in interactive rendering and, more in general, of Computer Graphics (Andersson, 2010). As well as transparency, other physical phenomenons such as absorption, emission and scattering processes identify energy transfers in the form of electromagnetic radiation. The equation of radiative transfer describes these interactions mathematically (Chandrasekhar, 1960) and in its simplest one-dimensional form it is known as

$$\frac{dI(s)}{ds} = \varepsilon(s) - \kappa(s)I(s) \tag{10}$$

where $\varepsilon(s)$ is the emissivity function (the source of light at each point along a sight path) and $\kappa(s)$ the extinction function (caused by light scattering or absorption). This differential equation has the general solution

$$I(s) = I(s_0)e^{-\int_{s_0}^{s}\kappa(\tilde{s})d\tilde{s}} + \int_{s_0}^{s} \mathcal{S}(\tilde{s})e^{-\int_{\tilde{s}}^{s}\kappa(\hat{s})d\hat{s}}d\tilde{s} \tag{11}$$

where $\mathcal{S}(s) := \varepsilon(s)/\kappa(s)$ is called the source function. For constant extinction $\kappa(s) = const.$ we can write the optical depth $\int_{s_0}^{s_1}\kappa(s)ds = \kappa(s_1 - s_0)$ leading to a simpler solution:

$$I(s) = I(s_0)e^{-\kappa(s-s_0)} + \int_{s_0}^{s} \mathcal{S}(\tilde{s})e^{-\kappa(s-\tilde{s})}d\tilde{s} \tag{12}$$

which for constant emissivity $\varepsilon(s) = const.$ (thus $\mathcal{S}(s) = const.$) reduces to the rendering equation used in computer graphics (Kajiya, 1986)

$$\begin{aligned} I(s) = I(s_0)e^{-\kappa(s-s_0)} + \mathcal{S}\int_{s_0}^{s} e^{-\kappa(s-\tilde{s})}d\tilde{s} &= I(s_0)\underbrace{e^{-\kappa(s-s_0)}}_{=:\alpha} + \mathcal{S}(1 - \underbrace{e^{-\kappa(s-s_0)}}_{\alpha}) \\ &= I(s_0)\alpha + \mathcal{S}(1-\alpha) \quad , \end{aligned} \tag{13}$$

also known as *alpha-blending*.

Alpha-blending has been introduced almost 30 years ago, and it is the *de facto* standard technique for transparency. Solving the rendering equation via alpha blending requires to handle object ordering, which means that if we want to correctly evaluate the rendering equation, we should assure that objects contribute to the final pixel color with the correct ordering. *Order-independent transparency* (OIT) denotes any technique that allows to render transparent objects without having to sort them before they are being rendered, thus sorting is performed by the technique itself (e.g. depth peeling (Everitt, 2001) and the A-Buffer (Carpenter, 1984)). OIT techniques are computationally intensive and complex. They often require extra efforts to be used with other techniques, e.g. antialiasing. For these reasons developers would avoid to implement OIT and would prefer a simpler and faster algorithm that does not require sorting. Of course, such techniques are non-physical, nevertheless sometimes they may still be sufficient to provide some insight into the data while avoiding the higher cost of a correct solution.

Bounded A-Buffer. The traditional image-based algorithm for fragment sorting is the Z-buffer (Catmull, 1974). However, for each pixel, only the fragment with the lowest (or greatest) depth is kept, and the others are discarded. The A-buffer (Carpenter, 1984) extends the Z-buffer by storing a list of rasterized fragments for each pixel, sorted by depth. Potentially, the A-buffer is a powerful tool to implement OIT techniques. Nevertheless, it requires an unbounded memory per pixel. Recently, Callahan et al. (2005) introduced the K-buffer, a per pixel fixed-size buffer of fragments that is maintained in GPU memory. There exist many proposals for implementing the K-Buffer on the GPU, for example by using the stencil routing algorithm (Myers & Bavoil, 2007) or by exploiting generic atomic operations available in OpenGL 4.0 (Yang et al., 2010). It requires only one geometric pass, thus being more efficient than techniques such as depth peeling (Everitt, 2001). Despite the technique can store only K fragments per pixel, a recent extension called Adaptive Transparency overcomes this limitation by using an approximated and adaptively compressed visibility representation (Salvi et al., 2011).

Moreover, even the data types involved in transparency drastically change the effectiveness of the technique used (e.g. particle vs volumetric vs mesh). While each of the previously mentioned methods is a more or less straightforward mapping of the given data to a rendering method using shaders in OpenGL, fusing them into an integrated visualization encompassing more than one such type is non-trivial if transparency is involved.

Combining volume and particle visualization. Astrophysical simulations are a classical visualization scenario where often data of different types are involved. An example is having both particle-based and volumetric data. The approaches used to visualize transparent particle-based (e.g. splatting approaches, see Westover, 1990), and the ones used for volume, (e.g. texture-based slicing, see Weiler et al., 2000) or GPU-assisted ray casting (Kaehler et al., 2007; Kruger & Westermann, 2003) work well alone. Again, if we relax our needs and we consider one of the two data sets in input as completely opaque, we can afford a combined visualization by using a two pass approach: the first pass for opaque objects, and the second for transparent ones, thus we combine results by exploting hardware depth-buffers and blending operations. Nevertheless, problems arise when both data sets are transparent. Kaehler et al. (2007) introduced an interesting approach for high-quality visualization of volume rendering of grid data and unstructured point set. Their approach is based on a GPU-accelerated raycasting algorithm, where point data is stored in a GPU-octree data structure (e.g. a 3d texture) in order to efficiently access it during ray traversal.

4. Results

4.1 Galaxy clusters

One main demand for a comprehensive data analysis is the visualization of both grid data (that comes from the hydrodynamical simulations) and the point data (coming from the N-body simulation and the SAM) at the same time. As expounded in the previous section, we are first and foremost interested in secondary effects of cosmic structure formation that influence the inter cluster medium. The analysis of the hydro-simulation should reveal, how material coming from galactic ram-pressure stripping and galactic winds behaves in this extremely hot ($10^7 - 10^9$K) and thin (10^{-3} particles per cm^3) ICM.

The density and temperature plots in figures 4 (a) and (b) reveal crosswise major in-falling regions of hot gas that is further (shock-)heated and condensed in the central region of the cluster. The dots represent the galactic halos of which structures partially coincide with peaks in gas temperature and density, but clearly not always. This unique combined visualization of these two types of data reveals that groups of galaxy halos can produce remarkable substructures that also show in the gaseous component of the cluster. Furthermore there is a considerable amount of galaxies that are not fast enough to shock-heat the ICM excessively while descending towards the cluster center.

Especially fascinating in this context are the metallicity plots that disentangle chemical enrichment caused by galactic winds (Fig. 4 (c)) and ram-pressure stripping (Fig. 4 (d)). The trails of metals are much longer for the wind component and they begin much earlier in time and in further outwards regions of the cluster. Ram-pressure only acts above a certain pressure threshold that can merely be achieved if both surrounding density of the ICM and the relative velocities of gas and galaxies are sufficient. Ram-pressure stripping however can in succession be especially effective and deprive the galaxy of a considerable fraction of its gas on cosmologically short time scales. Since galactic winds are predominantly driven by collective supernovae explosions as denoted in section 2.1.2, the metal yields are composed differently for wind respectively ram-pressure enriched material. This should lead to an inhomogeneous spatial distribution of chemical elements in a cluster which is however not yet observable with todays instruments and remains to be investigated also observationally in future projects.

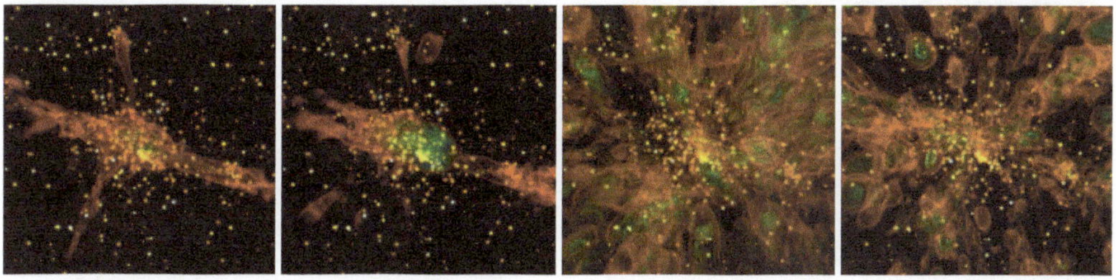

(a) Density (b) Temperature (c) Wind Metallicity (d) RAM Pressure

Fig. 4. Hydrodynamic quantities on the galaxy cluster.

4.2 Magnetic fields in clusters of galaxies

The technique outlined in section 3.4.2 is very well suited to investigate the correlations of plasma properties in the intra-cluster medium (ICM) from simulations of clusters of galaxies. The simulations cover the evolution of the intra-cluster medium (ICM) from an early redshift $z \sim 100$ to the present time ($z = 0$) and include cosmic expansion, dark matter, baryonic matter, magnetic fields, (magneto-) hydrodynamics and radiative cooling.

The correlations between certain plasma properties can give important information on the dynamics of the whole system that might not be detectable or quantifiable when looking at those properties separately. A good example for this is the correlation between the magnetic pressure p_{mag} and the density ρ in the ICM which gives an insight into the importance of turbulence and dynamo effects during the cluster creation. Fig. 5 shows a Dual Volume Rendering of magnetic pressure and density of the ICM for a cluster with a total mass of $M_{\text{tot}} = 2.72 \times 10^{14} M_{\odot}$ at redshift $z = 0$. In Fig. 5(d) the dense isocontours of the magnetic pressure together with the high density values (blue-green) but low density gradients clearly indicate a region with high magnetic turbulence, which is otherwise only visible in a log p_{mag}-log ρ correlation diagram where features cannot be localized in the 3D domain. Likewise in Fig. 5(e), the purple region marks increased magnetic pressure caused by the inflow of matter into the center of the cluster. It is important to note that a high density does not necessarily correspond to a high magnetic pressure, as the blue colored density isocontours in Fig. 5(e) demonstrate.

We find that the Dual Volume Rending technique is especially well suited for investigating correlations between plasma properties that are not obvious from the values alone, but also take into account the gradients of aforementioned properties. This visualization method

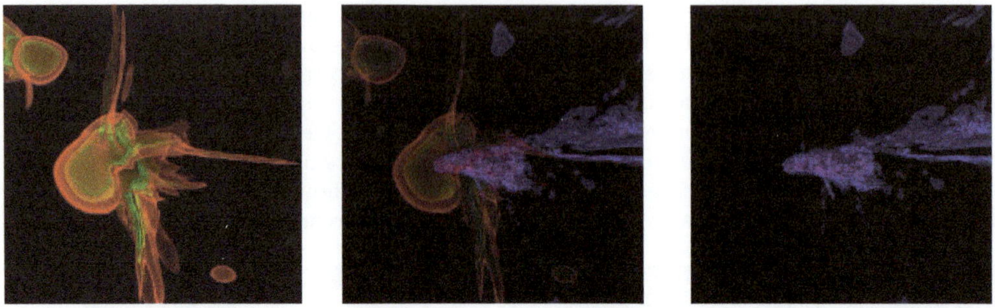

(a) Gas density (b) Gas density and magnetic pressure overlaid (c) Magnetic pressure

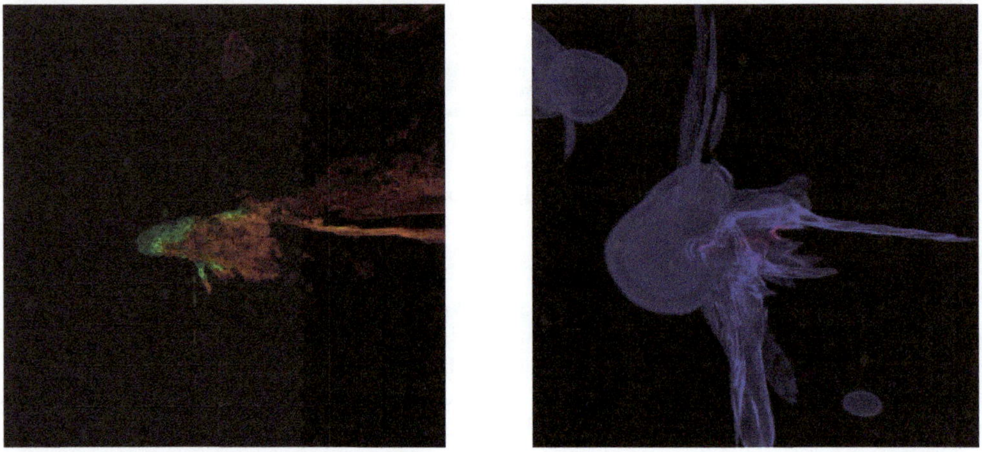

(d) Density per constant magnetic pressure (e) Magnetic pressure per constant density

Fig. 5. Results from a MHD simulation of a cluster of galaxies at redhift $z = 0$. A classical volume rendering is shown in (a) to (c). The overlaid rendering (b) works very well to visualize the distribution of gas and the large scale structure of the magnetic field strength. In regions of rapid changing values or close to the highest concentrations of density and magnetic field strength however the overlaid plots conceal each other's features. This can be omitted by using Dual Volume Rendering, where one quantity is color mapped onto the isocontours of another quantity. In (d) the gas density (red - blue / low - high) is mapped onto the isobars of the magnetic field pressure, while in (e) the magnetic field pressure (blue - purple / low - high) is mapped onto the isocontours of the gas density.

allows to follow even steep gradients of both quantities that are being displayed together (compare (a) – (c) in Fig. 5 to the dual volume renderings (d) and (e)).

4.3 Ram-pressure stripping of disk galaxies

The simulations reveal many interesting results on ram-pressure stripped galaxies. The stripping of the gaseous disk is in well agreement with the Gunn & Gott criterion. The ram pressure is calculated as $p_{ram} = \rho_{ICM} v_{rel}^2$. So, even low gas densities can deplete the gas mass in the disk if the relative velocity is higher. In the case of a high ram pressure, using a gas density of 10^{-27}g cm^{-3} and a relative velocity of $v_{rel} = 1000\,\mathrm{km\,s}^{-1}$, the gas disk is almost completely stripped.

Moreover, the simulations show that ram pressure enhances the star formation rate. In the previously described scenario, the star formation rate can be enhanced up to eight times compared to galaxies evolved in isolation. The reason can be found in the compression of the gas disk due to ram pressure and the formation of dense gas knots in the stripped gas wake of the galaxy. Also in these gas clumps stars are formed as already has been seen by (Kapferer et al., 2008). Hence, ram-pressure stripping can contribute to the intra-cluster stellar population (see e.g. Mihos et al., 2005).

Visualizations are crucial to study morphology and dynamics of the stripped ISM in detail. For this purpose, density plots and velocity fields of the different components of the model galaxies are shown in the following sections.

4.3.1 Scalar visualization

In Fig. 8 the density of the newly formed stars (bluish) and the ISM (reddish-greenish) is shown in a face-on (right) and edge-on view (left) for five different time steps. It can clearly be seen in the face on view that the gas disk gets compressed by ram pressure, with just a bunch of dense gas remaining in the disk. The stripping process first involves the outer parts of the gas disk which leads to a diffuse gas in the tail remaining at the starting point of the galaxy. From the inner parts of the spiral structured disk, a filamentary structure is formed in the wake which finally leads to the formation of the dense gas clumps, depicted in detail in Fig. 6 (b). These clumps are stable throughout a few hundred million years. In Fig. 6 (a) it is apparent, that the gas knots are affected by ram pressure as some gas gets stripped. Nevertheless, due to self-gravity and the pressure of the ICM, only a very small amount of gas is stripped and the gas loss is almost completely due to star formation. As can clearly be seen in Fig. 7, the new stars are formed in the dense gas clumps and are falling towards the disk as they do not feel the ram pressure anymore.

4.3.2 Vectorial visualization

A part from the morphology of the distribution of stars and gas, also the dynamic of these components is highly interesting. In Fig. 9, the method of Doppler Speckles is used to show the movement of the gas. From this plot, it is obvious that the angular momentum of the gas disk is being conserved also in the stripped gas tail. A detailed view is shown in Fig. 10. A differential movement of the gas can be seen. With respect to the galaxy itself, the gas knots in the wake are slowed down due to ram pressure. On the other hand it is very interesting that the gas knots themselves rotate as well as the filamentary dumbbell-like structure shows in Fig. 10 (b). This can explain the formation of the gas clumps. To this end, a detailed view on the dumbbell structure is shown in Fig. 11. The rotation of the filament leads to the formation of a gas clump at both ends. Furthermore, the part in between is more affected by ram pressure due to the lower density, hence it is pushed backwards with respect to the formed gas clumps. Finally, the two clumps are dismembered.

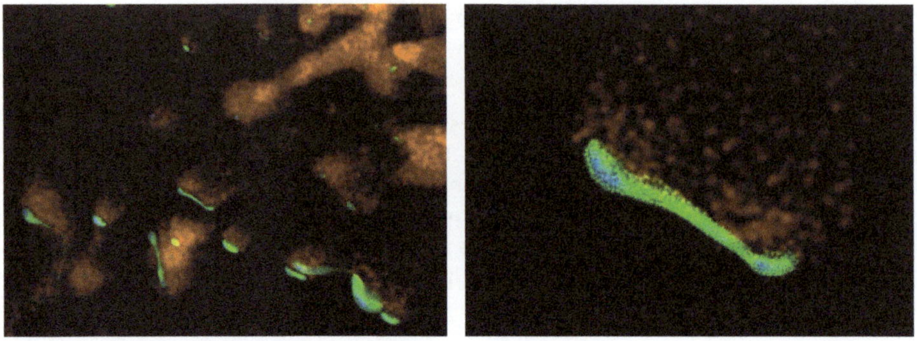

(a) (b)

Fig. 6. Detail of high-density regions in the stripped gas tail.

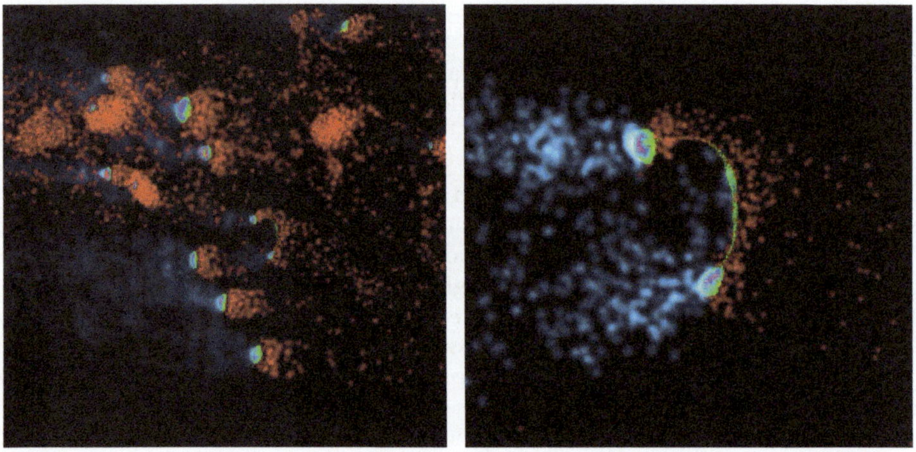

(a) (b)

Fig. 7. Detail of star formation area and high-density regions

4.3.3 Tensorial visualization

Computing and visualizing the point distribution tensor allows to identify regions of predominantly linear, planar and isotropic distributions immediately.

4.4 Cosmological evolution

In a cosmological evolution of a galaxy cluster we simulated the behavior of dark matter from initial conditions up to presence. The simulation was performed using 2,097,152 particles. Of particular interest in cosmological simulations is the formation of voids, filaments and clusters of galaxies, as depicted in Fig. 14. While a slight elliptical shape of the central class can be glimpsed from the intensity distribution (Fig. 14(f)) the tensor display clearly depicts the anisotropy of the distribution (Fig. 14(g)).

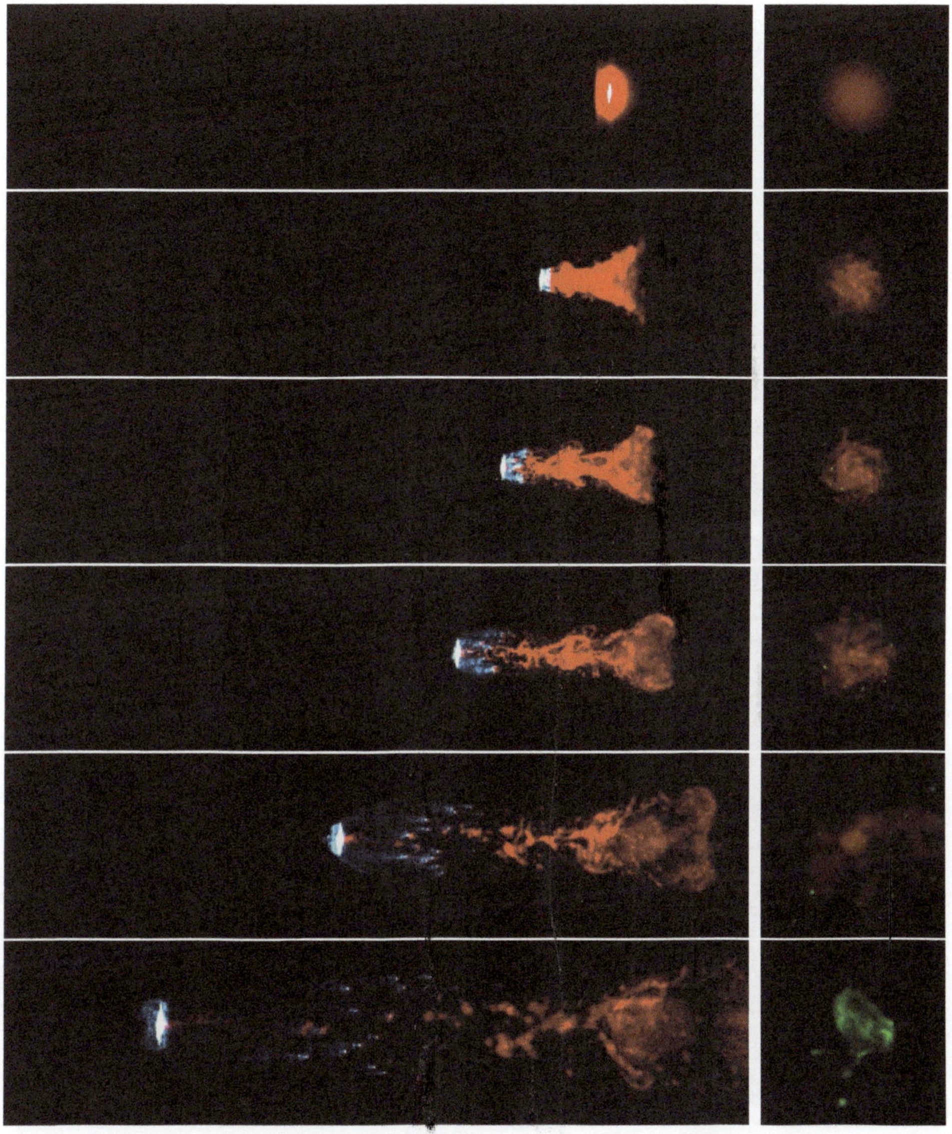

Fig. 8. Evolution of newly formed stars and gas for time steps T=10,100,150,200,300 and 400 Myr.

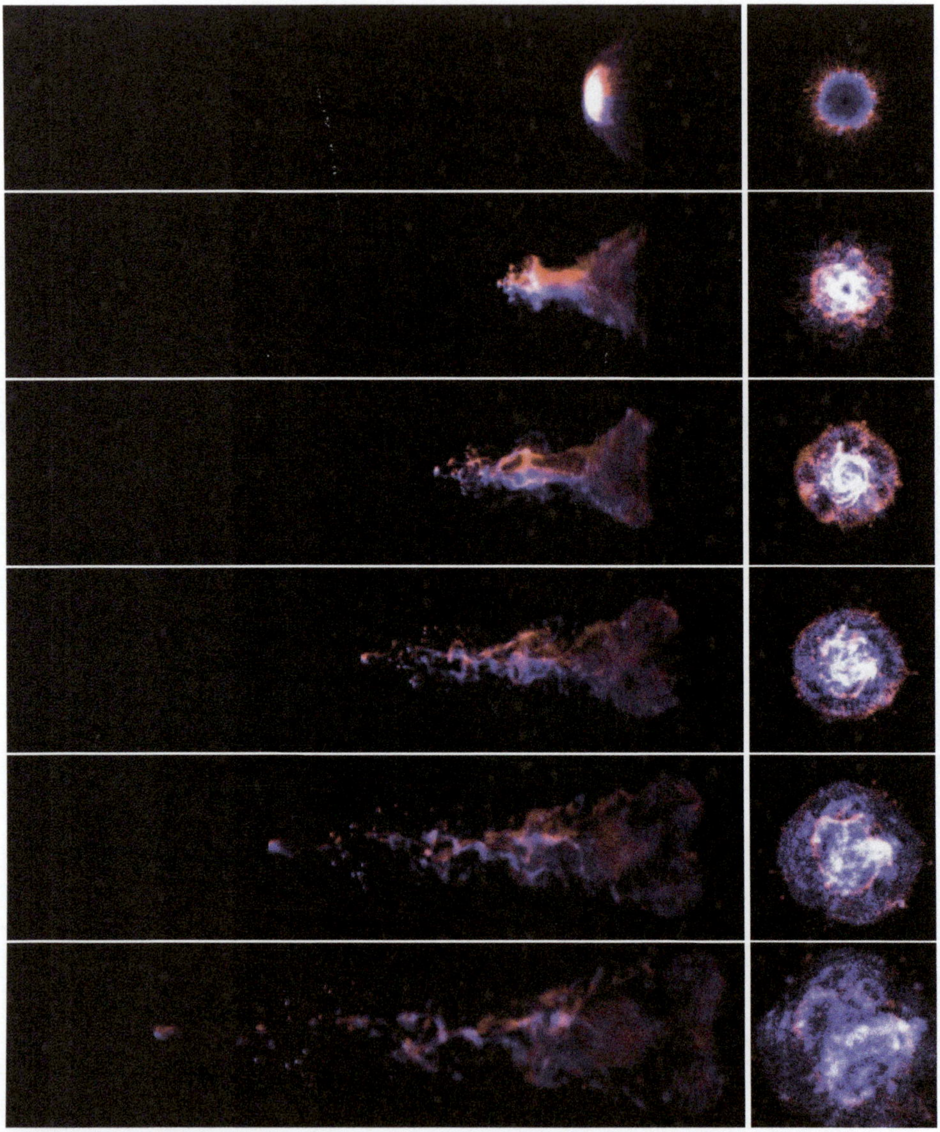

Fig. 9. Velocity of the gas, edge-on and front view for time steps T=30,100,150,200,250 and 300 Myr. Rendered using the method of Doppler Speckles.

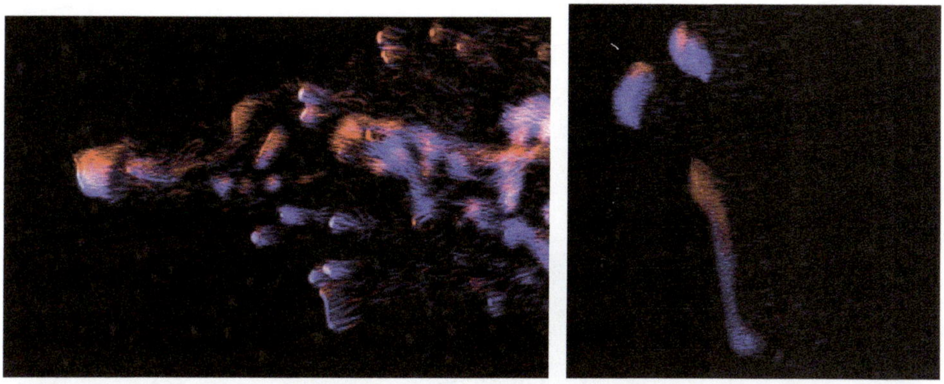

Fig. 10. Detail of velocity in high-density regions

(a) T=240 Myr (b) T=270 Myr (c) T=290 Myr

Fig. 11. Detailed evolution of a particular high-density region, showing velocity, density and newly formed stars.

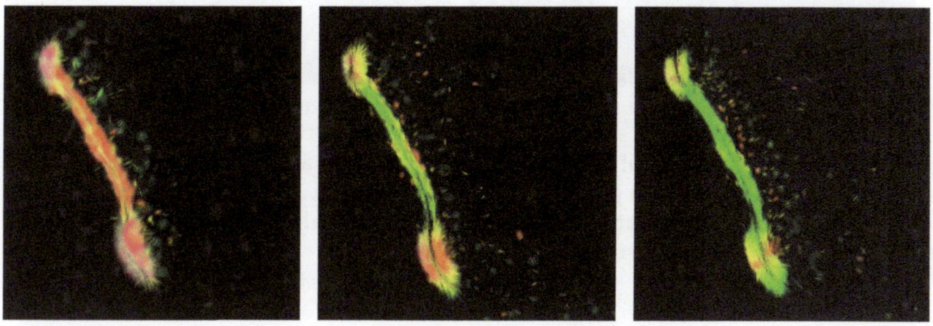

(a) $R = 0.4$ (b) $R = 0.8$ (c) $R = 1.2$

Fig. 12. Closeup of the dumbbell structure at $T = 25$ Myr. The images illustrate the influence of the radius of the chosen neighborhood on the resulting point distribution tensor. The geometric properties are best illustrated in (b) where the linear ribbon connnects the two high density regions. If the radius is increased the whole dumbbell is becoming linear.

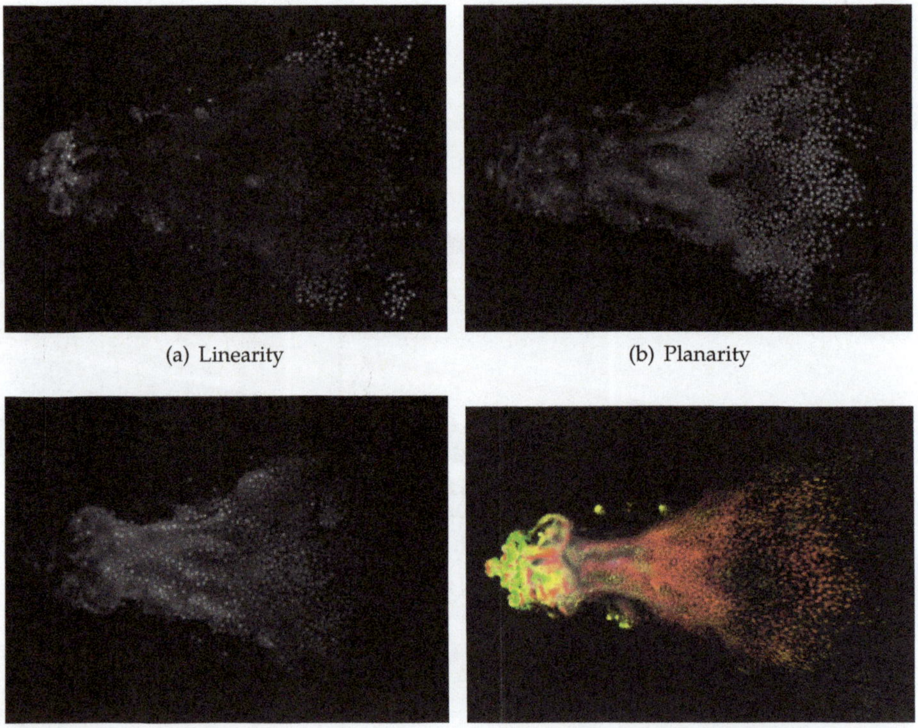

(a) Linearity (b) Planarity

(c) Sphericity (d) Point Distribution Tensor

Fig. 13. Geometric properties of the gas particle distribution at $T = 10$ Myr. The 1280 nearest neighbors are selected for each particle. The distribution has a cone-like planar structure on the right side and forms a linear ring on the left. The greyscale scalar fields and the direct tensor splats illustrate these structures.

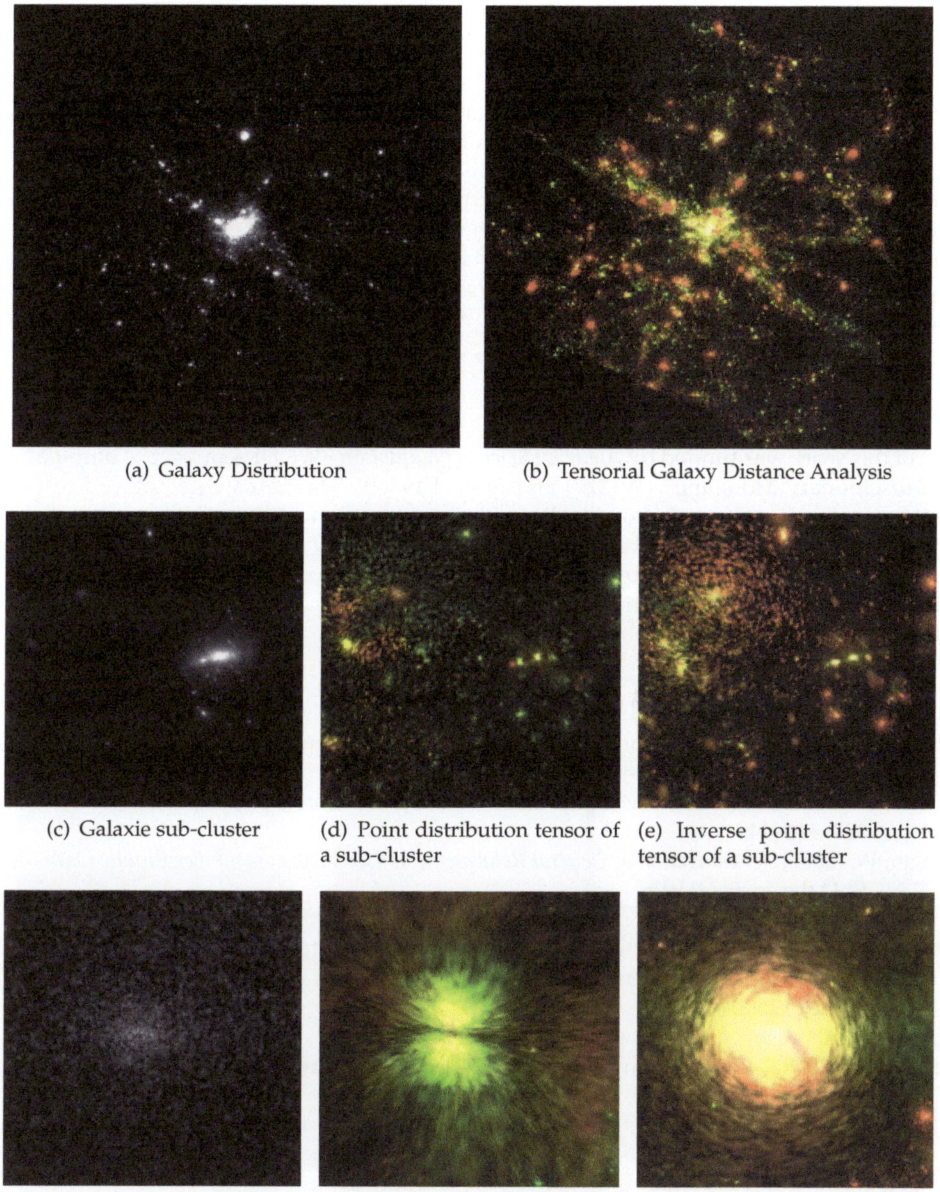

(a) Galaxy Distribution
(b) Tensorial Galaxy Distance Analysis
(c) Galaxie sub-cluster
(d) Point distribution tensor of a sub-cluster
(e) Inverse point distribution tensor of a sub-cluster
(f) Locations in central cluster
(g) Point distribution tensor of central cluster
(h) Inverse point distribution tensor of central cluster

Fig. 14. Tensorial point distribution distance (2000 points considered for neighborhood computation) allows to recognize distribution features within a large volume of galaxies.

5. Conclusion

In this article we have presented a systematic way to study data sets from numerical astrophysics originating from simulations exploring open questions. We review a model to organize data for scientific visualization supporting high reusability of visualization algorithms by avoiding problem-specific data structures. This model is based on the mathematics of fiber bundles and we discuss the mapping of astrophysical simulation data sets into this framework. We investigate the available data fields in the fiber space, exploring scalar-, vector- and tensor- fields that are given or extractable from the given data sets. We also discussed how visualization techniques face the different challenges of these data sets, such as visualizing transparency of combined data sets and exploiting the computational power and new features available in current GPU hardware. Results include (magneto-) hydrodynamic quantities of galaxy cluster, ram-pressure stripped galaxies with scalar, vectorial and tensorial visualization, and cosmological evolution of galaxy cluster.

6. Acknowledgements

Part of this work was funded by *Austrian Science Foundation FWF* DK+ project *"Computational Interdisciplinary Modeling"* (W1227) and grant P19300. This research employed resources of the Center for Computation & Technology at Louisiana State University, which is supported by funding from the Louisiana legislature's Information Technology Initiative. This work was supported by the Austrian Ministry of Science BMWF as part of the UniInfrastrukturprogramm of the Forschungsplattform Scientific Computing at LFU Innsbruck.

7. References

Andersson, J. (2010). Five Major Challenges in Interactive Rendering, ACM SIGGRAPH 2010 Beyond Programmable Shading Course.

Barnes, J. & Hut, P. (1986). A hierarchical O(N log N) force-calculation algorithm, *Nature* 324: 446–449.

Benger, W. (2004). *Visualization of General Relativistic Tensor Fields via a Fiber Bundle Data Model*, PhD thesis, FU Berlin.
URL: `http://www.lob.de/isbn/3865411088`

Benger, W. (2008). Colliding galaxies, rotating neutron stars and merging black holes - visualising high dimensional data sets on arbitrary meshes, *New Journal of Physics* 10(12): 125004–+.
URL: `http://stacks.iop.org/1367-2630/10/125004`

Benger, W. (2009). On safari in the file format djungle - why can't you visualize my data?, *Computing in Science & Engineering* 11(6): 98–102.
URL: `http://www.computer.org/cms/Computer.org/ComputingNow/homepage/2009/1109/rW_CS_SafariFileFormatJungle.pdf`

Benger, W., Bartsch, H., Hege, H.-C., Kitzler, H., Shumilina, A. & Werner, A. (2006). Visualizing Neuronal Structures in the Human Brain via Diffusion Tensor MRI, *International Journal of Neuroscience* 116(4): pp. 461–514.

Benger, W., Hamilton, A., Folk, M., Koziol, Q., Su, S., Schnetter, E., Ritter, M. & Ritter, G. (2010). Using geometric algebra for navigation in riemannian and hard disc space, *in* V. Skala & D. Hildebrand (eds), *GraVisMa 2009 - Computer Graphics, Vision and Mathematics*

for Scientific Computing, UNION Agency, Na Mazinach 9, CZ 322 00 Plzen, Czech Republic.
URL: http://gravisma.zcu.cz/GraVisMa-2009/

Benger, W. & Hege, H.-C. (2003). The tensor splats rendering technique, *Technical Report ZIB03-17*, Zuse Institute Berlin.

Benger, W. & Hege, H.-C. (2006). *Strategies for Direct Visualization of Second-Rank Tensor Fields*, Springer, chapter 11, pp. 191–214.
URL: http://www.springerlink.com/content/h264176262502825/

Benger, W., Heinzl, R., Hildenbrand, D., Weinkauf, T., Theisel, H. & Tschumperle, D. (2011). *Differential Methods for Multi-Dimensional Visual Data Analysis*, Springer Science + Business Media LLC, chapter 50, pp. 1533–1595.
URL: http://www.springerlink.com/content/r4j504485j73jk16/

Benger, W., Ritter, G. & Heinzl, R. (2007). The Concepts of VISH, *4th High-End Visualization Workshop, Obergurgl, Tyrol, Austria, June 18-21, 2007*, Berlin, Lehmanns Media-LOB.de, pp. 26–39.

Benger, W., Ritter, G., Su, S., Nikitopoulos, D. E., Walker, E., Acharya, S., Roy, S., Harhad, F. & Kapferer, W. (2009). Doppler speckles - a multi-purpose vectorfield visualization technique for arbitrary meshes, *CGVR'09 - The 2009 International Conference on Computer Graphics and Virtual Reality*.

Benger, W. & Ritter, M. (2010). Using Geometric Algebra for Visualizing Integral Curves, *in* E. M. Hitzer & V. Skala (eds), *GraVisMa 2010 - Computer Graphics, Vision and Mathematics for Scientific Computing*, Union Agency - Science Press.

Benson, A. J. (2010). Galacticus: A Semi-Analytic Model of Galaxy Formation, *ArXiv e-prints* .

Berger, M. J. & Oliger, J. (1984). Adaptive mesh refinement for hyperbolic partial differential equations, *J. Comput. Phys.* 53: 484–512.

Bertschinger, E. (1995). Cosmics: Cosmological initial conditions and microwave anisotropy codes.

Bertschinger, E. (1998). Simulations of structure formation in the universe, *ANNUAL REVIEW OF ASTRONOMY AND ASTROPHYSICS* 36: 599–654.

Blanton, E. L., Sarazin, C. L. & McNamara, B. R. (2003). Chandra Observation of the Cooling Flow Cluster Abell 2052, *The Astrophysical Journal* 585: 227–243.

Butler, D. M. & Pendley, M. H. (1989). A visualization model based on the mathematics of fiber bundles, *Computers in Physics* 3(5): 45–51.

Callahan, S. P., Ikits, M., Member, S., Comba, J. L. D. & Silva, C. T. (2005). Hardware-assisted visibility sorting for unstructured volume rendering, *IEEE Transactions on Visualization and Computer Graphics* 11: 2005.

Carpenter, L. (1984). The a-buffer, an antialiased hidden surface method, *Proceedings of the 11th annual conference on Computer graphics and interactive techniques*, SIGGRAPH '84, ACM, New York, NY, USA, pp. 103–108.
URL: http://doi.acm.org/10.1145/800031.808585

Catmull, E. E. (1974). *A Subdivision Algorithm for Computer Display of Curved Surfaces*, PhD thesis, Dept. of CS, U. of Utah.

Chandrasekhar, S. (1960). *Radiative Transfer*, Dover Publications Inc.

Crawfis, R. & Max, N. (1993). Textured splats for 3d scalar and vector field visualization, *in* Nielson & Bergeron (eds), *Visualization '93, San Jose*, IEEE CS Press, pp. 261–266.

Cullip, T. J. & Neumann, U. (1994). Accelerating Volume Reconstruction With 3D Texture Hardware, *Technical report*, Chapel Hill, NC, USA.

de Oliveira-Costa, A., Tegmark, M., Zaldarriaga, M. & Hamilton, A. (2004). Significance of the largest scale cmb fluctuations in wmap, *Phys. Rev. D* 69: 063516.
URL: http://link.aps.org/doi/10.1103/PhysRevD.69.063516

Domainko, W., Kapferer, W., Mair, M., Schindler, S., Kampen, E. V., Kronberger, T., Moll, R., Kimeswenger, S., Ruffert, M. & Mangete, O. E. (2008). Metal Enrichment of the ICM due to Ram-Pressure Stripping of Cluster Galaxies, *in* B. Aschenbach, V. Burwitz, G. Hasinger, & B. Leibundgut (ed.), *Relativistic Astrophysics Legacy and Cosmology - Einstein's*, pp. 300–+.

Dougherty, M. T., Folk, M. J., Bernstein, H. J., Bernstein, F. C., Eliceiri, K. W., Benger, W., Zadok, E. & Best, C. (2009). Unifying biological image formats with hdf5, *Communications of the ACM* 52(10): 42–47.

Everitt, C. (2001). Interactive order-independent transparency, *Technical report*, NVidia Technical Report.

Finoguenov, A., David, L. P. & Ponman, T. J. (2000). An ASCA Study of the Heavy-Element Distribution in Clusters of Galaxies, *The Astrophysical Journal* 544: 188–203.

Gingold, R. A. & Monaghan, J. J. (1977). Smoothed particle hydrodynamics - Theory and application to non-spherical stars, *mnras* 181: 375–389.

Gunn, J. E. & Gott, III, J. R. (1972). On the Infall of Matter Into Clusters of Galaxies and Some Effects on Their Evolution, *apj* 176: 1–+.

Hestenes, D. (1999). *New Foundations for Classical Mechanics, 2nd ed.*, Springer Verlag.

Hester, J. A., Seibert, M., Neill, J. D., Wyder, T. K., Gil de Paz, A., Madore, B. F., Martin, D. C., Schiminovich, D. & Rich, R. M. (2010). IC 3418: Star Formation in a Turbulent Wake, *apj* 716: L14–L18.

Kaehler, R., Abel, T. & Hege, H.-C. (2007). Simultaneous GPU-Assisted Raycasting of Unstructured Point Sets and Volumetric Grid Data , pp. 49–56.
URL: http://www.eg.org/EG/DL/WS/VG/VG07/049-056.pdf

Kajiya, J. T. (1986). The rendering equation., *SIGGRAPH'86*, pp. 143–150.

Kapferer, W., Kronberger, T., Ferrari, C., Riser, T. & Schindler, S. (2008). On the influence of ram-pressure stripping on interacting galaxies in clusters, *mnras* 389: 1405–1413.

Kapferer, W., Sluka, C., Schindler, S., Ferrari, C. & Ziegler, B. (2009). The effect of ram pressure on the star formation, mass distribution and morphology of galaxies, *A&A* 499: 87–102.

Katz, N., Weinberg, D. H. & Hernquist, L. (1996). Cosmological Simulations with TreeSPH, *apjs* 105: 19–+.

Knebe, A., Knollmann, S. R., Muldrew, S. I., Pearce, F. R., Aragon-Calvo, M. A., Ascasibar, Y., Behroozi, P. S., Ceverino, D., Colombi, S., Diemand, J., Dolag, K., Falck, B. L., Fasel, P., Gardner, J., Gottlöber, S., Hsu, C.-H., Iannuzzi, F., Klypin, A., Lukić, Z., Maciejewski, M., McBride, C., Neyrinck, M. C., Planelles, S., Potter, D., Quilis, V., Rasera, Y., Read, J. I., Ricker, P. M., Roy, F., Springel, V., Stadel, J., Stinson, G., Sutter, P. M., Turchaninov, V., Tweed, D., Yepes, G. & Zemp, M. (2011). Haloes gone MAD: The Halo-Finder Comparison Project, *Monthly Notices of the Royal Astronomical Society* 415: 2293–2318.

Kniss, J., Premoze, S., Hansen, C. & Ebert, D. (2002). Interactive Translucent Volume Rendering and Procedural Modeling, *In Proceedings of IEEE Visualization 2002*, pp. 109–116.

Kronberger, T., Kapferer, W., Ferrari, C., Unterguggenberger, S. & Schindler, S. (2008). On the influence of ram-pressure stripping on the star formation of simulated spiral galaxies, *A&A* 481: 337–343.

Kruger, J. & Westermann, R. (2003). Acceleration techniques for gpu-based volume rendering, *Proceedings of the 14th IEEE Visualization 2003 (VIS'03)*, VIS '03, IEEE Computer Society, Washington, DC, USA, pp. 38–.
URL: http://dx.doi.org/10.1109/VIS.2003.10001

Lorensen, W. E. & Cline, H. E. (n.d.). Marching cubes: A high resolution 3d surface construction algorithm, *Computer Graphics 21(4)* 21(4): 163–169.

Lovisari, L., Kapferer, W., Schindler, S. & Ferrari, C. (2009). Metallicity map of the galaxy cluster A3667, *Astronomy & Astrophysics* 508: 191–200.

Lucy, L. B. (1977). A numerical approach to the testing of the fission hypothesis, *aj* 82: 1013–1024.

Mihos, J. C., Harding, P., Feldmeier, J. & Morrison, H. (2005). Diffuse Light in the Virgo Cluster, *apj* 631: L41–L44.

Mo, H. J., Mao, S. & White, S. D. M. (1998). The formation of galactic discs, *mnras* 295: 319–336.

Monaghan, J. J. & Lattanzio, J. C. (1985). A refined particle method for astrophysical problems, *Astronomy & Astrophysics* 149: 135–143.

Myers, K. & Bavoil, L. (2007). Stencil Routed K-Buffer, *Technical report*, NVidia Technical Report.

Navarro, J. F., Frenk, C. S. & White, S. D. M. (1996). The Structure of Cold Dark Matter Halos, *apj* 462: 563–+.

Ostriker, J. & Steinhardt, P. J. (1995). Cosmic concordance.

Ritter, M. (2011). *Geodesics in Numerical Space Times*, Verlag Dr. Müller.

Salpeter, E. E. (1955). The Luminosity Function and Stellar Evolution., *apj* 121: 161–+.

Salvi, M., Montgomery, J. & Lefohn, A. (2011). Adaptive transparency, *Proceedings of the ACM SIGGRAPH Symposium on High Performance Graphics*, HPG '11, ACM, New York, NY, USA, pp. 119–126.
URL: http://doi.acm.org/10.1145/2018323.2018342

Springel, V. (2005). The cosmological simulation code GADGET-2, *Monthly Notices of the Royal Astronomical Society* 364: 1105–1134.

Springel, V. & Hernquist, L. (2003). Cosmological smoothed particle hydrodynamics simulations: a hybrid multiphase model for star formation, *mnras* 339: 289–311.

Stöckl, J., Steinhauser, D., Haider, M. & Riser, T. (2010). Improved visualisation of 3d volumetric datathrough pointwise phong shading based on normal mapping, *in* W. Benger, A. Gerndt, S. Su, W. Shoor, M. Koppitz, W. Kapferer, H. P. Bischof & M. Di Pierro (eds), *Proceedings of the 6th High-End Visualization Workshop*, Vol. 6 of *Proceedings of the High-End Visualization Workshop*, Lehmanns Media, Berlin, 2010, pp. 107–121.

Treinish, L. A. (1997). Data explorer data model, http://www.research.ibm.com/people/l/lloydt/dm/dx/dx_dm.htm.

van Kampen, E., Jimenez, R. & Peacock, J. A. (1999). Overmerging and mass-to-light ratios in phenomenological galaxy formation models, *Monthly Notices of the Royal Astronomical Society* 310: 43–56.

Weiler, M., Westermann, R., Hansen, C., Zimmermann, K. & Ertl, T. (2000). Level-of-detail volume rendering via 3d textures, *Proceedings of the 2000 IEEE symposium on Volume*

visualization, VVS '00, ACM, New York, NY, USA, pp. 7–13.
URL: `http://doi.acm.org/10.1145/353888.353889`

Westin, C., Peled, S., Gudbjartsson, H., Kikinis, R. & Jolesz, F. (1997). Geometrical diffusion measures for mri from tensor basis analysis, *Proceedings of ISMRM, Fifth Meeting, Vancouver, Canada*, p. 1742.
URL: `http://spl.bwh.harvard.edu:8000/pages/ppl/westin/papers/smr97/htmlversion.html`

Westover, L. (1990). Footprint evaluation for volume rendering, *Proceedings of the 17th annual conference on Computer graphics and interactive techniques*, ACM Press, NY, USA, pp. 367–376.

White, S. D. M. & Frenk, C. S. (1991). Galaxy formation through hierarchical clustering, *The Astrophysical Journal* 379: 52–79.

Yang, J. C., Hensley, J., Grün, H. & Thibieroz, N. (2010). Real-time concurrent linked list construction on the gpu, *Computer Graphics Forum* 29(4): 1297–1304.

14

A Comparison of Non Negative Blind Source Separation Methods for Identifying Astrophysical Ice Compounds

Jorge Igual and Raul llinares
Universidad Politecnica de Valencia
Spain

1. Introduction

The analysis of astrophysical ices and the determination of the compounds that are present in the molecular clouds play a fundamental role in order to predict the future evolution of the cloud, e.g., its transformation to protostellar bodies or the appearance of new radicals and molecules. Because of the difficulties of obtaining satellite data, the process is simulated first in the laboratory generating ice analogs under controlled conditions. In this case, the ice mixture is carried on allocating the different components in the appropriate concentrations in a deposit chamber with a substrate and recording the spectrum of the aggregated ice when the chamber is filled through a gas inlet with the corresponding compounds (see Figure 1). This process tries to simulate the real process of forming ice mantles under the environmental conditions of the Interstellar Medium. The spectrum is obtained analyzing the transfer function of the ice when it is excited by a source beam with a known spectrum and measuring with a detector the output spectrum after crossing the ice.

The spectrum of each ice can be modelled as the linear instantaneous superposition of the spectrum of the different compounds, so a Source Separation approach is proper. We review and compare in this chapter a set of Source Separation algorithms that approach to the problem in different ways.

Initially, the problem can be addressed as a classical Blind Source Separation problem. In this case, nothing is assumed about the statistical distribution of the compounds present in the ices or how the mixture of these compounds is produced in the ice. The goal is the identification of the elements present in it after recovering the spectrum of the components that there exist in the ice analog. In addition, if we are able to obtain the demixing matrix, we will get a whole description of the mixing process, obtaining the abundances of every component in every ice. But the Blind Source Separation can be slightly modified if we introduce in the modelling of the problem some constraints based on physical properties of the spectrum and mixing process. The priors involve two characteristics: on the one hand, the non-negativeness of the spectrum and the abundances; on the other hand, the sparseness of the spectrum and sometimes the sparseness of the mixing matrix, depending on the kind of astrophysical ice. We will review these approaches and will present the family of algorithms that can be obtained when this information is exploited.

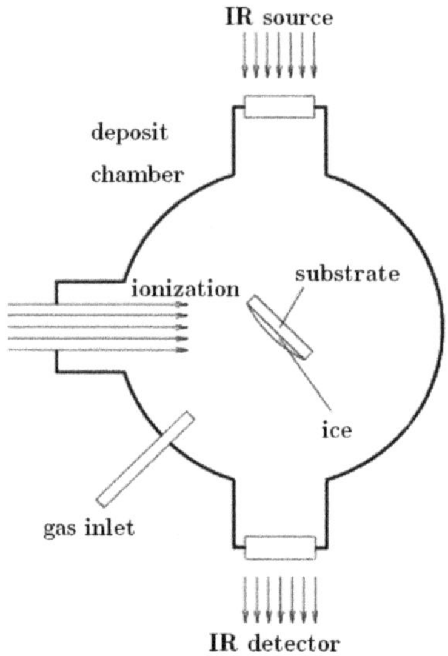

Fig. 1. Obtention of ices in the laboratory.

Blind Source Separation consists of the recovering of some independent sources up to a permutation, scaling and sign factor starting from instantaneous linear mixtures of them with the only assumption of the independence of the sources. The formulation is:

$$\mathbf{x} = \mathbf{A}\mathbf{s} \tag{1}$$

where **x** is the observed vector, **s** the source vector and **A** the mixing matrix.

To solve the Blind Source Separation problem, the statistical technique applied is the Independence Component Analysis (Oja et al., 2001; Cardoso, 1998 ; Comon, 1994), i.e., the generative model that try to decompose the data as a linear combination of statistically independent random variables. Traditional solutions are based on the minimization-maximization of a function that measures the statistical independence of the recovered signals. As we do not know anything about the distributions of the sources, a complete statistical analysis of the problem is difficult and it is not possible to guarantee exactly the independence of the recovered signals. In order to approximate the statistical independence of the sources, many approaches have been proposed. All of them use higher order statistics or nonlinear functions in order to approximate the independence criterion. In all these methods, no prior knowledge is available. However, in many applications, more information can be included in the model.

There are also different ways to modify the Blind Source Separation formulation to introduce this additional information. The best theoretical way to do it is to state the Bayesian formulation of the problem, where the prior information is combined with the likelihood function to obtain the posterior distribution. This approach has the advantage that it is statistically well grounded, since we do not obtain a point estimate but a distribution. But in practice it has some drawbacks. The most important is that it is difficult to implement even in the case when the distributions are approximated with some known expressions. We will implement it through the variational Bayesian approach. If we are interested in algorithms that do not require so high computational resources, another statement is necessary. The mathematical formulation is still based on the fact that the spectrum of the sources are decoupled, but some priors are imposed in the Independent Component Analysis solutions. The non-negativeness is enforced modifying classical Blind Source Separation solutions obtaining algorithms where some nonlinear function such as a threshold function (e.g., the unit step function) are used to assure that there are not negative values in the estimated spectra. The idea is that the sources must be projected into the space of feasible solution, i.e., the sources can not have negative values. But there is still another option to address the problem including the positive restriction. In this case, the independence assumption is relaxed and substituted by a matrix factorization where the observations are the product of two matrices; the restriction in this case is related to the sign of the values of these two matrices: they can not have negative entries. This factorization is referred to as Non Negative Matrix Factorization (Lee & Seung, 1999). Its main advantages are that it is relatively easy to obtain fast algorithms and to modify them in order to include some additional restrictions, such as the non-negativeness of the mixing matrix (the abundances or concentrations of the compounds that there exist in an ice). The problem is that the numerical analysis of these algorithms reveals problems in convergence issues; nevertheless, as we will see in the Results section, they work pretty well in the case of our application.

2. Laboratory simulations of astrophysical ice mixtures

The signature of each ice is its infrared absorption spectrum, where the absorption bands correspond to the specific vibrational mode of the molecules, each one with different atoms and bonds. The spectrum is relevant because we know that the frequencies corresponding to the middle infrared spectrum (4000-400 cm-1; 2.5-25 μm) span the same range as the vibrational frequencies of the adjacent atoms in molecules associated with the most common species. This is why the infrared spectroscopy is an adequate tool to detect the compounds combined in different concentrations in an ice. Specially, this technique is appropriate to detect substances formed by molecules that have dipolar moment, e.g., CO, CO_2, H_2O, NH_3 and CH_4. This dipolar moment can be permanent or induced, due to the presence of other molecules. The detection of the existence of methane in water or in beaches after fuel waste or other environmental disasters is a typical example of the importance of the detection of these kinds of molecules. Each molecule (called endmember) has its own signature, i.e., an unique spectrum. The infrared spectrum has indeed another interesting property. It not only allows to know which molecules are present; in addition, it provides other kind of information, as state of the sample (different spectra for the gas and the ice phase of the molecule), and specifically the interatomic bonds of the molecules.

The molecules are combined in different quantities, called concentrations or abundances depending on if we talk about percentages or absolute quantities. They are named "ice mixtures" or to abbreviate "ices" when they are found in the frozen state. The data are the different spectra recorded by a satellite, but due to the high cost of the equipment and the problems inherent to satellite recordings such as the time slot, focusing in the proper direction, sensors or noise, a lot of preprocessing must be carried out before the data are obtained and after that before they are ready to be used.

A solution is to simulate the real ISM processes in the laboratory. In the laboratory, we are interested in the analysis of the composition of different ices and their behaviour in specific conditions that reproduce situations in which these substances are found outside of the laboratory. For example: to raise the temperature gradually or to radiate the samples with ions that are similar to the ones originated outside of the planet Earth. This is the reason because in the deposit chamber used to obtain the ice (see Fig. 1) in the laboratory there is an entrance for the ion beam. In this way, we can obtain in the laboratory molecules of interest as carbonic acid (H_2CO_3) from simple molecules as CO by protons implantation in the same way that is supposed it happens in the cloud. The measured infrared spectrum is formed by absorption bands around some specific wavelengths determined by the atomic composition and bond structure of the ice; their peak position and width depend on the presence or absence of some molecules that can affect to their dipolar moment, in addition to temperature and particle shape. Besides, these bands usually have an area and a width related to the compound abundance in the ice.

3. Signal processing of laboratory ices

The analysis of laboratory simulations of astrophysical ice mixtures consists of the study of the spectra of the ices in order to establish the compounds present in them. The measured spectrum corresponds to the superposition of the different spectra of the molecules present in the ice; i.e., it describes the absorption features of the ice mixture as a linear combination of the features of the different compounds, e.g., the 2140 cm^{-1} C≡O stretching band. The basic model corresponds to the linear instantaneous mixture model of the Blind Source Separation problem (1).

The infrared absorption spectrum x_{ij} measured for $i = 1,...,M$ ice mixtures in the spectral band $j = 1,...,N$, typically corresponding to 4000 up to 400 cm^{-1} with resolutions 1 or 2 cm^{-1}, is the linear combination of the independent absorption spectra s_{kj} of the molecules (sources) $k = 1,...,K$ present in those ices. The concentration of molecule k in ice i is the mixing matrix entry a_{ik}. The concentrations a_{ik} and absorption spectra s_{kj} are non-negative, although some preprocessing tasks such as baseline removal, noise and complex physics in the measurement process can produce a negative x_{ij}. All these processes will be resumed in a noise term n_{ij} for ice i in wavelength j. In this case, we obtain the noisy instantaneous mixture Blind Source Separation model:

$$\mathbf{x} = \mathbf{A}\mathbf{s} + \mathbf{n} \tag{2}$$

The $Mx1$ data vector $\mathbf{x}$ (measured ices) is modeled as the linear instantaneous combination of a $Kx1$ source vector $\mathbf{s}$ (spectrum of the compounds), where $\mathbf{n}$ is an additive noise representing the error in the measurements and the goal is to recover the independent components of the source vector and/or estimate the MxK mixing matrix $\mathbf{A}$ (abundances of each molecule in every ice). In the case of $M=K$, we have the square problem and the mixing matrix can be inverted obtaining the demixing matrix; remember that the product of the mixing and demixing matrices is a permuted diagonal matrix, i.e., a matrix with one and only one non zero value in every row and column that corresponds to the sign, order and amplitude indetermination of the problem. The noise term is usually considered a centered, white and Gaussian random vector with a given diagonal covariance matrix $\mathbf{R}_n$.

3.1 Bayesian source separation

Given the set of observations $\mathbf{x}_j, j = 1 \ldots N$, the posterior distribution of the unknown interesting variables $\mathbf{A}, \mathbf{s}_j, j = 1 \ldots N$ of our generative model is, applying Bayes' Theorem:

$$p(\mathbf{A}, \mathbf{s}_j \mid \mathbf{x}_j) = p(\mathbf{A}, \mathbf{s}_j) \cdot \frac{p(\mathbf{x}_j \mid \mathbf{A}, \mathbf{s}_j)}{p(\mathbf{x}_j)} \tag{3}$$

where $p(\mathbf{A}, \mathbf{s}_j)$ is called the prior probability, $p(\mathbf{x}_j \mid \mathbf{A}, \mathbf{s}_j)$ the likelihood and $p(\mathbf{x}_j)$ the marginal likelihood or evidence.

The prior describes our knowledge before we obtain the observations, i.e., it encodes in advance the possible values of the model parameters, e.g., if we know before any observation that the sources follow a known distribution; the likelihood describes the goodness of the model, i.e., the probability that the observations follow the model; the evidence, in our case, because we assume the Blind Source Separation model, is just a normalization factor that can be dropped in the optimization step. Shortly, Bayes rule updates the prior after a new data is observed, obtaining the posterior probability. Because the spectrum of the molecules and the concentrations in the ices are independent, we can factorize the prior in two terms, obtaining the posterior:

$$p(\mathbf{A}, \mathbf{s}_{1 \ldots N} \mid \mathbf{x}_{1 \ldots N}) \propto p(\mathbf{A}) \cdot p(\mathbf{s}_{1 \ldots N}) \cdot p(\mathbf{x}_{1 \ldots N} \mid \mathbf{A}, \mathbf{s}_{1 \ldots N}) \tag{4}$$

Bayesian inference with the posterior density is intractable in a general framework. The first step consists of approximating the distributions using some parametric distributions. In our case, the prior distribution for the sources and the mixing matrix corresponds to a mixture of Gaussians with hyperparameters the mixing proportions, the mean and the variance value of the Gaussians (Igual & Llinares, 2007). This mixture model can be optimized attending to the non-negativeness of the spectrum. In this case, the distribution used is the truncated or positive Gaussian, i.e. a distribution with zero probability of taking negative values and two times the probability of the Gaussian for positive values.

The variational approach is a good solution to obtain an algorithm that can implement the parameterized posterior distribution. The objective function to be maximised is the negative free energy *E* (Choudrey & Roberts, 2001):

$$E = \langle \log p(\mathbf{x}_{1 \ldots N}, \boldsymbol{\Theta}) \rangle_{\hat{p}(\boldsymbol{\Theta})} + H(\hat{p}(\boldsymbol{\Theta})) \tag{5}$$

where $\boldsymbol{\Theta}$ is the set of all parameters: the mixing matrix, the noise covariance matrix, the variance of the mixing matrix entries, the sources, the mixing proportions, mean and variances of the mixture of Gaussian model used for the molecular ice distribution (sources); $\hat{p}(\boldsymbol{\Theta})$ is the approximating posterior. The first term in (5) is the expectation of the joint density with respect $\hat{p}(\boldsymbol{\Theta})$; the second one is the entropy. Maximizing (5) is equivalent to minimizing the Kullback-Leibler distance between the true and the approximating posteriors. The approximated pdf $\hat{p}(\boldsymbol{\Theta})$ is chosen such that it can be factorized over the set $\boldsymbol{\Theta}$. Therefore, the maximization can be done individually with coupled terms. The factorization we use is:

$$\hat{p}(\boldsymbol{\Theta}) = \hat{p}(\mathbf{A})\hat{p}(\mathbf{R})\hat{p}(\mathbf{s}_{1...N})\hat{p}(\boldsymbol{\pi})\hat{p}(\boldsymbol{\sigma}^2) \quad (6)$$

where $\{\boldsymbol{\pi}, \boldsymbol{\sigma}^2\}$ are the vectors defined by the mixing proportions and variances of the different mixture of truncated Gaussians used to model the source distribution. Their distribution is a product of symmetric Dirichlets for the mixing proportions and a product of inverted Gammas for the variance. The prior used for the noise variance is also a product of inverted Gammas. Maximizing (5) yields updating rules for the parameters of the posteriors. The parameters of the posteriors are updated versions of the priors. Unfortunately, the update equations are coupled and the implementation of the algorithm requires iterations; these are carried out starting with some initial value for the variables and iterating until convergence.

3.2 Non negative matrix factorization

Since the Bayesian approach involves the use of statistical distributions, it allows a physically well grounded exposition of the solution, at least satisfying the physical restrictions of the problem, in particular, the non-negativeness of the absorption spectrum and abundances. However, it has the drawback that in some point some approximations must be done in order to obtain an implementation of the algorithm. The good thing is that we work with distributions, so we can infer not only a point estimate, but a posterior distribution. This is the classical advantage of Bayesian approach. But there are more ways for obtaining an algorithm that enforce the physical restrictions.

One option is extending the Blind Source Separation model adding constraints about the non-negativeness. Another related approach is to model the observations as the product of two non-negative matrix, the so called Non Negative Matrix Factorization (Lee & Seung, 1999). It is closely related to Independent Component Analysis (Cichocki et al., 2006). Although it was not motivated originally by a Bayesian framework, it has been shown that both of them are the same under mild assumptions about the distributions of the sources (Igual & Llinares, 2008).

The NMF statement of the problem is (in matricial form): the infrared absorption spectrum $\mathbf{X}$ of the ices is factorized as the product of the matrix of concentrations $\mathbf{A}$ and the matrix of absorption spectra $\mathbf{S}$ of the molecules present in those ices:

$$\mathbf{X} = \mathbf{AS}, \mathbf{A} \geq 0, \mathbf{S} \geq 0 \quad (7)$$

Note that in this case we include in the formulation of the problem the whole matrices, i.e., we do not assume any generative model nor time series approach (in our case "time" corresponds to "wavelength"). In real measurements, as it was explained before, an additive noise term must be considered. Although in the laboratory the conditions are well controlled (a high signal to noise ratio), a perfect reconstruction such as (7) is not possible. For example, we experience difficulties to obtain pure mixtures of the different compounds and to control the temperature and pressure conditions in the vacuum chamber where the ice is aggregated. Therefore, in order to use a more complete model, we will assume the noisy model as we did in the Blind Source Separation in order to take into account all these sources of noise. It reads:

$$\mathbf{X} = \mathbf{AS} + \mathbf{V}, \quad \mathbf{A} \geq 0, \mathbf{S} \geq 0 \tag{8}$$

where $\mathbf{V} \in \Re^{M \times N}$ is the noise matrix. Thanks to this term, we will be able to explain measured spectra with negative absorptions, i.e., we will suppose that the negative values of our data are due to the noise vector. In addition, adding the noise term will also allow to understand (8) as a probabilistic generative model connected to (2) although this was not the original motivation of the Non Negative Matrix Factorization.

Assuming in the model the term $\mathbf{V}$ as an i.i.d. Gaussian noise, the maximum likelihood estimates of the matrices $\mathbf{A}$ and $\mathbf{S}$ are:

$$(\mathbf{A}^{ML}, \mathbf{S}^{ML}) = \arg\min_{\mathbf{A},\mathbf{S}}(-\log p(\mathbf{X} / \mathbf{A}, \mathbf{S})) = \arg\min_{\mathbf{A},\mathbf{S}} ||\mathbf{X} - \mathbf{AS}||^2, \mathbf{A} \geq 0, \mathbf{S} \geq 0 \tag{9}$$

Therefore, for a Gaussian assumption, the cost function to be minimized results rather intuitive: the ML estimate tries to minimize the squared Frobenius norm between the spectra of the ice mixtures and the factorization.

The original multiplicative version of the algorithm is:

$$\begin{aligned} \mathbf{A}_{i,k} &\leftarrow \mathbf{A}_{i,k} \frac{(\mathbf{XS}^T)_{i,k}}{(\mathbf{ASS}^T)_{i,k}} \\ \mathbf{S}_{k,j} &\leftarrow \mathbf{S}_{k,j} \frac{(\mathbf{A}^T\mathbf{X})_{k,j}}{(\mathbf{A}^T\mathbf{AS})_{k,j}} \end{aligned} \tag{10}$$

If we know that the spectra of the ices do not include negative values, the multiplicative version guarantees that the matrix of abundances and the spectra of the compounds are positive just initializing the algorithm with non negative matrices. However, if there exist negative values due to the noise term, the non-negativeness can not be guaranteed since all the spectral values of the ices are used in the updating step. If we do not take care of it, we can obtain for example compounds with negative absorptions, something physically impossible. When it happens, the rule is modified constraining the algorithm to enforce $\mathbf{A} \geq 0, \mathbf{S} \geq 0$ simply converting the negative or zero values of $\mathbf{A}$ and $\mathbf{S}$ to small positive numbers.

3.3 Sparseness of the spectrum

The algorithms explained in the previous section are based on the superposition of non negative signals. In the case of our application, and considering the typical molecules involved in the formation of ices, e.g., CO and CO_2, there exist another hypothesis that can be introduced in the statement of the problem: the sparseness of the signals, i.e., their value is zero excepting in some few wavelengths determined by the atomic composition and bond structure of the corresponding compound.

The consequence from a statistical signal processing point of view is that the spectra are characterized by deep narrow absorption peaks around some specific wavelengths, so they can be modeled from a flexible distribution with several parameters, such as a mixture of

Gaussians, to a simple supergaussian distribution such as the Laplacian one, with only one parameter. In the case of a mixture of Gaussians, it reads:

$$p(\mathbf{s}_j)=\prod_{k=1}^{K}p(\mathbf{S}_{k,j})=\prod_{k=1}^{K}\sum_{q_k=1}^{Q_k}\pi_{k,q_k}\frac{1}{\sqrt{2\pi\sigma^2_{k,q_k}}}\exp\left[-\frac{(\mathbf{S}_{k,j}-\mu_{k,q_k})^2}{2\sigma^2_{k,q_k}}\right] \tag{11}$$

where the pdf of the *k*-source at the wavelength j corresponds to a mixture of Q_k Gaussians with mixing proportions π_{k,q_k}, $\sum_{q_k=1}^{Q_k}\pi_{k,q_k}=1, k=1,\ldots,K$, and hyperparameters the mean μ_{k,q_k} and the variance σ^2_{k,q_k} for each Gaussian.

Note that this information is easily included in the Bayesian approach, since we can control the sparseness simply adjusting the corresponding parameters. In the case of the Non Negative Matrix Factorization, since we are not using the distributions, the sparseness must be incorporated in a different way. We have to modify the cost function, i.e., the squared Frobenius norm between the spectra of the ice mixtures and the factorization, to add a constraint in the optimization procedure that enforces the sparseness hypothesis.

The new cost function to be minimized is:

$$D(\mathbf{X}\,||\,\mathbf{AS})=\frac{1}{2}\sum_{i,j}(\mathbf{X}_{i,j}-(\mathbf{AS})_{i,j})^2+\gamma J(\mathbf{S}) \qquad \gamma,\mathbf{A}\geq 0,\mathbf{S}\geq 0 \tag{12}$$

where γ is a regularization parameter. With respect to the regularization function $J(\mathbf{S})$, to enforce the sparseness, it is defined such as $J(\mathbf{S})=\sum_{k,j}\mathbf{S}_{k,j}$, that is equivalent to model the source with a Laplacian distribution. We can calculate the derivatives and use a gradient descent algorithm that minimizes (12). We call this algorithm the regularized Non Negative Matrix Factorization estimate. The updating rules are:

$$\begin{aligned}
\mathbf{A}_{i,k} &\leftarrow \mathbf{A}_{i,k}\frac{(\mathbf{XS}^T)_{i,k}}{(\mathbf{ASS}^T)_{i,k}},\\
\mathbf{A}_{i,k} &\leftarrow \frac{\mathbf{A}_{i,k}}{\sum_{k=1}^{K}\mathbf{A}_{i,k}}\\
\mathbf{S}_{k,j} &\leftarrow \mathbf{S}_{k,j}\frac{(\mathbf{A}^T\mathbf{X})_{k,j}-\gamma}{(\mathbf{A}^T\mathbf{AS})_{k,j}}
\end{aligned} \tag{13}$$

In order to prevent that **A** and **S** can take negative values, in every iteration we apply the nonlinear function $\mathbf{A}_{i,k}=\max(\mathbf{A}_{i,k},\varepsilon)$, $\mathbf{S}_{k,j}=\max(\mathbf{S}_{k,j},\varepsilon)$, where ε is a very small value, i.e., the abundance and spectra are projected to the subspace of possible values as we did in previous sectionand it is typical to all implementations of NMF algorithms to prevent numerical problems.

4. Results

The data corresponds to the public ice analogs database of the University of Leiden. This database contains the infrared spectra of laboratory analogs of interstellar ices. Different mixtures of molecules (from one up to three components selected from H_2, H_2O, NH_3, CH_4, CO, H_2CO, CH_3OH, HCOOH, O_2, N_2 and CO_2) at different temperatures and UV radiation exposures were produced, the final spectrum being calculated rationing the measured and the background spectrum. The units of the data are absorbance and cm^{-1}. Figure 2 shows the spectra of the pure ices (molecules) used in the experiments.

The baseline was removed with Origin software and the useful wavelengths intervals were selected. Among all the preprocessing tasks, the most important one is the baseline removal, because in much attenuated absorption bands, a bad approximation of the baseline can mask some compounds or produce negative values of the optical depth.

The measure of performance used in this section is the Signal-to-Interference Ratio (SIR) defined as:

$$\mathrm{SIR_i} = 10\log\frac{\mathrm{E}\left[\mathbf{s}_i^{\,2}\right]}{\mathrm{E}\left[\left(\mathbf{s}_i - \hat{\mathbf{s}}_i\right)^2\right]} \tag{14}$$

where $\mathbf{s}_i$ represents the original source i and $\hat{\mathbf{s}}_i$ the corresponding recovered source, normalized to the same power. In this index, a high value means high quality results.

The spectra of the ices were mixed with a 10x5 uniform random positive mixing matrix in the range 0-1 in order to simulate nine real laboratory ices. Figure 3 shows an example of such a mixture. The first experiment involved the recovery of the five ices using the algorithms explained in previous sections: NMF (Non Negative Matrix Factorization - Eq. 10), RNMF (Regularized Non Negative Matrix Factorization - Eq. 13) and VBICA (Bayesian Source Separation). Figures 4 to 6 represent the ices recovered by the three algorithms.

Remember that the sources can be recovered in any order. The separation of CO and H_2O obtained by NMF algorithm is almost perfect. For the other sources, there is some remaining of the other molecules. The algorithm was not able to cancel in the rest of compounds the peak around 2300 cm^{-1} due to CO. The same problem occurs for the water, which is contaminating the spectrum of the estimated CO_2. In addition, the algorithm fails absolutely in the extraction of the CH_3OH, obtaining two HCOOH (third and fourth signals in Figure 4).

In the case of RNMF, the results improve. Only for the water (the fourth recovered source in the Figure 5) it is still visible the same noisy peak than in the case of the NMF algorithm: the very narrow absorption band around 2300 cm^{-1} due to CO. For the rest of compounds, the separation is excellent.

VBICA was able to recover all the sources but the CH_3OH (fourth source in Figure 6). It was not able to estimate correctly the right side of the spectrum of the CH_3OH. However, for the other sources, the algorithm worked very well, as we can see in the same Figure comparing with Figure 2.

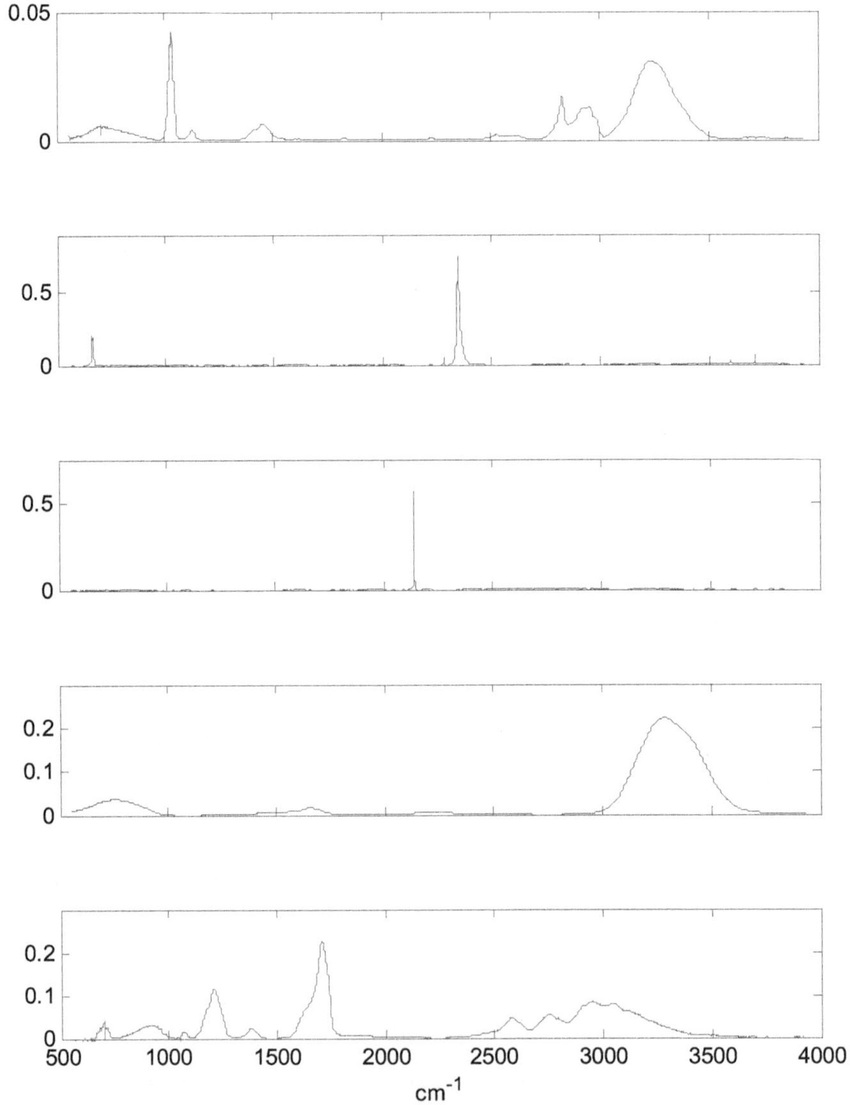

Fig. 2. Spectrum of the pure ices: CH_3OH, CO, CO_2, H_2O and HCOOH.

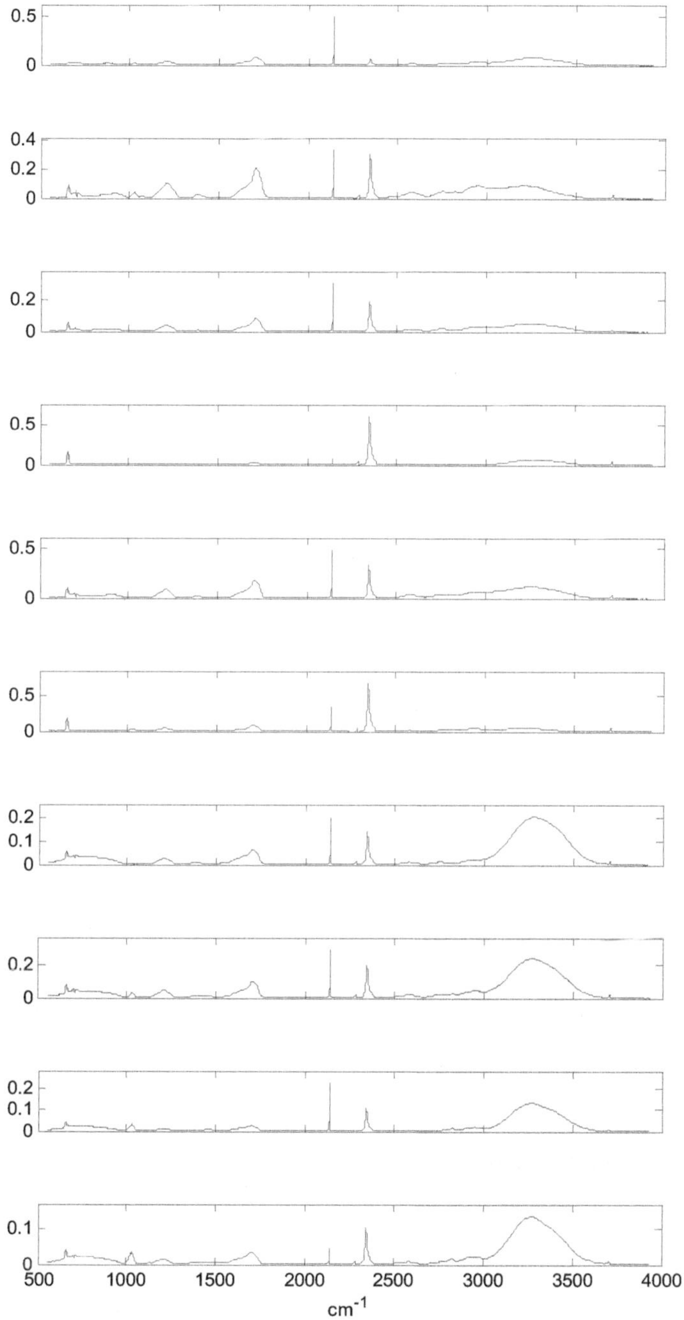

Fig. 3. A mixture of ices: CH_3OH, CO, CO_2, H_2O and HCOOH.

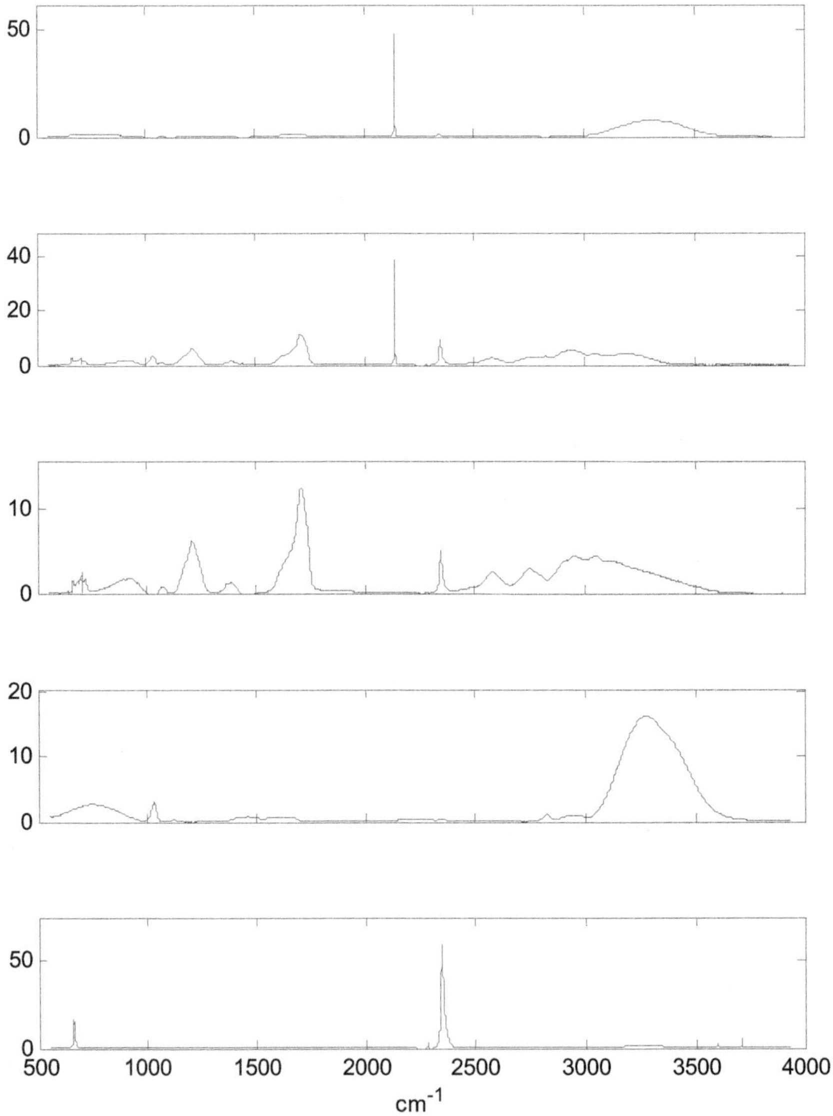

Fig. 4. Ices recovered by NMF. CO_2, HCOOH, HCOOH, H_2O, CO.

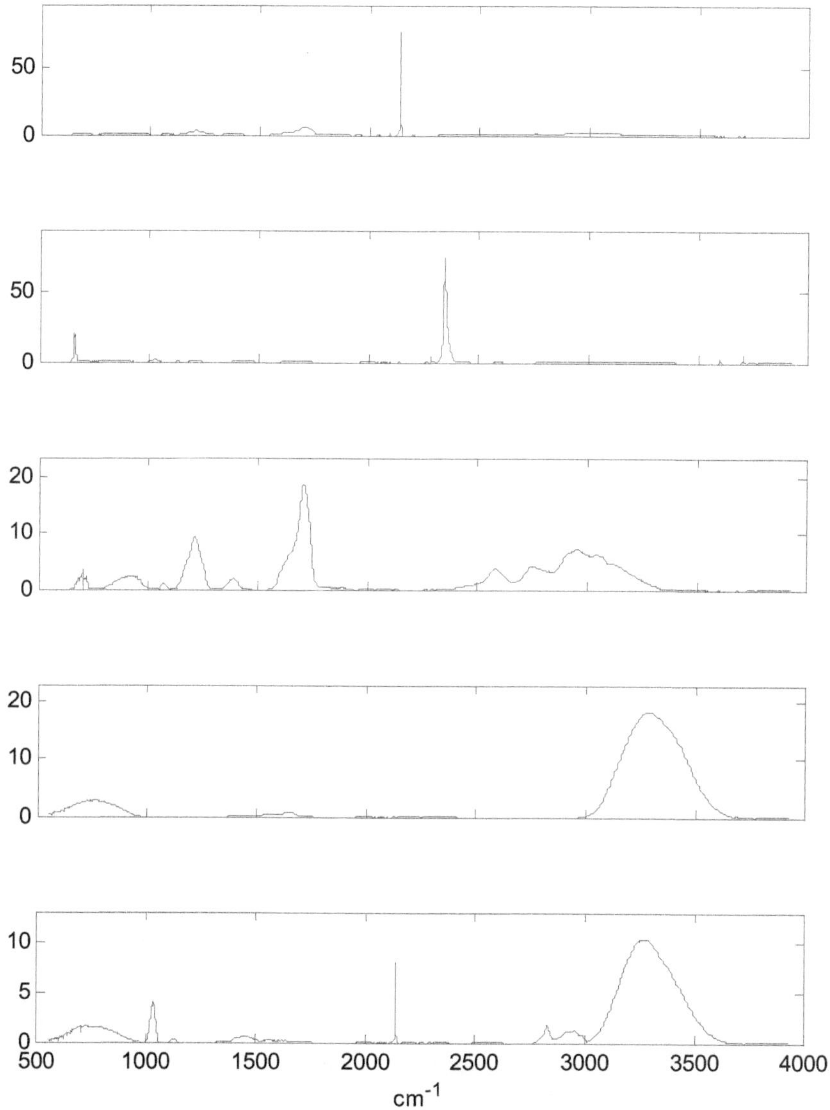

Fig. 5. Ices recovered by RNMF: CO_2, CO, HCOOH, H_2O, CH_3OH.

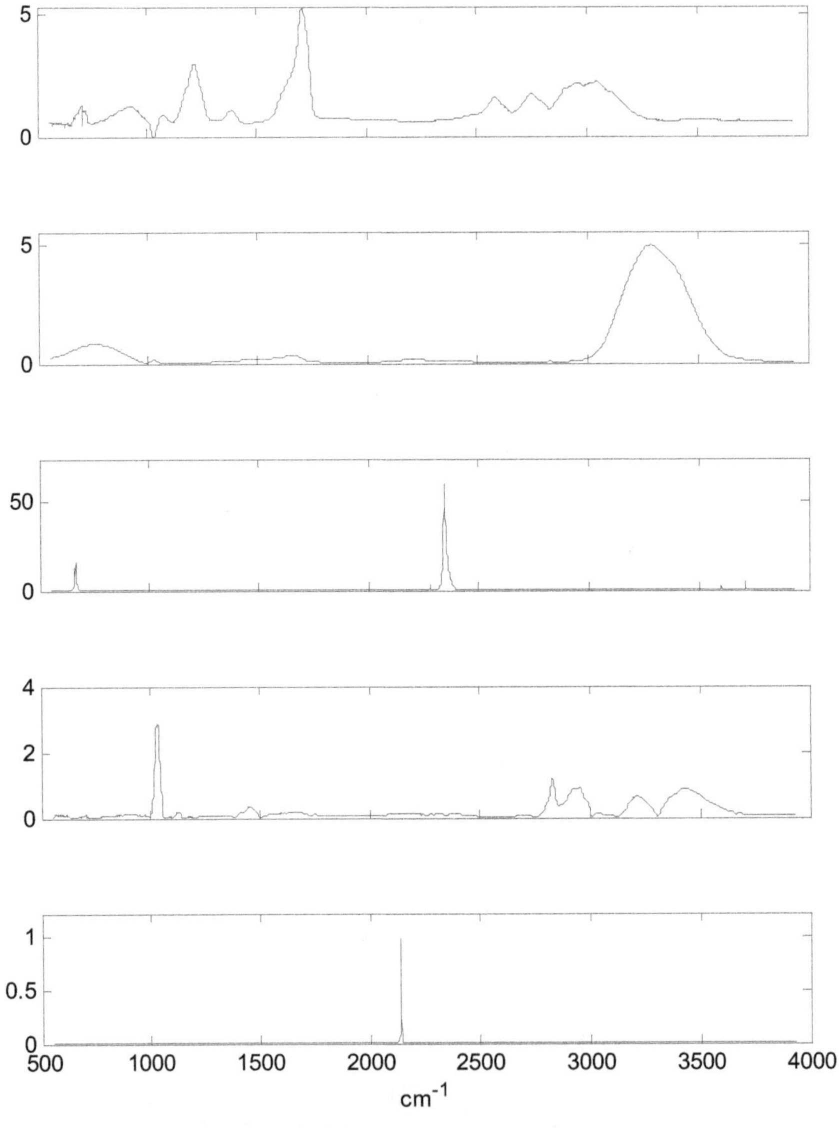

Fig. 6. Ices recovered by VBICA: HCOOH, H_2O, CO, CH_3OH and CO_2.

5. Conclusion

We have reviewed an application of Signal Processing techniques in Astrophysics: the study and analysis of ices obtained in the laboratories that try to simulate the astrophysical ices found in the outer space. We have exposed the problem from a Signal Processing point of view, showing the goodness of the Blind Source Separation approach to model the recorded absorption spectra as the combination of the spectra of the molecules that there exist in the obtained ice by aggregation of the molecules.

The problem can be addressed in a Bayesian framework, where the prior distributions about the spectrum of the compounds and the abundances of them in every ice can be treated in a proper way when they are approximated by models such as mixture of Gaussians. But these algorithms are very slow when compared to classical optimization methods based on a cost function that is maximized according to some gradient based algorithm. We have shown that the statistical approach can be relaxed when the Non Negative Matrix Factorization is introduced. The obtained algorithms are faster although they can suffer from convergence problems due to the initialization and adjustment of some parameters that control the performance of the algorithms.

Both procedures work, although attending to all the factors, we can conclude that the algorithms that enforce the non-negativeness condition in the optimization step obtain the best results, since they are fast and are able to cancel the remaining spectrum of the compounds demixing the spectrum of the ices. Nevertheless, the extraction is not always perfect, especially when some large narrow peaks are involved. On the other hand, when the sparseness restriction is included, the results improve, as we have seen with the RNMF algorithm, which obtained similar results to VBICA with a lower computational cost.

6. Acknowledgment

This chapter is in part supported by the Valencia Regional Government (Generalitat Valenciana) through project GV/2010/002 (Conselleria d'Educacio) and by the Universidad Politecnica de Valencia under grant no. PAID-06-09-003-382.

7. References

Cardoso, J.F.; Blind Signal Separation: statistical principles. *Proc.of the IEEE*, Vol. 86, No. 10, (1998), pp. 2009-2025

Choudrey, R.; Roberts, S.J.; Flexible Bayesian independent component analysis for blind source separation. *Proceedings of 3rd International Conference on Independent Component Analysis and Blind Signal Separation*, pp. 90-95, San Diego, USA, 2001

Cichocki, A.; Zdunek, R.; Amari, S.; New algorithms for non-negative matrix factorization in applications to blind source separation. *Proceedings of IEEE International Conference on Acoustics, Speech and Signal Processing ICASSP*, 2006

Comon, P.; Independent Component Analysis, a new concept? *Signal Processing*, Vol. 36, (1994), pp. 287-314.

Igual, J.; Llinares, R.; An informed source separation of astrophysical ice analogs. Digital Signal Procesing, Vol. 17 (2007), pp. 947–964, ISSN 1051-2004.

Igual, J. ; Llinares, R.; Nonnegative Matrix Factorization of Laboratory Astrophysical Ice Mixtures. IEEE Journal of Selected Topics in Signal Processing, Vol. 2, pp. 697 - 706, ISSN 1932-4553.

Lee, D.D.; Seung, S.; Learning the parts of objects by non-negative matrix factorization. Nature, Vol. 401, No. 6755 (1999), pp. 788-791.

Oja, E.; Hyvarinen, A.; Karhunen, J. (2001). Independent Component Analysis. John Wiley & Sons.

15

Energetic Charged Particles in the Heliosphere from 1-120 AU Measured by the Voyager Spacecraft

W.R. Webber
New Mexico State University, Department of Astronomy, Las Cruces, USA

1. Introduction

For nearly 35 years the Voyager Spacecraft have been pursuing their epic journey to the boundaries of the heliosphere and interstellar (IS) space beyond. The remarkable observations of Jupiter, Saturn, Uranus and Neptune and their moons by Voyager are now long past. The observations of the giant bubble produced in the interstellar plasma by the outward flowing solar wind are still continuing however and are, in fact, building in excitement as the Voyagers cross or approach important boundaries between our local heliosphere and interstellar space. At the end of 2004, V1 crossed the heliospheric termination shock (HTS) at a distance of 94 AU from the Sun at a location of ~33° North. This shock is produced at a point where the outward moving solar plasma pressure is balanced by that of the interstellar medium including the magnetic (B) field. V2 crossed the HTS in August, 2007, at a distance of 84 AU at a location of ~25° South. This North-South asymmetry in the distance to the HTS (Stone, et al., 2008) has been determined to be due to the direction of the interstellar magnetic field. Its magnitude is estimated to be 4-5 μG or even larger, in contrast to pre-Voyager estimates of only 2-3 μG. The direction of this field is also determined for the 1st time and is such that it pushes in on the South hemisphere of the heliosphere; the B field making an angle ~40° with the local equator and creating a very distorted and asymmetric heliosphere in the North-South dimension (Opher, 2009).

Beyond the HTS is a region called the heliosheath. This region is believed to extend to a distance ~1.5 times the distance to the HTS and may have further North-South asymmetries. It is a very turbulent region where considerable energetic (≥ 1 MeV) particle acceleration occurs. Near its outermost boundary the wavy heliospheric current sheet, which changes in polarity every 11 years, bends back across the polar regions creating a long heliospheric tail, much like that of the Earth's magnetosphere. The outer boundary of the heliosheath is the heliopause (HP), a surface discontinuity between the solar wind plasma and the local interstellar plasma.

In this paper we are interested in the ≥ 1 MeV charged particle environment of this overall region extending outward from the Sun to at least 120-150 AU. This study is based mainly on the results from the Cosmic Ray Science (CRS) experiment (Stone, et al., 1977). At 2012.0, Voyager 1, the outermost spacecraft is approaching 120 AU and V2 is at ~98 AU. A schematic side-on view of the heliosphere is shown in Figure 1 to give an overall view of the heliospheric dimensions and the location of V1 and V2.

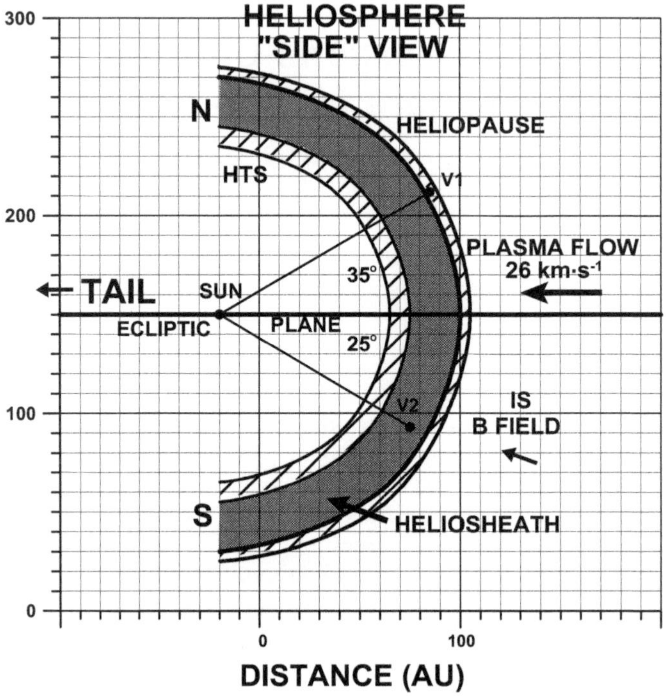

Fig. 1. Side on view of the solar wind dominated heliosphere showing the inner region, the HTS at about 90 AU, the heliosheath between 90-111 AU, and the outer transition region before the HP.

This environment includes galactic cosmic rays (nuclei and electrons) above a few MeV that enter the heliosphere from the galaxy where they populate interstellar space. Their intensity beyond the heliosphere is uncertain and their total energy density may even exceed that of the typical interstellar B fields and plasma. This cosmic ray component, which is the dominant one in the interplanetary region of the heliosphere, has a positive radial intensity gradient, indicating much larger intensities in interstellar space than at the Earth (for electrons, perhaps ~100 times).

There exist two other major components of energetic particles above ~1 MeV in the heliosphere region. These components are accelerated, either in the vicinity of the HTS (called Termination Shock Particles, TSP) or throughout the heliosheath and perhaps even near its outer transition region (called Anomalous Cosmic Rays, ACR). ACR's are only occasionally observed near the Earth. The details of these components have only been revealed as the Voyager approached and then crossed the HTS and explored the heliosheath region beyond.

The heliosheath is a giant region beyond the HTS of perhaps 20-40 AU in extent which is populated by protons, helium nuclei and oxygen nuclei with energies from ~1 MeV up to

~100 MeV (TSP and ACR). The power to accelerate these particles comes from the energy in the slowing down solar wind. This process may occur in several stages at various distances from the HTS to the outer boundary of the heliosheath. This region is bathed in massive intensities of these energetic particles, comparable to or larger than those found in the magnetospheres of the planets (e.g., the Earth, Jupiter, etc.) but in this case essentially surrounding the entire heliosphere outside of the HTS. This region, which will be described in detail in this paper, will certainly give pause to any space traveler whether they are leaving the solar system or coming to explore it, because of the severe radiation hazard over an extended period of time/distance that these energetic particles present.

2. Galactic cosmic rays

From its beginning over 70 years ago, one of the goals of the study of the very energetic galactic cosmic rays was to determine their intensity in the galaxy and also their origin. These goals have been hampered by the extensive modulation effects of the heliospheric plasma and magnetic fields. For the lowest energies these modulation effects are sufficient to reduce the intensities observed near the Earth by 10 to over 100 times from the interstellar intensities. Now at last the Voyagers spacecraft are pushing back this veil. In Figure 2 we show the total intensity of cosmic rays >70 MeV from the time of Voyager 1 launch in 1977 to the present time in late 2011, a time period of ~34 years. During this time the overall intensity of these particles has increased by a factor ~2.5 from that observed in 1977 which was a time of minimum modulation (highest intensity). The solar 11 year activity cycle is clearly evident in the data with successive intensity maximum in 1977, 1987, 1998 and 2009, at times of solar activity minima. The study of this solar modulation cycle as a function of radius is one of the ongoing areas of research for the Voyager Science Teams.

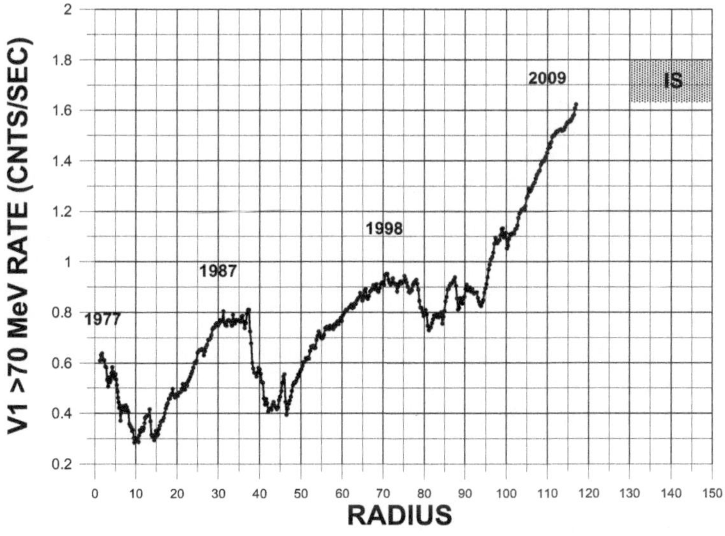

Fig. 2. Total intensity of cosmic rays >70 MeV at V1 from launch in 1977 to the present time. Times of maximum intensity (minimum solar activity) in the 11 year solar cycle are indicated. The estimated IS intensity is shown as a shaded region beyond 130 AU.

In late 2011, this integral intensity reached levels near the lower limit of the expected IS intensities based on models for propagation of cosmic ray nuclei in the galaxy (e.g., Webber and Higbie, 2009). However, as much as 10% of this integral intensity could be due to ACR's of energy >70 MeV and also to galactic cosmic ray electrons, not included in the models. With each passing AU this integral rate data becomes more and more constraining with regard to the IS energy density of these cosmic rays and their intensity and propagation in the galaxy.

In Figure 3 we present data for cosmic ray electrons of 6-14 MeV. These particles show a completely different radius vs. intensity profile than the nuclei. Near the Earth and out to perhaps ~40 AU these electrons are dominated by Jovian electrons. In fact, most of the electrons in this energy range measured near the Earth are not galactic electrons at all but are Jovian electrons propagating within the inner heliosphere (Ferreira, et al., 2001). These electrons peak at the time of the Jupiter encounter in 1979, with radiation levels that are dangerous to both man and the instruments (several instruments were affected). Beyond ~40 AU the electron intensities are dominated by galactic electrons, however. They show only weak modulation effects due to the solar 11 year cycle and at the time of the HTS crossing in late 2004 their intensity is within a factor ~2 of the intensity at the Earth after subtraction of the Jovian component at 1 AU.

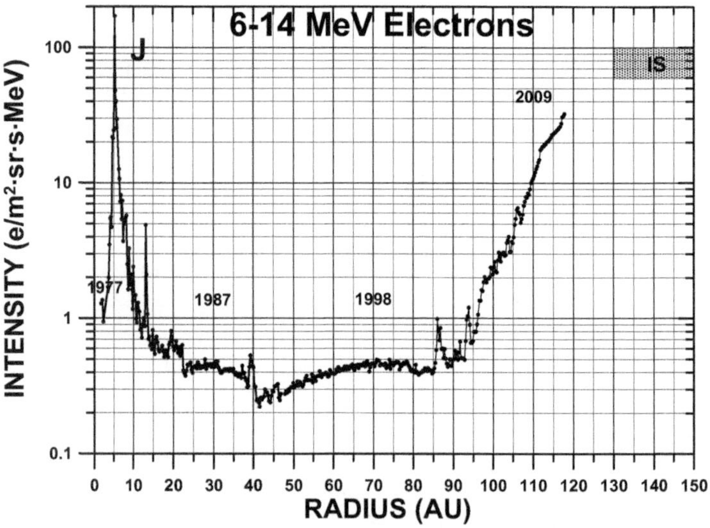

Fig. 3. Intensity of 6-14 MeV electrons at V1 from launch to the present time. The intensity peak at Jupiter encounter occurs in 1979. The intensity of these galactic electrons begins to increase only after crossing the HTS, reaching levels ~100 times those observed at the Earth. The estimated IS intensity is shown as a shaded region beyond 130 AU.

After crossing the HTS the 6-14 MeV electron intensity at V1 has risen continuously by a total factor ~60-80 by 2011.5, a period ~6.5 years. At 2009.7 and again at 2011.2 two sudden jumps in the electron intensity along with changes in their radial intensity gradient are observed. We believe that these times, which occur when V1 is at 111 AU and again at 116 AU, mark the times when V1 has passed significant structures in the outer heliosphere

(Webber, et al, 2011). These features may represent the fold back of the heliospheric wavy current sheet. This wavy current sheet has been observed by the Voyagers in the heliosphere inside the HTS and this layered structure may be carrying solar plasma and magnetic fields back to form the tail of the heliosphere (Florinski, 2011). This region is estimated to be 8-10 AU thick and its outer boundary may, in fact, be the heliopause, a 3-D surface separating the region of solar influence from the local interstellar medium (Washimi, et al., 2011). So in this scenario, V1 may be relatively close to observing the interstellar spectrum of these low energy electrons. It is seen in Figure 3 that the 6-14 MeV electron intensity at 2011.5 has almost reached the lower limits of the estimated IS electron intensities. These estimates are based on a combination of propagation models and the intensity of low frequency radio emission from these electrons in the galaxy. The overall uncertainty in the estimated IS electron spectrum is a factor ~2 (Webber and Higbie, 2008). So again with each AU that V1 travels outward, new limits are placed on the galactic propagation conditions at these low energies and also on the origin of these electrons and the distribution of their sources in the galaxy.

The changes in the spectrum of cosmic ray electrons at V1 as a function of time/radius are remarkable. These spectra from ~6-120 MeV are shown in Figure 4 from the time of the HTS crossing in one year intervals to 2011.5. As might be expected from Figure 3, the HTS spectrum/intensity is very similar to that at the Earth. Almost all of the intensity change occurs in these low energy electrons after the HTS crossing. The intensities increase by nearly a factor ~100 at the lowest energies and the spectrum quickly develops into a shape ~$E^{-1.7}$. This spectrum simply increases with increasing distance maintaining the same spectral exponent.

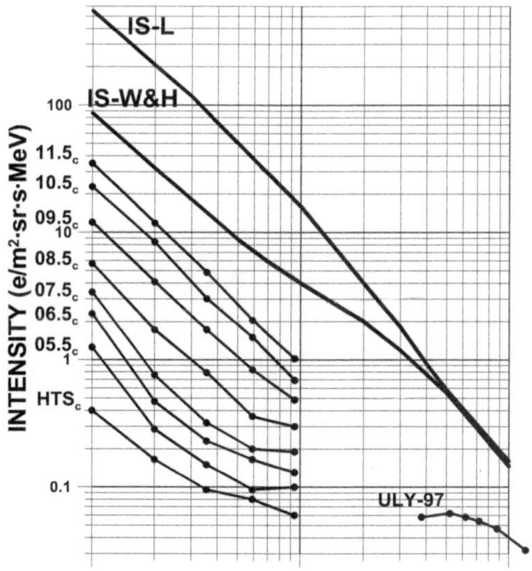

Fig. 4. Spectra of the galactic electron component from 10 MeV to 100 MeV. Voyager 1 measurements are shown at the time of the HTS crossing and yearly up to 2011.5 (117 AU). Estimated limits on the possible IS electron intensity are also shown. Ulysses measurements near the Earth at a time of minimum solar modulation are also shown at higher energies.

Two predictions of possible IS electron spectra are shown in the Figure. One is by Langner, et al., 2001, the other by Webber and Higbie, 2008. Both are based on the observed galactic radio synchrotron spectrum which results from the electrons propagating in the galactic magnetic fields, plus propagation/diffusion models. The electron spectrum measured at V1 at 117 AU is now only a factor ~2 below the lowest of these estimates. Measurements at energies above a few hundred MeV made at times of minimum modulation are shown from the Ulysses spacecraft. It is hoped that the new results from the PAMELA spacecraft which extend to lower energies, will help to constrain this higher energy part of the electron spectrum and overlap the Voyager measurements.

Next we present data for the individual nuclei, H, He and heavier nuclei such as C. Here we show the spectra of these nuclei measured by the CRS instrument on V1. The spectra for H nuclei are shown in Figure 5, for He nuclei in Figure 6 and for C nuclei in Figure 7. The spectra are shown for 3 times, 2009.5, 2010.5 and 2011.5 as V1 moves ever closer to the outer boundary of the heliosphere, at the rate of 3.6 AU/year. So these spectra are essentially snapshots of the intensities and spectra as these intensities approach the IS values. The H and He intensities are corrected for a background from ACR H and He nuclei which becomes important below ~100 MeV.

Also shown in all these figures are recently estimated IS intensities for these nuclei based on propagation models and estimated source spectra (Webber and Higbie, 2009). These estimates give the lowest intensities of the many historical estimates dating back over 30 years or more. It is seen that the intensities of all components, H, He and C are getting larger at the rate of 5-15% per year corresponding to radial distances of 110.3, 114.0 and 117.5 AU, but are still below these latest IS estimates.

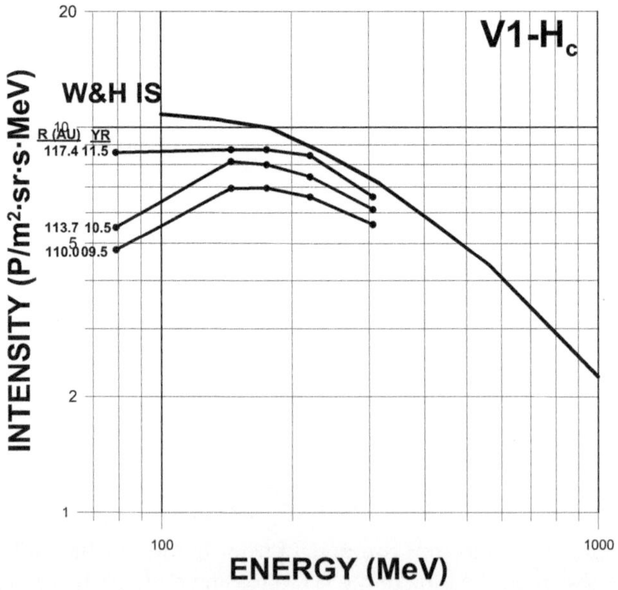

Fig. 5. Energy spectra of H nuclei measured at V1 at 2009.5, 2010.5 and 2011.5 (corrected for ACR background). Estimated IS spectrum for H is shown by a solid line.

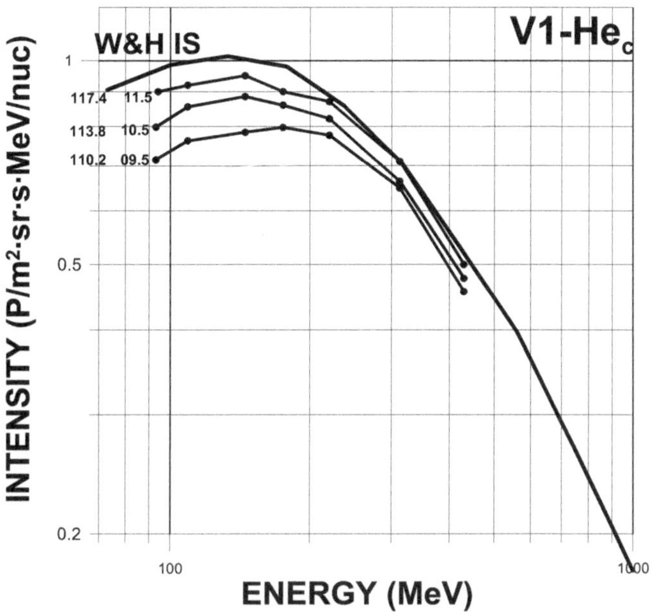

Fig. 6. Energy spectra of He nuclei measured at V1 at 2009.5, 2010.5 and 2011.5 (corrected for ACR background). Estimated IS spectrum for He is shown by a solid line.

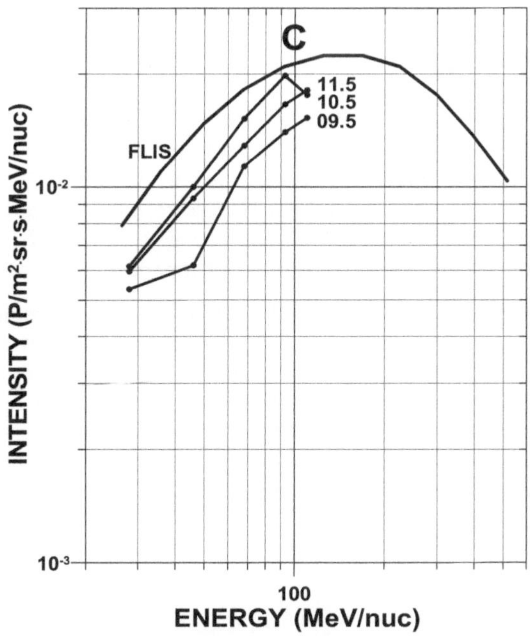

Fig. 7. Energy spectra of C nuclei measured at V1 at 2009.5, 2010.5 and 2011.5 (no correction for ACR needed). Estimated IS spectrum for C is shown by a solid line.

Using a simple "force field like" model which is commonly used to describe this solar modulation (e.g., Gleeson and Axford, 1968) and starting with the IS spectra given in Figures 5, 6 and 7, we find that a modulation parameter ϕ=53 MV fits the H data at 2009.5, a parameter = 39 MV fits the 2010.5 data and a parameter = 26 MV fits the 2011.5 data. The parameters that fit the He data are 47, 33 and 20 MV respectively. The spectra of all nuclei should be fit by the same value of the modulation parameter for each year. The slightly lower values of this modulation parameter for He relative to H could be explained by an increase in the IS He ~+3% relative to H.

As a point of reference the value of the modulation parameter required to reproduce the spectra of H and He nuclei that are observed at the Earth at the times of minimum modulation is ~400 MV, the same for both H and He. So the "amount" of modulation necessary to reproduce the H and He spectra observed at 2011.5 (117.5 AU), (which is between 20 and 26 MV), is 5-7% of that required to reproduce the spectra at the Earth. This data suggests that, at about 5-7 AU beyond the V1 location of 117.5 AU at 2011.5, the estimated IS spectrum will be reproduced by Voyager measurements at the current rate of intensity increase, e.g., the solar modulation will go to zero.

For C nuclei the spectrum can be measured from ~20-120 MeV/nuc using current analysis procedures. Fortunately this is the most interesting energy range, just below the peak in the differential spectrum. In the energy range below ~100 MeV/nuc the models used to describe the solar modulation predict a spectrum that is proportional to E, independent of the exact shape of the IS spectrum, as long as the modulation is "significant". In Figure 7 we note that the C spectrum at all times is very closely ~E and the "apparent" modulation is small as well.

Does this mean that the energy spectrum at V1 is close to the true IS spectrum at these energies or is the solar modulation larger at lower energies than that determined from the H and He data at higher energies so that, in fact, the modulation is producing the ~E spectrum? At ~100 MeV/nuc where the LIS spectrum can be calculated with reasonable accuracy (±10%) the apparent modulation for C (the value of ϕ) is ~45 MV at 2010.5 and ~27 MV at 2011.5, consistent with the modulation derived from the H and He spectra. So at ~100 MeV/nuc there is consistency between the relative IS H, He and C intensities and also consistency between the apparent solar modulations for these three components.

The ratio of IS He/C intensities at ~100 MeV/nuc is a critical parameter for any galactic propagation/source model for cosmic rays. This is so because the energy loss by ionization in the galaxy is large and becoming dominant at this energy and this E loss parameter is ~Z^2. The measured value for this ratio at V1 at 2011.5 is 46.5±3, a dramatic improvement over values for this ratio being used at the present time which have uncertainties of ±50% (e.g., Putze, Maurin and Donato, 2011). Note that small levels of solar modulation will not change this ratio because of the similar spectra and the identical A/Z ratio = 2.0 of these nuclei. These are the most important factors in the modulation calculation at these energies.

3. The anomalous components

ACR Oxygen nuclei were first discovered in 1974 by instruments on the Pioneer spacecraft just after their launch (McDonald, et al., 1974). This result was confirmed by the CRS instruments on Voyager. These particles were recognized as anomalous because of an unexpected upturn in the spectrum of low energy galactic O nuclei. In subsequent years it has been found that the ACR O component has a source spectrum between ~E^{-2} - E^{-3}, with the spectral index increasing with increasing energy at energies from a few MeV up to as high as 50 MeV/nuc. The time history of these nuclei from 1977 to the present as measured by V1 is shown in Figure 8. It can be seen that these nuclei almost vanished in 1980-81 at a time of the 11 year maximum in solar modulation. But the intensity recovered to still higher intensities at the next modulation minimum in 1987.

Theories for their origin have centered on the acceleration of IS neutral atoms, first ionized in the outer heliosphere and then accelerated (Fisk, Kozlovsky and Ramaty, 1974). But the apparent acceleration region has moved outward as the intensity has continued to increase with increasing radius. The intensity showed no sudden increase at the HTS, considered to be one of the possible acceleration locations. In fact, the intensity has continued to increase throughout the heliosheath, but now seems to have reached a maximum in 2009-2010, at about 110 AU, ~15 AU beyond the HTS. This distance of ~110 AU corresponds to the time at which the first of several sudden electron intensity increases, as was noted earlier, was seen at V1. A jump in magnetic field strength and a change of polarity of the field observed at this time (Burlaga and Ness, 2010) provides additional evidence that V1 crossed a significant structure at this time, perhaps moving into the 1st of several layers in a transition zone from solar dominated to interstellar dominated plasma conditions. This is also the time when the radial component of the solar wind speed becomes ~zero (Krimigis, et al., 2011).

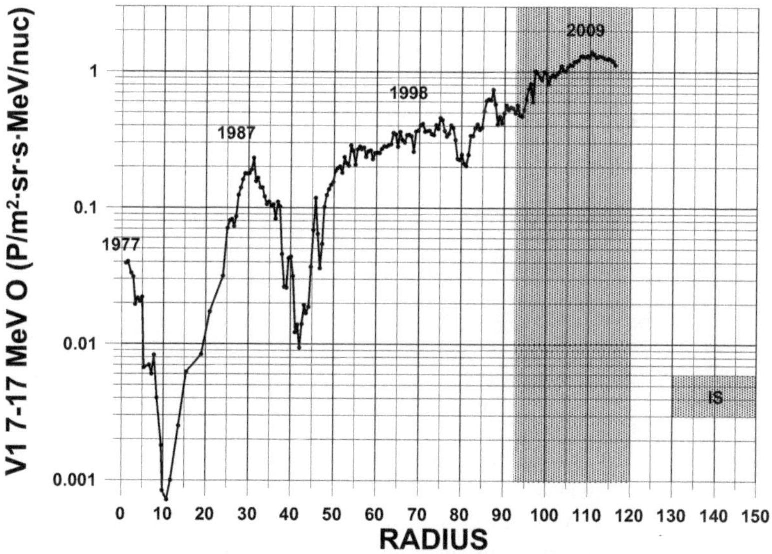

Fig. 8. Intensity of 7-17 MeV ACR-O nuclei from launch to the present time. The solar 11 year cycle and the extreme modulation of these ACR at times of maximum solar activity in 1980-1982 and 1990-1991 are evident.

At the time of maximum in 2009-2010 at between 110-115 AU, the intensity of 7-17 MeV/nuc ACR-O nuclei was ~50 times the maximum intensity observed at the Earth. The IS intensity of these low energy particles is completely unknown. A galactic component as part of the spectrum of cosmic ray O nuclei would be a factor of at least 500 below the peak intensity observed in the heliosheath (see Figure 8).

We next examine the other major components of ACR, the H and He nuclei, concentrating here on H nuclei. All of the ACR nuclei have very similar spectra, ~E^{-2} at the lowest energies below ~10-20 MeV, and steepening to E^{-3} at the higher energies and extending up to ~100 MeV for H (for more spectral information see Cummings, et al., 2011). The abundance of H nuclei is ~10 times that of He at a fixed energy (below ~20 MeV). We consider the highest energies first and then work down to lower energies.

Figure 9 presents the time history of 27-69 MeV H nuclei. At these energies the ACR spectrum sits atop the galactic spectrum of H nuclei. The estimated galactic H nuclei intensities at times of minimum modulation are shown as large dots which are only slightly below the total H nuclei intensities measured in the inner heliosphere. Even at the times of intensity maxima in 1977 and 1987, it was not clear that an anomalous H component was present. The existence of this component was clearly demonstrated at the time of the next maximum intensity in 1998 when the effects of solar modulation between V1 and the source became much less as V1 moved outward toward the source region. It is now recognized that H nuclei are the dominant component of ACR in the heliosheath energetically and intensity wise. At the time of peak intensity at ~110 AU, ACR H at an average energy ~50 MeV was ~20 times greater than the galactic H intensity at the same energy.

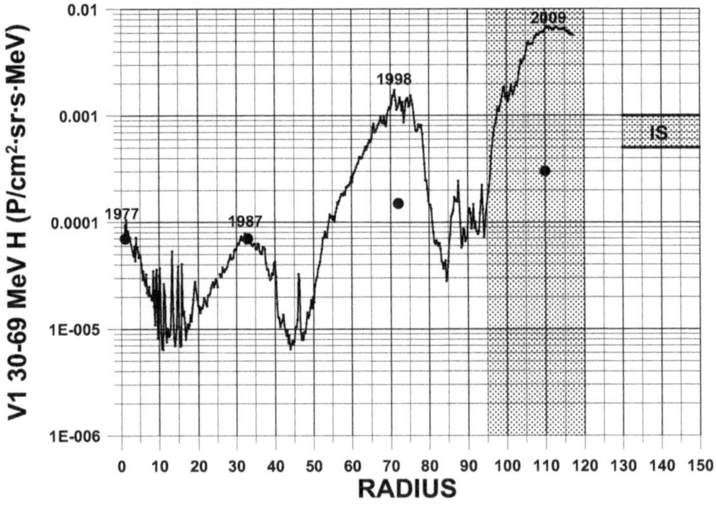

Fig. 9. Intensity of 27-69 MeV H nuclei, including both ACR and GCR from launch to the present time. Estimated galactic H nuclei contribution is shown as big dots at times of maximum intensity. The estimated IS intensity of galactic H nuclei at this energy is shown as a shaded region beyond 130 AU.

Note that there are several spikes in the intensity-time profile of 27-69 MeV H nuclei between 8-20 AU. This is a time of maximum solar activity between 1980-1982 and these spikes are the high energy tips of individual solar/interplanetary cosmic ray acceleration events propagating outward in the heliosphere.

In Figure 10 we present the time history of 10-21.5 MeV H nuclei, a factor ~3 lower in energy. This figure looks much like the higher energy H nuclei in Figure 9. At this lower energy the presence of ACR H again does not stand out in the inner heliosphere relative to an even lower intensity galactic component (the galactic intensity is scaled ~E). ACR-H nuclei become obvious again only at the time of the 1998 intensity maximum. The intensity maximum for these particles is observed beyond the HTS at between 110-115 AU and at that distance the intensity is ~100 times the estimated galactic intensity at this same energy.

Note the increased presence of spikes in the intensity-time profile between 8-20 AU in the 10-21.5 MeV channel. Again these spikes are the medium energy part of the spectra of individual solar/interplanetary cosmic ray acceleration events propagating outward through the heliosphere. The large spike at ~46 AU (at 1991.72) at V1 was from a solar (CME) event at 1991.42 at the Earth. The time difference of 0.30 year represents the propagation time of this disturbance (shock) between the Earth and V1.

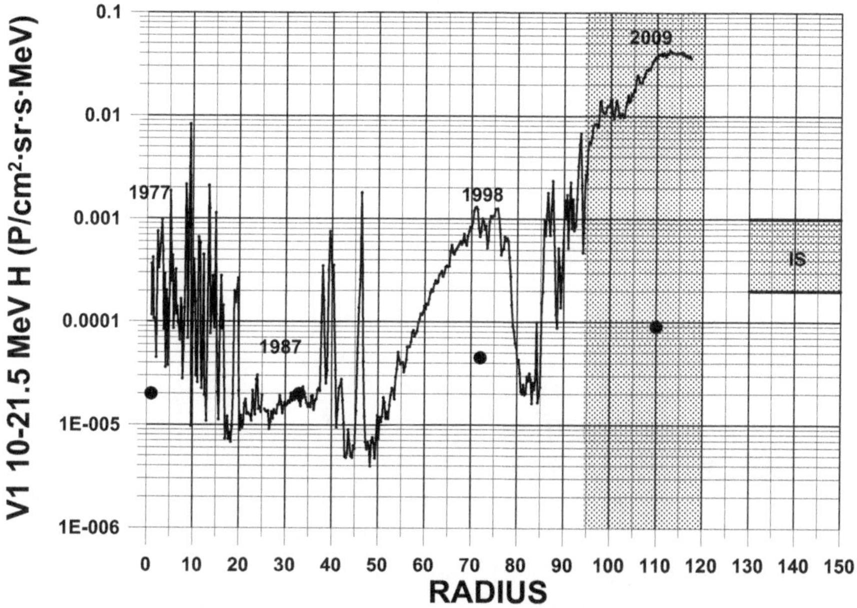

Fig. 10. Intensity of 10-21.5 MeV H nuclei, including both ACR/TSP and GCR and solar/interplanetary protons from launch to the present time. Estimated galactic contribution is shown as big dots at times of maximum intensity. The estimated IS intensity of galactic H nuclei in this energy range is shown as a shaded region beyond 130 AU. The presence of solar/interplanetary accelerated nuclei is an almost continuous excess intensity from launch to ~20 AU (1978-1982).

In Figure 11 we present the time history of 3-8 MeV H nuclei, another factor ~3 lower in energy than Figure 10. At this energy there is again no strong evidence of an ACR H component or even a galactic component before the 1998 intensity maximum. Most of the intensity variations in this channel up to ~50 AU are due to individual solar/interplanetary cosmic ray acceleration events. In fact these events, on average, increase the intensity in this energy channel by a factor >100 over the intensity due to that from galactic cosmic rays alone. Many of these events produce intensities $>10^3$ times the background intensity even at distances ~20 AU or greater from the Sun.

The actual 3-8 MeV heliosheath ACR/TSP "component" makes it first appearance at V1 about 10 AU before this spacecraft reaches the HTS. The intensity increases before the HTS crossing in a complicated series of increases, some of which are periodic. After the HTS crossing the intensity continues to increase rapidly and more smoothly, reaching levels ~10 times those at the time of the HTS crossing itself. Again, the intensity peaks at between 110-115 AU at levels between 10^2-10^3 times those expected for the estimated galactic spectrum in IS space and a factor $\sim 10^4$ times the quiet time intensity observed by V1 as it left the Earth.

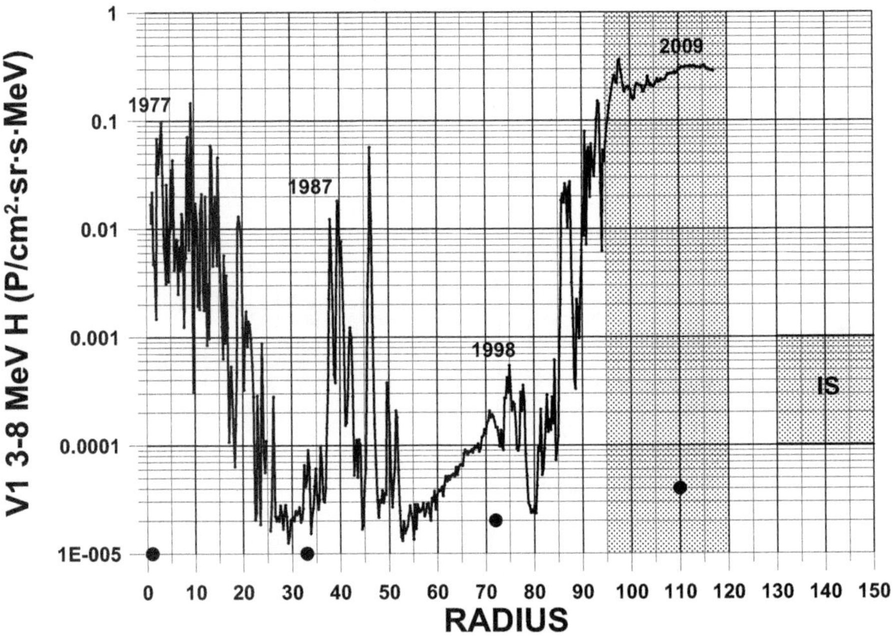

Fig. 11. The intensity of 3-8 MeV H nuclei from launch to the present time. These nuclei are mainly solar/interplanetary protons in the inner heliosphere during solar active periods and ACR/TSP H nuclei beyond ~50 AU, increasing rapidly just prior to the HTS and reaching an intensity maximum at ~110 AU, $\sim 10^3$ times a possible IS galactic component of H nuclei. Estimated galactic contribution is shown as big dots at times of maximum intensity. The estimated IS intensity of galactic H nuclei in this energy range is shown as a shaded region beyond 130 AU. The presence of solar/interplanetary accelerated nuclei increases the average intensity of 3-8 MeV H by a factor ~100 inside of 20 AU (1978-1982).

The final Figure, #12, shows the integral rate of all particles >0.5 MeV from launch to the present time. This rate is probably the best representation of the total radiation that would be received by an instrument or a person at that location. The large dots show the total rate that would be observed from galactic cosmic rays only at a time of minimum modulation. The solar 11 year modulation cycle is barely visible in the data. What are notable here again are the spikes of intensity observed in the inner heliosphere out to ~20 AU and beyond. As noted previously these are individual solar interplanetary acceleration events and they increase the average integral intensity by a factor 10-100 times the normal background level due to the galactic cosmic rays only.

The other feature in the >0.5 MeV data is the high intensity in the heliosheath, mainly due to TSP and ACR beginning, in fact, with discrete and large increases that begin to appear 5-10 AU prior to the HTS crossing. This intensity again peaks at a distance ~110-115 AU from the Sun.

The total integral intensities in the heliosheath are 10-20 times those expected from galactic cosmic rays which have an energy density ~1 eV/cm^3 in interstellar space.

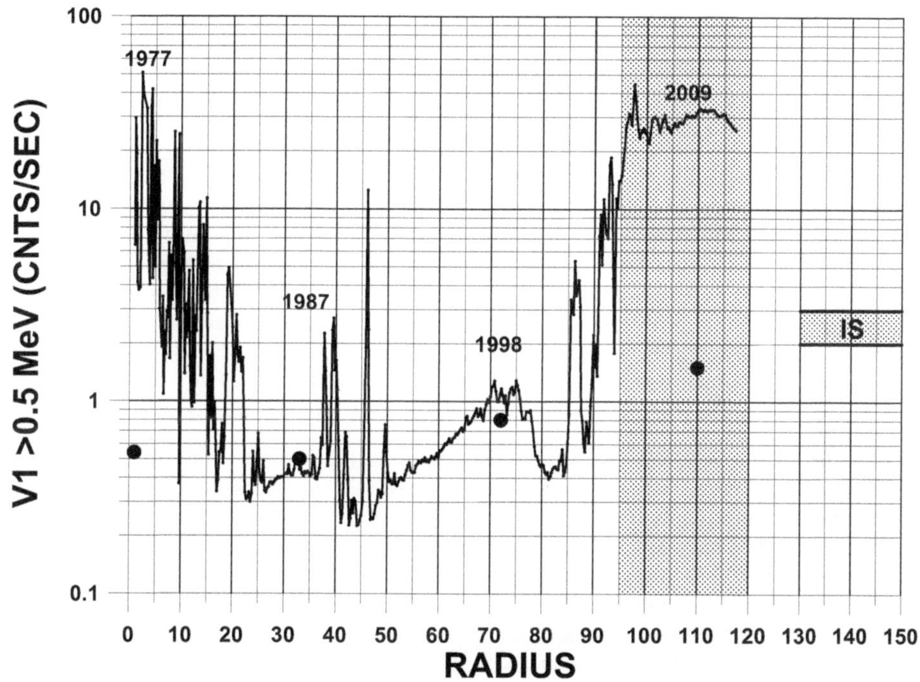

Fig. 12. The integral intensity of >0.5 MeV particles from launch to the present time. These particles again are mainly solar/interplanetary particles in the inner heliosphere and ACR/TSP in the outer heliosphere and in the heliosheath beyond the HTS. They increase to a maximum intensity at about 110 AU. The estimated galactic H nuclei contribution at times of minimum modulation is shown as big dots. The estimated IS (galactic) intensity of all H nuclei is shown as a shaded region beyond 130 AU.

4. Conclusions

The 34 year mission of the Voyagers to the outer heliosphere has led to nothing less than a complete reorganization of our understanding of the size and scope of the heliosphere surrounding the Sun and of the very considerable acceleration of particles within. This includes the global effects of the modulation process as the solar magnetic fields and plasma influence the energetic particles that enter this region from interstellar space beyond. The full extent of the reorganization of the modulation process is only just becoming apparent as this Voyager data is made available in articles such as this.

There is evidence that V1 (and maybe V2 as well) are nearing the outermost boundaries of the solar dominated heliosphere itself, thus lifting the veil that has existed and prevented access to the interstellar plasma, energetic particle intensities and the magnetic field. Already the intensities of the most important nuclei such as H, He and C, as well as the enigmatic electrons are approaching some of the lower and more recent estimates of the interstellar intensities at energies as low as ~100 MeV. In the next 3 years both V1 and V2 will travel a further 10 AU beyond their present locations of nearly 120 AU and 100 AU respectively. If in this time there is still no substantial evidence that we have reached full interstellar conditions, certainly the values and limits of key IS parameters such as the cosmic ray energy density, the magnetic field strength and indeed those parameters governing the origin and propagation of the great bulk of cosmic rays of energy ~1 GeV, or less, will be refined by an order of magnitude or more over the situation existing even today.

In the heliosphere itself, the most important result of the Voyager energetic particle studies has to be the observation of extraordinarily high intensities of energetic particles from 1-100 MeV in the outermost heliosphere. This includes both the large modulation effects observed between the inner and outer heliosphere and the massive intensity of MeV particles that is observed as the HTS is approached and the spacecraft enter the heliosheath beyond. The modulation effects, a factor of ~100 times or more at solar activity maxima, effectively shield the inner heliosphere from the source of these particles in the heliosheath while at the same time providing very sensitive tests of solar modulation theory. These tests will be more fully exploited as the V1 and V2 data are more fully disseminated.

The intensities of the 3-8 MeV ACR and TSP particles (H nuclei) in the heliosheath in the lower energy regime are perhaps 10^3 times the normal intensity of galactic particles at this energy. The same relative factor between ACR intensities and galactic intensities applies to ACR-He since the typical ACR H/He ratio (per nuc) is ~10, the same as the ratio for galactic cosmic rays. For ACR-O nuclei the H/O ratio at these low energies (per nuc) is ~300, compared to a ratio ~500 or more for galactic particles, so the ACR-O will make a relatively larger contribution. In terms of energy loss by ionization, which is $\sim Z^2$, if the contribution to the ionization of the IS medium from ACR-H is 1.0, then ACR-He is ~0.5 and ACR-O is ~0.2.

The overall intensity and energy supply for the ACR (and TSP) components must ultimately come from the solar wind plasma and magnetic fields as the solar wind is brought to a stop first near the HTS and then throughout the heliosheath by the interstellar plasma and magnetic fields. The ultimate acceleration process and its location have yet to be identified for the ACR. This source always seems to be just "beyond" the V1 location. The recently identified intensity peak at between 110-115 AU in the heliosheath observed at the location of V1 seems to be a most important step in this identifying process. Of the many models that have been suggested for this acceleration process, the one that seems to best reproduce the Voyager intensity vs. radius profiles, in our opinion, is the model by Ferreira, Potgieter and Scherer, 2007 which incorporates both stochastic and diffusive acceleration (so called Fermi II acceleration) throughout the heliosheath region beyond the HTS. These processes operate with different effectiveness at different locations in the heliosheath depending on the value of the diffusion coefficient.

For the TSP, which are important below ~10 MeV near the HTS, a separate acceleration source near the HTS is apparent just from the intensity-time profiles of these few MeV particles at V1 and particularly at V2 as they cross the termination shock. This acceleration process may be enhanced by the interaction of strong interplanetary shocks with the HTS itself producing periodic intensity-time profiles that are multiples or sub-multiples of the solar 27 day rotation period driven by the heliospheric wavy current sheet.

5. Acknowledgements

The author wishes to acknowledge his colleagues for over a generation now on the Voyager CRS team. First is Ed Stone, truly a captain of the "Starship Voyager" in the Captain Kirk tradition. Not to be outdone is Frank McDonald, our most senior advisor. In fact the original "team of three" is without doubt the oldest scientifically productive team in history for a major experiment with an average age of 81.0 years. And Alan Cummings would have been part of the original team, except that he was a graduate student at the time of the 1st proposal. Now he is ready to retire, but, of course, nobody working on Voyager data would even think of that. And Nand Lal, the guru of the data system, has been around almost as long (have you ever tried the Voyager web site, http://voyager.gsfc.nasa.gov). There is no better one for any space experiment. And Bryant Heikilla, my former student, who can provide any data you need usually within 48-72 hours real time, including the ~2 hour light travel time to Earth. And last but not least, Tina, who made this Chapter possible.

6. References

Burlaga, L.F. and Ness, N.F., (2010), Sectors and Large Scale Magnetic Field Strength Fluctuations in the Heliosheath Year 110 AU Voyager 1, *Ap.J.*, 725, 1306-1316, doi:10.108/0004-637X/725/1/1306

Cummings, A.C., et al., (2011), Voyager Observations of Anomalous Cosmic Rays in the Outer Heliosphere, *32nd ICRC*, Beijing,

Fisk, L.A., Kozlovsky, B. and Ramaty, R., (1974), An Interpretation of the Observed Oxygen and Nitrogen Enhancements in Low-Energy Cosmic Rays, *Ap.J. Lett.*, 190, L3508

Florinski, V., (2011), Transport of Cosmic Rays in the Distant Heliosheath, Advances in Space Research, 48, 308-313

Ferreira, S.E.S., et al., (2001), Modulation of Jovian and Galactic Electrons in the Heliosphere, *J. Geophys. Res.*, 106, 29313

Ferreira, S.E.S., Potgieter, M.S. and Scherer, K., (2001), Transport and Acceleration of Anomalous Cosmic Rays in the Inner Heliosheath, *J. Geophys. Res.*, 112, 11101, doi:10.1029/2007JA012477

Gleeson, L.J., and Axford, W. I., (1968), Solar Modulation of Galactic Cosmic Rays, *Ap. J.*, 154, 1011

Langner, U.W., de Jager, O.C. and Potgieter, M.S. (2001), Local Interstellar Spectra for Cosmic Ray Electrons, *Adv. Space Res.*, 27, 517

Krimigis, S.M., Roelof, E.C., Decker, R.B. and Hill, M.E., (2011), Zero Outward Flow Velocity for Plasma in a Heliosheath Transition Layer, *Nature*, 474, 359, doi:10.1038/nauture10115

McDonald, F.B., Teegarden, B.J., Trainor, J.H. and Webber, W.R., (1974), The Anomalous Abundance of Cosmic-Ray Nitrogen and Oxygen Nuclei at Low Energies, *Ap.J. Letters*, 187, 105

Opher, M., et al., 2009), A Strong, Highly Tilted Interstellar Magnetic Field Near the Solar System, *Nature*, 462, 1036, doi:10.1038/nature08567

Putze, A., Maurin, D. and Donato, F., (2011), P, He, C and Fe in the Cosmic Ray Primary Fluxes in Diffusion Models, *Astron and Ap.*,518, 1

Stone, E.C., et al., (1977), Cosmic Ray Investigations for the Voyager Missions: Energetic Particle Studies in the Outer Heliosphere- and Beyond, *Space Sci. Rev.*, 21, 355, doi:10.1077/BF00211546

Stone, E.C., et al., (2008), An Asymmetric Solar Wind Termination Shock, *Nature*, 454, 71, doi:10.1038/nature07022

Washimi, H., et al., (2011), Realistic and Time-Varying Outer Heliospheric Modeling by Three Dimensional MHD Stimulation, *Mon. Not. R. Astron. Soc.*, 416, 1475-1485 doi:10.111/j.1365-2966.2011.19144.x

Webber, W.R. and Higbie, P.R., (2008), Limits on the Interstellar Cosmic Ray Electron Spectrum Below 1-2 GeV Derived from the Galactic Polar Radio Spectrum, *J. Geophys. Res.*, 113, A11106, doi:10.1029/ 2008JA013386

Webber, W.R. and Higbie, P.R., (2009), Galactic Propagation of Cosmic Ray Nuclei in a Model with an Increasing Diffusion Coefficient at Low Rigidities, *J. Geophys. Res.*, 114, A02103, doi:10.1029/2009JA013689

Webber, W.R. et al., (2011), Sudden Intensity Increases and Radial Gradient Changes of cosmic Ray Electrons and Protons Observed at Voyager 1 Beyond 111 AU in the Heliosheath, submitted to *J. Geophys Res.*, Sept.

Part 3

Cosmology

16

Putting Einstein to Test – Astrometric Experiments from Space, Fundamental Physics and Local Cosmology

Alberto Vecchiato
INAF (National Institute for Astrophysics), Astronomical Observatory of Torino
Italy

1. Introduction

First experimental evidences of General Relativity (GR) came from Astrometry. Its measurements not only had put to evidence the existence of the effect of an excess of perihelion precession in the motion of the Solar System objects which couldn't be explained in terms of the Newtonian Mechanics (Le Verrier, 1859) but also provided a striking confirmation of Einstein's theory through the experiment conducted at Sobral and Principe by Dyson, Eddington & Davidson (1920) during the Solar Eclipse of 1919, which measured the light deflection foreseen by GR with an accuracy of $\sim 10\%$.

GR, however, has never been the only theory claiming to provide a detailed description of the Gravitational interaction beyond the Newtonian limit. Although GR is still the most favoured alternative, either because of the experimental results or of its simplicity, several other theories have been proposed. Examples of such theories are those where the field equations contain not only the metric tensor of GR, but also a scalar field coupled with the metric itself (Will, 2006), or the so-called fourth-order theories of gravity, where the scalar of curvature R in the field equations is replaced by a more complex function of this quantity, $f(R)$.

Tests of gravity theories within the Solar System are usually analysed in the framework of the so-called *Parametrized Post-Newtonian framework* which enables the comparison of several theories through the estimation of the value of a limited number of parameters. Among these parameters, γ and β are of particular importance for astrometry since they are connected with the classical astrometric phenomena of the light deflection and of the excess of perihelion precession in the orbits of massive objects. Besides its immediate implication for the fundamental physics problem of characterizing the best gravity theory, a precise estimation of these parameters has important consequences on the interpetation of observational evidences at different scales in space and time up to cosmological scales. This is the reason why these Solar System astrometric experiments can also be intended as a kind of "local cosmology" tests which will be addressed from space by planned missions like Gaia (Perryman et al., 2001) and, in the future, by other projects presently under study.

In the next section we will birefly recall the basic concepts standing at the foundations of the current gravity theories, and we will try to give a general overview of the "what" and the "why", that is which are the alternatives available on the "theoretical market" and their motivations. In Section 3 we will show the "how", i.e. the theoretical methods which have been developed to make the comparison among the various gravity theories easier, while in the following one we will concentrate on the problem of how it is possible to model a generic astrometric observable in a relativistic consistent way. Section 5 will adress the experimental side of this problem giving examples of data reduction techniques and of some experiments planned or under study, and finally the next and last section will come to the conclusions.

2. Theoretical background

The basic tenets of the so-called *calssical field theories* could be summarized, very shortly and roughly, as:

1. there exists something (the field source) producing a field;
2. this field tells to the particles how to move.

Mathematically, these two simple principles correspond to the *field equations* and to the *equations of motion* respectively.

The first and simplest theory of this kind is the newtonian theory of gravity, in which the Poisson equation

$$\nabla^2 U = 4\pi G\rho \tag{1}$$

is the field equation where the gravitational field U originates from a mass density ρ, while the Newtonian law of dynamics

$$\frac{\mathrm{d}^2 x^k}{\mathrm{d}t^2} = -\frac{\partial U}{\partial x^k} \tag{2}$$

are the equations of motion.

However, many other classical field theories aiminig at explaining how gravity works have been formulated. In GR the field equations write

$$R_{\alpha\beta} - \frac{1}{2} g_{\alpha\beta} R = \frac{8\pi G}{c^4} T_{\alpha\beta} \tag{3}$$

where the metric $g_{\alpha\beta}$ plays the role of the potential U and the stress-energy tensor $T_{\alpha\beta}$ is the relativistic counterpart of the matter density ρ.[1] This statement can be better understood if one notices that:

1. under the assumption of weak gravitational fields and slow motion, the relativistic equations of motion approximate to

$$\frac{\mathrm{d}^2 x^k}{\mathrm{d}t^2} = -\frac{c^2}{2} \frac{\partial g_{00}}{\partial x^k}, \tag{4}$$

and the requirement that, within these hypothesis, they have to reproduce the newtonian ones implies that

$$g_{00} \simeq -1 + \frac{2U}{c^2}; \tag{5}$$

[1] The definition of the stress-energy tensor is more general and takes into account that in GR any form of energy can contribute to the curvature of the space-time.

2. the quantities $R_{\alpha\beta}$ and R basically depend on the derivatives of the metric, so that the Poisson equation can be derived from Eq. (3) under the same weak-field and slow-motion assumptions.

Another example of relativistic field theory of gravity is the Brans-Dicke theory. (Brans & Dicke, 1961) This theory is the simplest example of the class of the so-called *scalar-tensor* theories, in which the field equation is characterized by the presence of a scalar field ϕ coupled with gravity in addition to the metric tensor

$$\phi\left(R_{\alpha\beta}-\frac{1}{2}g_{\alpha\beta}R\right)+g_{\alpha\beta}\Box\phi-\phi_{,\alpha;\beta}=\frac{8\pi G}{c^4}T_{\alpha\beta}+\frac{\omega}{\phi}\left(\phi_{,\alpha}\phi_{,\beta}-\frac{1}{2}g_{\alpha\beta}\phi_{,\sigma}\phi^{,\sigma}\right). \tag{6}$$

The quantity ω is a coupling constant[2] which gives a measure of the deviations of this theory from GR. From this equation, in fact, it can be shown that the relation between ϕ and the trace T of the stress-energy tensor is

$$\Box\phi=\frac{8\pi G}{(2\omega+3)\,c^4}T \tag{7}$$

so that Eq. (6) reduces to the GR field equation (3) when $\omega\rightarrow\infty$. The maini motivation at the basis of this theory was to have a gravity theory which could fully incorporate the Mach principle, according to which the inertial phenomena were caused by the relative motion of the mass in the Universe. (Ciufolini & Wheeler, 1995)

The importance of scalar-tensor theories comes also from other more complex models introduced as a way to explain inflationary cosmological scenarios (like e.g. in Damour & Nordtvedt (1993)), and also in particle-physics-inspired models of unification such as string theory (Damour, Piazza & Veneziano, 2002).

We finally want to mention the so-called $f(R)$ theories. This class of theories stem from the idea that the lagrangian $\mathcal{L}$ from which the field equations can be derived can be more general with respect to that of GR. The latter, in fact, assumes that it is simply equal to the Ricci scalar, i.e. $\mathcal{L}=R$, while these theories relax this condition and take a generic dependence $\mathcal{L}=f(R)$. From a mathematical point of view, a consequence of this hypothesis is that the field equations are no more second-order differential equations, but fourth-order. From a physical point of view, the additional terms emerging into the field equation can be interpreted also as sources, rather than fields, and take the same role of $T_{\alpha\beta}$. From a phenomenological point of view, finally, these theories have been invoked to explain several observational phenomena like cosmological acceleration, galactic dynamics, gravitational lensing, etc. without the need to resort to non-barionic Dark Matter or to Dark Energy. A more detailed review of this kind of theories can be found in Capozziello & Faraoni (2011).

3. The theoretical tools of the gravity arena

As we have just pointed out in the previous section, several other theories alternative to GR has been formulated in order to describe how gravity works. This situation naturally calls for a framework in which it is possible to model the observations taking into account many viable theories at the same time. In this section we will briefly explain two of these frameworks: the

[2] In more general versions of scalar-tensor theories ω is not a constant but a function of ϕ.

Parametrized Post-Newtonian (PPN) and the Mansouri-Sexl. More details about these two subjects can be found in Will (2006) and in Mansouri & Sexl (1977a;b;c).

3.1 The PPN framework

As pointed out in Will (2006) a viable theory of gravity must be complete, self-consistent, relativistic, and with a correct Post-Newtonian limit. The Post-Newtonian limit of a gravity theory is the approximation of such theory under the assumptions of slow-motion and weak gravity field, i.e. when $v/c \sim U/c^2 \sim \epsilon \ll 1$.[3]

The PPN formalism is a powerful tool which simplifies the confrontation among the Post-Newtonian limits of the different theories of gravity. This is particularly useful also because these conditions hold in the Solar System, where local experiments are confined.

It can be shown that, in the slow motion and weak-field limit,

$$g_{\alpha\beta} \simeq \eta_{\alpha\beta} + h_{\alpha\beta}, \tag{8}$$

i.e. that the metric can be written as an expansion about the Minkowskian metric of Special Relativity where the perturbations $h_{\alpha\beta}$ are functions of the different smallness parameters $\sim \epsilon$ like U, v, etc.

The noticeable thing is that the expansions predicted by nearly every theory of gravity have the same structure apart from the coefficients multiplying each kind of perturbation.

In the PPN formalism these coefficients are replaced by appropriate parameters in such a way that it is possible to reproduce any specific theory simply by assigning specific values to them. Currently the number of parameters of the PPN formalism is set to 10 (Table 1), suitably chosen to give informations on general properties of the selected theory of gravity. On the other side, the formulation of the experimental measurements in the PPN formalism can be translated to the estimation of the values of these parameters, and thus in a direct selection of the viable theories.

The two most important parameters of the PPN formalism are γ and β, which are both 1 in GR. A precise estimation of these two parameters, has important *theoratical and observational implications.* It could help to *fulfill theoretical needs* because they are the phenomenological "trace" of a scalar field coupled with gravity of scalar-tensor theories which, as explained in the previous section, is related to:

- theories fully compatible with the Mach principle (Brans & Dicke, 1961);
- cosmological scenarios with inflationary stage (Damour & Nordtvedt, 1993);
- theories aiming to provide a formulation of a quantum theory of gravity (Damour, Piazza & Veneziano, 2002).

For example, there are formulations of scalar-tensor theories in which the scalar field evolves with time toward a theory close to GR leaving little relic deviations at present times (Damour & Esposito-Farèse, 1992). Such deviations from GR today range from 10^{-5} to a few times 10^{-7} for $|\gamma - 1|$, depending on the cosmological model (Damour & Nordtvedt, 1993).

[3] Actually it is also required that the internal energy Π and the pressure-over-density ratio p/ρ are $\sim \epsilon$.

Parameter	Meaning	Value in			
		GR	Brans-Dicke	Vector-Tensor	$f(R)$
γ	How much space-curvature is produced by unit rest mass?	1	$\frac{1+\omega}{2+\omega}$	γ'	γ'
β	How much nonlinearity in the superposition law for gravity?	1	1	β'	β'
ξ	Preferred-location effects?	0	0	0	0
α_1	Preferred-frame effects?	0	0	α'_1	0
α_2		0	0	α'_2	0
α_3		0	0	0	0
ζ_1	Violation of conservation of total momentum?	0	0	0	0
ζ_2		0	0	0	0
ζ_3		0	0	0	0
ζ_4		0	0	0	0

Table 1. List of the PPN parameters, and of their meanings and values in some theories of gravity.

Experiment	Effect	Technique	$\|\gamma-1\|$ lower bound
HIPPARCOS	Light deflection	Global Astrometry	$3 \cdot 10^{-3}$
VLBI	Light deflection	Radio Interferometry	$4.5 \cdot 10^{-4}$
Cassini	Shapiro time delay	Round-trip travel time of radar signals	$2.3 \cdot 10^{-5}$

Table 2. Current best estimations for the γ parameter.

On the front of the $f(R)$ theories, instead, it has been argued that the current estimations of the γ and β PPN parameters are not sufficient to provide serious constraints on such theories, which claim of being able to explain *observational evidences* about several astrophysical and cosmological problems without any need for Dark Matter (DM) or Dark Energy (DE) like, e.g. Capozziello, de Filippis & Salzano (2009):

- DE dynamics (acceleration of cosmological expansion);
- DM dynamics (galactic rotation curves, galaxy cluster masses);
- observational data from gravitational lensing;
- Tully-Fisher relation.

Again, it seems that the desired experimental accuracy can be set at level of 10^{-7} to 10^{-8} for $|\gamma - 1|$, and from 10^{-5} for $|\beta - 1|$ (Capozziello, Stabile & Troisi, 2006).

The most stringent experimental limit for γ available so far was set by the Cassini experiment (Bertotti, Iess & Tortora, 2003) which used the relativistic time delay effect on the propagation of radar signals (Table 2). This result will be overcome by future astrometric measurements

Experiment	Effect/Technique	Combination with γ	$\lvert\beta-1\rvert$ lower bound
Radar observations of Mercury (1966-1990)	Perihelion shift excess of Mercury	$2\gamma-\beta$	$3\cdot10^{-3}$
Lunar Laser Ranging	Nordtvedt effect	$4\beta-\gamma$	$\sim 10^{-4}$
Radar observations of inner bodies (1963-2003)	"Grand fits" of Solar System equations of motion	$2\gamma-\beta$	$1\div2\cdot10^{-4}$

Table 3. Current best estimations for the β parameter.

from space (Vecchiato et al., 2003). The estimation of β depends on that of γ, since the former always appear in combination with the latter, and its present bounds are provided by different techniques, involving e.g. Lunar Laser Ranging measurements (Williams, Turyshev & Boggs, 2004) or "Grand fits" of the Solar System equations of motion obtained from the reduction of radar observations (Pitjeva, 2005; 2010) (Table 3). Once again, astrometric techniques could be used to improve on these limits (Vecchiato, Bernardi & Gai, 2010).

3.2 Beyond the PPN framework

Other interesting phenomena which can be used to test the gravity theories at a fundamental level are the higher-order quadrupole contribution to the light deflection, and the possible violations of the Local Lorentz Invariance (LLI). Both can be put to test with convenient astrometric experiments.

The first one is foreseen by GR and other theories of gravity when the light is deflected by perturbing bodies with non-spherically symmetric distributions of the mass. The coefficients of $g_{\alpha\beta}$, in fact, in this case does not depend only on the mass, but also on the higher order multipoles of the gravity field of the perturbing body. The deflection can then be described as a vectorial quantity with two components $\mathbf{n}$ and $\mathbf{m}$

$$\Delta\psi = \Delta\psi_1\mathbf{n} + \Delta\psi_2\mathbf{m}$$

with respect to a reference triad $(\mathbf{n}, \mathbf{m}, \mathbf{z})$ where the $\mathbf{z}$-axis is orthogonal to the celestial sphere.

The two components can be modelled as functions of a parameter ε whose value is 1 in GR like in the following formulae (Crosta & Mignard, 2006)

$$\Delta\psi_1 = \frac{2(1+\gamma)M}{b}\left[1+\varepsilon J_2\frac{R^2}{b^2}\left(1-2(\mathbf{n}\cdot\mathbf{z})^2-(\mathbf{t}\cdot\mathbf{z})^2\right)\right] \tag{9}$$

$$\Delta\psi_2 = \frac{4(1+\gamma)M\varepsilon J_2R^2\gamma}{b^3}(\mathbf{m}\cdot\mathbf{z})(\mathbf{n}\cdot\mathbf{z}) \tag{10}$$

where M is the mass of the perturbing body, J_2 is the quadrupole component of its gravitational field, and b the impact parameter, i.e. the distance of maximum approach of the light path to the perturbing body.

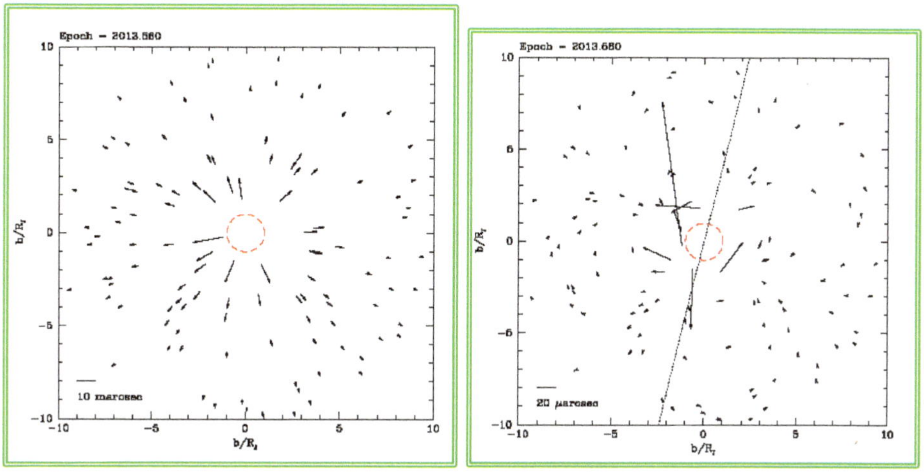

(a) Spherically symmetic case.

(b) Deflection pattern with $J_2 \neq 0$.

Fig. 1. A non-spherically symmetric distribution of the mass of a perturbing body produces specific patterns in the nearby light deflection.

This asymmetric perturbation on the light path induces specific patterns in the nearby light deflection (Fig. 1) which have never been mesuared up to now because of the smallness of this effect.

Tests of possible violations of the LLI are motivated by several theoretical models encompassing a large number of different subjects, from quantum gravity to varying speed of light cosmologies. A complete review of the tests and of the motivations linked to the LLI is out of scope here, and can be found in Mattingly (2005). Here we will limit ourselves to cite the Robertson-Mansouri-Sexl (RMS) formalism for its possible application to astrometry. A violation of the LLI, in fact, will show itself as a breaking of the Lorentz transformations. The RMS formalism is a way to describe this hypothetical breaking in a kinematical way.

In analogy to the PPN formalism, the RMS framework is developed under the assumption that $v \ll c$, and can be expressed as a generalization of the Lorentz transformations

$$
\begin{aligned}
T &= \frac{(t - \boldsymbol{\epsilon} \cdot \mathbf{x})}{a} \\
\mathbf{X} &= \frac{\mathbf{x}}{d} - \left(\frac{1}{d} - \frac{1}{b}\right) \frac{\mathbf{v}(\mathbf{v} \cdot \mathbf{x})}{v^2} + \frac{\mathbf{v}}{a} t
\end{aligned}
\tag{11}
$$

depending on a set of arbitrary parameters (a, b, d, ϵ). The potential impact of this formalism on astrometric measurements comes from the fact that those LLI violations depending on $f = d/b$, show themselves as an aberration effect which therefore puts to test the same properties of the Michelson-Morley experiment. It is probable that violations depending on such aberration effect are out of reach for the present planned space experiments like Gaia (Klioner, 2008) but possible applications to new and more sensitive experiments has not been investigated yet.

4. Modelling the observations

4.1 Basics of the relativistic Theory of measurements

In order to understand why Astrometry can be used to set limits to the PPN parameters and therefore to put to test the different theories of gravity, it has to be shown how these parameters enter in the astrometric observable. The most basic measure in Astrometry, ideally, is the angle ψ_{12} between two observing directions, which in the usual Euclidean geometry reads

$$\cos\psi_{12} = \frac{\mathbf{r}_1 \cdot \mathbf{r}_2}{|\mathbf{r}_1|\,|\mathbf{r}_2|}, \tag{12}$$

and can be represented as in Fig. 2 as the projection on a unit sphere of the two vectors $\mathbf{r}_1$ and $\mathbf{r}_2$ connecting the observer and the objects P and Q.

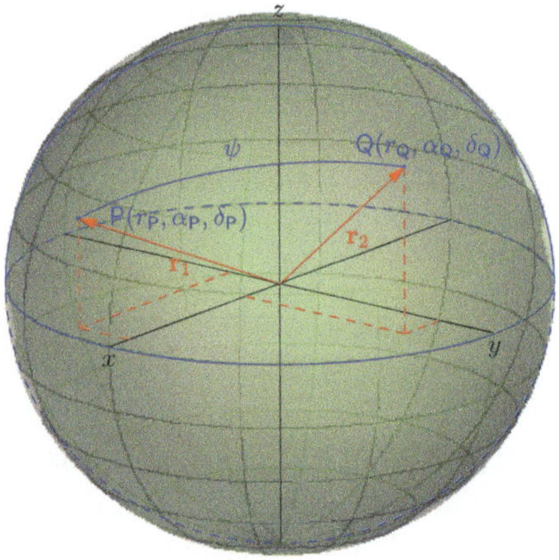

Fig. 2. Representation of the angular distance ψ on the unit sphere between two objects P and Q.

In order to give a correct interpretation of the experimental results, however, the measurements must be written in a proper relativistic way, i.e. following the prescription of the *theory of measurements*. The details of this formalism are out of scope in this article, and are fully developed in de Felice & Bini (2010). We therefore will give here only a brief overview of the main concepts needed to follow the present exposition.

The starting point is the definition of a physical observer, whose history, as shown in Fig. 3, is identified by his/her worldline $\Gamma(w)$ in the four-dimensional differential manifold representing the space-time. The four-vector tangent to this curve (i.e. the *four-velocity* of Γ) is $u^\alpha \equiv \dot{\Gamma}$, where the dot stands for the derivative with respect to the parameter w of the worldline, and if the four-velocity is unitary with respect to the metric $g_{\alpha\beta}$ of the space-time, that is if $g_{\alpha\beta}u^\alpha u^\beta = -1$, then u^α represents the observer at a given point of the worldline. The parameter τ of the curve Γ that satisfies the unitarity condition of u^α has the physical meaning of the *proper time* of the observer, so that u^α can be interpreted as the representation of the observer at each instant of his/her proper time.

With a certain degree of approximation, it can be said that at each point of Γ, u^α splits the space-time into the 1D and 3D subspaces parallel and orthogonal to itself, which are the *time*

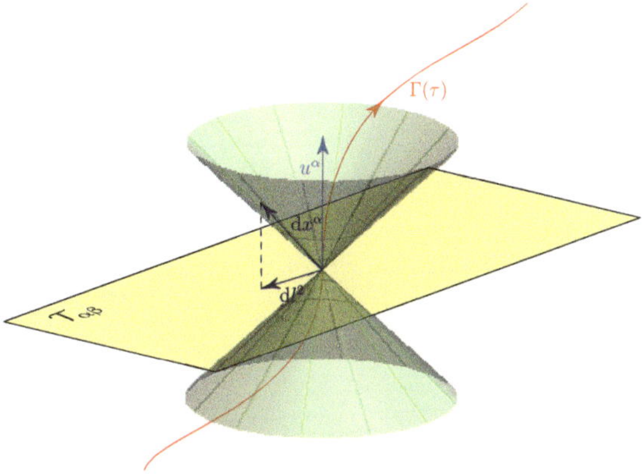

Fig. 3. Any time-like curve $\Gamma(\tau)$ in the space-time can represent the history of an observer, which is identified by its tangent vector u^α and whose proper time τ is the parameter of the worldline. Locally, u^α induces a $3+1$ splitting of the space-time where the two subspaces represent the space and the time associated to this observer.

direction and the *space* relative to the observer u^α respectively. More precisely, we can always define two operators $\mathcal{P}_{\alpha\beta} = -u_\alpha u_\beta$ and $\mathcal{T}_{\alpha\beta} = g_{\alpha\beta} + u_\alpha u_\beta$ which project any four-vector parallely and orthogonally to the observer. Quite obviously, these operators are called the *parallel* and *transverse* projector respectively, and any interval of space-time $\mathrm{d}s^2 = g_{\alpha\beta}\mathrm{d}x^\alpha \mathrm{d}x^\beta$ can be written as

$$\mathrm{d}s^2 = -c^2\left(-\frac{1}{c^2}\mathcal{P}_{\alpha\beta}\mathrm{d}x^\alpha \mathrm{d}x^\beta\right) + \left(\mathcal{T}_{\alpha\beta}\mathrm{d}x^\alpha \mathrm{d}x^\beta\right), \tag{13}$$

where $\mathrm{d}\tau^2 = -c^{-2}\mathcal{P}_{\alpha\beta}\mathrm{d}x^\alpha \mathrm{d}x^\beta$ has the physical meaning of the (square of the) infinitesimal time interval measured by the observer u^α, while similarly $\mathrm{d}l^2 = \mathcal{T}_{\alpha\beta}\mathrm{d}x^\alpha \mathrm{d}x^\beta$ is the measured spatial distance associated to $\mathrm{d}s$.

4.2 Linking the observable with the unknown parameters

With the above indications in mind, and remembering the spatial nature of angular measurements, it is easy to understand why the relativistic counterpart of Eq. (12) can be written as

$$\cos\psi_{12} = \frac{\mathcal{T}_{\alpha\beta}k_1^\alpha k_2^\beta}{\sqrt{\mathcal{T}_{\alpha\beta}k_1^\alpha k_1^\beta}\sqrt{\mathcal{T}_{\alpha\beta}k_2^\alpha k_2^\beta}} \tag{14}$$

where, as shown in Fig. 4 k_1^α and k_2^α are the two observing directions (de Felice & Clarke, 1990). It has to be noticed, however, that the expression "observing direction" has not to be intended as the result of a measure. The observable is the angle at the left-hand-side of the equation, while the four-vectors $k_{1,2}^\alpha$ have the mathematical meaning of the tangents to the path of the incoming light rays connecting the positions of the observed objects to the observer, i.e. of the null geodesics of the observed photons. In other words, if s is the parameter of the null geodesic x^α, then

$$k^\alpha = \frac{\mathrm{d}x^\alpha}{\mathrm{d}s}$$

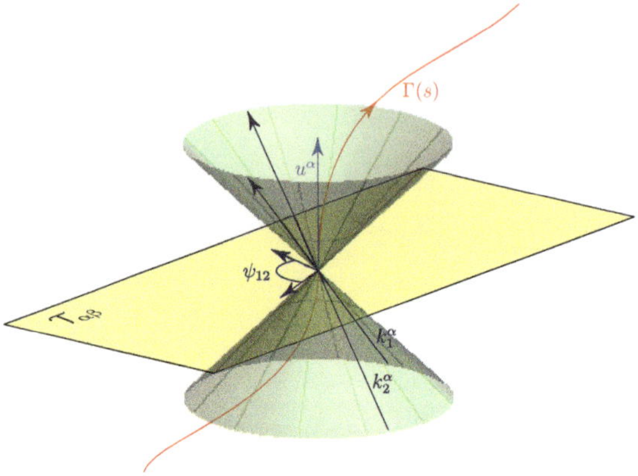

Fig. 4. The formula expressing the arc measured between two directions can be found considering that the relativistic viewing directions k_1^α and k_2^α – i.e. the tangents to the null geodesics connecting the observed objects to the observer and computed at the observation event – have to be projected onto the *spatial* hypersurface relative to u^α.

and this four-vector can be obtained by solving the geodesic equation (i.e. the relativistic equation of motion)

$$\frac{d^2x^\alpha}{ds^2} + \Gamma^\alpha_{\beta\gamma}\frac{dx^\beta}{ds}\frac{dx^\gamma}{ds} = 0. \tag{15}$$

This formula has to be expressed as a function of the desired astrometric unknowns, namely the positions and proper motions of the observed objects. Moreover, as it is known from classical astrometry (see e.g. Smart, 1965), measurements of the same object change in time not only because of its intrinsic motion, but also because of the motion of the observer itself, with the aberration and parallax effects. In addition to this, a relativistic formulation of the astrometric measurements has to properly take into account the geometry of the space-time.

The most important thing to be noticed in Eq. (14) is that it is natural to choose a coordinate system of the space-time metric coincident with that of the desired catalog, and that this helps to express automatically the observer as a function of the needed parameters introducing immediately and naturally the observer-related effects of classical astrometry like the aberration and the parallactic displacement. If we imagine, just as an example, to use the Schwarzschild metric, its coordinates are already centered on the Sun, therefore the aberration is automatically taken into account in the observation equation since the transverse projector

$$\mathcal{T}_{\alpha\beta} = g_{\alpha\beta} + u_\alpha u_\beta$$

contains the (four-)velocity u^α expressed with respect to the origin of the coordinate system.

It's easy to understand that the dependence on the positions of the observed objects enters in the equation through the k_i^α, while the relativistic formulation of the observable, together with the correct solution of Eq. (15), can take into account of all the "relativistic effects" such as the light deflection, i.e. the geometry of the space-time. However, many ways to address the problem of the integration of the null geodesics have been developed so far, therefore the actual expression of Eq. (14) can vary according to the accuracy level of the model and/or to the specific technique used for the integration of the equation of motion. A sub-mas $(v/c)^2$ model is described in Vecchiato et al. (2003) and reference therein, while de Felice et al. (2004) is an example of a μas-level $(v/c)^2$ model which takes into account the gravitational influence of all the bodies of the Solar System. More accurate sub-μas models can be found in Klioner (2003) and references therein, Kopeikin & Schäfer (1999), and de Felice et al. (2006). Finally, Teyssandier & Le Poncin-Lafitte (2008) have applied the method of the Synge's World Function to obtain a relativistic expression of the basic astrometric observables without integrating directly the geodesic equations.

Whatever is the explicit method used for the integration of the null geodesic, the PPN parameters such as γ and β can be easily included in the formula of the observable by chosing an appropriate PPN expression of the metric tensor, however there exists an important difference between them. If we consider, e.g., the PPN approximation of the Schwarzschild metric[4] we can see that its expression in isotropic coordinates at the first Post-Newtonian order (1PN) can be written as

$$ds^2 = -\left(1 - 2U + 2\beta U^2\right) dt^2 + (1 + 2\gamma U)\,\delta_{ij} dx^i dx^j. \tag{16}$$

This shows that β enter the metric at the second order in the gravitational potential U, while γ is at the first order, and since the relativistic deflection of light is in general

$$\Delta\psi = f\left((g_{00})^{-1}, g_{\mu\nu}\right), \tag{17}$$

this explains why astrometric measurements exploiting the light deflection are particularly suited for the estimation of γ but not for that of β. The latter, actually, can also be addressed by astrometric measurements able to detect the well-known effect of the excess of perihelion shift

$$\Delta\omega = \frac{6\pi m}{a\left(1-e^2\right)}\left[\frac{1}{3}\left(2 + 2\gamma - \beta\right) + f\left(\alpha_1, \alpha_2, \alpha_3, \zeta_2, J_2\right)\right] \tag{18}$$

in the orbit of massive objects. If we neglect the contribution from the other PPN parameters and from the quadrupole of the gravitational source, which in the case of the Sun is not important, we can see that this observation is affected by γ and β at the same order, and therefore β can be estimated given a prior knowledge of γ, usually known at a better order because of the above considerations.

4.3 Relativistic reference systems in astrometry

If the required measure is the observed direction of the object, instead of an angle between two directions, then one has to deal with a relativistic theory of reference systems. The Equivalence Principle states that for any observer it is always possible to find a locally inertial reference system where Special Relativity holds. This reference system is called *tetrad*, and it is represented by a set of 4 orthonormal four-vectors $\left(E^{\alpha}_{\hat{0}}, E^{\alpha}_{\hat{1}}, E^{\alpha}_{\hat{2}}, E^{\alpha}_{\hat{3}}\right)$ which can be built at

[4] The Schwarzschild metric considers the spherical non-rotating Sun as the only source of gravity.

any point P of the space-time by imposing the conditions

$$\begin{aligned} E^{\alpha}_{\hat{0}} &\equiv u^{\alpha} \\ \left(g_{\alpha\beta} E^{\alpha}_{\hat{a}} E^{\beta}_{\hat{b}}\right)_{\mathsf{P}} &= \eta_{\hat{a}\hat{b}} \quad \forall\, a, b = 1, 2, 3 \end{aligned} \tag{19}$$

where u^{α} is the four-velocity of the observer. In other words (see Fig. 5) the first axis is by definition the four-velocity of the observer and represents the direction of the proper time, while the other three axes are projected onto the spatial hypersurface of u^{α} and define a reference frame for the 3D space locally seen by the observer at each instant. The above conditions imply that the spatial axes of a tetrad are defined except for three arbitrary spatial rotations.

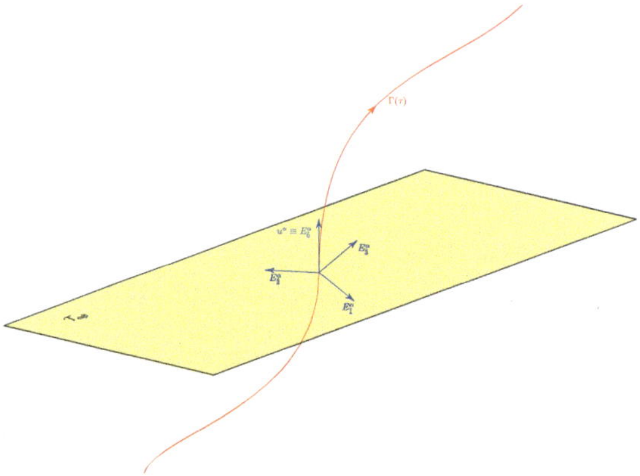

Fig. 5. A tetrad is a set of four orthonormal axes in the 4D space-time which locally represent a Minkowskian reference frame.

The formula which gives the direction cosine seen by an observer u^{α} on the a-th axis of the tetrad is

$$\cos \psi_{\hat{a},k} = \frac{\mathcal{T}_{\alpha\beta} E^{\alpha}_{\hat{a}} k^{\beta}}{\sqrt{\mathcal{T}^{\alpha}_{\alpha\beta} k^{\alpha} k^{\beta}}} \tag{20}$$

which can be easily obtained from Eq. (14) by substituting one observing direction k^{α} with the required tetrad axis and considering that, by construction, tetrad axes are such that $\mathcal{T}_{\alpha\beta} E^{\alpha}_{\hat{a}} E^{\beta}_{\hat{a}} = 1$ for $a = 1, 2, 3$.

5. Methods of data reduction and experimental overview

5.1 Estimation of the γ parameter

The position and motion of the objects in the sky can be determined by astrometric measurements using the two differents approaches of the global and relative astrometry.

5.1.1 The global astrometry approach

Global astrometric techniques have been used in the HIPPARCOS space mission to estimate the γ parameter at the 10^{-3} accuracy level Froeschle, Mignard & Arenou (1997). The same techniques, with a more advanced technology, will be applied by the Gaia space mission (Perryman et al., 2001) to push this measurement to a higher level of accuracy.

The main goal of the Gaia mission is the determination of a 6D map of $\sim 10^9$ objects of the Milky Way. To achieve this objective the observations can't be made in pointed mode. The satellite will rather scan continuously the celestial sphere, measuring the angular positions of the observed objects with respect to a local reference frame associated to the satellite's attitude, and accumulating several billions of observations on the entire sky. Each observation can be modelled as indicated in Section 4 and corresponds to an equation like the (20) which is highly non-linear. In general this equation can depend not only on the astrometric parameters and the PPN γ, but also on other kind of parameters, like those defining the satellite attitude or other instrumental ones needed for the calibration, which cannot always be determined *a priori* with sufficient accuracy[5]

$$\cos\psi \equiv F\left(\underbrace{\alpha_*,\delta_*,\varpi_*,\mu_{\alpha*},\mu_{\delta*}}_{\text{Astrometric parameters}},\underbrace{a_1^{(j)},a_2^{(j)},\ldots}_{\text{Attitude}},\underbrace{c_1,c_2,\ldots}_{\text{Instrument}},\underbrace{\gamma,\ldots}_{\text{Global}}\right). \tag{21}$$

Several billions of observations are accumulated during the five years of the mission lifetime. The final result therefore will be a big and sparse system of equations (up to $\sim 10^{10} \times 10^8$ in the case of Gaia, see Lammers & Lindegren, 2011). Since the solution of such systems of non-linear equations is not practically feasible, the observation equations (21) are linearized about a convenient starting point represented by the current best estimation of the required unknowns. In this case the problem is converted into that of the solution of a big and sparse system of linear equations

$$\begin{aligned} -\sin\psi\,\mathrm{d}\psi = & \underbrace{\left.\frac{\partial F}{\partial\alpha_*}\right|_{\bar\alpha_*}\delta\alpha_* + \left.\frac{\partial F}{\partial\delta_*}\right|_{\bar\delta_*}\delta\delta_* + \left.\frac{\partial F}{\partial\varpi_*}\right|_{\bar\varpi_*}\delta\varpi_* + \cdots}_{\text{Astrometric parameters}} \\ & \underbrace{\sum_{ij}\left.\frac{\partial F}{\partial a_i^{(j)}}\right|_{\bar a_i^{(j)}}\delta a_i^{(j)}}_{\text{Attitude}} + \underbrace{\sum_i\left.\frac{\partial F}{\partial c_i}\right|_{\bar c_i}\delta c_i}_{\text{Instrument}} + \underbrace{\left.\frac{\partial F}{\partial\gamma}\right|_{\bar\gamma}\delta\gamma + \cdots}_{\text{Global parameters}} \end{aligned} \tag{22}$$

whose unknowns are the *corrections* δx to the starting catalog values. Despite this simplification, the system is still too large to be solved with direct methods, and iterative methods are then used. Since the number of observations is much larger than the number of unknowns, the system is overdetermined and can be solved in the least-squares sense (Fig.6). The complexity of the observation equations and the dimensions of the problem are only two of the components contributing to the challenge of such global astrometry experiments. Another one comes from the correlations among some of the unknowns which has to be determined. As shown in Fig. 7, one of them involves the γ parameter, the angle between the two viewing directions (basic angle) and the parallax of the observed objects.

[5] The PPN γ, contrary to the largest part of the unknowns, appears on every equation of the system. Conventionally such kind of unknowns are called *global parameters*.

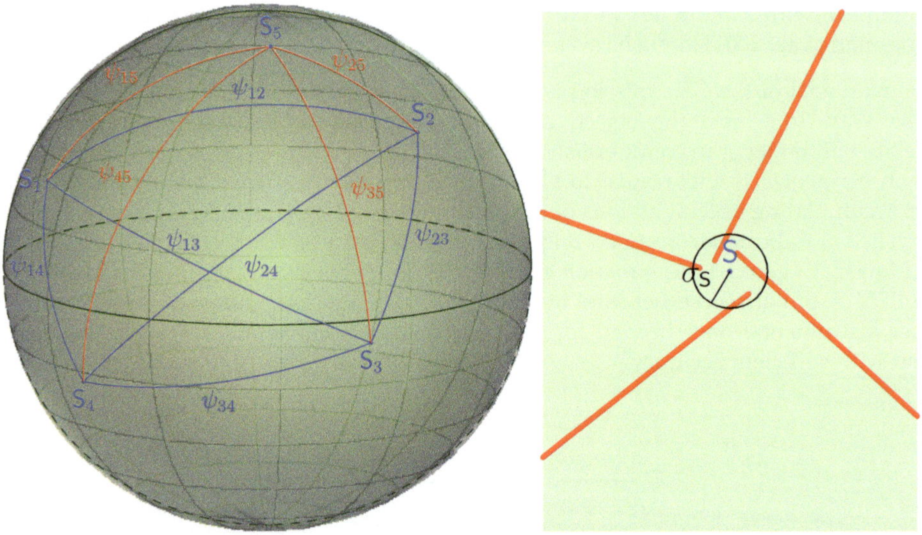

(a) Measured arcs accumulate faster than the number of unknowns in global sphere reconstruction.

(b) Conceptual representation of least-squares fit of measurement errors.

Fig. 6. Concept of sphere reconstruction with global astrometric measurements in the case of arc measurements. The pictures show how repeated observations of the same stars can accumulate a number of measurements greater than the number of unknowns (a) thus allowing to find them by solving an overdetermined system of equations in the least-squares sense (b).

When all of these complications are solved, the γ parameter can thus be obtained as a by-product of the system solution. The foreseen accuracy for the Gaia mission is at the $10^{-6} - 10^{-7}$ level (Vecchiato et al., 2003) i.e. one or two orders of magnitude better than the present estimations.

5.1.2 The relative astrometry approach

Other proposals have been made in order to attempt a high-precision estimation of the γ parameter with techiniques of relative astrometry.

Under many aspects this approach is the opposite of the global one examined in the previous section. Global astrometry is specially fitted to deduce positions and motions of stellar objects with respect to a common global inertial reference frame, and it resorts on the measurements of large angles on the sky. Relative astrometry, instead, is targeted to the high-precision estimation of the *relative* position and motion of celestial objects, i.e. with respect to some distant background objects, using small-field measurements.

The design of the proposed medium-class mission LATOR (Turyshev et al., 2004) is based on a three-spacecraft configuration involving two micro-satellites moving on a trailing-Earth orbit and the International Space Station (ISS). The two satellites are pushed away from the Earth until they reach the opposite side of the Sun. Each satellite is equipped with a laser

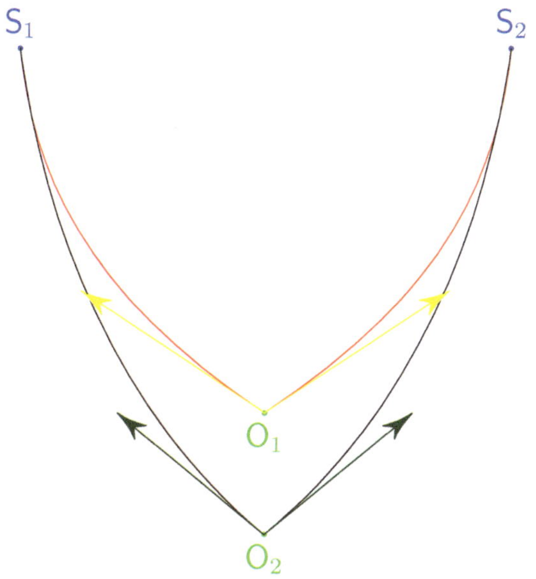

Fig. 7. The γ parameter is highly correlated with parallax, in the sense that any variation in the arc measured by two stars can be interpreted with almost no distinction as a variation of the distance of the stars or of the PPN parameter as well. In any Gaia-like global astrometric sphere reconstruction this parameter is also correlated with the variations of the basic angle, i.e. of the angle between the two viewing directions.

beam which can be detected by a long baseline interferometer housed in the ISS. In this way it is possible to measure the angular distance of these two "artificial stars". These measures adds to those done with an additional laser-ranging system onboard of each spacecraft which is able to measure the distances among them. Gravity should deform the shape of the large triangle formed by the two satellites and the ISS, making LATOR able to reach the 10^{-8} level of accuracy on the estimation of γ.

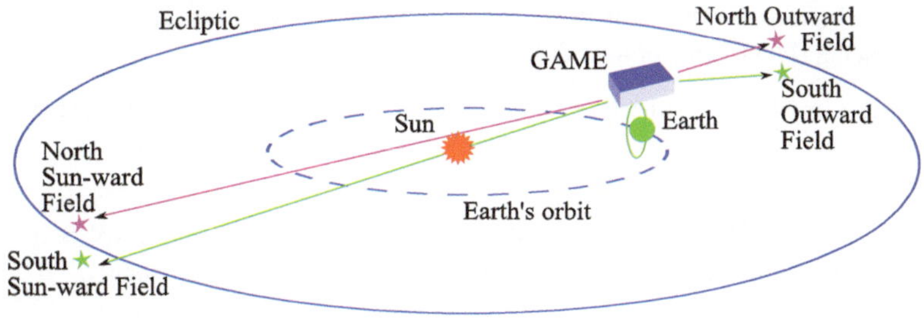

Fig. 8. Schematic depiction of the GAME mission concept.

The estimation of the same PPN parameters is the main scientific goal of GAME, another astrometric mission whose mission concept is sketched in Fig. 8.

The instrument concept is based on a multiple-field, multiple-aperture Fizeau interferometer, observing simultaneously several sky regions close to the solar limb. A beam combiner folds the telescope line of sight on different directions on the sky, separated by a base angle of a few degrees ($\simeq 4^{\circ}$).

This satellite will measure repeatedly the arcs between selected star pairs, placed in the different fields of view of the telescope, when they are close to the Sun. It will observe again the same arcs after some months, when the Sun is at a large angular distance. In the first configuration the arcs are larger because of the gravitaional pull of the Sun, so that their comparison makes it possible to estimate the γ parameter.

The GAME mission and instrument concepts are highly scalable and the final performances vary according to the selected configuration. A small-mission version with two fields of view, capable to reach a final accuracy between 10^{-6} and 10^{-7} in its 2-yr lifetme, was proposed in Gai et al. (2009); Vecchiato et al. (2009). At present both smaller versions eligible to be hosted in high-altitude baloon flights, and larger medium-class implementations of the same concept are under study. The former have the capability of performing in few days quick and cheap tests at the same level of the Cassini experiment (i.e. 10^{-5}), while the latter is able to reach the 10^{-8} level of accuracy thanks to a longer duration (five years) and a more complex instrument configuration with up to 4 fields of view and simultaneous front and rear observations which enable a better control of the systematic errors.

5.2 Estimation of the β parameter

As explained in Section 4.2, the astrometric attempts for the β parameter estimation, are not based on the light deflection, but rather on high precision reconstruction of the orbits of massive bodies in the Solar System. Since the perihelion shift effect increases with the eccentricity of the orbit and decreases with the distance from the Sun, the most convenient targets are Mercury and the Near-Earth Objects.

Gaia will not be able to estimate the value of β from Mercury since this planet cannot be observed by the satellite because it is too close to the Sun. It will instead revert to the observation of NEOs, which are tipically in less favourable conditions, but can somewhat compensate because of their large ($\sim 10^{5}$) number. Numerical simulations, however, have shown that Gaia will not be able to improve on the accuracy of this parameter, and will rather reach the current best estimations at the 10^{-4} level using the satellite observations for some 1300 of these objects (Hestroffer et al., 2009).

On the other side, simple order-of-magnitude calculations suggest that with GAME, especially in its medium-class mission implementation, the β parameter can be estimated at an unprecedented level of accuracy.

The shift excess for Mercury is $\Delta\omega \simeq 0.104''/\text{orbit}$. If we consider the simplified scenario in which Mercury is observed only close to superior conjunction[6] the "observed" displacement from the predicted newtonian position is

$$\Delta\alpha \simeq 0.27\Delta\omega \simeq 28\,\text{mas}. \tag{23}$$

[6] This is the most favourable condition because the displacement due to $\Delta\omega$ is largest from the point of view of the observer, except for the inferior conjuction where, however, the planet has the faintest magnitude. See Fig. 9

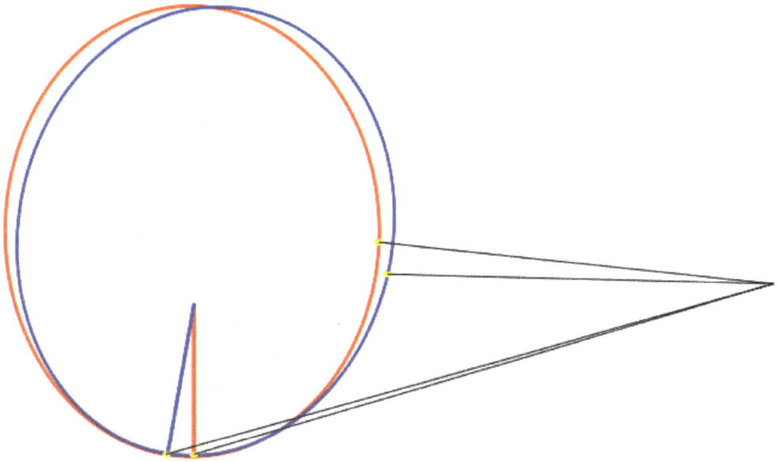

Fig. 9. In principle observations at quadrature are not good for estimating the β parameter.

The integrated magnitude of Mercury in this geometry is $V_{\rm merc} \gtrsim -1.5$, and thus its position can be estimated with a formal relative accuracy of $\sigma_{\Delta\alpha}/\Delta\alpha \simeq 0.5$ mas with a single exposure of 0.1 s by the medium-class GAME. Therefore, from Eqs. (18) and (23) it easy to deduce that

$$\frac{\sigma_\beta}{\beta} \simeq 3\frac{\sigma_\Gamma}{\Gamma} = 3\frac{\sigma_{\Delta\alpha}}{\Delta\alpha} \sim 5 \cdot 10^{-2}$$

with this single observation. If, for the sake of simplicity, we consider the case in which Mercury is observed for one day each orbit in these conditions, after 5 years it is possible to collect some $5 \cdot 10^8$ observations, which would bring the final accuracy to $\approx 2.5 \cdot 10^{-6}$ on β.

5.3 Estimation of the quadrupole effect

The quadrupole light deflection effect (q-effect) has its preferred targets on the stars which can be observed close to the largest and most oblate planets of the Solar System, like Jupiter or Saturn, where the effect is larger. The maximum amount of deflection which can be ascribed to the J_2 component of the gravitational field is in fact 240 μas and 95 μas for these two planets respectively, and this quantity has to be compared with the 10 μas for the next largest contribution coming from Neptune, or with the ~ 1 μas at the solar limb.

Similarly to the case of γ, the ε parameter of the quadrupole deflection can be estimated with a *global* approach, as the result of a sphere reduction in which this effect has been appropriately modeled, or with a *differential* approach, in which the configurations of suitable stellar fields are compared when Jupiter/Saturn is in between, and some time later, when the planet has moved away.

At the moment only the performances of Gaia in the differential case and for Jupiter has been investigated in some detail. The results depend obviously on the initial condition of the scanning law of the satellite, because they determine if a favourable configuration of bright stars around Jupiter can be detected. With the present scanning law, the most positive situation will happen in mid-2017, when there will be the possibility to determine the ε parameter at the 10σ-level.

It has to be added, however, that GAME, as a pointed mission is probably in a more convenient condition for this kind of measurements.

6. Conclusions

In this paper we have given a brief overview of the motivations at the basis of the quest for the determination of the most reliable gravity theory, and of the possible tests inspired by the theoretical scenario. It has also been shown how astrometry, probably the most ancient experimental discipline, can play an important role in these tests. What have eventually emerged, is that it is likely that in the next one or two decades some crucial experiment will be successfully conducted, the outcome of which will be able to revolutionize our view of the physics and of cosmology at a fundamental level.

7. Acknowledgements

This work has been partially funded by ASI under contract to INAF I/058/10/0 (Gaia Mission - The Italian Participation to DPAC).

8. References

Bertotti, B., Iess, L. & Tortora, P. (2003). A test of general relativity using radio links with the Cassini spacecraft, *Nature* 425: 374–376.

Brans, C. & Dicke, R. H. (1961). Mach's Principle and a Relativistic Theory of Gravitation, *Physical Review* 124: 925–935.

Capozziello, S., de Filippis, E. & Salzano, V. (2009). Modelling clusters of galaxies by f(R) gravity, *Monthly Notices of the Royal Astronomical Society* 394: 947–959.

Capozziello, S. & Faraoni, V. (2011). *Beyond Einstein Gravity*, Springer.

Capozziello, S., Stabile, A. & Troisi, A. (2006). Fourth-Order Gravity and Experimental Constraints on Eddington Parameters, *Modern Physics Letters A* 21: 2291–2301.

Ciufolini, I. & Wheeler, J. A. (1995). *Gravitation and Inertia*, Princeton University Press.

Crosta, M. T. & Mignard, F. (2006). Microarcsecond light bending by Jupiter, *Class. Quantum Grav.* 23: 4853–4871.

Damour, T. & Esposito-Farèse, G. (1992). Tensor-multi-scalar theories of gravitation, *Class. Quantum Grav.* 9: 2093–2176.

Damour, T. & Nordtvedt, K. (1993). General relativity as a cosmological attractor of tensor-scalar theories, *Phys. Rev. Lett.* 70: 2217–2219.

Damour, T., Piazza, F. & Veneziano, G. (2002). Violations of the equivalence principle in a dilaton-runaway scenario, *Phys. Rev. D* 66.

de Felice, F. & Bini, D. (2010). *Classical Measurements in Curved Space-Times*, Cambridge Univeristy Press.

de Felice, F. & Clarke, C. J. S. (1990). *Relativity on curved manifolds*, Cambridge University Press.

de Felice, F., Crosta, M. T., Vecchiato, A., Lattanzi, M. G. & Bucciarelli, B. (2004). A General Relativistic Model of Light Propagation in the Gravitational Field of the Solar System: The Static Case, *Astrophys. J.* 607: 580–595.

de Felice, F., Vecchiato, A., Crosta, M. T., Bucciarelli, B. & Lattanzi, M. G. (2006). A General Relativistic Model of Light Propagation in the Gravitational Field of the Solar System: The Dynamical Case, *Astrophys. J.* 653: 1552–1565.

Dyson, F. W., Eddington, A. S. & Davidson, C. (1920). A Determination of the Deflection of Light by the Sun's Gravitational Field, from Observations Made at the Total Eclips of May 29, 1919, *Phil. Trans. R. Soc. A* 220: 291–333.

Froeschle, M., Mignard, F. & Arenou, F. (1997). Determination of the PPN Parameter gamma with the HIPPARCOS Data, *in* R. M. Bonnet, E. Høg, P. L. Bernacca, L. Emiliani, A. Blaauw, C. Turon, J. Kovalevsky, L. Lindegren, H. Hassan, M. Bouffard, B. Strim, D. Heger, M. A. C. Perryman, & L. Woltjer (ed.), *Hipparcos - Venice '97*, Vol. 402 of *ESA Special Publication*, pp. 49–52.

Gai, M., Vecchiato, A., Lattanzi, M. G., Ligori, S. & Loreggia, D. (2009). Gamma Astrometric Measurement Experiment (GAME) - Implementation and performance, *Advances in Space Research* 44: 588–596.

Hestroffer, D., Mouret, S., Mignard, F., Tanga, P. & Berthier, J. (2009). Gaia and the Asteroids: Local Test of GR, *American Astronomical Society, IAU Symposium #261. Relativity in Fundamental Astronomy: Dynamics, Reference Frames, and Data Analysis 27 April - 1 May 2009 Virginia Beach, VA, USA, #9.09; Bulletin of the American Astronomical Society, Vol. 41, p.884* 261: 909–+.

Klioner, S. A. (2003). A Practical Relativistic Model for Microarcsecond Astrometry in Space, *Astronomical Journal* 125: 1580–1597.

Klioner, S. A. (2008). Testing Relativity with Space Astrometry Missions, *in* H. Dittus, C. Lammerzahl, & S. G. Turyshev (ed.), *Lasers, Clocks and Drag-Free Control: Exploration of Relativistic Gravity in Space*, Vol. 349 of *Astrophysics and Space Science Library*, pp. 399–+.

Kopeikin, S. M. & Schäfer, G. (1999). Lorentz covariant theory of light propagation in gravitational fields of arbitrary-moving bodies, *Physical Review D* 60(12): 124002–+.

Lammers, U. & Lindegren, L. (2011). News on Seeking Gaia's Astrometric Core Solution with AGIS, *EAS Publications Series*, Vol. 45 of *EAS Publications Series*, pp. 123–126.

Le Verrier, U. J. (1859). Theorie DU mouvement de Mercure, *Annales de l'Observatoire de Paris* 5: 1–+.

Mansouri, R. & Sexl, R. U. (1977a). A test theory of special relativity. I - Simultaneity and clock synchronization. II - First order tests, *General Relativity and Gravitation* 8: 497–513.

Mansouri, R. & Sexl, R. U. (1977b). A test theory of special relativity: II. First order tests., *General Relativity and Gravitation* 8: 515–524.

Mansouri, R. & Sexl, R. U. (1977c). A test theory of special relativity: III. Second-order tests., *General Relativity and Gravitation* 8: 809–814.

Mattingly, D. (2005). Modern tests of lorentz invariance, *Living Reviews in Relativity* 8(5). URL: *http://www.livingreviews.org/lrr-2005-5*

Perryman, M. A. C., de Boer, K. S., Gilmore, G., Høg, E., Lattanzi, M. G., Lindegren, L., Luri, X., Mignard, F., Pace, O. & de Zeeuw, P. T. (2001). GAIA: Composition, formation and evolution of the Galaxy, *Astron. Astrophys.* 369: 339–363.

Pitjeva, E. V. (2005). High-Precision Ephemerides of Planets – EPM and Determination of Some Astronomical Constants, *Solar System Research* 39: 176–186.

Pitjeva, E. V. (2010). EPM ephemerides and relativity, *in* S. A. Klioner, P. K. Seidelmann, & M. H. Soffel (ed.), *IAU Symposium*, Vol. 261 of *IAU Symposium*, pp. 170–178.

Smart, W. M. (1965). *Text-book on spherical astronomy*, Cambridge: University Press.

Teyssandier, P. & Le Poncin-Lafitte, C. (2008). General post-Minkowskian expansion of time transfer functions, *Classical and Quantum Gravity* 25(14): 145020–+.

Turyshev, S. G., Shao, M. & Nordtvedt, Jr., K. (2004). New concept for testing General Relativity: the Laser Astrometric Test Of Relativity (LATOR) mission, *Astronomische Nachrichten* 325: 267–277.

Vecchiato, A., Bernardi, G. & Gai, M. (2010). Measurement of the PPN-beta parameter in the GAME mission., *38th COSPAR Scientific Assembly*, Vol. 38, pp. 3840–+.

Vecchiato, A., Gai, M., Lattanzi, M. G., Crosta, M. & Sozzetti, A. (2009). Gamma astrometric measurement experiment (GAME) - Science case, *Advances in Space Research* 44: 579–587.

Vecchiato, A., Lattanzi, M. G., Bucciarelli, B., Crosta, M., de Felice, F. & Gai, M. (2003). Testing general relativity by micro-arcsecond global astrometry, *Astron. Astrophys.* 399: 337–342.

Will, C. M. (2006). The confrontation between general relativity and experiment, *Living Reviews in Relativity* 9(3).
URL: *http://www.livingreviews.org/lrr-2006-3*

Williams, J. G., Turyshev, S. G. & Boggs, D. H. (2004). Progress in Lunar Laser Ranging Tests of Relativistic Gravity, *Physical Review Letters* 93(26): 261101–+.

17

Graviton Emission in the Bulk and Nucleosynthesis in a Model with Extra Dimension

Mikhail Z. Iofa
Moscow State University, Skobeltsyn Institute of Nuclear Physics, Moscow, Russia

1. Introduction

Brane-world scenarios with a 3-brane identified with the observable Universe which is embedded in a higher-dimensional space-time provide an alternative to the standard 4D cosmology (Binetruy, Deffayet & Langlois 2000; Binetruy et al., 2000; Collins & Holdom, 2000; Kraus, 1999; Shiromizu; Maeda & Sasaki, 2000), reviews (Durrer, 2005; Maartens, 2004; Rubakov, 2001). A necessary requirement on these models is that they should reproduce the main observational cosmological data, the age of the Universe, abundances of elements produced in primordial nucleosynthesis, etc.

A general property of the models with extra dimensions is that gravity propagates in the extra dimensions independent of whether the ordinary matter is confined to the brane or not. This entails a peculiar property of the models with extra dimensions which is absent in the standard cosmology: gravitons which are produced in reactions of particles of matter on the brane can escape from the brane and propagate in the bulk (Hebecker & March-Russel, 2001; Langlois; Sorbo & Rodriguez-Martinez, 2002; Langlois & Sorbo, 2003; Maartens, 2004; Tanaka & Himemoto, 2003). As a consequence the Einstein equations contain terms accounting for the graviton emission. Cosmological evolution of matter on the brane is also affected by this process.

Roughly the energy loss due to the process $a + b \to G + X$ can be estimated (Gorbunov & Rubakov, 2008) as

$$\frac{d\hat{\rho}}{dt} = - < n_a n_b \sigma_{a+b\to G+X} v E_G >,$$

where in the radiation-dominated period of the evolution of the Universe $n_a, n_b \sim T^3$ and $E_G \sim T$. This yields

$$\frac{d\hat{\rho}}{dt} \sim -\kappa^2 T^8.$$

Here T is temperature of the Universe and $\kappa^2 = 8\pi/M^3$, is the 5D gravitational constant, where M is the 5D Planck mass,

Below we consider the problem of graviton emission to the bulk in a model with one 3-brane embedded in the bulk with one infinite extra dimension (Randall-Sundrum type II model

(Randall & Sundrum, 1999) with matter). In a class of metrics defined below, the 5D model can be treated in two alternative approaches. In the first approach the brane is moving in the static 5D space-time, constructed by attaching two AdS spaces to the brane (Birmingham, 1999; Chamblin & Reall, 1999; Collins & Holdom, 2000; Kraus, 1999). In the second approach the brane is located at a fixed position in the extra dimension, and the metric is time-dependent (Binetruy, Deffayet & Langlois 2000; Binetruy et al., 2000). We review a method to connect both approaches.

The Einstein equations including the terms due to graviton production are solved in the perturbative approach. In the leading order we neglect the graviton production, and include it in the next order. We perform our calculations in a picture in which the metric is time-dependent and the brane is located at a fixed position in the extra dimension. In the leading order solution of the system of 5D Einstein equations is the warped extension of the metric on the brane to the bulk. Restriction of the 5D metric to the brane has the form of the FRW metric $ds^2 = -dt^2 + a^2(t)\eta_{\mu\nu}dx^\mu dx^\nu$ where the scale factor $a(t)$ is determined from the generalized Friedmann equation (Binetruy, Deffayet & Langlois 2000; Binetruy et al., 2000; Collins & Holdom, 2000; Kraus, 1999; Maartens, 2004; Mukohyama, Shiromizu & Maeda). The generalized Friedmann equation is obtained by solving the system of the 5D Einstein equations. In the 5D models the generalized Friedmann equation contains the terms linear and quadratic in energy density on the brane. In the leading order the Friedmann equation does not contain terms due to graviton production which are included in the next order.

Evolution of the energy density on the brane is determined by the Boltzmann equation. The collision term in the Boltzmann equation accounts for the graviton emission from the brane resulting from annihilation of the Standard model particles to gravitons. To calculate the collision term, solving the field equations for fluctuations over the background metric, we find the spectrum of the tower of Kaluza-Klein gravitons.

Explicit form of the bulk energy-momentum tensor is obtained by identifying the collision term in the Boltzmann equation which accounts for the loss of energy density on the brane with the component of the energy-momentum tensor in the 5D conservation equation which also represents the energy flow from the brane to the bulk (Langlois & Sorbo, 2003).

Graviton emission changes cosmological evolution of matter on the brane. Time (temperature) dependence of of the Hubble function determined from the Friedmann equation which includs the components of the bulk graviton energy-momentum tensor is different from that in the standard cosmological model. This, in turn, results in a change of abundances of light elements produced in primordial nucleosynthesis (Steigman; Walker & Zentner, 2001). Perturbatively solving the system of the Friedmann and 5D conservation equations, we find the difference of abundances of ^4He produced in primordial nucleosynthesis calculated in the models with and without graviton production. Calculations are performed in the period of late cosmology, in which in the Friedmann equation the term linear in matter energy density is dominant. We find that the difference of abundances of ^4He calculated in both models is a small number, much smaller than that estimated in (Hebecker & March-Russel, 2001; Langlois; Sorbo & Rodriguez-Martinez, 2002; Langlois & Sorbo, 2003). We make an estimate of production of ^4He in the period of early cosmology, in which in the Friedmann equation the quadratic term in matter energy density is dominant. In this period it is important to account for multiple bounces of gravitons back to the brane (Hebecker & March-Russel, 2001; Langlois & Sorbo, 2003). Under the assumption that the component of the graviton energy-momentum tensor representing the bouncing gravitons cancels the large terms contained in the energy-momentum tensor of the emitted gravitons, we find that production of ^4He is consistent with the result of the standard cosmological model.

The plan of the paper is as follows. After briefly reviewing in Sect. 2 two approaches to the 5D model, in Sects. 3 and 4 we establish explicit connection between the two pictures. In Sect. 5, starting from the system of the Einstein equations containing the terms due to graviton emission to the bulk, we obtain the generalized Friedmann equation. In Sect. 8, using the results of Sect. 7, we calculate the collision integral in the Boltzmann equation and find the components of the energy-momentum tensor of the gravitons emitted to the bulk. In Sects. 9 and 10 we calculate the effect of the graviton emission on abundance of ^4He produced in primordial nucleosynthesis. In Sect. 11 we find condition for bouncing of the emitted gravitons back to the brane.

2. Two pictures of 3-brane in the 5D space-time

We consider the 5D model with one 3D brane embedded in the bulk. Matter is confined to the brane, gravity extends to the bulk. In the leading approximation we neglect gravitation emission from the brane to the bulk. The action is taken in the form

$$S_5 = \frac{1}{2\kappa^2}\left[\int_{\Sigma} d^5x\sqrt{-g^{(5)}}(R^{(5)} - 2\Lambda) + 2\int_{\partial\Sigma} K\right] - \int_{\partial\Sigma} d^4x\sqrt{-g^{(4)}}\,\hat{\sigma} - \int_{\partial\Sigma} d^4x\sqrt{-g^{(4)}}L_m, \quad (1)$$

where $x_4 \equiv y$ is coordinate of the infinite extra dimension, $\kappa^2 = 8\pi/M^3$.

The 5D model can be treated in two alternative approaches. Each approach proves to be useful for certain problems discussed below. In the first approach metric is non-static, and the brane is located at a fixed position in the extra dimension (Binetruy, Deffayet & Langlois 2000; Binetruy et al., 2000). We consider the class of metrics of the form [1]

$$ds_5^2 = g_{ij}^{(5)}dx^i dx^j = -n^2(y,t)dt^2 + a^2(y,t)\eta_{ab}dx^a dx^b + dy^2 \equiv dy^2 + g_{\mu\nu}dx^\mu dx^\nu. \quad (2)$$

The brane is spatially flat and located at $y = 0$. Making use of the freedom in parametrization of t, we can make $n(0,t) = 0$. Reduction of the metric (2) to the brane is

$$ds^2 = dt^2 + a^2(0,t)\eta_{ab}dx^a dx^b. \quad (3)$$

The energy-momentum tensor of matter on the brane is taken in the form

$$\hat{T}_\mu^\nu = diag\,\delta(y)\{-\hat{\rho},\hat{p},\hat{p},\hat{p}\}. \quad (4)$$

For the following it is convenient to introduce the normalized expressions for energy density, pressure and cosmological constant on the brane which all have the same dimensionality $[GeV]$

$$\mu = \sqrt{-\frac{\Lambda}{6}}, \quad \sigma = \frac{\kappa^2\hat{\sigma}}{6}, \quad \rho = \frac{\kappa^2\hat{\rho}}{6}, \quad p = \frac{\kappa^2\hat{p}}{6}. \quad (5)$$

[1] The indices i,j run over 0,...,4, the Greek indices are 0,...,3, and $a,b = 1,2,3$

In the leading approximation the system of 5D Einstein equations and junction conditions admits a solution (Binetruy et al., 2000)

$$a^2(y,t) = \frac{a^2(0,t)}{4}\left[e^{2\mu|y|}\left(\left(\frac{\rho+\sigma}{\mu}-1\right)^2+\frac{\rho_w}{\mu}\right)+e^{-2\mu|y|}\left(\left(\frac{\rho+\sigma}{\mu}+1\right)^2+\frac{\rho_w}{\mu}\right)\right. \tag{6}$$

$$\left.-2\left(\left(\frac{\rho+\sigma}{\mu}\right)^2-1+\frac{\rho_w}{\mu}\right)\right],$$

$$n(y,t) = \frac{\dot{a}(y,t)}{\dot{a}(0,t)}. \tag{7}$$

The function $a(t) = a(0,t)$ is solution of the generalized Friedmann equation (Binetruy, Deffayet & Langlois 2000; Binetruy et al., 2000)

$$H^2(t) = -\mu^2 + (\rho+\sigma)^2 + \mu\rho_w(t). \tag{8}$$

Here we introduced the Hubble function $H(t)$ and the so-called Weyl radiation term $\rho_w(t)$ (Maartens, 2004; Shiromizu; Maeda & Sasaki, 2000)

$$H(t) = \frac{\dot{a}(0,t)}{a(0,t)}, \qquad \rho_w(t) = \rho_{w0}\left(\frac{a(0,t_0)}{a(0,t)}\right)^4. \tag{9}$$

In the second approach the brane separates two static 5D AdS spaces attached to both sides of the brane (Birmingham, 1999; Chamblin & Reall, 1999; Collins & Holdom, 2000; Kraus, 1999). The bulk actions are as the bulk part of the action (1). The metrics, which are solutions of the Einstein equations, are

$$ds^2 = -f_i(R)dT^2 + \frac{dR^2}{f_i(R)} + \mu_i^2 R^2 dx^a dx_a, \tag{10}$$

where

$$f_i(R) = \mu_i^2 R^2 - \frac{P_i}{R^2}.$$

Below we consider the case $\mu_1 = \mu_2$ and $P_1 = P_2$. Reduction of the 5D metric to the brane is

$$ds^2 = -dt^2 + R_b^2(t)dx^a dx_a. \tag{11}$$

Trajectory of the moving brane is defined through the proper time t on the brane as

$$R = R_b(t),\ T = T_b(t), \text{where} - f(R_b)\dot{T}_b^2 + f^{-1}(R_b)\dot{R}_b^2 = -1.$$

The junction conditions on the brane are (Binetruy, Deffayet & Langlois 2000; Binetruy et al., 2000; Chamblin & Reall, 1999; Collins & Holdom, 2000; Kraus, 1999; Shiromizu; Maeda & Sasaki, 2000)

$$[h_i^k \nabla_k n_j] = \tau_{ij} - \frac{1}{3}\tau h_{ij}. \tag{12}$$

Here $h_{ij} = g_{ij} - n_i n_j$ is the induced metric on the brane, $v^i = \pm(\dot{T}_b, \dot{R}_b, 0)$ and $n_i = \pm(-\dot{R}_b, \dot{T}_b, 0)$ are velocity and normal vector to the brane, $\tau_{ij} = (\rho + p)v_i v_j + p h_{ij}$. [2] [X]

[2] Here and below prime and dot denote differentiation over y and t.

denotes the difference of expressions calculated at the opposite sides of the brane. From the spatial components $(a, b = 1, 2, 3)$ of the junction conditions follows the generalized Friedmann equation

$$\left(\frac{\dot{R}_b}{R_b}\right)^2 = -\mu^2 + (\rho + \sigma)^2 + \frac{P}{R_b^4}. \tag{13}$$

Equation (13) is of the same form as (8). $a^2(0,t)$ can be identified with $\mu^2 R_b^2(t)$, and the term $\rho_w(t) = \rho_{w0}/a^4(t)$ can be identified with the term $P/\mu R_b^4(t)$. Below we set $\sigma = \mu R_b^2(t)$ and $a^2(0,t)$ are interpreted as scale factors of the Universe. [3] .

3. Geodesic equations in the picture with static metric

To prepare necessary relations for establishing connection between the two pictures, we consider the geodesic equations in the picture with the static metric and moving brane. Let y be a parameter along a geodesic. Geodesic equations are

$$\frac{d^2T}{dy^2} + 2\Gamma^T_{TR}\frac{dT}{dy}\frac{dR}{dy} = 0 \tag{14}$$

$$\frac{d^2x^a}{dy^2} + 2\Gamma^a_{bR}\frac{dx^b}{dy}\frac{dR}{dy} = 0 \tag{15}$$

$$\frac{d^2R}{dy^2} + \Gamma^R_{RR}\left(\frac{dR}{dy}\right)^2 + \Gamma^R_{TT}\left(\frac{dT}{dy}\right)^2 + \Gamma^R_{ab}\frac{dx^a}{dy}\frac{dx^b}{dy} = 0, \tag{16}$$

where the Christoffel symbols calculated with the metric (10) are

$$\Gamma^T_{TR} = \frac{f'}{2f}, \quad \Gamma^R_{RR} = -\frac{f'}{2f}, \quad \Gamma^R_{TT} = \frac{1}{2}ff', \quad \Gamma^R_{ab} = -\eta_{ab}f\mu^2 R, \quad \Gamma^a_{Rb} = \frac{\delta^a_b}{R}.$$

Here $(T, R) \equiv (T^{\pm}, R^{\pm})$ are coordinates in the AdS spaces at the opposite sides of the brane. Integrating the geodesic equations, we obtain

$$\frac{dT^{\pm}}{dy} = \frac{E^{\pm}}{f(R)}, \qquad \frac{dx^a}{dy} = \frac{C^a}{\mu^2 R^2}, \qquad \left(\frac{dR^{\pm}}{dy}\right)^2 = f(R)(C^{R\pm})^2 + E^{\pm 2} - \frac{C^{a2}f}{\mu^2 R^2}, \tag{17}$$

where $(E^{\pm}, C^a, C^{R\pm})$ are integration parameters. $(dT/dy, dR/dy, dx^a/dy)$ are the components of the tangent vector to the geodesic which we normalize to unity. Imposing normalization condition

$$\frac{dx^i}{dy}\frac{dx^j}{dy}g_{ij} = 1, \qquad i = T, R, a$$

we obtain that $(C^{R\pm})^2 = 1$.

Let us consider the foliation of the hypersurface $(T, R, x^a = 0)$ by geodesics $(T(y,t), R(y,t))$ that intersect the trajectory of the brane $(T_b(t), R_b(t))$ and at the intersection point are orthogonal to it (cf. (Mukohyama, Shiromizu & Maeda)). The geodesics are subject to the

[3] From the fit of the cosmological data in the leading approximation of the present model it follows that $\sigma^2 = \mu^2(1 + O(H_0^2/\mu^2))$, where H_0 is the present-time Hubble parameter (Iofa, 2009a;b). For $\mu \sim 10^{-12} GeV$ correction is $\sim 10^{-60}$.

initial conditions at $y=0$: $R^{\pm}(0,t)=R_b(t),\ T^{\pm}(0,t)=T_b(t)$. We consider solutions of the geodesic equations even in y: $T^+(y)=T^-(-y),\ R^+(y)=R^-(-y)$

$$\frac{dT^{\pm}(y)}{dy}=\frac{E\varepsilon(y)}{f(R)},\qquad \frac{dR^{\pm}(y)}{dy}=\alpha\varepsilon(y)\left(f(R)+E^2\right)^{1/2}, \tag{18}$$

where $\alpha=\pm1$, and $E^+(t)=E^-(t)=E(t)$.

The normalized velocity of the brane and the normal vector to the brane are

$$v_b^i=(\dot{T}_b,\dot{R}_b),\qquad n_b^{i\pm}=\eta\varepsilon(y)\left(\frac{\dot{R}_b}{f(R_b)},f(R_b)\dot{T}_b\right), \tag{19}$$

where $\eta=\pm1$. From (17), setting $C^a=0$, we obtain the tangent vector to the geodesic

$$u^{i\pm}=\left(\frac{E\varepsilon(y)}{f(R)},\alpha\varepsilon(y)\sqrt{f(R)+E^2}\right), \tag{20}$$

By construction, at the intersection point with the trajectory of the brane the tangent vector to a geodesic is parallel to the normal to the trajectory of the brane,

$$u^i\,|_{y=0}||n_b^i. \tag{21}$$

Substituting $\dot{T}_b=\xi\sqrt{f(R_b)+\dot{R}_b^2}/f(R_b)$, where $\xi=\pm1$, we have

$$E=\eta\dot{R}_b,\qquad \alpha=\xi\eta. \tag{22}$$

Integrating Eq.(18) for R, we express coordinate R in the hyperplane $C^a=0$ through coordinates (y,t)

$$R^2(y,t)=R_b^2(t)\left[\cosh(2\mu y)+\frac{H^2}{2\mu^2}(\cosh(2\mu y)-1)\pm\sqrt{1+\frac{H^2}{\mu^2}-\frac{P}{R_b^4\mu^2}}\sinh(2\mu|y|)\right]. \tag{23}$$

Here $H^2=(\dot{R}_b(t)/R_b(t))^2$. Substituting H^2 from the Friedmann equation (13) with $\sigma=\mu$, we transform (23) to the form (6) with $\sigma=\mu$. Thus, we can identify

$$\mu^2R^2(y,t)=a^2(y,t). \tag{24}$$

4. Integration of geodesic equations and connection between two forms of the metric

In the following we consider the metric (10) with $P=0$, i.e. $f(R)=\mu^2R^2$. In the hyperplane $x^a=const$ we set $C^a=0$. Integrating Eq. (18)

$$\frac{dR^{\pm}(y,t)}{dy}=\alpha\varepsilon(y)\sqrt{\mu^2R^2(y,t)+E^2} \tag{25}$$

we obtain

$$R^{\pm}(y,t)=R_b(t)\cosh\mu y+\alpha\sqrt{\frac{\dot{R}_b^2}{\mu^2}+R_b^2}\,\sinh\mu|y|. \tag{26}$$

Using the Friedmann equation

$$H^2 = \dot{R}_b^2/R_b^2 = \rho^2 + 2\mu\rho \tag{27}$$

omitting ($\pm$) we rewrite (26) as

$$R(y,t) = R_b(t)\left(\cosh\mu y + \alpha\beta\left(1+\frac{\rho}{\mu}\right)\sinh\mu|y|\right), \tag{28}$$

where $\beta \equiv sign(R_b(t))$. To identify $\mu^2 R^2(y,t)$ with

$$a^2(y,t) = \frac{a^2(0,t)}{4}\left[e^{2\mu|y|}\left(\frac{\rho}{\mu}\right)^2 + e^{-2\mu|y|}\left(\frac{\rho}{\mu}+2\right) - 2\frac{\rho}{\mu}\left(\frac{\rho}{\mu}+2\right)\right], \tag{29}$$

(cf. (6)) we set

$$\alpha\beta = -1.$$

and obtain

$$R(y,t) = \frac{R_b(t)}{2}\left[e^{-\mu|y|}\left(\frac{\rho}{\mu}+2\right) - e^{\mu|y|}\frac{\rho}{\mu}\right]. \tag{30}$$

Also we have

$$\frac{dR(y,t)}{dy} = -\varepsilon(y)\frac{\mu R_b}{2}\left[e^{-\mu|y|}\left(\frac{\rho}{\mu}+2\right) + e^{\mu|y|}\frac{\rho}{\mu}\right]. \tag{31}$$

Introducing y_0, such that

$$e^{\mu y_0} = \left(\frac{\rho}{\rho+2\mu}\right)^{1/2} \tag{32}$$

we express $R(y,t)$ and $R'(y,t)$ as

$$R(y,t) = -\frac{HR_b(t)}{\mu}\sinh(\mu|y| + \mu y_0), \qquad R'(y,t) = -\varepsilon(y)HR_b(t)\cosh(\mu|y| + \mu y_0). \tag{33}$$

Substituting in Eq. (18) for $T(y,t)$ expression (33) for $R(y,t)$, and integrating the equation, we have

$$T^{\pm}(y,t) = -\frac{1}{\mu E}\frac{\cosh(\mu|y| + \mu y_0)}{\sinh(\mu|y| + \mu y_0)} + C^{\pm}(t). \tag{34}$$

Taking $C^+(t) = C^-(t) = C(t)$, we obtain that the limits $y = 0$ of $T^{\pm}(y,t)$ from both sides of the brane are the same .

To determine $C(t)$, first, we consider transformation of the metric (10) from coordinates $R(y,t)$, $T(y,t)$ to coordinates y, t. We have

$$ds^2 = dy^2\left(-\mu^2R^2T'^2 + \frac{R'^2}{\mu^2R^2}\right) + 2dtdt\left(-\mu^2R^2\dot{T}T' + \frac{\dot{R}R'}{\mu^2R^2}\right) \tag{35}$$

$$+dt^2\left(-\mu^2R^2\dot{T}^2 + \frac{\dot{R}^2}{\mu^2R^2}\right) + \mu^2R^2dx^{a2}.$$

On solutions of the geodesic equations (18) the coefficient at dy^2 is $\varepsilon^2(y)$. The coefficient at $dydt$ is zero, if

$$\dot{T} = \frac{\dot{R}\,R'}{\mu^4R^4\,T'} = \frac{\dot{R}\,R'}{\mu^2R^2(\varepsilon(y)E)}, \tag{36}$$

where in the second equality we have used (18). Substituting (36) in the coefficient at dt^2 and using (18) and (33), we obtain $\dot{R}^2/\dot{R_b}^2 = n^2(y,t)$ (cf. (6)).

Writing (34) as

$$T(y,t) = -\frac{R'(y,t)/\varepsilon(y)}{\mu^2 R(y,t)E} + C(t) \tag{37}$$

and taking the time derivative, we obtain

$$\dot{T} = \frac{\dot{R}R'}{\mu^2 R^2(\varepsilon(y)E)} - \frac{1}{\mu^2 R}\frac{d}{dt}\left(\frac{R'/\varepsilon(y)}{E}\right) + \dot{C} \tag{38}$$

The first term in the rhs of (38) is the same as (36). Remarkably, substituting explicit expressions (33) for R and R', we find that the second term in the right-hand side of (38) is independent of y

$$\frac{1}{\mu^2 R}\frac{d}{dt}\left(\frac{R'\varepsilon(y)}{E}\right) = \frac{\eta\dot{y}_0(t)}{\dot{R}_b(t)}. \tag{39}$$

Choosing

$$\dot{C} = \eta\frac{\dot{y}_0}{\dot{R}_b},$$

we obtain $\dot{T}$ in the form (36). To conclude, we have transformed the metric (10) to the form (2).

5. Generalized Friedmann equation with the graviton emission terms included

In this section, keeping in mind application of the results of this section to calculation of primordial nucleosynthesis, we derive the generalized Friedmann equation containing the bulk energy-momentum tensor due to the graviton emission from the brane to the bulk. We use formulation based on the metric (2). Dynamics of the model is contained in the system of 5D Einstein equations

$$G_{ij} \equiv R_{ij} - \frac{1}{2}g_{ij}R = \kappa^2 T^{(5)}_{ij} - g_{ij}\Lambda - \delta^{\mu\nu}_{ij}\frac{\sqrt{-g^{(4)}}}{\sqrt{-g^{(5)}}}\delta(y)g_{\mu\nu}\kappa^2\hat{\sigma}. \tag{40}$$

Here $T^{(5)}_{ij}$ is the sum of the energy-momentum tensor of matter confined to the brane $\hat{T}_{ij}$ (11) and the bulk energy-momentum tensor $\check{T}_{ij}$.

The components of the Einstein tensor $G_{ij}(y,t)$ are

$$G_{00} = 3\left[\frac{\dot{a}^2}{a^2} - n^2\left(\frac{a''}{a} + \frac{a'^2}{a^2}\right)\right] \tag{41}$$

$$G_{44} = 3\left[\left(\frac{a'^2}{a^2} + \frac{a'n'}{a\,n}\right) - \frac{1}{n^2}\left(\frac{\dot{a}^2}{a^2} - \frac{\dot{a}\dot{n}}{a\,n} + \frac{\ddot{a}}{a}\right)\right] \tag{42}$$

$$G_{04} = 3\left(\frac{n'}{n}\frac{\dot{a}}{a} - \frac{\dot{a}'}{a}\right). \tag{43}$$

The components $G_{ij}(y,t)$ satisfy the relations (cf. (Binetruy et al., 2000))

$$G_0^0 - G_4^0 \frac{\dot{a}}{a'} = \frac{3}{2a'a^3} F' \quad (44)$$

$$G_{44} - G_{04} \frac{a'}{\dot{a}} = \frac{3}{2\dot{a}a^3} \dot{F}, \quad (45)$$

where

$$F = (a'a)^2 - \frac{(\dot{a}a)^2}{n^2} \quad (46)$$

The functions $a(y,t)$ and $n(y,t)$ satisfy junction conditions on the brane [4]

$$\frac{a'(0,t)}{a(0,t)} = -\sigma - \rho(t), \qquad \frac{n'(0,t)}{n(0,t)} = 2\rho(t) + 3p(t) - \sigma \quad (47)$$

Reparametrization of t allows to set $n(0,t) = 1$, i.e. t is the proper time on the brane.

Eq. (45) can be rewritten as

$$\dot{F} = -\mu^2 (\dot{a^4}) - \frac{\kappa^2}{6} (a^4)' \check{T}_{04} + \frac{\kappa^2}{6} (\dot{a^4}) \check{T}_{44}. \quad (48)$$

On the brane, at $y = 0$, using junction conditions and setting $\sigma \simeq \mu$, we have

$$\dot{F} = \mu^2 (\dot{a^4}) + \frac{2\kappa^2 a^4}{3} (\rho + \mu) \check{T}_{04} + \frac{2\kappa^2 a^3 \dot{a}}{3} \check{T}_{44}. \quad (49)$$

Integrating (48) in the interval $(t,\ t_l)$, where the initial time t_l is defined below, we obtain

$$F(0,t) = \mu^2 a^4(0,t) + \frac{2\kappa^2}{3} \int_{t_l}^{t} dt' \check{T}_{04}(t') (\rho(t') + \mu) a^4(0,t') + \frac{2\kappa^2}{3} \int_{t_l}^{t} dt' \check{T}_{44}(t') \dot{a}(0,t') a^3(0,t') - C, \quad (50)$$

where C is an integration constant. Substituting expression (46) for F and using the junction conditions, we rewrite (50) in a form of the generalized Friedmann equation (cf. (Binetruy, Deffayet & Langlois 2000; Binetruy et al., 2000))

$$H^2(t) = \rho^2(t) + 2\mu\rho(t) + \mu\rho_w(t) - \frac{2\kappa^2}{3a^4(0,t)} \int_{t_l}^{t} dt' \left[\check{T}_{04}(t') (\rho(t') + \mu) + \check{T}_{44}(t') H(t') \right] a^4(0,t'). \quad (51)$$

On the brane, at $y = 0$, substituting the expressions for n'/n and a'/a from the junction conditions (47), we transform the (04) component of the Einstein equations (43) to the form

$$\dot{\rho} + 3H(\rho + p) = \frac{\kappa^2 \check{T}_{04}}{3}. \quad (52)$$

On the other hand, the same equation, which is the generalization of the conservation equation for the energy-momentum tensor of the matter confined to the brane to the case with the energy-momentum flow in the bulk, is obtained by integration of the 5D conservation law $\nabla_i T_0^i = 0$ across the brane (Tanaka & Himemoto, 2003).

[4] We assume invariance $y \leftrightarrow -y$.

6. Period of late cosmology

The generalized Friedmann equation contains the terms linear and quadratic in the energy density ρ. The period, in which the term linear in energy density is dominant, i.e. $\rho/\mu < 1$ is called the period of late cosmology. Graviton production by hot matter is sufficiently intensive in the radiation-dominated period of cosmology. In the radiation-dominated phase of the Universe condition of late cosmology is

$$\frac{\rho_r(T)}{\mu} = \frac{\kappa^2\hat{\rho}_r(T)}{6\mu} = \frac{4\pi^3 g_*(T)T^4}{90\mu M^3} \simeq \frac{4\pi^3 g_*(T)T^4}{90(\mu M_{pl})^2} < 1, \tag{53}$$

where we used that $\mu M_{pl}^2/M^3 \simeq 1$ Iofa (2009a;b). Taking $\mu \sim 10^{-12}GeV$, we find that the approximation of late cosmology is valid up to the temperatures of order $5 \cdot 10^2 GeV$.

Without the Weyl radiation term, the function $a^2(y,t)$ (29) has the minimum equal to zero at the point $|\bar{y}|$

$$e^{2\mu|\bar{y}|} = 1 + \frac{2\mu}{\rho}. \tag{54}$$

In the region $0 < |y| < |\bar{y}|$ and for $\rho/\mu \ll 1$ the functions $a(y,t)$ and $n(y,t) = \dot{a}(y,t)/\dot{a}(0,t)$ can be approximated as

$$\begin{aligned} a(y,t) &\simeq a(0,t)e^{-\mu|y|} \\ n(y,t) &\simeq e^{-\mu|y|}, \end{aligned} \tag{55}$$

and the approximate 5D metric is

$$ds^2 \simeq dy^2 + e^{-2\mu|y|}(-dt^2 + a^2(0,t)\eta_{ab}dx^a dx^b). \tag{56}$$

7. Fluctuations of the background metric

in the models with extra dimensions, in interactions of particles of the hot plasma on the brane, are produced not only massless gravitons, but the whole Kaluza-Klein tower of gravitons. Gravitons are fluctuations over the background metric. To calculate the energy loss from the brane due to graviton emission we need the spectrum of gravitons. The part of the action quadratic in fluctuations is

$$\begin{aligned} I = \frac{1}{2}\int d^5x\sqrt{-g^{(5)}}\Big[(R-\Lambda_{(5)})\left(-\frac{1}{2}h_i^j h_j^i + \frac{1}{4}h^2\right) - R_i^j h_j^i h + 2R_i^j h_j^k h_k^i \\ + \frac{1}{2}\left(2h_{qi;k}h^{ik;q} - h_{ik;q}h^{ik;q} + h_{,q}h^{,q} - 2h_{,i}h^{ik}{}_{;k}\right)\Big]. \end{aligned} \tag{57}$$

The action I is invariant under the gauge transformations

$$\tilde{h}_{kl} = h_{kl} - (\nabla_k\xi_l + \nabla_l\xi_k), \tag{58}$$

where ∇ is defined with respect to the background metric. The gauge freedom allows to set the components h_{4i} to zero. There remain residual gauge transformations, allowing for subsequent simplifications.

With the full metric $g_{\mu\nu}(y,t)$ (2) it is complicated to solve the equations for $h_{\mu\nu}$ analytically. To solve the equations, instead of the exact metric, in the regions $0 < y < \bar{y}$ and $y > \bar{y}$ we use the approximate metrics with the separated dependence on y and t. In the region $0 < y < \bar{y}$, we use the approximate metric (56). In the region $y > \bar{y}$ the approximate metric is

$$ds^2 = dy^2 + e^{2\mu y}\left(-dt^2 + \frac{a^2(0,t)}{4}\left(\frac{\rho}{\mu}\right)^2 \eta_{ab}dx^a dx^b\right). \tag{59}$$

In the metrics (56) and (59), using the residual gauge transformations and field equations for h_{ij}, it is possible to transform h_{ij} to the traceless, transverse form $h_\mu{}^\mu = 0$, $D^\mu h_{\mu\nu} = 0$ (Iofa, 2011).

Equations for the eigenmodes are considered in the next subsection. We show that the norm of the function $h_m^>$ is smaller than that of $h_m^<$. Effectively, in the period of late cosmology, this allows to consider the contribution from the region $0 < y < \bar{y}$ only.

We obtain the spectrum

$$m_n \simeq \mu e^{-\mu\bar{y}}\left(n\pi + \frac{\pi}{2}\right) \tag{60}$$

and the normalized eigenmode $h_m(0)$

$$h_m(0) \simeq (\mu e^{-\mu\bar{y}})^{1/2}. \tag{61}$$

For the following we need the sum $\sum_n h_{m_n}^2(0)$, where m_n is determined by (60). Because of a narrow spacing between the levels, we change summation to integration and obtain

$$\sum_n h_{m_n}^2(0) \simeq \int \frac{dm\, e^{\mu\bar{y}}}{\mu\pi}\mu e^{-\mu\bar{y}} = \int \frac{dm}{\pi}. \tag{62}$$

The integral (62) is independent of $\bar{y}$. The same measure of integration was obtained in Langlois; Sorbo & Rodriguez-Martinez (2002), where the authors used the graviton modes of the Randal-Sundrum II model (Randall & Sundrum, 1999) without matter, in which case the integration over y extends to infinity and the spectrum is continuous. Similarity of the results can be traced to the fact that we performed calculations in the period of late cosmology neglecting the terms of order $O(\rho/\mu)$ as compared to unity.

7.1 Equations for eigenmodes

In the background of the approximate metric (56), in the region $0 < y < \bar{y}$, in the gauge $D^\mu h_{\mu\nu}(y,x) = 0$, $h_\mu^\mu(y,x) = 0$, the $(\mu\nu)$ components of the field equations for fluctuations are

$$h''_{\mu\nu} - 4\mu^2 h_{\mu\nu} + b^{-1}(y) D_\rho D^\rho h_{\mu\nu} + \delta(y) 4\mu h_{\mu\nu} = 0. \tag{63}$$

We expand the functions $h_\nu^\mu(x,y)$ as

$$h_\nu^\mu(x,y) = \sum_m \phi_{(m)\nu}^\mu(x) h_m(y),$$

where the functions $h_m^<(y)$ satisfy the equation [5]

$$h''_m(y) - 4\mu^2 h_m(y) + e^{2\mu|y|} m_<^2 h_m(y) + \delta(y) 4\mu h_m(y) = 0. \tag{64}$$

[5] Wherever it does not lead to ambiguity, we omit the (sub)superscripts $<$ and $>$.

Solution of the equation

$$h''_m(y) - 4\mu^2 h_m(y) + e^{2\mu|y|} m^2_< h_m(y) = 0 \tag{65}$$

is

$$h^<_m(y) = C_1 J_2(\tilde{m} e^{\mu|y|}) + C_2 N_2(\tilde{m} e^{\mu|y|}), \tag{66}$$

where

$$\tilde{m} = \frac{m_<}{\mu}.$$

The terms with $\delta(y)$ are taken into account by the boundary condition

$$\left[\frac{dh_m(y)}{dy} + \frac{2}{\tilde{m}} h_m(y)\right]_{y=0_+} = 0$$

which yields the relation $C_1 J_1(\tilde{m}) + C_2 N_1(\tilde{m}) = 0$. The eigenfunctions (66) take the form

$$h^<_{m_<}(y) = C\left[(N_1(\tilde{m}) J_2(\tilde{m} e^{\mu|y|}) - J_1(\tilde{m}) N_2(\tilde{m} e^{\mu|y|})\right]. \tag{67}$$

Introducing the functions

$$f_k(y) = N_1(\tilde{m}) J_k(\tilde{m} e^{\mu|y|}) - J_1(\tilde{m}) N_k(\tilde{m} e^{\mu|y|})$$

we obtain the norm of h_m

$$||h^<_{m_<}||^2 = 2C^2 \int_0^{\bar{y}} e^{2\mu y} f_2^2(y) dy = \frac{C^2}{\mu}\left[e^{2\mu\bar{y}}\left(f_2^2(\bar{y}) - f_1(\bar{y}) f_3(\bar{y})\right) - \left(f_2^2(0) - f_1(0) f_3(0)\right)\right]. \tag{68}$$

Typical masses (energies) of the emitted Kaluza-Klein gravitons are of order of temperature of the Universe T. In the case of small $\mu \sim 10^{-12}$ GeV we have $m/\mu \gg 1$. Substituting the asymptotics of the Bessel functions, we obtain

$$||h^<_{m_<}||^2 = \frac{C^2}{\mu}\left(\frac{2}{\pi\tilde{m}}\right)^2 (e^{\mu\bar{y}} - 1). \tag{69}$$

In the region $y > \bar{y}$ we use the metric in (59) with the increasing exponent. The equations for the eigenmodes $h^>$ is

$$h''_m(y) - 4\mu^2 h_m(y) + e^{-2\mu|y|} m^2_> h_m(y) = 0. \tag{70}$$

The eigenfunctions are

$$h^>_{m_>}(y) = \tilde{C}_1 J_2(\tilde{m} e^{-\mu y}) - \tilde{C}_2 N_2(\tilde{m} e^{-\mu y}), \tag{71}$$

where $\tilde{m} = m_>/\mu$. For large y, such that $\tilde{m} e^{-\mu y} \ll 1$, the function $N_2(\tilde{m} e^{-\mu y}) \sim (\tilde{m} e^{-\mu y})^{-2}$ rapidly increases, and to have normalizable eigenfunctions we set $\tilde{C}_2 = 0$. The norm of the function $h^>_m(y)$ is

$$||h^>_{m_>}||^2 = \tilde{C}_1^2 \int_{\bar{y}}^{\infty} e^{-2\mu y} J_2^2(\tilde{m} e^{-\mu y}) dy = \tilde{C_1}^2 \frac{e^{-2\mu\bar{y}}}{2\mu} [J_2^2(\tilde{m} e^{-\mu\bar{y}}) - J_1(\tilde{m} e^{-\mu\bar{y}}) J_3(\tilde{m} e^{-\mu\bar{y}})]. \tag{72}$$

At temperatures $T \gg \mu$ at which production of gravitons is sufficiently intensive, the argument of the Bessel function

$$\tilde{m}e^{-\mu\bar{y}} \sim \frac{T}{\mu}\left(\frac{\rho}{\mu}\right)^{1/2} \sim \frac{T}{\mu}g_*^{1/2}(T)\frac{T^2}{\mu M_{pl}}$$

is large and we can substitute the asymptotics of the Bessel functions

$$h^{>}_{\tilde{m}_>}(y) = \tilde{C}_1\sqrt{\frac{2}{\pi\tilde{m}}}e^{\mu\bar{y}/2}\cos(\tilde{m}(e^{\mu\bar{y}}-1)). \tag{73}$$

Instead of sewing the oscillating functions $h^{<}_{\tilde{m}}(y)$ and $h^{>}_{\tilde{m}}(y)$, we sew the envelopes of their asymptotics

$$\frac{2C}{\pi\tilde{m}_<}e^{-\mu\bar{y}/2} = \tilde{C}_1\left(\frac{2}{\pi\tilde{m}_>}\right)^{1/2}e^{\mu\bar{y}/2} \tag{74}$$

giving

$$\tilde{C}_1 = Ce^{-\mu\bar{y}}\sqrt{\frac{2}{\pi}}\frac{\sqrt{\tilde{m}_>}}{\tilde{m}_<}$$

The norm (72) is smaller than (69). Effectively, we neglect the contribution from the region $y > \bar{y}$ and impose the condition $h_m(\bar{y}) = 0$, which yields us the spectrum (60). Using the norm (69), we obtain the normalized eigenmode $h_m(0)$ (61).

8. Production of Kaluza-Klein gravitons

In this section we calculate the rate of production of Kaluza-Klein gravitons in interactions of particles of the hot matter on the brane in the radiation-dominated period. The leading contribution to this process is given by the annihilation reactions $\psi^i + \bar{\psi}^i \to G$, where ψ and $\bar{\psi}$ are the standard model particles on the brane (vector, spinor, scalar) and G is a state of mass m_n from the graviton Kaluza-Klein tower. Production of Kaluza-Klein gravitons is calculated with the interaction Lagrangian

$$I = \kappa\int d^4x\sqrt{-\bar{g}}\,h_{\mu\nu}(0,x)T^{\mu\nu}(x),$$

where $T^{\mu\nu}$ is the energy-momentum of particles on the brane. Evolution of energy density of matter on the brane is determined from the Boltzmann equation (Gorbunov & Rubakov, 2008; Kolb & Turner, 1990)

$$\frac{d\hat{\rho}}{dt} + 4H\hat{\rho} = -\sum_n\sum_i\int\frac{d^3p}{(2\pi)^3}\int\frac{d^3k_1}{(2\pi)^3 2E_1}\frac{d^3k_2}{(2\pi)^3 2E_2}f_1^i(E_1)f_2^i(E_2)|M_n^i|^2(2\pi)^4\delta^4(k_1+k_2-p). \tag{75}$$

Here f^i are the Bose/Fermi distributions of colliding particles and M_n^i is the amplitude of the annihilation reaction. The square of the annihilation amplitude is (Langlois & Sorbo, 2003)

$$|M_n^i|^2 = A_i\frac{\kappa^2}{8}h_n^2(0)s^4, \tag{76}$$

where $A_i = A_s, A_v, A_f = 2/3, 4, 1$ for scalars, vectors and fermions and $s^2 = (k_1+k_2)^2$. The sum over the graviton states is transformed to the integral as in (62). Integrating over the angular variables of the momenta of interacting particles, we obtain Boltzmann equation in the form

$$\frac{d\hat{\rho}}{dt}+4H\hat{\rho}=-\frac{\kappa^2\sum_i A_i}{8(2\pi)^4}\int\frac{dme^{\mu\bar{y}}}{\mu}h_m^2(0)m^4\int dk_1dk_2f(k_1)f(k_2)(k_1+k_2)\theta\left(1-\left|1-\frac{m^2}{2k_1k_2}\right|\right). \tag{77}$$

Integrating over m, we obtain (cf. (Hebecker & March-Russel, 2001; Langlois & Sorbo, 2003))

$$\frac{d\hat{\rho}}{dt}+4H\hat{\rho}=-\frac{\kappa^2\sum_i A_i}{8(2\pi)^4}\int_0^{(2k_1k_2)^{1/2}} dmm^4\int dk_1dk_2f(k_1)f(k_2)(k_1+k_2) \tag{78}$$

$$=-\frac{\kappa^2A315\zeta(9/2)\zeta(7/2)T^8}{\pi^3 2^8},$$

where the sum extends over relativistic degrees of freedom

$$A=\sum_i A_i=\frac{2g_s}{3}+g_f(1-2^{-7/2})(1-2^{-9/2})+4g_v. \tag{79}$$

In the high-energy period, when all the standard model degrees of freedom are relativistic, $A=166,2$.

Eq.(78) has the same form as Eq.(52) and the right hand side of (78) can be identified with the component $\check{T}_{04}$ of the bulk energy-momentum tensor (Langlois & Sorbo, 2003).

The energy-momentum tensor of the emitted gravitons is taken in the form of free radiation of massless particles (Langlois & Sorbo, 2003)

$$\check{T}^{ij}(x)=\int d^5p\sqrt{-g}\delta(p_ip^i)\tilde{f}(x,p)p^ip^j, \tag{80}$$

where $\tilde{f}(x,p)$ is the phase space density of the distribution function of gravitons.

Expanding the graviton momentum p^i in the orthonormal basis v_b^i, n_b^i (19), we have

$$p^i=Ev_b^i+mn_b^i+p^ae_a^i,$$

where $p_ip^i=-E^2+m^2+p_ap^a$. Substituting these expressions in (80), we obtain

$$\check{T}_{nu}=u^in^jT_{ij}=-\frac{1}{2}\int d^3pdm\,mf(E,m,p^a)\bigg|_{E=\sqrt{m^2+p^2}} \tag{81}$$

The interaction tern in (75) can be rewritten as

$$\frac{1}{(2\pi)^5}\int d^3p\frac{dm}{\pi}\int\frac{d^3k_1}{2E_1}\frac{d^3k_2}{2E_2}f_1^i(E_1)f_2^i(E_2)\sum_i|M_n^i|^2\delta^4(k_1+k_2-p) \tag{82}$$

Comparing (81) and (82), one determines $f(E,m,p^a)$ and calculates the other components of $\check{T}_{ij}$

$$\check{T}_{nn}(0,t)=\frac{3A\zeta(9/2)\zeta(7/2)}{2\pi^4}\kappa^2T^8,\qquad \check{T}_{uu}(0,t)=\frac{21\,A\zeta(9/2)\zeta(7/2)}{8\pi^4}\kappa^2T^8. \tag{83}$$

In the basis (19), the components $\check{T}_{nu}$ and $\check{T}_{nn}$ are equal to $\check{T}_{04}$ and $\check{T}_{44}$.

9. Graviton emission in the bulk in the period of late cosmology and nucleosynthesis

To estimate the effect of the graviton emission on nucleosynthesis we solve perturbatively the system of the generalized Friedmann equation (51) and the 5D conservation equation (52). In the period of late cosmology the leading approximation of the Friedmann equation coincides with that in the standard cosmological model. Supposing that the terms with the bulk energy-momentum tensor do not alter the leading-order results significantly, we treat these terms perturbatively. The Weyl radiation term and the term quadratic in radiation energy density can be treated perturbatively also.

The Friedmann equation can be written as

$$H^2 \simeq 2\mu\rho - I_{04} - I_{44}. \tag{84}$$

Using the expressions (83) for the components $\check{T}_{ij}$, we obtain

$$I_{04} = \frac{2\kappa^2}{3a^4(0,t)} \int_{t_l}^{t} dt' \check{T}_{04}(0,t')(\mu+\rho(t'))a^{-4}(0,t') = \tag{85}$$
$$\mu\rho(t)A_{04}\left(\frac{1}{12}\left(\frac{1}{\mu t_l}-\frac{1}{\mu t}\right)+\frac{1}{288}\left(\frac{1}{(\mu t_l)^3}-\frac{1}{(\mu t)^3}\right)\right),$$

and

$$I_{44} = \frac{2\kappa^2}{3a^4(0,t)} \int_{t_l}^{t} dt' \check{T}_{44}(t')(2\mu\rho(t'))^{1/2}a^{-4}(0,t') = \frac{\mu\rho(t)A_{44}}{48}\left(\frac{1}{(\mu t_l)^2}-\frac{1}{(\mu t)^2}\right), \tag{86}$$

Here we substituted

$$\kappa^2\check{T}_{04} = A_{04}\rho^2 = -\frac{315\,A\,\zeta(9/2)\zeta(7/2)}{2^9\pi^3}\left(\frac{180}{g_*\pi^2}\right)^2\rho^2(t) \tag{87}$$

$$\kappa^2 T_{44} = A_{44}\rho^2 = \frac{3\,A\,\zeta(9/2)\zeta(7/2)}{4\pi^4}\left(\frac{180}{g_*\pi^2}\right)^2\rho^2(t) \tag{88}$$

The integrals have a strong dependence on the value of the lower limit. Therefore, in the integrand the slowly varying functions can be taken at the times when all the Standard model degrees of freedom are relativistic. In this period $A = 166.2$, $g_*(T) = 106.7$ and $A_{04} \simeq -0.126$. In the leading approximation $\rho(t) \simeq 1/8\mu t^2$ and $1/\mu t_l \simeq (8\rho(t_l)/\mu)^{1/2}$ Taking $\rho(t_l)/\mu \sim 0.1 \div 0.001$ and $\mu \sim 10^{-12} GeV$, we have $1/\mu t_l \simeq 0.9 \div 0.09$ and $T_l \sim (5.1 \div 1.6)\cdot 10^2 GeV$. For $\rho(t_l)/\mu \sim 0.1$ we obtain

$$I_{04} \simeq -2\mu\rho(t)\cdot 0.0048 \tag{89}$$
$$I_{44} \simeq 2\mu\rho(t)\cdot 0.00094. \tag{90}$$

In the radiation-dominated period,the 5D conservation equation is

$$\dot{\rho} + 4H\rho = -\frac{A_{04}\rho^2}{3} \tag{91}$$

Let $\bar{\rho}$ and $\bar{H}$ be the energy density and the Hubble function in the leading order

$$\bar{H}^2 = 2\mu\bar{\rho}.$$

Defining

$$\rho = \bar{\rho} + \rho_1, \qquad H = \bar{H} + H_1,$$

and separating in (84) and (91) the leading-order terms, we obtain

$$2\bar{H}H_1 = 2\mu\rho_1 - \frac{2\bar{\rho}}{3}\int dt' \left[A_{04}(\mu+\bar{\rho}) + A_{44}\bar{H}\right]\bar{\rho}(t') \tag{92}$$

$$\dot{\rho}_1 + 4\bar{H}\rho_1 + 4\bar{\rho}H_1 = \frac{A_{04}\bar{\rho}^2}{3}. \tag{93}$$

Substituting in the above system $\bar{\rho}(t) \simeq 1/8\mu t^2$ and performing integration, we have

$$H_1(t) \simeq \frac{\mu}{\bar{H}}\rho_1 - \frac{\bar{H}A_{04}}{48\mu}\left[\frac{1}{t_l} - \frac{1}{t} + \frac{1}{24\mu^2}\left(\frac{1}{t_l^3} - \frac{1}{t^3}\right)\right] - \frac{\bar{H}A_{44}}{192\mu^2}\left(\frac{1}{t_l^2} - \frac{1}{t^2}\right). \tag{94}$$

Substituting in (92) expression (94) for H_1 and noting that $\bar{H} = 1/2t$, we obtain

$$\dot{\rho}_1 + \frac{3}{t}\rho_1 = \frac{A_{04}}{192\mu^2 t^3}\left[\frac{1}{t_l} + \frac{1}{24\mu^2}\left(\frac{1}{t_l^3} - \frac{1}{t^3}\right)\right] + \frac{A_{44}}{768\mu^3 t^3}\left(\frac{1}{t_l^2} - \frac{1}{t^2}\right).$$

Solving this equation, we find

$$\rho_1(t) = \frac{C_1}{t^3} + \frac{A_{04}}{192\mu^2 t^3}\left[\frac{t-t_l}{t_l} + \frac{1}{24\mu^2}\left(\frac{t-t_l}{t_l^3} - \frac{1}{2}\left(\frac{1}{t_l^2} - \frac{1}{t^2}\right)\right)\right] + \frac{A_{44}}{768\mu^3 t^3}\left(\frac{t}{t_l^2} - \frac{2}{t_l} + \frac{1}{t}\right) \tag{95}$$

The constant C_1 can be determined by sewing solutions of Friedmann equation in the periods of early and late cosmologies. For a moment we set $C_1 = 0$, i.e. we look for a contribution from the period of late cosmology. For H_1 we obtain

$$H_1(t) = -\frac{A_{04}}{1536\mu^3 t^2}\left(\frac{1}{t_l^2} - \frac{1}{t^2}\right) + \frac{A_{44}}{192\mu^2 t^2}\left(-\frac{1}{t_l} + \frac{1}{t}\right) \simeq -\bar{\rho}(t)\left(\frac{A_{04}}{192\mu^2 t_l^2} + \frac{A_{44}}{24\mu t_l}\right). \tag{96}$$

The expressions (95) and (96) show that corrections to the leading terms are small, i.e. the perturbative approach is justified. Note that the leading term proportional to $A_{04}/\mu t_l$ was canceled in H_1.

The mass fraction of ^4He produced in primordial nucleosynthesis is (Gorbunov & Rubakov, 2008; Kolb & Turner, 1990)

$$X_4 = \frac{2(n/p)}{(n/p)+1},$$

where the ratio n/p is taken at the end of nucleosunthesis. Characteristic temperature at the onset of the period of nucleosynthesis (freezing temperature T_n of the reaction $n \leftrightarrow p$), estimated as the temperature at which the reaction rate $\sim G_F T^5$ is approximately equal to the Hubble parameter $G_F T_n^5 \sim H$ (Gorbunov & Rubakov, 2008; Kolb & Turner, 1990), is $T_n \sim 10^{-3} GeV$. The difference of the freezing temperatures in the models with and without the account of the graviton emission is

$$\frac{\delta T_n}{T_n} \simeq \frac{H_1}{5\bar{H}} \simeq \frac{1}{5}\sqrt{\frac{\bar{\rho}(t_n)}{2\mu}}\left(\frac{A_{04}}{192\mu^2 t_l^2} + \frac{A_{44}}{24\mu t_l}\right), \tag{97}$$

where

$$\frac{\bar{\rho}(t_n)}{\mu} = \frac{4\pi^3 g_*(T_n)}{90} \frac{T_n^4}{\mu M^3}. \tag{98}$$

Substituting $T_n \sim 10^{-3} GeV$ and $\mu M^3 \sim (\mu M_{pl})^2 \sim 10^{14} GeV^4$, we find that the ratio $\delta T_n / T_n$ is very small. The equilibrium value of the $n - p$ ratio at the freezing temperature is

$$\left(\frac{n}{p}\right)_n = \exp\left[-\frac{(m_n - m_p)}{T_n}\right].$$

Substituting $\delta(n/p)_n = (n/p)\ln(n/p)_n \delta T_n / T_n$, we obtain variation of X_4 under variation of the freezing temperature

$$\delta X_4 \simeq \frac{2}{(n/p+1)^2} \ln\left(\frac{n}{p}\right)\left(\frac{n}{p}\right)_n \frac{\delta T_n}{T_n} \tag{99}$$

which is also a very small number (Iofa, 2011).

10. Estimate of the graviton emission in the bulk in the period of early cosmology

In the period of early cosmology in the Friedmann equation the ρ^2 term is dominant, i.e. $\rho(t)/\mu > 1$. For the value of $\mu \sim 10^{-12} GeV$ assumed in the present study the characteristic temperatures of the period of early cosmology are above $5 \cdot 10^2 GeV$. Not much is known about physics at such temperatures. To make a crude estimate of the effect of the graviton emission on the nucleosynthesis we adopt a conservative point of view assuming that the collision integral in the Boltzmann equation and the expressions for $\check{T}_{ij}$ calculated in the period of late cosmology remain qualitatively valid in the early cosmological period.

The new phenomenon in the early cosmological period is that some of the emitted gravitons can return to the brane and be again reflected in the bulk with a different momentum. These gravitons do not contribute to the component $\check{T}_{04}$, because they are not produced, but reflected, but contribute to the component $\check{T}_{44}$. The new ingredient in the Friedmann equation is the term $\check{T}_{44}^{(b)}$ representing the energy-momentum tensor of gravitons bouncing back to the brane (Hebecker & March-Russel, 2001; Langlois & Sorbo, 2003)

$$\begin{aligned} H^2(t) &= \rho^2(t) + 2\mu\rho(t) + \mu\rho_w(t) \\ &- \frac{2\kappa^2}{3a^4(0,t)} \int_{t_c}^{t} dt' \left[\check{T}_{04}(t')(\rho(t') + \mu) + \check{T}_{44}(t')H(t') - \check{T}_{44}^{(b)}(t')H(t')\right] a^4(0,t'). \end{aligned} \tag{100}$$

Here the initial time t_c is the time of reheating. Numerical estimates and considerations from the Vaidya model (Langlois; Sorbo & Rodriguez-Martinez, 2002; Langlois & Sorbo, 2003) suggest that at the period of early cosmology the dominant contributions from T_{04}, T_{44} and $\check{T}_{44}^{(b)}$ mutually cancel. Below we solve the system of the Friedmann and 5D conservation equations assuming that this cancellation takes place.

Let $\bar{\rho}$ and $\bar{H}$ be the energy density of matter on the brane and the Hubble function calculated in the model without the graviton emission in the period of early cosmology $\bar{H}^2 \simeq \bar{\rho}^2(t)$. For the equation of state of the hot plasma $\bar{p} = \bar{\rho}/3$, we have $\bar{\rho}(t) \simeq 1/4t$. Defining

$$\rho = \bar{\rho} + \rho_2, \qquad H = \bar{H} + H_2,$$

we obtain

$$2\bar{H}H_2 \simeq 2\bar{\rho}\rho_2 - \frac{2\bar{\rho}}{3}\int_{t_c}^{t} dt' \left[(A_{04}(\mu+\bar{\rho}) + A_{44}\bar{H})\,\bar{\rho}(t') - \kappa^2 \check{T}_{44}^{(b)}(t') \right] \tag{101}$$

$$\dot{\rho}_2 + 4\bar{H}\rho_2 + 4\bar{\rho}H_2 = \frac{A_{04}\bar{\rho}^2}{3}. \tag{102}$$

The equation for ρ_2 is

$$\dot{\rho}_2 + \frac{2}{t}\rho_2 - \frac{1}{t}\left[\frac{A_{04}}{12}\left(\mu \ln\frac{t}{t_c} + \frac{1}{4}\left(\frac{1}{t_c} - \frac{1}{t}\right)\right) + \frac{A_{44}}{48}\left(\frac{1}{t_c} - \frac{1}{t}\right)\right] + \frac{1}{12t}\int_{t_c}^{t} \kappa^2 \check{T}_{44}^{(b)}(t')dt' = \frac{A_{04}}{48t^2}, \tag{103}$$

Integrating (102) with the initial condition $\rho_2(t_c) = 0$, we obtain

$$\rho_2(t) = \frac{A_{04}\mu}{24}\left(\ln\frac{t}{t_c} - \frac{1}{2} + \frac{t_c^2}{2t^2}\right) + \frac{A_{04}+A_{44}}{96t_c}\left(1 - \frac{t_c^2}{t^2}\right) - \frac{A_{44}}{48}\left(\frac{1}{t} - \frac{t_c}{t^2}\right) \tag{104}$$
$$- \frac{1}{12t^2}\int_{t_c}^{t} dyy \int_{t_c}^{y} dx\kappa^2 \check{T}_{44}^{(b)}(x)$$

The time t_c of reheating is estimated for the reheating temperature $T_R \sim 5 \cdot 10^6 GeV$ (Mielczarek, 2009). Using the relation (53) and substituting $\bar{\rho}(t_c)/\mu \simeq 1/4\mu t_c$, we have

$$\frac{1}{4\mu t_c} \simeq \frac{4\pi^3 g_*(T_R)T_R^4}{90\mu M^3} \sim 9 \cdot 10^{14}.$$

It follows that $t_l/t_c \sim 1/\mu t_c \simeq 3.5 \cdot 10^{15}$.

In (104) there is a large term $(A_{04} + A_{44})/t_c$. Omitting the small terms, we have

$$\rho_2(t) \simeq \frac{A_{04}\mu}{24}\left(\ln\frac{t}{t_c} - \frac{1}{2}\right) + \frac{A_{04}+A_{44}}{96t_c} - \frac{A_{44}}{48}\frac{1}{t} - \frac{1}{12t^2}\int_{t_c}^{t} dyy \int_{t_c}^{y} dx\kappa^2 \check{T}_{44}^{(b)}(x) \tag{105}$$

On dimensional grounds at small t the term $\kappa^2 \check{T}_{44}^{(b)}(t)$ has the following structure

$$\kappa^2 \check{T}_{44}^{(b)}(t) = \frac{b_2}{t^2} + \frac{b_1\mu}{t} + b_0\mu^2 + \cdots \tag{106}$$

Performing integration of the last term in (105) and taking $t \sim t_l$, we obtain

$$\rho_2(t_l) \simeq \frac{A_{04}\mu}{24}\left(\ln\frac{t_l}{t_c} - \frac{1}{2}\right) - \frac{A_{44}}{48t_l} + \frac{A_{04}+A_{44}}{96t_c} - \frac{b_2}{24t_c} + \frac{b_2}{12t_l} + \frac{b_1\mu}{24}\ln\frac{t_l}{t_c}. \tag{107}$$

Next, we equate $\rho_2(t_l)$ and C_1/t_l^3 in (95). At the times $t \sim t_l$, where $\mu t_l \sim 1$, we have

$$C_1\mu^3 \simeq \mu\left(\frac{A_{04}}{24}\left(\ln\frac{1}{\mu t_c} - \frac{1}{2}\right) - \frac{A_{44}}{48}\right) + \frac{A_{04}+A_{44}}{96t_c} - \frac{b_2}{24t_c} + \frac{b_2\mu}{12} + \frac{b_1\mu}{24}\ln\frac{1}{\mu t_c}. \tag{108}$$

The term C_1/t^3 is

$$\frac{C_1}{t^3} \simeq \frac{1}{\mu^2 t^3}\left[\mu\left(\frac{A_{04}}{24}\left(\ln\frac{1}{\mu t_c} - \frac{1}{2}\right) - \frac{A_{44}}{48}\right) + \frac{A_{04}+A_{44}}{96t_c} - \frac{b_2}{24t_c} + \frac{b_2\mu}{12} + \frac{b_1\mu}{24}\ln\frac{1}{\mu t_c}\right]. \tag{109}$$

In the period of late cosmology the term C_1/t^3 generates in Eq.(94) for H_1 the contribution $\Delta H_1 = \mu C_1/\bar{H}t^3$

$$\Delta H_1 = 16\bar{\rho}\left[\mu\left(\frac{A_{04}}{24}\left(\ln\frac{1}{\mu t_c}-\frac{1}{2}\right)-\frac{A_{44}}{48}\right)+\frac{A_{04}+A_{44}}{96t_c}-\frac{b_2}{24t_c}+\frac{b_2\mu}{12}+\frac{b_1\mu}{24}\ln\frac{1}{\mu t_c}\right]. \tag{110}$$

Here $\bar{\rho}(t) \simeq 1/8\mu t^2$ is the matter energy density in the period of late cosmology.

If the term $(A_{04}+A_{44})/96t_c \sim 3\cdot 10^{-4}/t_c$ was not canceled, it would produce in ρ_1 the contribution

$$\Delta\rho_1 = \frac{A_{04}+A_{44}}{96t_c}\frac{1}{(\mu t)^3} \simeq \frac{3\cdot 10^{-4}}{(\mu t)^3 t_c}.$$

From (94) we would have

$$\frac{\Delta H_1}{\bar{H}} = \frac{\mu}{\bar{H}^2}\Delta\rho_1 \simeq \frac{1.2\cdot 10^{-3}}{(\mu t_c)(\mu t)}. \tag{111}$$

At time of the nucleosynthesis

$$\frac{1}{8(\mu t_n)^2} \simeq \frac{4\pi^3 g_*(T_n)}{90}\frac{T_n^4}{(\mu M_{pl})^2}.$$

For $T_n \sim 10^{-3}GeV$, we have $\mu t_n \sim 10^{12}$. From (111) we obtain $\Delta H_1(t_n)/\bar{H}(t_n) \simeq 4$, which is too large a value, and would contradict the experimental data.

Assuming that the large terms in (110) cancel, we have

$$\Delta H_1 = \frac{2\bar{\rho}(t_n)}{3}\left[A_{04}\left(\ln\frac{1}{\mu t_c}-\frac{1}{2}\right)+\frac{A_{04}}{2}+b_1\ln\frac{1}{\mu t_c}\right]. \tag{112}$$

Because in the period of nucleosynthesis $\Delta H_1/\bar{H} \sim (\bar{\rho}(t_n)/\mu)^{1/2}$ is a small number, contribution from the early cosmology would result in a small variation of $\delta X_4/X_4$.

11. Gravitons bouncing to the brane

In this section, in the framework of the brane moving in the static space-time we present arguments that in the period of early cosmology gravitons can bounce back to the brane.

Using Friedmann equation, $(\dot{R}_b/R_b)^2 = \rho^2 + 2\mu\rho$ the equation for the brane trajectory can be written as (Langlois & Sorbo, 2003)

$$\frac{dR_b}{dT_b} = \frac{\mu^2 R_b^2 \dot{R}_b}{\xi\sqrt{\mu^2 R_b^2 + \dot{R}_b^2}} = \xi\beta\mu^2 R_b^2 \frac{H}{\sqrt{\mu^2+H^2}}, \tag{113}$$

where $\beta = sign(R_b(t))$. Expanding the right hand side of (113) in powers of $\mu/\rho < 1$, we obtain

$$\frac{dR_b}{dT_b} \simeq \xi\beta\mu^2 R_b^2\left(1-\frac{\mu^2}{2\rho^2}\right). \tag{114}$$

Using that $\rho(t) = \rho(t_1)(R_b(t_1)/R_b(t))^4$, we have

$$\left(1 - \frac{\mu^2}{2\rho^2}\right)^{-1} \simeq 1 + \frac{\mu^2 R_b^8(t)}{2\rho^2(t_1)R_b^8(t_1)}.$$

Integrating Eq. (113) with the initial conditions $R_b = R_b(t_1),\ T_b = T_b(t_1)$, we obtain

$$\xi\beta\mu^2(T_b(t) - T_b(t_1)) = -\frac{1}{R_b(t)} + \frac{1}{R_b(t_1)} + \frac{\mu^2}{14(\rho(t_1)R_b^4(t_1))^2}\left(R_b^7(t) - R_b^7(t_1)\right). \tag{115}$$

Gravitons propagate along the null geodesics which are found from the geodesic equations (14)-(16). The tangent vectors to a null geodesic satisfy the relation

$$g_{ij}\frac{dx^i}{d\lambda}\frac{dx^j}{d\lambda} = 0, \tag{116}$$

where, to distinguish the case of the null geodesics, we relabeled the affine parameter from y to λ. Substituting in (116) solution (17), we obtain that $C^R = 0$. The tangent vectors to a null geodesic are

$$\frac{d\tilde{T}}{d\lambda} = \frac{C^T}{f(\tilde{R})}, \qquad \frac{d\tilde{x}^a}{d\lambda} = \frac{C^a}{\mu^2\tilde{R}^2}, \qquad \frac{d\tilde{R}}{d\lambda} = \nu|C^T|\,(1-\gamma)^{1/2}, \tag{117}$$

where *tilde* indicates that the point is at the null geodesic, $\nu = \pm 1$. From the equations (117) we obtain

$$\frac{d\tilde{R}}{d\tilde{T}} = \nu\kappa(1-\gamma)^{1/2}\mu^2\tilde{R}^2, \tag{118}$$

where $\kappa \equiv sign(C^T)$. Integrating this equation with the initial conditions $\tilde{R} = R_b(t_1),\ \tilde{T} = T_b(t_1)$, we have

$$\frac{1}{R_b(t_1)} - \frac{1}{\tilde{R}} = \nu\kappa(1-\gamma)^{1/2}\mu^2(\tilde{T} - T_b(t_1)). \tag{119}$$

To find, if at a time t the graviton trajectory returns to the brane world sheet, i.e. $\tilde{R} = R_b(t)$ and $\tilde{T} = T_b(t)$, we combine equations (115) and (119). In the case $\xi\beta = \nu\kappa$ we obtain

$$[(1-\gamma)^{-1/2} - 1]\left(\frac{1}{R_b(t_1)} - \frac{1}{R_b(t)}\right) = \frac{\mu^2}{14\rho^2(t_1)}\frac{1}{R_b(t_1)}\left(\left(\frac{R_b(t)}{R_b(t_1)}\right)^7 - 1\right). \tag{120}$$

Eq. (120) means that graviton emitted from the brane at the time t_1 has bounced back on the brane at the time t. Setting $z \equiv R_b(t_1)/R_b(t)$, and $\hat{\gamma}/2 \equiv [(1-\gamma)^{-1/2} - 1]$, and noting that in the period of early cosmology $\mu/\rho(t) \simeq 4\mu t \ll 1$, we rewrite (120) as

$$\frac{(1-z^7)}{z^7(1-z)} = \frac{14\hat{\gamma}\rho^2(t_1)}{\mu^2} \simeq \frac{14\hat{\gamma}}{(4\mu t_1)^2} \tag{121}$$

The function $(1-z^7)z^{-7}(1-z)^{-1}$ is monotone decreasing with the minimum at $z = 1$, and Eq. (121) has a unique solution provided $2\hat{\gamma}/(4\mu t_1)^2 > 1$.

From these relations, using that in the period of early cosmology $z = R_b(t_1)/R_b(t) \simeq (t_1/t)^{1/4}$, is calculated the time of the bounce. The time of the bounce is small, if $z \sim 1$, or $\gamma/(4\mu t)^2 \sim 1$, that is if gravitons are emitted at small angles to the brane.

12. Conclusion

In a model of 3-brane embedded in 5D space-time we calculated graviton emission of interacting hot matter on the brane in the bulk. Reliable calculations can be performed in the period of late cosmology, when $\rho(T)/\mu < 1$, where $\rho(T)$ is the normalized radiation energy density of matter on the brane, $\mu = \sqrt{-\Lambda/6}$ is the scale of the warping factor in the metric, Λ is the 5D cosmological constant. For $\mu \sim 10^{-12} GeV$, which we adopted in this paper, the limiting temperatures of the Universe at which the approximation of late cosmology is valid are of order $T_l \sim 5 \cdot 10^2 GeV$. In the period of late cosmology it was possible to make a number of approximations, which enabled us to obtain the analytic expression for the energy loss from the brane to the bulk.

The 5D model of the present paper can be treated in two alternative approaches - with moving brane in the static 5D bulk and with static brane in 5D bulk with time-dependent metric. Each picture proves to be useful for particular problems. We establish explicit connection between two pictures.

From the system of the Einstein equations containing the terms due to graviton emission we obtained the generalized Friedmann equation. The Einstein equations and Friedmann equation were solved perturbatively, the leading-order solution being that without the graviton emission. We have shown that in the period of late cosmology corrections to the leading-order approximation are small.

Graviton emission changes cosmological evolution of matter on the brane and thus production of light elements in primordial nucleosynthesis. Solving the system of the generalized Friedmann and the 5D energy conservation equations, which included the graviton emission terms, we found the difference of the mass fractions of ^4He produced in primordial nucleosynthesis calculated in the models with and without the graviton emission

$$\frac{\delta X_4}{X_4} \sim C\sqrt{\frac{\rho(T_n)}{\mu}}.$$

Here $\rho(T)$ is is the radiation energy density of matter on the brane, $T_n \sim 10^{-3} GeV$ is the freezing temperature of the reaction $p \leftrightarrow n$, and $C \ll 1$ is a small number. The ratio $\rho(T_n)/\mu \sim 10^{-26}$ is a small number, and thus correction to abundance of ^4He due to the graviton emission is small.

To estimate the effect of the graviton emission in the period of early cosmology on Helium production, we assumed that the collision integral in the Boltzmann equation obtained for the period of late cosmology qualitatively retains its form in the period of early cosmology. As the upper limit of temperatures of matter on the brane at which we estimate the graviton production, we take the reheating temperature $\sim 10^6 GeV$. It appears that straightforward calculation of abundance of ^4He produced in the early cosmology is too large to comply with the data. In the period of early cosmology a new effect becomes important (Hebecker & March-Russel, 2001; Langlois & Sorbo, 2003): gravitons emitted from the brane can bounce back to the brane. Assuming that the large terms in the components of the energy-momentum tensors from the emitted and bouncing gravitons mutually cancel, we find that the graviton emission in the period of early cosmology yields a small correction to abundance of ^4He calculated in the standard cosmological model.

13. References

Binetruy, P., Deffayet, C. & Langlois, D. (2000). Non-conventional cosmology from a brane-universe, *Nucl. Phys.* B565, 269-287. [hep-th/9905012]

Binetruy, P. et al. (2000). Brane cosmological evolution in a bulk with cosmological constant, *Phys. Lett.* B477, 285-291. [hep-th/9910219]

Birmingham, D. (1999). Topological Black Holes in Anti-de Sitter Space. *Class. Quant. Grav.* 16, 1197-1205. [hep-th/9808032],

Chamblin, H.A. & Reall, H.S. (1999) Dynamic Dilatonic Domain Walls. *Nucl. Phys.* B562, 133-157. [hep-th/9903225],

Collins, H. & Holdom, B. (2000). Brane Cosmologies without Orbifolds, *Phys. Rev.* D62, 105009. [hep-ph/0003173]

Durrer, R. (2005). Braneworlds, *XI Brazilian School of Cosmology and Gravitation, AIP Conf. Proc.* 782. [hep-th/0507006]

Gorbunov, D. & Rubakov, V. (2008). *Introduction to the theory of early Universe,* ISBN 978-5-382-00657-4, Moscow.

Hebecker, A. & March-Russel, J. (2001) Randall-Sundrum II Cosmology, AdS/CFT, and the Bulk Black Hole, *Nucl. Phys.* B608, 375-393. [hep-ph/0103214]

Iofa, M. (2009)a. Constraints on parameters of models with extra dimension from primordial nucleosynthesis. *Mod. Phys. Lett.* A24, 2307-2318. [0907.0547]

Iofa, M. (2009)b. Cosmological constraints on parameters of one-brane models with extra dimension. *JCAP,* 11, 023. [0907.4039]

Iofa, M. (2011). Graviton emission from the brane in the bulk in a model with extra dimension *JCAP,* 06, 025. [1012.2445]

Kolb, E. & Turner, M. (1990). *The Early Universe.* Addison-Wesley, Frontiers in Physics, Redwood City.

Kraus, P. (1999) Dynamics of Anti-de Sitter Domain Walls. *JHEP* 9912 (1999) 011, [hep-th/9910149]

Langlois, D.; Sorbo, L. & Rodriguez-Martinez, M. (2002). Cosmology of a brane radiating gravitons into the extra dimension. *Phys. Rev. Lett.* 89, 171301. [hep-th/0206146]

Langlois, D. & Sorbo, L. (2003) Bulk gravitons from a cosmological brane. *Phys. Rev.* D68, 084006. [hep-th/0306281]

Maartens, R. (2004). Brane-world gravity. *Liv. Rev. Rel.* 7, 7. [gr-qc/0312059]

Mielczarek, J. (2009) Reheating temperature from the CMB. *Phys. Rev.* D83, 023502. [astro-ph/1009.2359]

Mukohyama, S., Shiromizu,T. & Maeda, K-i (2000). Global structure of exact cosmological solutions in the brane world. *Phys. Rev.* D62 (2000) 024028. [arXiv:hep-th/9912287]

Randall, L. & Sundrum, R. (1999) An Alternative to Compactification. *Phys. Rev. Lett.* B83, 4690-4693. [arXiv:hep-th/9906064]

Rubakov, V. (2001). Large and infinite extra dimensions. *Phys. Usp.* 44, 871-893. [hep-ph/0104152]

Shiromizu, T; Maeda, K-i. & Sasaki, M. (2000) The Einstein equations on the 3-brane world. *Phys. Rev.* D62, 024012. [gr-qc/9910076]

Steigman, G.; Walker, T. & Zentner, A. (2001) Global Constraints On Key Cosmological Parameters. [astro-ph/0012149]

Tanaka, T. & Himemoto, Y. (2003) Generation of dark radiation in the bulk inflaton model, *Phys. Rev.* D67 (2003) 104007 [gr-qc/0301010]

BBN as Probe of Fundamental Physics

L. A. Popa and A. Caramete
ISS/Institute for Space Sciences Bucharest-Magurele, Romania

1. Introduction

The primordial Universe provides a laboratory to probe fundamental physics at high energies. Relics from those early epochs, such as the light elements synthesized during primordial Big Bang Nucleosynthesis (BBN) (Burles et al., 2001; Eidelman et al., 2004; Olive et al., 2000; Wagoner et al., 1967) when the Universe was only a few minutes old, and the Cosmic Microwave Background (CMB) photons, last scattered when the protons and electrons recombined some 400 thousand years later, represent powerful probes of the high energy phenomena pointing beyond the standard models of cosmology, particle physics and general relativity (Boesgaard & Steigman, 1985; Steigman et al., 1977; Steigman, 2007).

During its earlier evolution the Universe was hot and dense. The combination of high temperature and density ensures that collision rates are very high during early epochs, guaranteeing that all particles were in equilibrium at sufficiently early times. As the Universe expands and cools, interaction rates decline and, depending on the strength of their interactions, different particles depart from equilibrium at different epochs.

At a temperature of few MeV, the neutrino interaction rates become lower than the Hubble expansion rate, decoupling from the CMB photons and $e^{\pm}$ pairs present at that time. However, electron neutrinos (and antineutrinos) continue to interact with the baryons through the weak interactions until the Universe where few seconds old and the temperature has dropped below MeV. The interactions among neutrons, protons and $e^{\pm}$ continue to influence the ratio of neutron to proton number densities, tracking its equilibrium value. After electrons, neutrons and protons combine to form neutral atoms at recombination, the CMB photons propagate freely. This occurs when the Universe is some 400 thousand years old. The relic photons from this epoch redshifted to the currently observed CMB black body temperature $T_{cmb} = 2.725$ K.

Predictions of the abundances of the light elements as D,^{3}He, ^{4}He and ^{7}Li synthesized at the BBN epoch are in good agreement with the primordial abundances inferred from observational data, which validates the Standard Big Bang Nucleosynthesis (SBBN).

The primordial abundances of the relic nuclei produced during SBBN depend on the baryon number density and on the Hubble expansion rate of the Universe. At the same time, the shape and amplitude of the CMB anisotropy angular power spectra at the time of recombination depend on the same parameters. One of the key cosmological tests in understanding the cosmological dynamics between BBN and recombination is the determination of baryon energy density mass fraction, $\Omega_b h^2$, at these two different epochs.

The abundance of baryons, η_{10}, is related to the number density of baryons, n_B, and the number density of CMB photons, n_γ, through: $\eta_{10} = 10^{10}(n_B/n_\gamma)$. The baryon energy density mass fraction influences the growth rate of the density perturbation through its impact on the Hubble expansion rate. It is convenient to express η_{10} in terms of this two parameters as: $\eta_{10} = 274\,\Omega_b h^2$, where h is the reduced Hubble constant at the present time ($h = H_0/100$ km s^{-1} Mpc^{-1}). During expansion of the Universe, n_B and n_γ decrease but remain unchanged in a given comoving volume. Therefore, the value of η_{10} measured at BBN epoch and at recombination epoch should be the same.

Among the nuclides synthesized at the BBN epoch, deuterium is the better indicator of the primordial baryon abundance since no significant amounts of deuterium are synthesized after BBN (Epstein et al., 1976). The measurement of the QSO Absorption Lines Systems (QSOALS) leads to the determination of the primordial deuterium abundance (Pettini et al., 2008) from which the infered baryon abundance is $\eta_{10}(BBN) = 5.80^{+0.27}_{-0.28}$ ($\Omega_b h^2 = 0.0212 \pm 0.0010$).

The analysis of WMAP 7-year data (Komatsu et al., 2009) provides an independent determination of the baryon abundance $\eta_{10}(CMB) = 6.190 \pm 0.145$ ($\Omega_b h^2 0.02260 \pm 0.00053$).

The two independent determinations are in good agreement, differing by only 1.3σ.

BBN also provides powerful constraints on possible deviations from the standard cosmology and on new physics beyond the Standard Model (SM) particle physics (Sarkar, 1996). Many non-SBBN models introduce new free parameters in addition to the baryon energy density parameter. Most of these models assume either non-standard contribution to the total energy density, or a lepton asymmetry. This paper aims to place constraints on parameters of two types of non-SBBN models by using most of the present cosmological data complemented with the BBN predictions on the ^{4}He abundance.

These models include: i) the leptonic asymmetric cosmological models, challenging the standard neutrino sector and ii) the Higgs inflation models, challenging the electroweak sector of the SM of particle physics.

2. Challenging the neutrino physics

The radiation budget of the Universe relies on a strong theoretical prejudice: apart from the CMB photons, the relativistic background would consist of neutrinos and of possible contributions from other relativistic relicts. The main constraints on the radiation energy density come either from the very early Universe, where the radiation was the dominant source of energy, or from the observation of cosmological perturbations which carry the information about the time equality between matter and radiation.

The primordial abundance of the light elements depends also on the radiation energy density at the BBN epoch (energy density of order MeV^4), usually parametrized by the effective number of relativistic neutrino species, N_{eff}. Meanwhile, the number of active neutrino flavors have been fixed by Z^0 boson decay width to $N_\nu = 2.944 \pm 0.012$ (Eidelman et al., 2004), while the combined study of the incomplete neutrino decoupling and the QED corrections indicate that the number of relativistic neutrino species is $N_{eff} = 3.046$ (Mangano et al., 2002). Any departure of N_{eff} from this last value would be due to non-standard neutrino features or to the contribution of other relativistic relics, having as main effect the modification of the competition between the nuclear reaction rates and the Hubble expansion rate. Since the primordial ^{4}He mass fraction, Y_P is largely determined by the neutron to proton ratio, Y_P is quite sensitive to the competition between the weak interaction rates and the expansion rate.

The most natural phenomenological extension of the standard neutrino sector is the consideration of the leptonic asymmetry (Freese et al., 1983; Ruffini et al., 1983; 1988), parametrized by the neutrino degeneracy parameter $\xi_\nu = \mu_\nu / T_{\nu_0}$ [μ_ν is the neutrino chemical potential and T_{ν_0} is the present temperature of the neutrino background, $T_{\nu_0}/T_{\rm cmb} = (4/11)^{1/3}$]. Although the standard model of particle physics predicts the value of leptonic asymmetry of the same order as the value of the baryonic asymmetry, $B \sim 10^{-10}$, there are many particle physics scenarios in which a leptonic asymmetry much larger can be generated (Chu & Cirelli, 2006; Smith et al., 2006).
One of the cosmological implications of a larger leptonic asymmetry is the possibility to generate baryonic asymmetry of the Universe through the sphaleron processes (Buchmuller et al., 2004; Falcone & Tramontano, 2001; Kuzmin et al., 1985). Therefore, distinguishing between a vanishing and non-vanishing ξ_ν at the BBN epoch is a crucial test of the standard assumption that sphaleron effects equilibrating the cosmic lepton and baryon asymmetries.

The measured neutrino mixing parameters implies that neutrinos reach the chemical equilibrium before BBN (Abazajian et al., 2002; Dolgov et al., 2002; Wong, 2002), so that all neutrino flavors are characterized by the same degeneracy parameter, ξ_ν, at this epoch. The most important impact of the leptonic asymmetry on BBN is the shift of the beta equilibrium between protons and neutrons and the increase of the radiation energy density parametrized by:

$$\Delta N_{eff}(\xi_\nu) = 3\left[\frac{30}{7}\left(\frac{\xi_\nu}{\pi}\right)^2 + \frac{15}{7}\left(\frac{\xi_\nu}{\pi}\right)^4\right]. \quad (1)$$

The total extra energy density can be splitted in two distinct uncorrelated contributions, first due to net lepton asymmetry of the neutrino background, $\Delta N_{eff}(\xi_\nu)$, and second due to the extra contributions from other unknown processes, ΔN_{eff}^{oth}:

$$\Delta N_{eff} = \Delta N_{eff}(\xi) + \Delta N_{eff}^{oth}. \quad (2)$$

2.1 Present bounds on lepton asymmetry and radiation energy density

The BBN constraints on N_{eff} have been analyzed by comparing the theoretical predictions and experimental data on the primordial abundances of light elements, by using the baryon abundance derived from the WMAP 3-year CMB temperature and polarization measurements (Hinshaw et al., 2007; Page et al., 2007; Spergel et al., 2007): $\eta_B = 6.14 \times 10^{-10}(1.00 \pm 0.04)$. In particular, the ^{4}He abundance, Y_p, is quite sensitive to the value of N_{eff}. The conservative error analysis of helium abundance, $Y_P = 0.249 \pm 0.009$ (Olive & Skillman, 2004), yielded to $N_{eff} = 3.1^{+1.4}_{-1.2}$ (95% CL) in good agreement with the standard value (Mangano et al., 2007), but still leaving some room for non-standard values.

More stringent error bars of helium abundance , $Y_p = 0.2516 \pm 0.0011$ (Izotov et al., 2007), leaded to $N_{eff} = 3.32^{+0.23}_{-0.24}$ (95% CL) and a degeneracy parameter $-0.04 < \xi_\nu < 0.07$ (68% CL) (Ichikawa et al., 2007; Serpico & Raffelt, 2005).

The CMB anisotropies and LSS matter density fluctuations power spectra carry the signature of the energy density of the Universe at the time of matter-radiation equality (energy density of order eV^4), making possible the measurement of N_{eff} through its effects on the growth of cosmological perturbations. The number of relativistic neutrino species influences the

CMB power spectrum by changing the time of matter-radiation equality that enhances the integrated Sachs-Wolfe effect, leading to a higher first acoustic Doppler peak amplitude. Also, the temperature anisotropy of the neutrino background (the anisotropic stress) acts as an additional source term for the gravitational potential (Hu et al., 1995; Trotta & Melchiorri, 2005), changing the CMB anisotropy power spectrum at the level of $\sim 20\%$.

The delay of the epoch of matter-radiation equality shifts the LSS matter power spectrum turnover position toward larger angular scales, suppressing the power at small scales.

In particular, the non-zero neutrino chemical potential leads to changes in neutrino free-streaming length and neutrino Jeans mass due to the increase of the neutrino velocity dispersion (Ichiki et al., 2007; Lattanzi et al., 2005).

A lower limit to $N_{eff} > 2.3$ (95% CL) was recently obtained from the analysis of the WMAP 5-year (WMAP5) data alone (Dunkley et al., 2009), while the combination of the WMAP5 data with distance information from baryonic acoustic oscillations (BAO), supernovae (SN) and Hubble constant measured by Hubble Space Telescope (HST), leaded to $N_{eff} = 4.4 \pm 1.5$ (68% CL), fully consistent with the standard value (Komatsu et al., 2009).

The bounds on the radiation content of the Universe and neutrino properties obtained from the analysis of the WMAP-5 year CMB measurements complemented with most of the existing CMB and LSS measurements, with self-consistent constraints on the primordial helium abundance from BBN leaded to a mean value (at 68% CL) of the effective number of relativistic neutrino species of $N_{eff} = 3.256^{+0.607}_{-0.641}$ and a neutrino degeneracy parameter of $-0.216 \leq \xi_\nu \leq 0.226$ (Popa & Vasile, 2008).

2.2 New bounds on neutrino properties: BBN and CMB constraints

In this section we revisit the constraints on the lepton asymmetry and radiation energy density by using the latest cosmological and astrophysical measurements: the WMAP 7-year CMB measurements (Komatsu et al., 2011; Larson et al., 2011) complemented with geometric probes from the Type Ia supernovae (SN) distance-redshift relation, the baryon acoustic oscillations (BAO) and the BBN predictions on Y_p.

The SN distance-redshift relation has been studied in detail in the recent unified analysis of the published heterogeneous SN data sets the Union Compilation08 (Kowalski et al., 2008; Riess et al., 2009).

The BAO in the distribution of galaxies are extracted from Two Degree Field Galaxy Redshift Survey (2DFGRS)the Sloan Digital Sky Surveys Data Release 7 (Percival et al., 2010).

The BBN predicted values of Y_P are obtained by using the PArthENoPE code (Pisanti et al., 2008). Starting from nuclear statistical equilibrium conditions, the code determines Y_P as function of $\Omega_b h^2$, ΔN^{oth}_{eff} and ξ_ν.

We use these measurements especially because we are testing models deviating from the standard Friedmann expansion. These datasets properly enables us to account for any shift of the CMB angular diameter distance and of the expansion rate of the universe. Hereafter, we will denote WMAP7+BAO+SNIa+BBN data set as WMAP7+BBN+All. We perform our analysis in the framework of the extended ΛCDM cosmological model described by $6+3$ free parameters:

$$\Theta = \{\underbrace{\Omega_b h^2, \Omega_{dm} h^2, H_0, z_{re}, n_s, A_s,}_{standard} \Omega_\nu h^2, \xi_\nu, \Delta N^{oth}_{eff}\}\,. \quad (3)$$

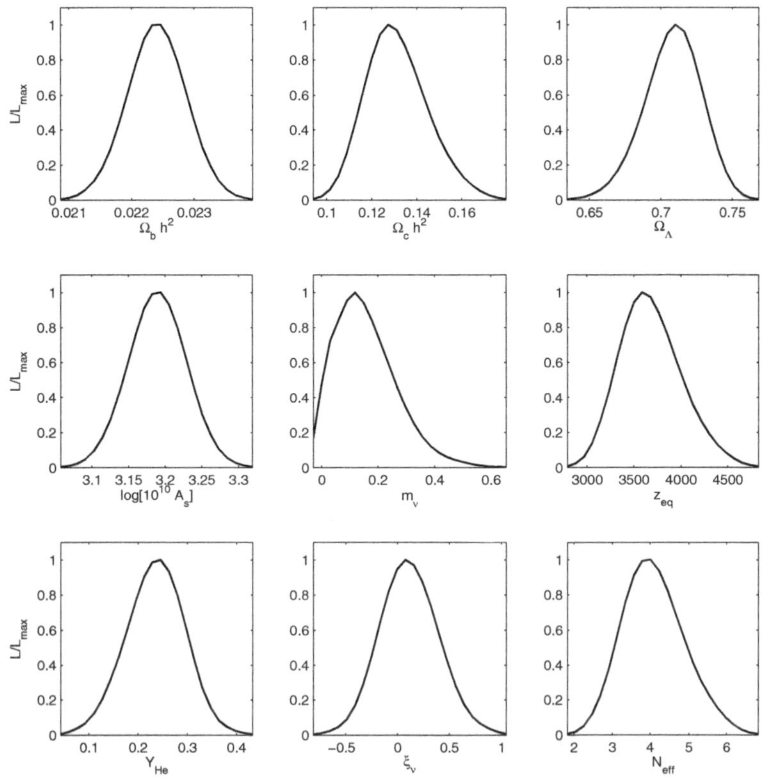

Fig. 1. The marginalized posterior likelihood probability distributions of the main cosmological parameters obtained from the fit of the leptonic asymmetric cosmological models to the WMAP7+BBN+All data set.

Here $\Omega_b h^2$ and $\Omega_{dm} h^2$ are the baryon and cold dark matter energy density parameters, H_0 is the Hubble expansion rate, z_{re} is the redshift of reionization, n_s is the scalar spectral index of the primordial density perturbation power spectrum and A_s is its amplitude at the pivot scale $k_* = 0.002\,\mathrm{hMpc}^{-1}$. The additional three parameters denote the neutrino energy density $\Omega_\nu h^2$, the neutrino degeneracy parameter ξ_ν and the contribution of extra relativistic degrees of freedom from other unknown processes ΔN_{oth}^{eff}.

The likelihood probabilities are evaluated by using the public packages COSMOMC and CAMB (Lewis et al., 2000; Lewis & S. Briddle, 2002) modified to include the formalism for the leptonic asymmetric cosmological models (Popa & Vasile, 2008).

We assume uniform prior probability on parameters Θ and compute the cumulative distribution function $C(\theta) = \int_{\Theta_{min}}^{\Theta} \mathcal{L}(\Theta) d\Theta / \int_{\Theta_{min}}^{\Theta_{max}} \mathcal{L} d\Theta$, quoting the upper and lower intervals at 68% CL.

Figure 1 presents the marginalized likelihood probabilities of the main cosmological parameters as obtained from the fit of the leptonic asymmetric cosmological models to the WMAP7+BBN+All data set and Figure 2 presents the degeneracies among them.

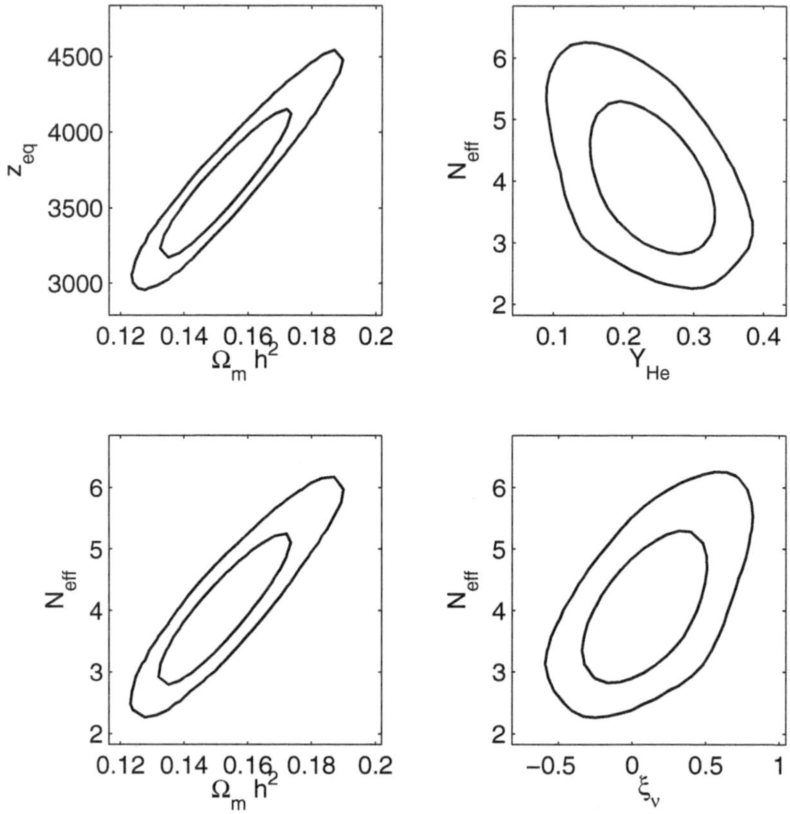

Fig. 2. Two dimensional joint probability distributions (68% and 95% CL) showing the degeneracy between different cosmological parameters as obtained from the fit of the leptonic asymmetric cosmological models to the WMAP7+BBN+All data set.

As neutrinos with eV mass decouple when they are still relativistic ($T_{dec} \sim 2$ MeV), the main effect of including ΔN_{eff} is the change of relativistic energy density. This changes the redshift of matter-radiation equality, z_{eq}, that affects the determination of $\Omega_m h^2$ from CMB measurements because of its linear dependence on N_{eff} (Komatsu et al., 2009):

$$1 + z_{eq} = \frac{\Omega_m h^2}{\Omega_\gamma h^2} \frac{1}{1 + 0.2271 N_{eff}} \,. \tag{4}$$

Here $\Omega_\gamma h^2$=2.469 $\times 10^{-5}$ is the present photon energy density parameter for $T_{cmb} = 2.725$ K. As a consequence, N_{eff} and $\Omega_m h^2$ are linearly correlated, with the width of degeneracy line given by the uncertainty in the determination of z_{eq}.

In Table 1 we compare the mean values and the absolute errors on the main cosmological parameters obtained from the analysis of WMAP5+BBN+All and WMAP7+BBN+All data sets. For all parameters, except m_ν, we quote the errors at 68% CL. For m_ν we give the upper limits at 95% CL.

Parameter	WMAP5+BBN+All	WMAP7+BBN+All
$\Omega_b h^2$	$0.02246^{+0.00063}_{-0.00072}$	$0.02240^{+0.00022}_{-0.00045}$
$\Omega_{dm} h^2$	$0.1115^{+0.0089}_{-0.0093}$	$0.1135^{+0.0061}_{-0.0084}$
z_{re}	$11.31^{+1.92}_{-2.11}$	$10.96^{+0.59}_{-0.83}$
n_s	0.965 ± 0.018	0.969 ± 0.008
$\ln[10^{10} A_s]$	3.265 ± 0.056	$3.189^{+0.018}_{-0.019}$
m_ν(eV)	≤ 0.535	≤ 0.412
ξ_ν	0.051 ± 0.221	$0.037^{+0.123}_{-0.143}$
ΔN_{eff}	$0.256^{+0.607}_{-0.641}$	$0.209^{+0.332}_{-0.401}$
Y_p	$0.2487^{+0.0451}_{-0.0484}$	$0.2356^{+0.0271}_{-0.0284}$
z_{eq}	3124^{+120}_{-128}	3132^{+64}_{-87}

Table 1. The table shows the mean values and the absolute errors on the main cosmological parameters obtained from the analysis of WMAP5+BBN+All and WMAP7+BBN+All data sets. For all parameters, except m_ν, we quote the errors at 68% CL. For m_ν we give the upper limits at 95% CL.

From the analysis of WMAP7+BBN+All data we find a mean value of $N_{eff} = 3.21^{+0.332}_{-0.401}$, bringing an improvement over the similar result obtained from WMAP5+BBN+ALL data, $N_{eff} = 3.26^{+0.638}_{-0.690}$. We also obtain improved values for ^{4}He mass fraction, $Y_p = 0.2356^{+0.0271}_{-0.0284}$, and neutrino degeneracy parameter, $-0.123 \leq \xi_\nu \leq 0.143$.

We find also a significantly reduced upper limit of the neutrino mass, $m_\nu < 0.412$ eV.

3. Challenging the standard model of particle physics

The primary goal of particle cosmology is to obtain a concordant description of the early evolution of the universe, establishing a testable link between cosmology and particle physics, consistent with both unified field theory and astrophysical and cosmological measurements.

Inflation is the most simple and robust theory able to explain the astrophysical and cosmological observations, providing at the same time self-consistent primordial initial condition mechanisms for the quantum generation of scalar (curvature) and tensor (gravitational waves) perturbations. In the simplest class of inflationary models, inflation is driven by a single scalar field ϕ (or inflaton) with some potential $V(\phi)$ minimally coupled to the Einstein gravity. The perturbations are predicted to be adiabatic, nearly scale-invariant and Gaussian distributed, resulting in an effectively flat universe.

The possibility that the Standard Model (SM) of particle physics with an additional non-minimally coupled term of the Higgs field to the gravitational Ricci scalar can give rise to inflation have been recently investigated by a number of authors (Barvinsky et al., 2008; Bezrukov & Shaposhnikov, 2008; Bezrukov et al., 2009; De Simone et al., 2009). This scenario is based on the observation that the problem of the very small value of Higgs quadratic coupling required by the CMB anisotropy data can be solved if the Higgs inflaton has a large coupling to gravity. The resultant Higgs inflaton effective potential in the inflationary domain

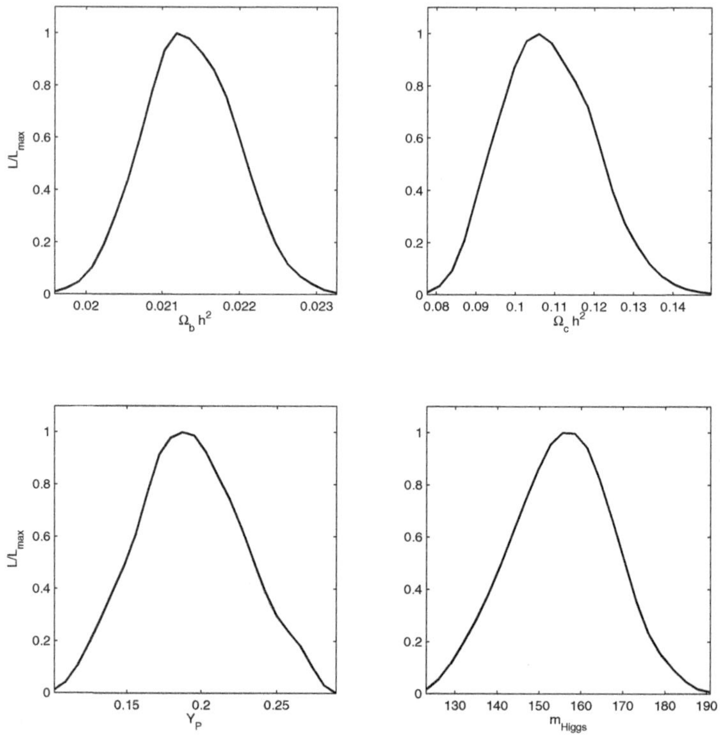

Fig. 3. The marginalized posterior likelihood probability distributions of the main cosmological parameters obtained from the fit of the Higgs inflation model to the WMAP7+BBN+All data set.

is effectively flat and can result in successful inflation for values of non-minimally coupling constant $\xi \sim 10^3 - 10^4$, allowing for cosmological values for Higgs boson mass in a window in which the Higgs vacuum expectation value $< v >= 246.22$ GeV is the minimum of the inflationary potential (Espinosa et al., 2008). The fluctuations of the electroweak vacuum expectation value might change the dynamics of the Higgs field during inflation, leading to modification of the main inflationary parameters and of the Higgs boson mass.

3.1 BBN and the Higgs vacuum expectation value

A modification of the Higgs vacuum expectation value during BBN leads to variations of the the neutron-proton mass difference, Δm_{np}, the Fermi constant, G_F, the deuterium binding energy, ϵ_D, and the electron mass, m_e.

The ^{4}He abundance is very sensitive to the parameters that fixed the neutron-to-proton ratio. In thermal equilibrium, this ratio is given by:

$$\frac{Y_n}{Y_p} = e^{-\Delta m_{np}/T}, \tag{5}$$

where Y_n and Y_p are the neutron and proton abundances and T is the temperature in MeV. The neutron-proton mass difference is affected by a change in the Higgs vacuum expectation value according to (Bergstrom et al., 1999):

$$\frac{\delta \Delta m_{np}}{\Delta m_{np}} = 1.587 \frac{\Delta < v >}{< v >} . \tag{6}$$

An increase in the Higgs vacuum expectation value $< v >$ leads to an increase in Δm_{np}. This produces a smaller neutron-to-proton equilibrium ratio and a smaller abundance of ^{4}He (Yoo & Scherrer, 2003). On the other hand, a larger Higgs vacuum expectation value during BBN results in: i) a smaller value of the Fermi coupling constant, $G_F = 1/\sqrt{2} < v >^2$, leading to earlier freeze-out of the weak reactions and producing more ^{4}He; ii) an increase in the electron mass, m_e, and a decrease of the weak reaction rates, producing also more ^{4}He (Landau et al., 2008).
The dependence of the deuterium binding energy, ϵ_D, on the Higgs vacuum expectation value is extremely model dependent and can be approximated as (Yoo & Scherrer, 2003):

$$\frac{\Delta \epsilon_D}{(\epsilon_D)_0} \simeq \kappa \frac{\Delta < v >}{< v >_0} , \tag{7}$$

where κ is a model dependent constant (e.g. $\kappa \simeq -6.230$ for a Reid potential). An increase in the Higgs vacuum expectation value results in a decrease in the deuterium binding energy, leading to a smaller initial deuterium abundance (Landau et al., 2008):

$$Y_d = \frac{Y_n Y_p e^{11.605 \epsilon_D / T_9}}{0.471\, T_9^{3/2}} , \tag{8}$$

where T_9 is the temperature in units of 10^9K, and ϵ_D is in MeV. The production of ^{4}He begins later, leading to a smaller helium abundance and an increase in the deuterium abundance (Yoo & Scherrer, 2003).

The electron mass is proportional to the Higgs vacuum expectation value:

$$\frac{\Delta m_e}{m_e} = \frac{\Delta < v >}{< v >} . \tag{9}$$

A value of electron mass during BBN different from the present one translates into changes of electron and positron energy densities, leading to modifications of the Hubble expansion rate. Also, an increase of the electron mass during BBN slowers the neutron-proton intreaction rates, leading to a higher ^{4}He abundance. It is important to note that the modification of the primordial abundances due to the changes of the weak interaction rates are dominant over those due to the changes in the Hubble expansion rate (Yoo & Scherrer, 2003).

3.2 Bounds on the Higgs boson properties and the inflationary parameters: BBN and CMB constraints

For a robust interpretation of upcoming high precision temperature and polarization CMB anisotropies (Mandolesi et al., 2010), it is imperative to understand how the inflationary dynamics of a non-minimally coupled Higgs scalar field (ξ-inflation) may affect the degeneracy of the inflationary observables, the determination of the Higgs boson mass and of the vacuum expectation value.

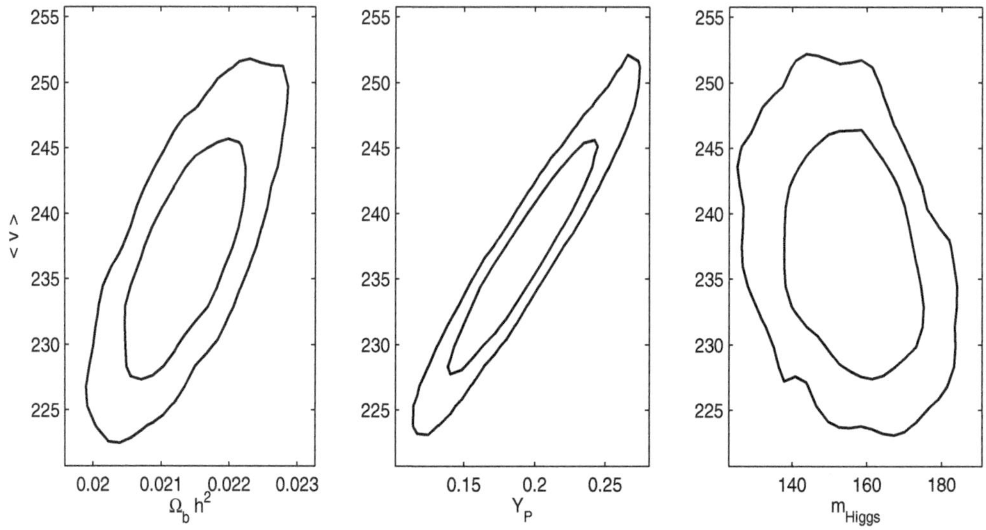

Fig. 4. Two dimensional joint probability distributions (68% and 95% CL) showing the degeneracy between different cosmological parameters as obtained from the fit of Higgs inflation model to the WMAP7+BBN+All data set.

Model Parameter	Standard Inflation	Higgs Inflation
$100\Omega_b h^2$	2.259±0.054	2.257±0.051
$\Omega_c h^2$	0.113±0.003	0.114±0.003
τ	0.088±0.015	0.086±0.013
θ_s	1.038±0.002	1.037±0.002
Y_P	0.245±0.014	0.247±0.091
m_{Higgs}(GeV)	-	155.372±3.851
$< v >$(GeV)	-	243.131±5.912

Table 2. The mean values from the posterior distributions of the parameters obtained from the fit of the standard inflation model and Higgs inflation model with m_{Top}=171.3 GeV to the WMAP7+BBN+All dataset. The errors are quoted at 68% CL. All parameters are computed at the Hubble radius crossing k_*=0.002 Mpc^{-1}.

We place bounds on these parameters by using WMAP7+BBN+All data set, in the context of the non-minimally coupled Higgs inflaton field with gravity.

We obtain the CMB temperature anisotropy and polarization power spectra by integrating the coupled Friedmann equation and equation of motion of the Higgs scalar field with respect to the conformal time (Popa & Caramete, 2010; Popa, 2011).

The quantum corrections due to the interaction effects of the SM particles with Higgs boson through quantum loops modify Higgs potential from its classical expression. We consider the quantum corrections to the Higgs potential by including in the computation the running of the different coupling constants: the SU(2) $\times$ SU(1) gauge couplings $\{g', g\}$, the $SU(3)$ strong coupling g_s, the top Yukawa coupling y_t, and the Higgs quadratic coupling λ (Espinosa et al., 2008).

At the top quark mass scale m_{Top}, the Higgs quadratic coupling $\lambda(0)$ and the top Yukawa coupling $y_t(0)$ are determined by the corresponding Higgs boson and top quark pole masses and the vacuum expectation value $< v >$:

$$\lambda(0) = \frac{m^2_{Higgs}}{2 < v >^2}\left[1 + 2\Delta_H(m_{Higgs})\right], \qquad y_t(0) = \frac{\sqrt{2} m_{Top}}{< v >}\left[1 + \Delta_T(m_{Top})\right], \quad (10)$$

where $\Delta_H(m_{Higgs})$ and $\Delta_T(m_{Top})$ are the corrections to Higgs and top quark mass respectively.

The likelihood probabilities are evaluated by using the public packages COSMOMC and CAMB (Lewis et al., 2000; Lewis & S. Briddle, 2002) modified to include the formalisms for the SM Higgs driven ξ-inflation.

Our fiducial model is the ΛCDM cosmological model described by the following set of parameters receiving uniform priors:

$$\Theta = \{\Omega_b h^2, \ \Omega_c h^2, \ \theta_s, \ \tau, \ m_{Higgs}, \ < v > \ \} \quad (11)$$

where: $\Omega_b h^2$ is the physical baryon density, $\Omega_c h^2$ is the physical dark matter density, θ_s is the ratio of the sound horizon distance to the angular diameter distance, τ is the reionization optical depth, m_{Higgs} is the Higgs boson pole mass, $< v >$ is the Higgs vacuum expectation value. As in the previous case, we use the PArthENoPE code (Pisanti et al., 2008) to obtain the ^{4}He mass fraction as function of $\Omega_b h^2$ and $< v >$.

Figure 3 presents the marginalized posterior likelihood probability distributions of the main cosmological parameters and Figure 4 presents the two-dimensional joint probability distributions showing the degeneracy in the planes $< v >$ - $\Omega_b h^2$, $< v >$ - Y_P and $< v >$ - m_{Higgs}, as obtained from the fit of the Higgs ξ-inflation model to the WMAP7+BBN+All data set.

Table 2 presents the mean values from the posterior distributions of the parameters obtained from the fit of the standard inflation model and Higgs inflation model with m_{Top}=171.3 GeV to the WMAP7+BBN+All dataset. The errors are quoted at 68% CL.

From this analysis we obtain the vacuum expectation value and the Higgs boson mass in the limits expected from the collider experiments, while the cosmological parameters are compatible to the standard inflation scenario (Popa & Caramete, 2009).

4. Acknowledgments

This work was partially supported by CNCSIS Contract 539/2009 and by ESA/PECS Contract C98051.

5. References

Abazajian, K. N.; Beacom, J. F.; Bell, N. F. (2002). Stringent Constraints on Cosmological Neutrino-Antineutrino Asymmetries from Synchronized Flavor Transformation. *Phys. Rev.* D 66, 013008 [astro-ph/0203442]

Barvinsky, A. O.; Kamenshchik, A. Yu.; Starobinsky A. A. (2008). Inflation scenario via the Standard Model Higgs boson and LHC. *JCAP* 11, 021 [arXiv:0809.2104]

Bergstrom, L.; Iguri, S.; Rubinstein, H. (1999). Constraints on the variation of the fine structure constant from big bang nucleosynthesis. *Phys. Review* D 60, 045005 [arXiv:astro-ph/9902157]

Bezrukov, F. & Shaposhnikov, M. (2008). The Standard Model Higgs boson as the inflaton. *Phys. Lett.* B 659, 703 [arXiv:0710.3755]

Bezrukov, F. L.; Magnin, A.; Shaposhnikov, M. (2009). Standard Model Higgs boson mass from inflation. *Phys. Lett.* B 675, 88 [arXiv:0812.4950]

Bezrukov, F.; Gorbunov, D.; Shaposhnikov, M. (2009). On initial conditions for the Hot Big Bang. *JCAP* 06, 029 [arXiv:0812.3622]

Boesgaard, A. M.; Steigman, G. (1985). Big bang nucleosynthesis - Theories and observations. *Ann. Rev. Astron. Astrophys.*, 23, 319

Buchmuller, W.; Di Bari, P.; Plumacher, M. (2004). Some Aspects of Thermal Leptogenesis. *New Jour. Phys.* 6, 105 [hep-ph/0406014]

Burles, S.; Nollett, K. M.; Turner, M. S. (2001). Big-Bang Nucleosynthesis Predictions for Precision Cosmology. *Astrophys. J.* 552, L1 [astro-ph/0010171]

Chu, Y. Z. & Cirelli, M. (2006). Sterile neutrinos, lepton asymmetries, primordial elements: How much of each? *Phys. Rev.* D 74, 085015

de Simone, A.; Hertzberg, M. P.; Wilczek, F. (2009). Running Inflation in the Standard Model. *Phys. Lett.* B 678, 1 [arXiv:0812.4946]

Dunkley, J. et al. (2009). Five-Year Wilkinson Microwave Anisotropy Probe (WMAP) Observations: Likelihoods and Parameters from the WMAP data. *Astrophys. J. Supp.* 180, 306 [arXiv:0803.0586]

Dolgov, A. D.; Hansen, S. H.; Pastor, S.; Petcov, S. T.; Raffelt, G.; Semikoz, D. V. (2002). Cosmological bounds on neutrino degeneracy improved by flavor oscillations. *Nucl.Phys.* B 632, 363 [hep-ph/0201287]

Eidelman S. et al. (2004). Review of particle physics. *Phys. Lett.* B 592, 1

Epstein, R.; Lattimer, J. & Schramm, D. N. (1976). The origin of deuterium. *Nature,* 263, 198

Espinosa, J. R.; Giudice, G. F.; Riotto, A. (2008). Cosmological implications of the Higgs mass measurement. *JCAP* 05, 002 [arXiv:0710.2484]

Falcone, D. & Tramontano, F. (2001). Neutrino oscillations and neutrinoless double beta decay. *Phys. Rev.* D 64, 077302 [hep-ph/0102136]

Freese, K.; Kolb, E. W.; Turner, M. S. (1983). Massive, degenerate neutrinos and cosmology. *Phys. Rev.* D27, 1689

Hinshaw et al. (2007). Three-Year Wilkinson Microwave Anisotropy Probe (WMAP) Observations: Temperature Analysis. *Astrophys. J. Suppl.* 170, 288 [astro-ph/0603451]

Hu, W.; Scott, D.; Sugiyama, N.; White, M. (1995). The Effect of Physical Assumptions on the Calculation of Microwave Background Anisotropies. *Phys. Rev.* D 52, 5498 [astro-ph/9505043]

Ichikawa, K.; Kawasaki, M.; Nakayama, K.; Senami, M., Takahashi, F. (2007). Increasing the effective number of neutrinos with decaying particles. *JCAP* 05, 008 [hep-ph/0703034]

Ichiki, K.; Yamaguchi, M.; Yokoyama, J. (2007). Lepton asymmetry in the primordial gravitational wave spectrum. *Phys.Rev* D 75, 084017 [hep-ph/0611121]

Izotov, Y. I. & Thuan, T. X. (2004) Systematic effects and a new determination of the primordial abundance of 4He and dY/dZ from observations of blue compact galaxies. *Astrophys. J.* 602, 200 [astro-ph/0310421]

Izotov, Y. I.; Thuan, T. X.; Stasinska, G. (2007). The primordial abundance of 4He: a self-consistent empirical analysis of systematic effects in a large sample of low-metallicity HII regions. *Astrophys. J.* 662, 15 [astro-ph/0702072]

Komatsu E. et al. (2009). Five-Year Wilkinson Microwave Anisotropy Probe (WMAP) Observations: Cosmological Interpretation. *Astrophys. J. Suppl.* 180, 330 [arXiv:0803.0547 [astro-ph]].

Komatsu, E. et al. (2011). Seven-Year Wilkinson Microwave Anisotropy Probe (WMAP) Observations: Cosmological Interpretation. *Astrophys. J. Suppl.* 192, 18 [arXiv:1001.4538 [astro-ph.CO]]

Kowalski, M. et al. (2008). Improved Cosmological Constraints from New, Old and Combined Supernova Datasets. *ApJ* 686, 749 [arXiv:0804.4142 [astro-ph]]

Kuzmin, V.; Rubakov, V.; Shaposhnikov, M. (1985). On anomalous electroweak baryon-number non-conservation in the early universe. *Phys. Lett.* B 155, 36

Landau, S. J.; Mosquera, M. E.; Scoccola, C. G.; Vucetich, H. (2008). Early universe constraints on time variation of fundamental constants. *Phys. rev.* D 78, 083527 [arXiv:0809.2033]

Larson, D. et al. (2011). Seven-Year Wilkinson Microwave Anisotropy Probe (WMAP) Observations: Power Spectra and WMAP-Derived Parameters. *Astrophys. J. Suppl.* 192, 16 [arXiv:1001.4635 [astro-ph.CO]]

Lattanzi, M.; Ruffini, R.; Vereshchagin, G. V. (2005) Joint constraints on the lepton asymmetry of the Universe and neutrino mass from the Wilkinson Microwave Anisotropy Probe. *Phys. Rev.* D 72, 063003 [astro-ph/0509079]

Lewis, A.; Challinor, A.; Lasenby, A. (2000). Efficient Computation of Cosmic Microwave Background Anisotropies in Closed Friedmann-Robertson-Walker Models. *ApJ* 538, 473 [1] [arXiv:astro-ph/9911177]

Lewis, A. & Briddle S. (2002). Cosmological parameters from CMB and other data: a Monte-Carlo approach. *Phys. Rev.* D 66, 103511 [2] [arXiv:astro-ph/0205436]

Mandolesi, N. et al. (Planck Collaboration) (2010). Planck pre-launch status: The Planck-LFI programme. *Astron. & Astrophys.* 520, A3 [arXiv:1001.2657 [astro-ph.CO]]

Mangano, G.; Miele, G.; Pastor, S.; Peloso, M. (2002). A precision calculation of the effective number of cosmological neutrinos. *Phys. Lett.* B 534, 8 [astro-ph/0111408]

Mangano, G.; Melchiorri, A.; Mena, O.; Miele, G.; Slosar, A. (2007). Present bounds on the relativistic energy density in the Universe from cosmological observables. *JCAP* 03, 006 [astro-ph/0612150]

Olive, K. A.; Steigman, G.; Walker, T. P. (2000). Primordial Nucleosynthesis: Theory and Observations. *Phys. Rep.* 333, 389 [astro-ph/9905320]

Olive, K. A. & Skillman, E. D. (2004). A Realistic Determination of the Error on the Primordial Helium Abundance: Steps Toward Non-Parametric Nebular Helium Abundances. *Astrophys. J.* 617, 29 [astro-ph/0405588]

Page, L. et al. (2007) Three Year Wilkinson Microwave Anisotropy Probe (WMAP) Observations: Polarization Analysis. *Astrophys. J.Suppl.* 170, 335 [astro-ph/0603450]

W. J. Percival, et al. (2010). Baryon Acoustic Oscillations in the Sloan Digital Sky Survey Data Release 7 Galaxy Sample. *MNRAS* 401, 2148 [arXiv:0907.1660 [astro-ph]]

Pettini, M.; Zych, B. J.; Murphy, M. T.; Lewis, A.; Steidel, C. C. (2008). Deuterium abundance in the most metal-poor damped Lyman alpha system: converging on $O_{b,0}h^2$. *MNRAS* 391, 1499 [arXiv:0805.0594 [astro-ph.CO]]

Pisanti, O.; Cirillo, A.; Esposito, S.; Iocco, F.; Mangano, G.; Miele, G.; Serpico, P. D. (2008). PArthENoPE: Public algorithm evaluating the nucleosynthesis of primordial elements. *Computer Physics Communications* 178, 956 [arXiv:0705.0290 [astro-ph.CO]]

Popa, L. A. & Vasile, A. (2008). WMAP 5-year constraints on lepton asymmetry and radiation energy density: Implications for Planck. *JCAP* 06, 028 [arXiv:0804.2971 [astro-ph.CO]]

Popa, L.A. & Caramete, A. (2009). From WMAP to Planck: Exact Reconstruction of Four- and Five-dimensional Inflationary Potential from High-precision Cosmic Microwave Background Measurements. *ApJ* 706 , 1008 [arXiv:0907.5558 [astro-ph.CO]]

Popa, L.A. & Caramete, A. (2010). Cosmological Constraints on the Higgs Boson Mass. *ApJ* 723 , 803 [arXiv:1009.1293 [astro-ph.CO]]

Popa, L.A. (2011). Observational consequences of the Standard Model Higgs inflation variants. *JCAP* 10, 025 [arXiv:1107.3436 [astro-ph.CO]]

Riess et al. (2009). A Redetermination of the Hubble Constant with the Hubble Space Telescope from a Differential Distance Ladder. *ApJ* 699, 539 (2009) [arXiv:0905.0695 [astro-ph.CO]]

Ruffini, R.; Song, D. J.; Stella, L. (1983). On the statistical distribution off massive fermions and bosons in a Friedmann universe. *Astron. Astrophys.* 125, 265

Ruffini, R.; Song, D. J.; Taraglio, S. (1988). The 'ino' mass and the cellular large-scale structure of the universe. *Astron. Astrophys.* 190, 1

Sarkar, S. (1996). Big bang nucleosynthesis and physics beyond the standard model. *Rep. Prog. Phys.* 59, 1493 [arXiv:hep-ph/9602260]

Serpico, P. D. & Raffelt, G. G. (2005) Lepton asymmetry and primordial nucleosynthesis in the era of precision cosmology. *Phys.Rev.* D 71, 127301 [astro-ph/0506162]

Smith, C. J.; Fuller, G. M.; Kishimoto, C. T.; Abazajian, K. N. (2006). Light Element Signatures of Sterile Neutrinos and Cosmological Lepton Numbers. *Phys. Rev.* D 74, 085008 [astro-ph/0608377]

Spergel, D. N. et al. (2007). Wilkinson Microwave Anisotropy Probe (WMAP) Three Year Results: Implications for Cosmology. *Astrophys. J. Suppl.* 170, 377 [astro-ph/0603449]

Steigman, G.; Schramm, D. N.; Gunn, J. E. (1977). Cosmological limits to the number of massive leptons. *Phys. Lett.* B 66, 202

Steigman, G. (2007). Primordial Nucleosynthesis in the Precision Cosmology Era. *Ann. Rev. Nucl. Part. Sci.* 57, 463

Trotta, R. & Melchiorri, A. (2005). Indication for Primordial Anisotropies in the Neutrino Background from WMAP and SDSS. *Phys. Rev Lett.* 95, 011305 [astro-ph/0412066]

Yoo, J. J. & Scherrer, R. J. (2003). Big bang nucleosynthesis and cosmic microwave background constraints on the time variation of the Higgs vacuum expectation value. *Phys. Rev.* D 67, 043517 [arXiv:astro-ph/0211545]

Wagoner R. V.; Fowler W. A.; Hoyle F. (1967). On the Synthesis of Elements at Very High Temperatures. *Astrophys. J.* 148, 3

Wong, Y. Y. Y. (2002). Analytical treatment of neutrino asymmetry equilibration from flavour oscillations in the early universe. *Phys. Rev.* D 66, 025015 [hep-ph/0203180]

Permissions

The contributors of this book come from diverse backgrounds, making this book a truly international effort. This book will bring forth new frontiers with its revolutionizing research information and detailed analysis of the nascent developments around the world.

We would like to thank İbrahim Küçük, for lending his expertise to make the book truly unique. He has played a crucial role in the development of this book. Without his invaluable contribution this book wouldn't have been possible. He has made vital efforts to compile up to date information on the varied aspects of this subject to make this book a valuable addition to the collection of many professionals and students.

This book was conceptualized with the vision of imparting up-to-date information and advanced data in this field. To ensure the same, a matchless editorial board was set up. Every individual on the board went through rigorous rounds of assessment to prove their worth. After which they invested a large part of their time researching and compiling the most relevant data for our readers. Conferences and sessions were held from time to time between the editorial board and the contributing authors to present the data in the most comprehensible form. The editorial team has worked tirelessly to provide valuable and valid information to help people across the globe.

Every chapter published in this book has been scrutinized by our experts. Their significance has been extensively debated. The topics covered herein carry significant findings which will fuel the growth of the discipline. They may even be implemented as practical applications or may be referred to as a beginning point for another development. Chapters in this book were first published by InTech; hereby published with permission under the Creative Commons Attribution License or equivalent.

The editorial board has been involved in producing this book since its inception. They have spent rigorous hours researching and exploring the diverse topics which have resulted in the successful publishing of this book. They have passed on their knowledge of decades through this book. To expedite this challenging task, the publisher supported the team at every step. A small team of assistant editors was also appointed to further simplify the editing procedure and attain best results for the readers.

Our editorial team has been hand-picked from every corner of the world. Their multi-ethnicity adds dynamic inputs to the discussions which result in innovative outcomes. These outcomes are then further discussed with the researchers and contributors who give their valuable feedback and opinion regarding the same. The feedback is then collaborated with the researches and they are edited in a comprehensive manner to aid the understanding of the subject.

Apart from the editorial board, the designing team has also invested a significant amount of their time in understanding the subject and creating the most relevant covers. They scrutinized every image to scout for the most suitable representation of the subject and create an appropriate cover for the book.

The publishing team has been involved in this book since its early stages. They were actively engaged in every process, be it collecting the data, connecting with the contributors or procuring relevant information. The team has been an ardent support to the editorial, designing and production team. Their endless efforts to recruit the best for this project, has resulted in the accomplishment of this book. They are a veteran in the field of academics and their pool of knowledge is as vast as their experience in printing. Their expertise and guidance has proved useful at every step. Their uncompromising quality standards have made this book an exceptional effort. Their encouragement from time to time has been an inspiration for everyone.

The publisher and the editorial board hope that this book will prove to be a valuable piece of knowledge for researchers, students, practitioners and scholars across the globe.

List of Contributors

Wei Wang
National Astronomical Observatories, Chinese Academy of Sciences, Beijing, China

G. Gosselin, V. Méot and P. Morel
CEA, DAM, DIF, Arpajon, France

P. Mohr
Diakonie-Klinikum, Schwäbisch Hall, Germany
ATOMKI, Debrecen, Hungary

Boris Pritychenko
National Nuclear Data Center, Brookhaven National Laboratory, Upton, NY, USA

Marcel Arnould and Stephane Goriely
Université Libre de Bruxelles, Belgium

M. L. Pumo
Università degli Studi di Padova, Dip. di Fisica e Astronomia "G. Galilei", Padova, Italy
INAF-Osservatorio Astronomico di Padova, Padova, Italy
Bonino-Pulejo Foundation, Messina, INAF-Osservatorio Astrofisico di Catania, Catania, Italy

İbrahim Küçük
Erciyes University, Science Faculty, Department of Astronomy and Space Sciences, Kayseri, Turkey

Chaudry Masood Khalique
International Institute for Symmetry Analysis and Mathematical Modelling, Department of Mathematical Sciences, North-West University, Mmabatho, Republic of South Africa

S. Capozziello, L. Consiglio, M. De Laurentis, G. De Rosa and C. Di Donato
Dipartimento di Scienze Fisiche, Université di Napoli "Federico II", Compl. Univ. di Monte S. Angelo, Edificio G, Napoli, INFN Sezione di Napoli, Compl Univ. di Monte S. Angelo, Edificio G, Napoli, Italy

Francesca Sammarruca
University of Idaho, USA

A.B. Aleksandrov, N.G. Polukhina and N.I. Starkov
Lebedev Physical Institute, Russian Academy of Sciences, Moscow, Russia

V.I. Sotnikov, R.W. Deming and L. Perlovsky
Air Force Research Laboratory, Sensors Directorate, Hanscom AFB, MA, USA

Sergey Bastrukov
State Key Laboratory of Nuclear Physics and Technology, School of Physics, Peking University, Beijing, China
Joint Institute for Nuclear Research, Dubna, Russia

Renxin Xu and Junwei Yu
State Key Laboratory of Nuclear Physics and Technology, School of Physics, Peking University, Beijing, China

Irina Molodtsova
Joint Institute for Nuclear Research, Dubna, Russia

Hsiang-Kuang Chang
Institute of Astronomy and Department of Physics, National Tsing Hua University, Hsinchu, Taiwan

Werner Benger
Department of Astro- and Particle Physics, Distributed and Parallel Systems Group, University of Innsbruck, Innsbruck, Austria
Center for Computation & Technology at Louisiana State University, Baton Rouge, USA

Markus Haider, Harald Höller, Dominik Steinhauser, Josef Stöckl, Biagio Cosenza and Marcel Ritter
Department of Astro- and Particle Physics, Distributed and Parallel Systems Group, University of Innsbruck, Innsbruck, Austria

Jorge Igual and Raul llinares
Universidad Politecnica de Valencia, Spain

W.R. Webber
New Mexico State University, Department of Astronomy, Las Cruces, USA

Alberto Vecchiato
INAF (National Institute for Astrophysics), Astronomical Observatory of Torino, Italy

Mikhail Z. Iofa
Moscow State University, Skobeltsyn Institute of Nuclear Physics, Moscow, Russia

L. A. Popa and A. Caramete
ISS/Institute for Space Sciences Bucharest-Magurele, Romania